Finite Mixture Models

Finite Mixture Models

GEOFFREY McLACHLAN

DAVID PEEL

Department of Mathematics
The University of Queensland

A Wiley-Interscience Publication
JOHN WILEY & SONS, INC.
New York • Chichester • Weinheim • Brisbane • Singapore • Toronto

This text is printed on acid-free paper. ♾

For ordering and customer service, call 1-800-CALL-WILEY.

Library of Congress Cataloging in Publication Data

McLachlan, Geoffrey J., 1946–
Finite mixture models / G.J. McLachlan and D. Peel.
p. cm. — (Wiley series in probability and statistics. Applied probability and statistics section)
Includes bibliographical references and indexes.
ISBN 0-471-00626-2 (cloth : alk. paper)
1. Mixture distribution (Probability theory). I. Peel, D. (David). II. Title. III. Series.
Wiley series in probability and statistics. Applied probability and statistics.
QA273.6.M395 2001
519.2—dc21 00-043324

Printed in the United States of America.

10 9 8 7 6 5 4 3 2

To
Beryl, Jonathan, and Robbie

Samantha

Contents

Preface

The importance of finite mixture models in the statistical analysis of data is underscored by the ever-increasing rate at which articles on mixture applications appear in the statistical and general scientific literature. The aim of this monograph is to provide an up-to-date account of the theory and applications of modeling via finite mixture distributions. Since the appearance of the monograph of McLachlan and Basford (1988) on finite mixtures, the literature has expanded enormously to the extent that another monograph on the topic is apt. In the past decade the extent and the potential of the applications of finite mixture models have widened considerably. Because of their flexibility, mixture models are being increasingly exploited as a convenient, semiparametric way in which to model unknown distributional shapes. This is in addition to their obvious applications where there is group-structure in the data or where the aim is to explore the data for such structure, as in a cluster analysis.

In this book the more recent work is surveyed against the background of the existing literature. The widespread use of mixture models in recent times is demonstrated by the fact that of the 800 or so references in this book, almost 40% of them have been published since 1995. A comprehensive account of the major issues involved with modeling via finite mixture distributions is provided. They include identifiability problems, the actual fitting of finite mixtures through use of the EM algorithm, the properties of the maximum likelihood estimators so obtained, the assessment of the number of components to be used in the mixture, and the applicability of asymptotic theory in providing a basis for the solutions to some of these problems. The intent is to provide guidelines to users of mixture models on these various issues. The emphasis is on the applications of mixture models, not only in mainstream statistical analyses, but also in other areas such as unsupervised pattern recognition, speech recognition, and medical imaging.

With the advent of inexpensive, high-speed computers and the simultaneous rapid development in posterior simulation techniques such as Markov chain Monte Carlo

(MCMC) methods for enabling Bayesian estimation to be undertaken, practitioners are increasingly turning to Bayesian methods for the analysis of complicated statistical models. In this book, we consider the latest developments in Bayesian estimation of mixture models.

New topics that are covered in this book include the scaling of the EM algorithm to allow mixture models to be used in data mining applications involving massively huge databases. In the same spirit, there is also an account of the use of the sparse/incremental EM algorithm and of multiresolution kd-trees for speeding up the implementation of the standard EM algorithm for the fitting of mixture models. Another topic concerns the use of hierarchical mixtures-of-experts models as a powerful new approach to nonlinear regression that is a serious competitor to well-known statistical procedures such as MARS and CART. Other recent developments covered include the use of mixture models for handling overdispersion in generalized linear models and proposals for dealing with mixed continuous and categorical variables. In other recent work, there is the proposal to use t components in the mixture model to provide a robust approach to mixture modeling. A further topic is the use of mixtures of factor analyzers which provide a way of fitting mixture models to high-dimensional data. As mixture models provide a convenient basis for the modeling of dependent data by allowing the component-indicator variables to have a Markovian structure, there is also coverage of the latest developments in hidden Markov models, including the Bayesian approach to this problem. Another problem considered is the fitting of mixture models to multivariate data in binned form, which arises in some important medical applications in practice.

The book also covers the latest developments on existing issues with mixture modeling, such as assessing the number of components to be used in a mixture model and the associated problem of determining how many clusters there are in clustering applications with mixture models.

In presenting these latest results, the authors have attempted to draw together the statistical literature with the machine learning and pattern recognition literature.

It is intended that the book should appeal to both applied and theoretical statisticians, as well as to investigators working in the many diverse areas in which relevant use can be made of finite mixture models. It will be assumed that the reader has a fair mathematical or statistical background. The main parts of the book describing the formulation of the finite mixture approach, detailing its methodology, discussing aspects of its implementation, and illustrating its application in many simple statistical contexts should be comprehensible to graduates with statistics as their major subject. The emphasis is on the practical applications of mixture models; and to this end, numerous examples are given.

Chapter 1 begins with a discussion of mixture models and their applications and gives a brief overview of the current state of the area. It also includes a brief history of mixture models.

Chapter 2 focuses on the maximum likelihood fitting of mixture models via the EM algorithm. It covers issues such as the choice of starting values, stopping criteria, the calculation of the observed information matrix, and the provision of standard errors either by information-based methods or by the bootstrap.

Chapter 3 specializes the results in Chapter 2 to mixtures of normal components. Given the tractability of the multivariate normal distribution, it is not surprising that mixture modeling of continuous data is invariably undertaken by normal mixtures. This chapter also discusses the occurrence and identification of so-called spurious local maximizers of the likelihood function, which is an issue with the fitting of normal mixture models with no restrictions on the component-covariance matrices. A number of illustrative examples are presented.

In Chapter 4, we consider the Bayesian approach to the fitting of mixture models. Estimation in a Bayesian framework is now feasible using posterior simulation via recently developed MCMC methods. Bayes estimators for mixture models are well-defined so long as the prior distributions are proper. One main hindrance is that improper priors yield improper posterior distributions. We discuss the use of "partially proper priors," which do not require subjective input for the component parameters, yet the posterior is proper. We also discuss ways of handling other hindrances, including the effect of label switching, which arises when there there is no real prior information that allows one to discriminate between the components of a mixture model belonging to the same parametric family.

In Chapter 5, we consider the fitting of mixture models with nonnormal component densities, including components suitable for mixed feature variables, where some are continuous and some are categorical. The maximum likelihood fitting of commonly used discrete components such as the binomial and Poisson are undertaken within the wider framework of a mixture of generalized linear models (GLMs). The latter also has the capacity to handle the regression case, where the response is allowed to depend on the value of a vector of covariates. The use of mixtures of GLMs for handling overdispersion in a single GLM component is discussed. In work related to mixtures of GLMs, the mixtures-of-experts model is considered, along with its extension, the hierarchical mixtures-of-experts model. This approach which combines aspects of finite mixture models and GLMs provides a comparatively fast learning and good generalization for nonlinear regression problems, including classification.

Chapter 6 is devoted to the estimation of the order of a mixture model. It covers the two main approaches. One way is based on a penalized form of the log likelihood whereby the likelihood is penalized by the subtraction of a term that "penalizes" the model for the number of parameters in it. The other main way for deciding on the order of a mixture model is to carry out a hypothesis test, using the likelihood ratio as the test statistic.

In Chapter 7, we consider the fitting of mixtures of (multivariate) t distributions, as proposed in McLachlan and Peel (1998a) and Peel and McLachlan (2000). The t distribution provides a longer-tailed alternative to the normal distribution. Hence it provides a more robust approach to the fitting of normal mixture models, as observations that are atypical of a component are given reduced weight in the calculation of its parameters. Also, the use of t components gives less extreme estimates of the posterior probabilities of component membership of the mixture model.

In Chapter 8, we consider mixtures of factor analyzers from the perspective of both (a) a method for model-based density estimation from high-dimensional data, and hence for the clustering of such data, and (b) a method for local dimensionality reduction. We also discuss the close link of mixtures of factor analyzers with mixtures of probabilistic principal component analyzers. The mixtures of factor analyzers model enables a normal mixture model to be fitted to high-dimensional data. The number of free parameters is controlled through the dimension q of the latent factor space. It allows thus an interpolation in model complexities from isotropic to full covariance structures without any restrictions.

In Chapter 9, we consider the fitting of finite mixture models to binned and truncated multivariate data by maximum likelihood via the EM algorithm. The solution for an arbitrary number of dimensions of the feature vector is specialized to the case of bivariate normal mixtures.

Chapter 10 is on the use of mixture distributions to model failure-time data in a variety of situations, which occur in reliability and survival analyses. The focus is on the use of mixture distributions to model time to failure in the case of competing risks or failures.

In Chapter 11, a case study is provided to illustrate the use of mixture models in the analysis of multivariate directional data. Mixtures of Kent distributions are used as an aid in joint set identification.

In Chapter 12, we consider methods for improving the speed of the EM algorithm for the maximum likelihood fitting of mixture models to large databases that preserve the simplicity of implementation of the EM in its standard form. They include the incremental version of the EM algorithm, where only a partial E-step is performed before each M-step, and a sparse version, where not all the posterior probabilities of component membership are updated on each iteration. The use of multiresolution kd-trees to speed up the implementation of the E-step is also described. In addition, we consider how the EM algorithm can be scaled to handle very large databases with a limited memory buffer.

In Chapter 13, recent advances on hidden Markov models are covered. Hidden Markov models are increasingly being adopted in applications, since they provide a convenient way of formulating an extension of a mixture model to allow for dependent data. We discuss hidden Markov chain models in the one-dimensional case and hidden Markov random fields in two or higher dimensions.

A brief account of some of the available software for the fitting of mixture models is provided in the Appendix. This account includes a description of the program EMMIX (McLachlan et al., 1999).

The authors wish to thank Dr. Angus Ng for many helpful and insightful discussions on mixture models, and Katrina Monico for her constructive comments and suggestions on drafts of the manuscript. They would also like to acknowledge gratefully financial support from the Australian Research Council. Thanks are due too to the authors and owners of copyrighted material for permission to reproduce tables and figures.

Brisbane, Australia

Geoffrey J. McLachlan
David Peel

1

General Introduction

1.1 INTRODUCTION

1.1.1 Flexible Method of Modeling

Finite mixtures of distributions have provided a mathematical-based approach to the statistical modeling of a wide variety of random phenomena. Because of their usefulness as an extremely flexible method of modeling, finite mixture models have continued to receive increasing attention over the years, from both a practical and theoretical point of view. Indeed, in the past decade the extent and the potential of the applications of finite mixture models have widened considerably. Fields in which mixture models have been successfully applied include astronomy, biology, genetics, medicine, psychiatry, economics, engineering, and marketing, among many other fields in the biological, physical, and social sciences. In these applications, finite mixture models underpin a variety of techniques in major areas of statistics, including cluster and latent class analyses, discriminant analysis, image analysis, and survival analysis, in addition to their more direct role in data analysis and inference of providing descriptive models for distributions.

The usefulness of mixture distributions in the modeling of heterogeneity in a cluster analysis context is obvious. In another example where there is group structure, they have a very useful role in assessing the error rates (sensitivity and specificity) of diagnostic and screening procedures in the absence of a gold standard. But as any continuous distribution can be approximated arbitrarily well by a finite mixture of normal densities with common variance (or covariance matrix in the multivariate case), mixture models provide a convenient semiparametric framework in which to model unknown distributional shapes, whatever the objective, whether it be, say, density estimation or the flexible construction of Bayesian priors. For example, Priebe (1994) showed that with $n = 10,000$ observations, a log normal density can be well approximated by a mixture of about 30 normals. In contrast, a kernel density estimator uses a mixture of 10,000 normals. A mixture model is able to model quite complex distributions through an appropriate choice of its components to represent accurately the local areas of support of the true distribution. It can thus handle situations where a single parametric family is unable to provide a satisfactory model for local variations in the observed data. Inferences about the modeled phenomenon can be made without difficulties from the mixture components, since the latter are chosen for their tractability.

This flexibility allows mixture models to play a useful role in neural networks (Bishop, 1995, Section 5.9). For with neural networks formed using radial basis functions, the input data can be modeled by a mixture model (for example, a normal mixture). That is, the basis functions can be taken to be the components of this mixture model after estimation by maximum likelihood from the input data. The second-layer weights in the neural network can then be estimated from the input data and their known outputs.

1.1.2 Initial Approach to Mixture Analysis

One of the first major analyses involving the use of mixture models was undertaken just over 100 years ago by the famous biometrician Karl Pearson. In his now classic paper, Pearson (1894) fitted a mixture of two normal probability density functions with different means μ_1 and μ_2 and variances σ_1^2 and σ_2^2 in proportions π_1 and π_2 to some data provided by Weldon (1892, 1893). The latter paper may have been the first ever to advocate statistical analysis as a primary method for studying biological problems (Pearson, 1906); see Stigler (1986, Chapter 8) and Tarter and Lock (1993, Chapter 1) for a more detailed account.

The data set analyzed by Pearson (1894) consisted of measurements on the ratio of forehead to body length of $n = 1000$ crabs sampled from the Bay of Naples. These measurements, which were recorded in the form of $v = 29$ intervals, are displayed in Figure 1.1, along with the plot of the density of a single normal distribution fitted to them. Weldon (1893) had speculated that the asymmetry in the histogram of these data might be a signal that this population was evolving toward two new subspecies. Sensing that his own mathematical training was inadequate, Weldon turned to his colleague Karl Pearson for assistance.

Pearson's (1894) mixture model-based approach suggested that there were two subspecies present. This paper was the first of two monstrous memoirs in a series of "Contributions to the Mathematical Theory of Evolution" (Stigler, 1986, Chapter 10). In Figure 1.1 we have plotted the density of the two-component normal mixture, as obtained by using maximum likelihood to fit this model to the data in their original interval form. Pearson (1894) had used the method of moments to fit this mixture model to the mid-points of the intervals, which, for this data set, gives a fit very similar to that obtained with the more efficient method of maximum likelihood. It can be seen that this mixture of two normal heteroscedastic components (components with unequal variances) does the job that Pearson (1894) had intended it to do, and that was to accommodate the apparent skewness in the data, which cannot be modeled adequately by the symmetric normal distribution. Pearson (1894) obtained his moments-based estimates of the five parameters of his normal heteroscedastic mixture model as a solution of a ninth degree polynomial (nonic). The computational effort in fitting this model (that is, finding the roots of a nonic) must have been at the time a daunting prospect to potential users of this mixture methodology. Indeed, as Everitt (1996) noted in his historical review of the development of finite mixture models, Charlier (1906) was led to comment "The solution of an equation of the ninth degree, where almost all powers, to the ninth, of the unknown quantity are existing, is, however, a very laborious task. Mr. Pearson has indeed possessed the energy to perform this heroic task in some instances in his first memoir on these topics from the year 1894. But I fear that he will have few successors, if the dissection of the frequency curve into two components is not very urgent." Not surprisingly, various attempts were made over the ensuing years to simplify Pearson's (1894) moments-based approach to the fitting of a normal mixture model.

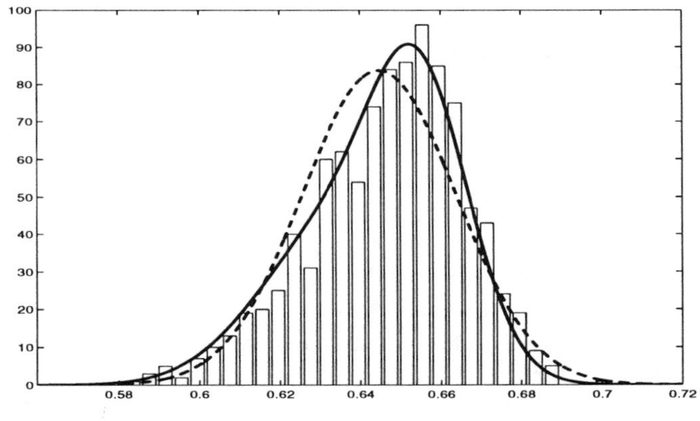

Fig. 1.1 Plot of forehead to body length data on 1000 crabs and of the fitted one-component (dashed line) and two-component (solid line) normal mixture models.

1.1.3 Impact of EM Algorithm

As to be elaborated on in Section 1.18, where we give a brief history of finite mixture models, it has only been in the last 20 or so years that considerable advances have been made in the fitting of finite mixture models, in particular by the method of maximum likelihood. Even with the advent of high-speed computers, there had been some reluctance in the past to fit mixture models to data of more than one dimension, possibly because of a lack of understanding of issues that arise with their fitting. They include the presence of multiple maxima in the mixture likelihood function and the unboundedness of the likelihood function in the case of normal components with unequal covariance matrices. But as the difficulties concerning these computational issues came to be properly understood and successfully addressed, it led to the increasing use of mixture models in practice.

In the 1960s, the fitting of finite mixture models by maximum likelihood had been studied in a number of papers, including the seminal papers by Day (1969) and Wolfe (1965, 1967, 1970). However, it was the publication of the seminal paper of Dempster, Laird, and Rubin (1977) on the EM algorithm that greatly stimulated interest in the use of finite mixture distributions to model heterogeneous data. This is because the fitting of mixture models by maximum likelihood is a classic example of a problem that is simplified considerably by the EM's conceptual unification of maximum likelihood (ML) estimation from data that can be viewed as being incomplete. As Aitkin and Aitkin (1994) noted, almost all the post-1978 applications of mixture modeling reported in the books on mixtures by Titterington, Smith, and Makov (1985) and McLachlan and Basford (1988) use the EM algorithm; see McLachlan and Krishnan (1997, Section 1.8). This also applies to the applications in this book.

1.2 OVERVIEW OF BOOK

The use of the EM algorithm for the fitting of finite mixture models, especially normal mixture models, has been demonstrated by McLachlan and Basford (1988) for the analysis of data arising from a wide variety of fields. For most commonly used parametric formulations of finite mixture models, the use of the EM algorithm to find a local maximizer of the likelihood function is straightforward. However, a number of issues remain. For example, as the likelihood function for mixture models usually has multiple local maxima, there is the question of which root of the likelihood equation corresponding to a local maximum of the likelihood function (that is, which local maximizer) to choose as the estimate of the vector of unknown parameters. Typically, the desired root corresponds to the global maximizer of the likelihood function in those situations where the likelihood function is bounded over the parameter space. But with mixtures of normal components with unequal variances in the univariate case or unequal covariance matrices in the multivariate case, the likelihood function is unbounded. In this case, the choice of root of the likelihood equation is not as obvious as in the bounded case and so requires careful consideration in practice. There is also the associated problem of how to select suitable starting values for the EM algorithm in the search of appropriate roots of the likelihood equation in the first instance. Another important consideration with the fitting of finite mixture models concerns the choice of the number of components g in the mixture model in those applications where g has to be inferred from the data.

In this book we give an extensive coverage of these problems, focusing on the latest developments. Indeed, almost 40% of the references in the book have been published since 1995.

Practitioners are increasingly turning to Bayesian methods for the analysis of complicated statistical models. This move is due in large part to the advent of inexpensive high speed computers and the simultaneous rapid development in posterior simulation techniques such as Markov chain Monte Carlo (MCMC) methods for enabling Bayesian estimation to be undertaken. In this book, we consider the latest developments in Bayesian estimation of mixture models.

One of the major problems of interest in mixture models concerns the choice of the number of components. In recent times, a number of new criteria have been suggested, some of which, like the integrated classification criterion, have given encouraging results in empirical studies designed to test their performance. In this book we discuss these various criteria, along with standard criteria such as the Bayesian information criterion (BIC) and the bootstrap likelihood ratio test.

As many applications of nonnormal mixtures are with components belonging to the exponential family, consideration is given to the family of mixtures of generalized linear models (GLMs). In this framework, the mixing proportions, as well as the component distributions, are allowed to depend on some associated covariates. A common way in which mixtures of GLMs arises in practice is in the handling of overdispersion in a single GLM. A convenient way to proceed in this case is to introduce a random effect into the linear predictor and to consider a mixture of such models with different intercepts in different proportions.

In work related to mixtures of GLMs, the mixtures-of-experts model is considered, along with its extension, the hierarchical mixtures-of-experts (HMEs) model. This approach, which combines aspects of finite mixture models and GLMs, provides a comparatively fast learning and good generalization for nonlinear regression problems, including classification. It is thus a serious competitor to well-known statistical procedures such as MARS and CART.

On other material concerning the use of nonnormal components, there is treatment of the case where the feature variables are mixed, with some being categorical and some continuous. There is also a separate chapter devoted to mixture distributions in modeling failure-time data with competing risks. A case study is presented on the use of mixtures of Kent distributions for the analysis of multivariate directional data.

The problem of fitting finite mixture models to binned and truncated data is also covered. The methodology is illustrated with a case study on the diagnosis of iron-deficient anemia by the mixture modeling of binned and truncated data on a patient in the form of volume and hemoglobin concentration of red blood cells, as measured by a cytometric blood cell counter.

This book contains a number of recent results on mixture models by the authors, including the use of mixtures of t distributions to provide a robust extension of normal mixture models, and the use of mixtures of factor analyzers to enable the fitting of normal mixture models to high-dimensional data. With the considerable attention being given to the analysis of large data sets, as in typical data mining applications, recent work on speeding up the implementation of the EM algorithm is discussed, including (a) the use of the sparse/incremental EM and of multiresolution kd-trees and (b) the scaling of the EM algorithm to massively large databases where there is a limited memory buffer.

Hidden Markov models are increasingly being used, as they provide a way of formulating an extension of mixture models to allow for dependent data. This book reviews the latest results on maximum likelihood and Bayesian methods of estimation for such models. It concludes with a concise summary of software available for the fitting of mixture models, including the EMMIX program (McLachlan et al., 1999).

Numerous examples of applications of mixture models are given throughout the book to demonstrate the methodology. Where available, the data sets considered in this book may be found on the World Wide Web at http://www.maths.uq.edu.au/~gjm.

1.3 BASIC DEFINITION

We let $\boldsymbol{Y}_1, \ldots, \boldsymbol{Y}_n$ denote a random sample of size n, where $\boldsymbol{Y}_j$ is a p-dimensional random vector with probability density function $f(\boldsymbol{y}_j)$ on $\mathbb{R}^p$. In practice, $\boldsymbol{Y}_j$ contains the random variables corresponding to p measurements made on the jth recording of some features on the phenomenon under study. We let $\boldsymbol{Y} = (\boldsymbol{Y}_1^T, \ldots, \boldsymbol{Y}_n^T)^T$, where the superscript T denotes vector transpose. Note that we are using $\boldsymbol{Y}$ to represent the entire sample; that is, $\boldsymbol{Y}$ is an n-tuple of points in $\mathbb{R}^p$. Where possible, a realization of a random vector is denoted by the corresponding lower-case letter. For example, $\boldsymbol{y} = (\boldsymbol{y}_1^T, \ldots, \boldsymbol{y}_n^T)^T$ denotes an observed random sample where $\boldsymbol{y}_j$ is the observed value of the random vector $\boldsymbol{Y}_j$.

Although we are taking the feature vector $\boldsymbol{Y}_j$ to be a continuous random vector here, we can still view $f(\boldsymbol{y}_j)$ as a density in the case where $\boldsymbol{Y}_j$ is discrete by the adoption of counting measure. We suppose that the density $f(\boldsymbol{y}_j)$ of $\boldsymbol{Y}_j$ can be written in the form

$$f(\boldsymbol{y}_j) = \sum_{i=1}^{g} \pi_i f_i(\boldsymbol{y}_j), \tag{1.1}$$

where the $f_i(\boldsymbol{y}_j)$ are densities and the π_i are nonnegative quantities that sum to one; that is,

$$0 \leq \pi_i \leq 1 \qquad (i = 1, \ldots, g) \tag{1.2}$$

and

$$\sum_{i=1}^{g} \pi_i = 1. \tag{1.3}$$

The quantities $\pi_1, \ldots, \pi_g$ are called the mixing proportions or weights. As the functions, $f_1(\boldsymbol{y}_j), \ldots, f_g(\boldsymbol{y}_j)$, are densities, it is obvious that (1.1) defines a density. The $f_i(\boldsymbol{y}_j)$ are called the *component densities* of the mixture. We shall refer to the density (1.1) as a g-component finite mixture density and refer to its corresponding distribution function $F(\boldsymbol{y}_j)$ as a g-component finite mixture distribution. Since we shall be focusing almost exclusively on finite mixtures of distributions, we shall usually refer to finite mixture models as just mixture models in the sequel.

In this formulation of the mixture model, the number of components g is considered fixed. But of course in many applications, the value of g is unknown and has to be inferred from the available data, along with the mixing proportions and the parameters in the specified forms for the component densities.

When the number of components is allowed to increase with the sample size n, the model is called a Gaussian mixture sieve; see Geman and Hwang (1982), Roeder

(1992), Priebe and Marchette (1993), Priebe (1994), and Roeder and Wasserman (1997).

1.4 INTERPRETATION OF MIXTURE MODELS

An obvious way of generating a random vector $\boldsymbol{Y}_j$ with the g-component mixture density $f(\boldsymbol{y}_j)$, given by (1.1), is as follows. Let Z_j be a categorical random variable taking on the values $1, \ldots, g$ with probabilities $\pi_1, \ldots, \pi_g$, respectively, and suppose that the conditional density of $\boldsymbol{Y}_j$ given $Z_j = i$ is $f_i(\boldsymbol{y}_j)$ $(i = 1, \ldots, g)$. Then the unconditional density of $\boldsymbol{Y}_j$ (that is, its marginal density) is given by $f(\boldsymbol{y}_j)$. In this context, the variable Z_j can be thought of as the component label of the feature vector $\boldsymbol{Y}_j$. In later work, it is convenient to work with a g-dimensional component-label vector $\boldsymbol{Z}_j$ in place of the single categorical variable Z_j, where the ith element of $\boldsymbol{Z}_j$, $Z_{ij} = (\boldsymbol{Z}_j)_i$, is defined to be one or zero, according to whether the component of origin of $\boldsymbol{Y}_j$ in the mixture is equal to i or not $(i = 1, \ldots, g)$. Thus $\boldsymbol{Z}_j$ is distributed according to a multinomial distribution consisting of one draw on g categories with probabilities $\pi_1, \ldots, \pi_g$; that is,

$$\text{pr}\{\boldsymbol{Z}_j = \boldsymbol{z}_j\} = \pi_1^{z_{1j}} \pi_2^{z_{2j}} \cdots \pi_g^{z_{gj}}. \tag{1.4}$$

We write

$$\boldsymbol{Z}_j \sim \text{Mult}_g(1, \boldsymbol{\pi}), \tag{1.5}$$

where $\boldsymbol{\pi} = (\pi_1, \ldots, \pi_g)^T$.

In the interpretation above of a mixture model, an obvious situation where the g-component mixture model (1.1) is directly applicable is where $\boldsymbol{Y}_j$ is drawn from a population G which consists of g groups, $G_1, \ldots, G_g$, in proportions $\pi_1, \ldots, \pi_g$. If the density of $\boldsymbol{Y}_j$ in group G_i is given by $f_i(\boldsymbol{y}_j)$ for $i = 1, \ldots, g$, then the density of $\boldsymbol{Y}_j$ has the g-component mixture form (1.1). In this situation, the g components of the mixture can be physically identified with the g externally existing groups, $G_1, \ldots, G_g$.

In biometric applications for instance, a source of the heterogeneity is often age, sex, species, geographical origin, and cohort status. For example, a population G may consist of two groups G_1 and G_2, corresponding to those members with or without a particular disease that is under study. The problem may be to estimate the disease prevalence (that is, the mixing proportion π_1 here) on the basis of some feature vector measured on a randomly selected sample of members of the population. In the case study of Do and McLachlan (1984), in which $p = 4$ variables were measured on the skulls of Malaysian rats collected from owl pellets, the components of the fitted mixture corresponded to $g = 7$ different species of rats. The aim of their study was to assess the rat diet of owls in terms of the proportion of each species of rat represented in the fitted mixture model.

We shall see in this book that there are many other examples in practice where the population is a mixture of g distinct groups that are known *a priori* to exist in some physical sense. However, there are also many examples involving the use of mixture models where the components cannot be identified with externally existing groups as above. In some instances, the components are introduced into the mixture model to allow for greater flexibility in modeling a heterogeneous population that is apparently unable to be modeled by a single component distribution. At the extreme end of this exercise, we obtain the nonparametric kernel estimate of a density if we fit a mixture of $g = n$ components in equal proportions $1/n$, where n is the size of the observed sample. For example, if $y_1, \ldots, y_n$ denote an observed (univariate) sample of size n, then we obtain the kernel estimate of the density of Y_j given by

$$\hat{f}(y_j) = \frac{1}{nh} \sum_{i=1}^{n} k((y_j - y_i)/h), \tag{1.6}$$

if in (1.1) we set $g = n$ and $\pi_i = 1/n$ and take

$$f_i(y_j) = h^{-1} k((y_j - y_i)/h)$$

for some kernel function $k(\cdot)$ and parameter h. Usually, the kernel $k(\cdot)$, which is a density, has its mode at the origin; see, for example, the monographs of Devroye and Györfi (1985), Silverman (1986), and Scott (1992) on nonparametric density estimation.

Thus for values of the number of components g between 1 and the sample size n, mixture models can be can be viewed as a semiparametric compromise between (a) the fully parametric model as represented by a single ($g = 1$) parametric family and (b) a nonparametric model as represented in the case of $g = n$ by the kernel method of density estimation. Although a single normal distribution is obviously inadequate for modeling a continuous skewed distribution, a mixture of two normal distributions may provide an adequate fit as, for example, considered in Section 1.6 on the modeling of the distribution of blood pressure in humans.

Thus it can be seen that mixture models occupy an interesting niche between parametric and nonparametric approaches to statistical estimation. As explained by Jordan and Xu (1995), mixture model-based approaches are parametric in that parametric forms $f_i(\boldsymbol{y}_j; \boldsymbol{\theta}_i)$ are specified for the component density functions, but that they can also be regarded as nonparametric by allowing the number of components g to grow. Hence mixture models have much of the flexibility of nonparametric approaches, while retaining some of the advantages of parametric approaches, such as keeping the dimension of the parameter space down to a reasonable size. Mixture models therefore provide a convenient method of density estimation that lies somewhere between parametric models and kernel density estimators.

Concerning the modeling of count data, the fitting of a single Poisson distribution often forces too much structure on the data leading to problems such as overdispersion. The use of a mixture model allows a compromise between the homogeneous Poisson model and nonparametric models which, although avoiding strong distributional assumptions, have other disadvantages including high-data dependency of model estimates (Böhning et al., 1994; Böhning, 1999). The problem of overdispersion in the modeling of count data is to be taken up further in Chapter 5.

1.5 SHAPES OF SOME UNIVARIATE NORMAL MIXTURES

1.5.1 Mixtures of Two Normal Homoscedastic Components

To illustrate some of the shapes taken by a univariate normal mixture density, we first consider a mixture of two univariate normal components with common variance σ^2 and means μ_1 and μ_2 in proportions π_1 and π_2, so that

$$f(y_j) = \pi_1\phi(y_j;\, \mu_1,\, \sigma^2) + \pi_2\phi(y_j;\, \mu_2,\, \sigma^2), \tag{1.7}$$

where

$$\phi(y_j;\, \mu,\, \sigma^2) = (2\pi)^{-\frac{1}{2}}\sigma^{-1}\exp\{-\tfrac{1}{2}(y_j - \mu)^2/\sigma^2\} \tag{1.8}$$

denotes the univariate normal density with mean μ and variance σ^2.

If the two component normal densities are sufficiently far apart, then one would expect the mixture density $f(y_j)$ to resemble two normal densities side by side, that is, a bimodal density. To demonstrate this, we have plotted this normal mixture density in Figure 1.2 for various values of Δ in the case where $\mu_1 = 0$, $\mu_2 = \Delta, \sigma^2 = 1$, and the proportions are equal $(\pi_1 = \pi_2 = 0.5)$. It can be seen that as Δ increases, the shape of the mixture density $f(y_j)$ changes from being unimodal to bimodal. The threshold for this change, in the present case of two univariate normal densities in equal proportions, is $\Delta = 2$ where, more generally,

$$\Delta = \mid \mu_1 - \mu_2 \mid / \sigma$$

is the Mahalanobis distance between the homoscedastic components of the normal mixture density; see Titterington et al. (1985, Section 5.5). This figure demonstrates how the graphical resolution of a mixture into its constituent components can be a straightforward task for widely separated components ($\Delta = 3$ and 4), but that it can be quite a challenge when the components are close together $(\Delta = 1)$.

If the means of the two component densities in the mixture model (1.7) are close enough together, then the overlap between the two component densities would tend to obscure the distinction between them and the result would be an asymmetric density if the components are not represented in equal proportions. To demonstrate this, we give in Figure 1.3 the plots of the mixture density $f(y_j)$ corresponding to those in Figure 1.2 but where now the components are mixed in the unequal proportions $\pi_1 = 0.75$ and $\pi_2 = 0.25$. In this case, it can be seen that the shape of the mixture density $f(y_j)$ changes from being bimodal to skewed in appearance for $\Delta = 4$ in Figure 1.3(d). The shape of the mixture density $f(y_j)$ for $\Delta = 3$ in Figure 1.3(c) demonstrates bitangentiality, which occurs when there are two distinct points at which there is a common tangent to the density. Thus, bitangentiality is implied by, but does not imply, bimodality. Informally, bimodality implies an extra hump, but bitangentiality merely an extra bump in departures from unimodality (Titterington et al., 1985, Section 3.3).

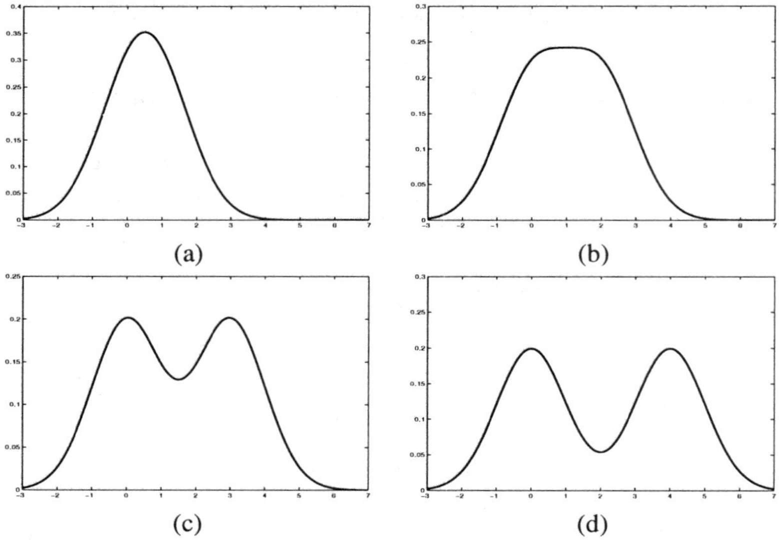

Fig. 1.2 Plot of a mixture density of two univariate normal components in equal proportions with common variance $\sigma^2 = 1$ and means $\mu_1 = 0$ and $\mu_2 = \Delta$ in the cases: (a) $\Delta = 1$; (b) $\Delta = 2$; (c) $\Delta = 3$; (d) $\Delta = 4$.

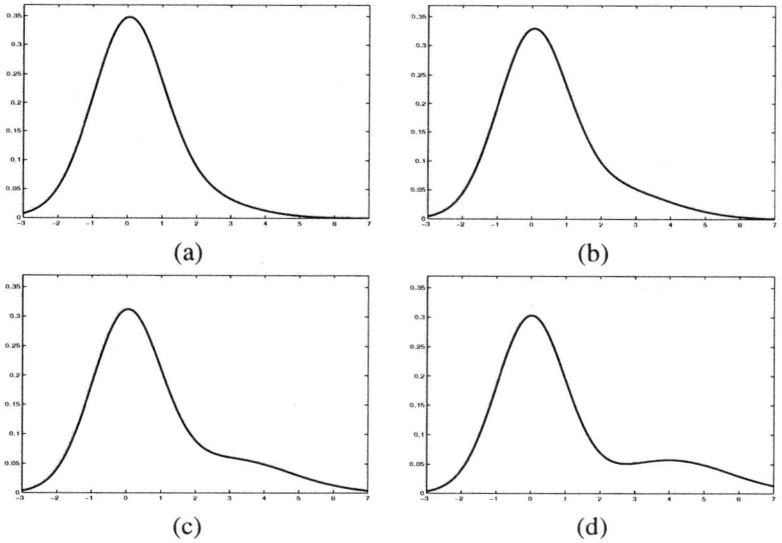

Fig. 1.3 Plot of a mixture density of two univariate normal components in proportions 0.75 and 0.25 with common variance $\sigma^2 = 1$ and means $\mu_1 = 0$ and $\mu_2 = \Delta$ in the cases: (a) $\Delta = 1$; (b) $\Delta = 2$; (c) $\Delta = 3$; (d) $\Delta = 4$.

For the univariate normal mixture density $f(y_j)$ given by (1.7), Preston (1953) obtained explicit expressions for its skewness γ_1 and kurtosis γ_2 in terms of the separation Δ between its components and the relative size of its mixing proportions, namely,

$$\gamma_1 = \frac{a(a-1)\Delta^3}{\{a\Delta^2 + (a+1)^2\}^{3/2}} \tag{1.9}$$

and

$$\gamma_2 = \frac{a(a^2 - 4a + 1)\Delta^4}{\{a\Delta^2 + (a+1)^2\}^2}, \tag{1.10}$$

where a is the ratio of the larger mixing proportion to the smaller.

For simplicity here, we have taken the normal mixture to have only $g = 2$ components with a common variance. Obviously, more flexibility is introduced by having more than two components and not necessarily constraining them to have the same variances, as illustrated in the next section.

1.5.2 Mixtures of Univariate Normal Heteroscedastic Components

A useful paper in characterizing the shape of the density of a mixture of two normal components with unrestricted variances, σ_1^2 and σ_2^2, is Eisenberger (1964). For instance, this density cannot be bimodal if

$$\Delta^2 < (27\sigma_2^2)/\{4(1+k)\},$$

where $k = \sigma_2^2/\sigma_1^2$. For

$$\Delta^2 > (27\sigma_2^2)/\{4(1+k)\},$$

a value of π_1 exists for which the density is bimodal.

As the family of g-component normal mixtures is very flexible, Marron and Wand (1992) used it to represent a wide variety of density shapes in their analytical study of the mean integrated squared error of the kernel density estimator. To demonstrate that the family of normal mixtures is indeed a very broad one, they gave fifteen examples of the (univariate) normal mixture density, corresponding to various combinations of the components, as listed in Table 1.1. The plots are reproduced here in Figure 1.4. The idea that any density can be approximated arbitrarily closely by a normal mixture is made visually clear in Figure 1.4.

Some idea of the range of shapes provided by mixtures of bivariate normal components may be obtained from Johnson (1987, Section 4.2), where contour plots are given for various selections of the parameters in the case of $g = 2$ components.

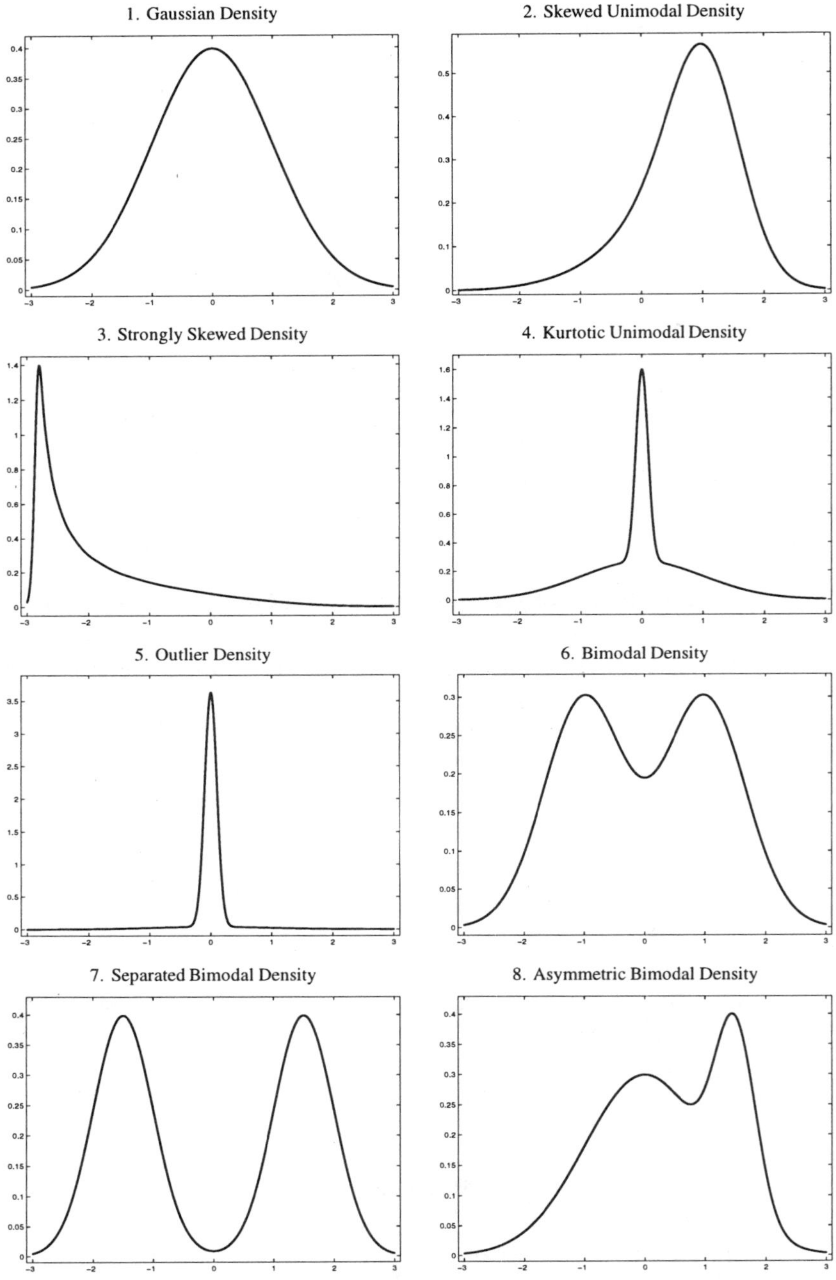

Fig. 1.4 Plots of normal mixture densities. From Marron and Wand (1992).

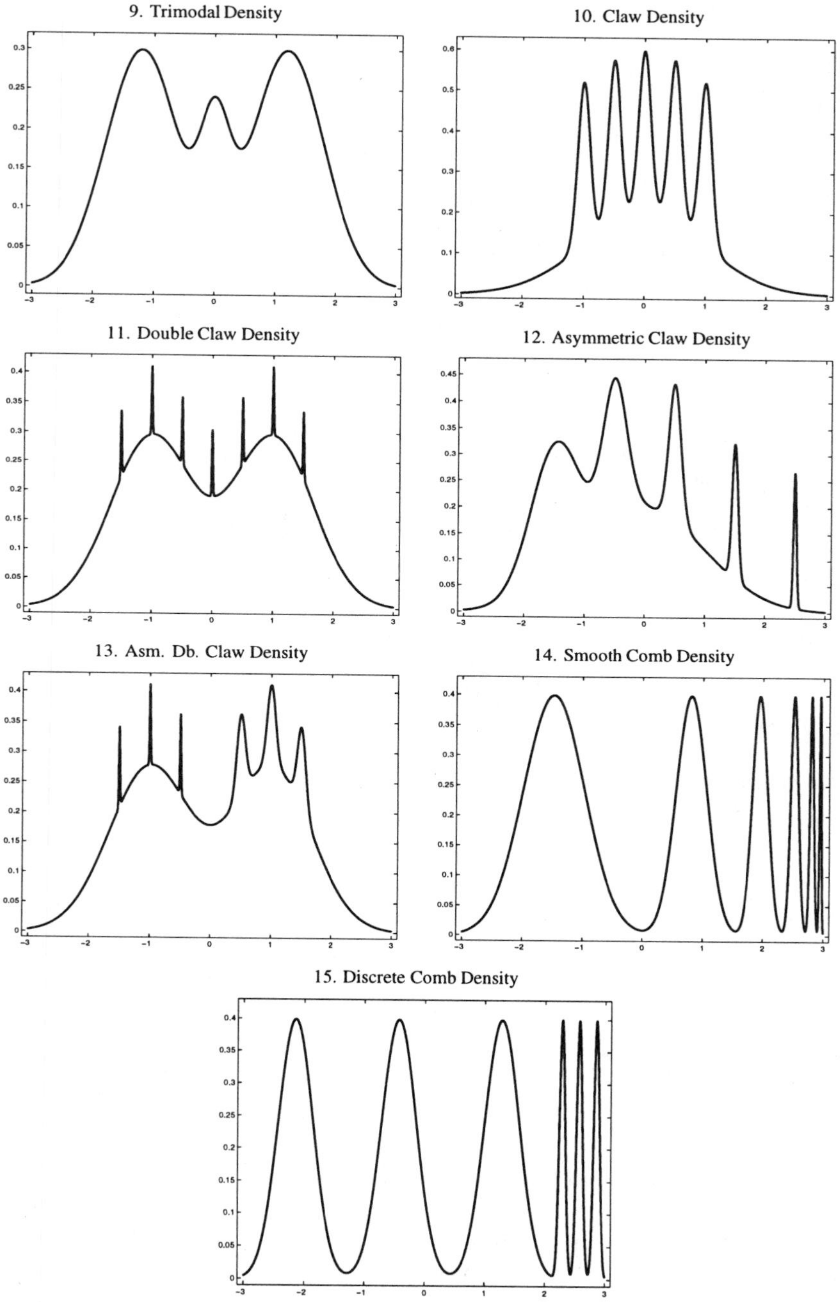

Fig. 1.4 Continued

Table 1.1 Parameters for Fifteen Examples of a Normal Mixture Density

Density	$f(y)$
1. Gaussian	$N(0,1)$
2. Skewed unimodal	$\frac{1}{5}N(0,1)+\frac{1}{5}N(\frac{1}{2},(\frac{2}{3})^2)+\frac{3}{5}N(\frac{13}{15},(\frac{5}{9})^2)$
3. Strongly skewed	$\sum_{i=0}^{7}\frac{1}{8}N(3\{(\frac{2}{3})^i-1\},(\frac{2}{3})^{2i})$
4. Kurtotic unimodal	$\frac{2}{3}N(0,1)+\frac{1}{3}N(0,(\frac{1}{10})^2)$
5. Outlier	$\frac{1}{10}N(0,1)+\frac{9}{10}N(0,(\frac{1}{10})^2)$
6. Bimodal	$\frac{1}{2}N(-1,(\frac{2}{3})^2)+\frac{1}{2}N(1,(\frac{2}{3})^2)$
7. Separated bimodal	$\frac{1}{2}N(-\frac{3}{2},(\frac{1}{2})^2)+\frac{1}{2}N(\frac{3}{2},(\frac{1}{2})^2)$
8. Skewed bimodal	$\frac{3}{4}N(0,1)+\frac{1}{4}N(\frac{3}{2},(\frac{1}{3})^2)$
9. Trimodal	$\frac{9}{20}N(-\frac{6}{5},(\frac{3}{5})^2)+\frac{9}{20}N(\frac{6}{5},(\frac{3}{5})^2)+\frac{1}{10}N(0,(\frac{1}{4})^2)$
10. Claw	$\frac{1}{2}N(0,1)+\sum_{i=0}^{4}\frac{1}{10}N(i/2-1,(\frac{1}{10})^2)$
11. Double claw	$\frac{49}{100}N(-1,(\frac{2}{3})^2)+\frac{49}{100}N(1,(\frac{2}{3})^2)$ $+\sum_{i=0}^{6}\frac{1}{350}N((i-3)/2,(\frac{1}{100})^2)$
12. Asymmetric claw	$\frac{1}{2}N(0,1)+\sum_{i=-2}^{2}(2^{1-i}/31)N(i+\frac{1}{2},(2^{-i}/10)^2)$
13. Asymmetric double claw	$\sum_{i=0}^{1}\frac{46}{100}N(2i-1,(\frac{2}{3})^2)$ $+\sum_{i=1}^{3}\frac{1}{300}N(-i/2,(\frac{1}{100})^2)$ $+\sum_{i=1}^{3}\frac{7}{300}N(i/2,(\frac{7}{100})^2)$
14. Smooth comb	$\sum_{i=0}^{5}(2^{5-i}/63)N((65-96(\frac{1}{2})^i)/21,(\frac{32}{63})^2/2^{2i})$
15. Discrete comb	$\sum_{i=0}^{2}\frac{2}{7}N((12i-15)/7,(\frac{2}{7})^2)+\sum_{i=8}^{10}\frac{1}{21}N(2i/7,(\frac{1}{21})^2)$

Source: Adapted from Marron and Wand (1992).

1.6 MODELING OF ASYMMETRICAL DATA

As seen in the previous section, normal mixture densities can play a useful role in modeling the distribution of data that have asymmetrical distributions, as recognized by Pearson (1894). Another way of proceeding with the modeling of skewed data is to first apply a transformation in an attempt to remove or reduce the asymmetry in the data. For this purpose the log transformation is often helpful. It is well known that the parameters of a normal mixture of two univariate normal homoscedastic components may be chosen so that its density is close in appearance to that of a log normal distribution. If the random variable Y_j has a normal distribution with mean μ and variance σ^2 on the log scale, then its density is given by

$$(2\pi)^{-\frac{1}{2}} y_j^{-1} \sigma^{-1} \exp\{-\tfrac{1}{2}(\log y_j - \mu)^2/\sigma^2\}. \tag{1.11}$$

To illustrate this point, Titterington et al. (1985) plotted the two-component univariate normal mixture density

$$f(y) = \pi_1\, \phi(y_j;\, \mu_1,\, \sigma^2) + \pi_2\, \phi(y_j;\, \mu_2,\, \sigma^2),$$

where $\pi_1 = 0.9, \mu_1 = 9.5, \mu_2 = 13.5$, and $\sigma^2 = 2.5$, and the log normal density (1.11), where $\mu = \log(10)$ and $\sigma^2 = 0.04$. Their plots are reproduced here in Figure 1.5.

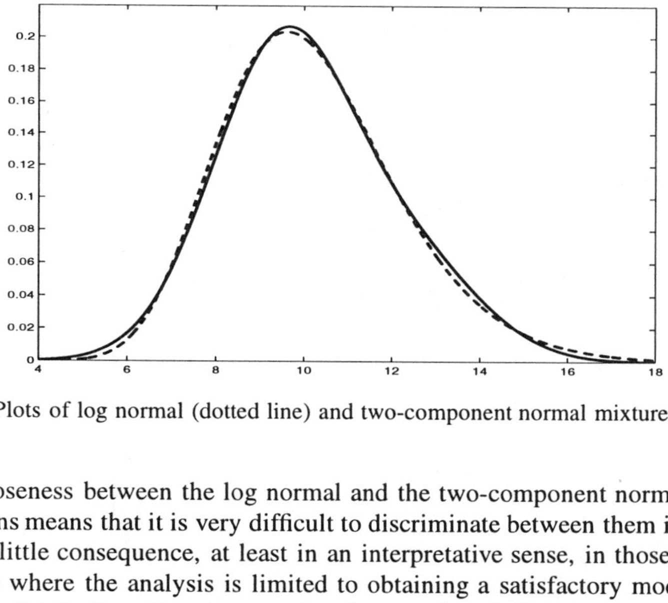

Fig. 1.5 Plots of log normal (dotted line) and two-component normal mixture (solid line) densities.

This closeness between the log normal and the two-component normal mixture distributions means that it is very difficult to discriminate between them in practice. This is of little consequence, at least in an interpretative sense, in those situations in practice where the analysis is limited to obtaining a satisfactory model for the underlying distribution. (In a computational sense, there is the consequence of some extra effort in fitting a five-parameter normal mixture model over a two-parameter log normal model.) However, in some situations, the choice between the log normal and normal mixture models is of much interest.

Pearson (1895, p. 394) recognized the problems in attempting to distinguish between inherently skewed distributions and mixtures, saying "... how [we are] to discriminate between a true curve of skew type and a compound [that is, a mixture] curve, supposing we have no reason to suspect our statistics *a priori* of mixture. I have at present been unable to find any general condition among the moments, which would be impossible for a skew curve and possible for a compound, and so indicate compoundness. I do not, however, despair of one being found."

As an example of these problems, there are the issues associated with the modeling of the distribution of blood pressure by a mixture of two normal homoscedastic components, which surfaced in the very heated debate in the 1950s and 1960s (the famous "Pickering/Platt" debate) about the pathophysiology of hypertension; see Swales (1985) and Schork, Weder, and Schork (1990) and the references therein. Pickering and Platt were two noted English internists with differing views on the etiology of essential hypertension. Platt (1963) claimed that hypertension was a "disease" and placed much emphasis on his personal observations that the distribution of blood pressure has a skewness that may be the manifestation of the effects of a Mendelian dominant gene (that is, the blood pressure distribution admits a (two-component) mixture as the consequence of the mixing of two groups corresponding to the "hypertensive" and "normotensive" subpopulations). Pickering (1968) staunchly opposed Platt's interpretation, arguing that the designation "hypertension" was entirely arbitrary, being merely a label assigned to those with blood pressure readings in the upper tail of the distribution. Many researchers, including Clark et al. (1968) and McManus (1983), have since then attempted to settle the dispute by fitting normal mixtures to large samples of blood pressure values, but results have been inconclusive; see Schork, Allison, and Thiel (1996) for a recent account. As Schork et al. (1990) have noted, investigations into this issue should be widened to include the use of mixtures of normal heteroscedastic components. To demonstrate the inconclusive nature of this issue, we have plotted in Figure 1.6 the two-component normal mixture and log normal models fitted by Schork et al. (1990) to systolic and diastolic blood pressures collected on 941 white male subjects participating in a random, statewide blood pressure screening program in Michigan. The data were first standardized to adjust for the effects of age, height, and weight.

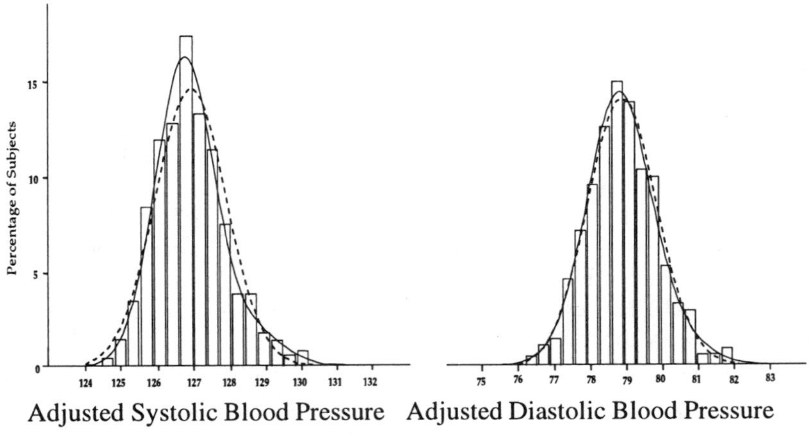

Fig. 1.6 Histograms of height-, weight- and age-adjusted systolic and diastolic blood pressures of 941 men. The superimposed solid and dotted curves represent the density of the fitted mixture of two normal homoscedastic components and single log normal component, respectively. From Schork et al. (1990).

The effect of skewness on hypothesis tests for the number of components in normal mixture models is to be considered further in Section 6.1.3. This problem has been considered by Maclean et al. (1976), Schork and Schork (1988), Gutierrez et al. (1995), and Ning and Finch (2000). In their work the Box–Cox (1964) transformation is employed initially in an attempt to obtain normal components.

1.7 NORMAL SCALE MIXTURE MODEL

Normal mixture models have been used in the investigation of the performances of certain estimators to departures from normality. In addition to this latter role of assessing the performance of estimators in nonnormal situations, normal mixture models have been used of course in the development of robust estimators. For example, under the contaminated normal family as suggested by Tukey (1960), the density of an observation is taken to be a mixture of two univariate normal densities with the same means but where the second component has a greater variance than the first. This family was introduced to model a population which follows a normal distribution except on those few occasions where a grossly atypical observation is recorded. That is,

$$f(y_j) = \pi_1 \phi(y_j;\, \mu,\, \sigma^2) + \pi_2 \phi(y_j;\, \mu,\, k\sigma^2), \tag{1.12}$$

where k is large and $\pi_2 = 1 - \pi_1$ is small, representing the small proportion of observations that have a relatively large variance. Huber (1964) subsequently considered more general forms of contamination of the normal distribution in the development of his robust M-estimators of a location parameter.

The normal scale mixture model (1.12) can be written as

$$f(y_j) = \int \phi(y_j; \mu; \sigma^2/u)\, dH(u), \tag{1.13}$$

where H is the probability distribution that places mass π_1 at the point $u = 1$ and mass $\pi_2 = 1 - \pi_1$ at the point $1/k$. If we replace H by the distribution of a chi-squared random variable on its degrees of freedom ν, we obtain the t distribution with ν degrees of freedom. The family of t distributions provides a heavy-tailed alternative to the normal family. The t distribution is considered further in Chapter 7 in the context of fitting mixtures of multivariate t components.

1.8 SPURIOUS CLUSTERS

We have seen in Section 1.5 that the shape of the density of a mixture of two normal components will be bimodal in appearance if the two components have sufficient separation between them. Hence, in practice, bimodality in a histogram of the feature data will obviously be suggestive of the possibility that the data have been drawn from a mixture distribution. However, bimodality in histograms of the data (or of linear combinations of the data if multivariate) does not always imply that the data have been sampled from a mixture distribution.

This point was illustrated in the seminal paper of Day (1969) on normal mixture models. To demonstrate the presence of spurious clusters in a data set, Day (1969) generated three random samples, each of size $n = 50$, from a spherically symmetric $p = 10$-dimensional normal distribution. He then plotted for each sample the histogram of the univariate projections $\hat{\boldsymbol{a}}^T \boldsymbol{y}_1, ..., \hat{\boldsymbol{a}}^T \boldsymbol{y}_n$, where

$$\hat{\boldsymbol{a}} = \hat{\boldsymbol{\Sigma}}^{-1}(\hat{\boldsymbol{\mu}}_1 - \hat{\boldsymbol{\mu}}_2),$$

and $\hat{\boldsymbol{\mu}}_1$, $\hat{\boldsymbol{\mu}}_2$, and $\hat{\boldsymbol{\Sigma}}$ are the estimates obtained in fitting a mixture of two ten-dimensional normal components with means $\boldsymbol{\mu}_1$ and $\boldsymbol{\mu}_2$ and common covariance matrix $\boldsymbol{\Sigma}$. That is, these univariate projections are the first canonical variates when two multivariate normal groups with means $\hat{\boldsymbol{\mu}}_1$ and $\hat{\boldsymbol{\mu}}_2$ and common covariance matrix $\hat{\boldsymbol{\Sigma}}$ are imposed on the data. The bimodal shape of these histograms suggested that the data were not generated from a unimodal distribution.

Following the approach of Day (1969), we generated a random sample of size $n = 50$ from a single $p = 10$-dimensional normal distribution with mean null vector and covariance matrix equal to the identity matrix and then plotted the univariate projections $\hat{\boldsymbol{a}}^T \boldsymbol{y}_1, ..., \hat{\boldsymbol{a}}^T \boldsymbol{y}_n$ of the generated data in histogram form. This plot is given in Figure 1.7. As with Day's (1969) simulated samples, the bimodal nature of the histogram suggests that the data have not come from a single normal distribution.

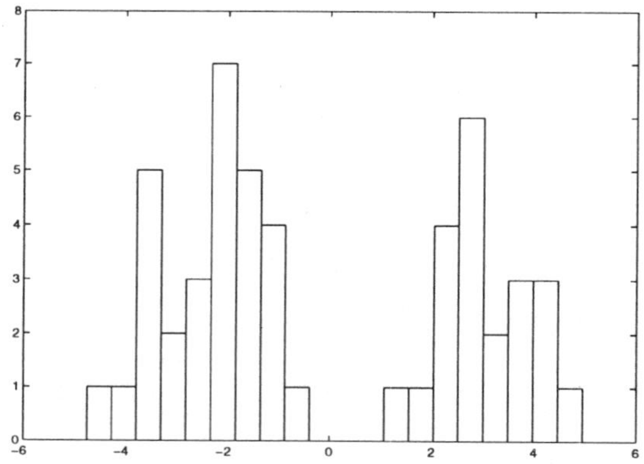

Fig. 1.7 Histogram of first canonical variate for ten-dimensional simulated normal data set of size $n = 50$.

It is of interest to see if this spurious clustering can be detected in practice. We now apply the likelihood ratio test statistic λ to the simulated data represented in Figure 1.7 to test the null hypothesis of a single normal component against the alternative of a two-component normal mixture with equal covariance matrices. The value of $-2 \log \lambda$ is 31.41. As is well known, regularity conditions do not hold for the likelihood ratio test statistic for this test to have its usual null distribution of chi-squared. Using a resampling approach to the assess the P-value, we generated $K = 199$ replications. As the realized value of 31.41 for $-2 \log \lambda$ is between the 93rd and 94th smallest replicated values of $-2 \log \lambda$, the P-value is assessed as approximately 47%. Hence the null hypothesis of a single normal component would be retained at any conventional level of significance.

The problem of testing for the number of components in a mixture model, including by the likelihood ratio test, is to be considered in Chapter 6.

1.9 INCOMPLETE-DATA STRUCTURE OF MIXTURE PROBLEM

In Section 1.4 we introduced the component-label vector $\boldsymbol{Z}_j$ of zero–one indicator variables to define the component in the mixture model (1.1) from which the feature random vector $\boldsymbol{Y}_j$ is viewed to have arisen. The concept of there being a label vector $\boldsymbol{Z}_j$ associated with each feature vector $\boldsymbol{Y}_j$ is a useful one, even though in a physical sense it may not always be appropriate to view the mixture model in this sense. It will be seen that this conceptualization of the mixture model in terms of $\boldsymbol{Y}_j$ and $\boldsymbol{Z}_j$ is most useful in that it allows the maximum likelihood estimate (MLE) of the mixture distribution to be computed via a straightforward application of the EM algorithm. It is also useful in implementing the MCMC methods in the fitting of mixture models in a Bayesian framework.

In this book the emphasis is on the estimation of mixture distributions on the basis of data, $\boldsymbol{y}_1, \ldots, \boldsymbol{y}_n$, usually available in the form of an observed random sample taken from the mixture density (1.1). That is, $\boldsymbol{y}_1, \ldots, \boldsymbol{y}_n$ are the realized values of n independent and identically distributed (i.i.d.) random vectors $\boldsymbol{Y}_1, \ldots, \boldsymbol{Y}_n$ with common density $f(\boldsymbol{y}_j)$. We write

$$\boldsymbol{Y}_1, \ldots, \boldsymbol{Y}_n \overset{\text{i.i.d.}}{\sim} F, \tag{1.14}$$

where $F(\boldsymbol{y}_j)$ denotes the distribution function corresponding to the mixture density $f(\boldsymbol{y}_j)$.

In the EM framework, the feature data $\boldsymbol{y}_1, \ldots, \boldsymbol{y}_n$ are viewed as being incomplete since their associated component-indicator vectors, $\boldsymbol{z}_1, \ldots, \boldsymbol{z}_n$, are not available. The complete-data vector is therefore declared to be

$$\boldsymbol{y}_c = (\boldsymbol{y}^T, \boldsymbol{z}^T)^T, \tag{1.15}$$

where

$$\boldsymbol{y} = (\boldsymbol{y}_1^T, \ldots, \boldsymbol{y}_n^T)^T \tag{1.16}$$

is the observed-data or incomplete-data vector and where

$$\boldsymbol{z} = (\boldsymbol{z}_1^T, \ldots, \boldsymbol{z}_n^T)^T \tag{1.17}$$

is the unobservable vector of component-indicator variables. It is assumed here that all the observations $\boldsymbol{y}_j$ have been completely recorded.

The component-label vectors $\boldsymbol{z}_1, \ldots, \boldsymbol{z}_n$ are taken to be the realized values of the random vectors $\boldsymbol{Z}_1, \ldots, \boldsymbol{Z}_n$, where, for independent feature data, it is appropriate to assume that they are distributed unconditionally as

$$\boldsymbol{Z}_1, \ldots, \boldsymbol{Z}_n \overset{\text{i.i.d.}}{\sim} \text{Mult}_g(1, \boldsymbol{\pi}). \tag{1.18}$$

The ith mixing proportion π_i can be viewed as the prior probability that the entity belongs to the ith component of the mixture ($i = 1, \ldots, g$), while the posterior probability that the entity belongs to the ith component with $\boldsymbol{y}_j$ having been observed on it, is given by

$$\begin{aligned}\tau_i(\boldsymbol{y}_j) &= \text{pr}\{\text{ entity} \in i\text{th component } \mid \boldsymbol{y}_j\}\\ &= \text{pr}\{Z_{ij} = 1 \mid \boldsymbol{y}_j\}\\ &= \pi_i f_i(\boldsymbol{y}_j)/f(\boldsymbol{y}_j) \quad (i = 1, \ldots, g;\ j = 1, \ldots, n).\end{aligned} \tag{1.19}$$

In Section 1.15.2 we shall consider the formation of an optimal rule of allocation in terms of these posterior probabilities of component membership $\tau_i(\boldsymbol{y}_j)$.

It can be seen that in this incomplete-data context, the mixture model arises because the component-label vectors are "missing" from the complete-data vector, and we have to estimate the mixture distribution on data available from the marginal distribution of $\boldsymbol{Y}_j$ only rather than from the joint distribution of the feature vector $\boldsymbol{Y}_j$ and its component label $\boldsymbol{Z}_j$. It will be seen in Section 2.8 that the EM algorithm exploits this reduced simplicity of working with the joint distribution of $\boldsymbol{Y}_j$ and $\boldsymbol{Z}_j$ to compute the MLEs on the basis of the observed (marginal) data $\boldsymbol{y}_j$. It forms the likelihood function on the basis of the complete-data vector $\boldsymbol{y}_c$ and then overcomes the fact that the label vectors $\boldsymbol{z}_j$ are unknown by iteratively working with the conditional expectation of the complete-data log likelihood given the observed data $\boldsymbol{y}$, which is effected using the current fit for the unknown parameters.

If the complete-data vector $\boldsymbol{y}_c$ were available, then estimation of the mixture distribution would be more straightforward than on the basis of the observed data $\boldsymbol{y}$, since each component density $f_i(\boldsymbol{y})$ could be estimated directly from the data known to have come from it; that is, from those feature data $\boldsymbol{y}_j$ with $z_{ij} = (\boldsymbol{z}_j)_i = 1$. This would be a trivial task if, say, the component densities were postulated to be multivariate normal. The only other parameters then to be estimated would be the mixing proportions which, in the case of a mixture sampling design for the classified data, can be estimated by the proportion of these data from each component, namely

$$\hat{\pi}_i = \sum_{j=1}^{n} z_{ij}/n \qquad (i = 1, \ldots, g).$$

The reader is referred to McLachlan (1992) for a comprehensive account of discriminant analysis.

Hence we shall not consider further the fitting of mixture models to data that are completely classified with respect to the components of the mixture model to be fitted unless the component densities themselves are specified to be finite mixtures. McLachlan and Gordon (1989) used such models in their development of a discriminant rule for the diagnosis of renal artery stenosis; see also McLachlan (1992, Section

7.8). More recently, Hastie and Tibshirani (1996) have exploited the flexibility of normal mixtures by proposing a discriminant rule where the group-conditional densities are modeled as finite mixtures of normal homoscedastic components.

In the standard case of independent data where (1.18) is valid, Titterington (1990) contrived the nomenclature *hidden multinomial* for the mixture model. The advantages of this nomenclature is that links can be made with two more general structures in the case of dependent data. If the $\boldsymbol{z}_1, \ldots, \boldsymbol{z}_n$ are assumed to follow a Markov chain, then the mixture model becomes the *hidden Markov chain model.* This model has become popular in the modeling of speech patterns (Rabiner, 1989). If the component-label vectors $\boldsymbol{z}_1, \ldots, \boldsymbol{z}_n$ correspond to some two-dimensional lattice for which a Markov random field is adopted, then the model can be described as a *hidden Markov random field model.* The application of the EM algorithm to these last two models for dependent data is to be considered in Chapter 13. As to be discussed there, the E-step can still be undertaken explicitly for the hidden Markov model, but exact calculations are not feasible for both the E- and M-steps in the case of the hidden Markov random field model. Qian and Titterington (1991) have considered approximations to the E- and M-steps and their links with the approaches of Besag (1986) and Geman and Geman (1984).

1.10 SAMPLING DESIGNS FOR CLASSIFIED DATA

As discussed in the previous section, in some applications of mixture models, the feature vector $\boldsymbol{Y}_j$ can be physically identified with having come from one of the g components. If this is the case and the associated component-label vector $\boldsymbol{Z}_j$ is known, then we shall say that $\boldsymbol{Y}_j$ is classified with respect to the components of the mixture model.

In this book we shall be primarily concerned with the fitting of mixture models to an observed random sample from the mixture (1.1), $\boldsymbol{y}_1, \ldots, \boldsymbol{y}_n$, which are unclassified with respect to the components of the mixture. However, we shall also consider the case where the available data are partially classified; that is, after an appropriate relabeling of the sample, the component-indicator vectors $\boldsymbol{z}_1, \ldots, \boldsymbol{z}_m$ are known for $\boldsymbol{y}_1, \ldots, \boldsymbol{y}_m$ $(0 < m < n)$.

There are two major sampling designs under which the classified data may be realized: (a) joint or mixture sampling and (b) $\boldsymbol{z}$-conditional or separate sampling. They correspond, respectively, to sampling from the joint distribution of $\boldsymbol{Y}_j$ and $\boldsymbol{Z}_j$ and to sampling from the distribution of $\boldsymbol{Y}_j$ conditional on $\boldsymbol{z}_j$. Mixture sampling is common in prospective studies and diagnostic situations. In a prospective study design involving a population of distinct groups, a sample of individuals from the population is followed until their group memberships (component labels) are determined. With separate sampling in practice, the feature vectors are observed for a sample of m_i entities taken separately from each group G_i corresponding to the ith component $(i = 1, \ldots, g)$. Hence it is appropriate to retrospective studies which are common in epidemiological investigations. For example, with the simplest retrospective case-control study of a disease, one sample is taken from the cases that occurred during the study period and the other sample is taken from the group of individuals who remained free of the disease. As many diseases are rare and even a large prospective study may produce few diseased individuals, retrospective sampling can result in important economies in cost and study duration. But of course with separate sampling, the observed proportions of entities from the groups (components of the mixture) do not provide estimates of the mixing proportions π_i.

1.11 PARAMETRIC FORMULATION OF MIXTURE MODEL

In many applications, the component densities $f_i(\boldsymbol{y}_j)$ are specified to belong to some parametric family. In this case, the component densities $f_i(\boldsymbol{y}_j)$ are specified as $f_i(\boldsymbol{y}_j;\,\boldsymbol{\theta}_i)$, where $\boldsymbol{\theta}_i$ is the vector of unknown parameters in the postulated form for the ith component density in the mixture. The mixture density $f(\boldsymbol{y}_j)$ can then be written as

$$f(\boldsymbol{y}_j;\,\boldsymbol{\Psi}) = \sum_{i=1}^{g} \pi_i f_i(\boldsymbol{y}_j;\,\boldsymbol{\theta}_i), \tag{1.20}$$

where the vector $\boldsymbol{\Psi}$ containing all the unknown parameters in the mixture model can be written as

$$\boldsymbol{\Psi} = (\pi_1,\,\ldots,\,\pi_{g-1},\,\boldsymbol{\xi}^T)^T, \tag{1.21}$$

where $\boldsymbol{\xi}$ is the vector containing all the parameters in $\boldsymbol{\theta}_1,\,\ldots,\,\boldsymbol{\theta}_g$ known *a priori* to be distinct. We let $\boldsymbol{\Omega}$ denote the specified parameter space for $\boldsymbol{\Psi}$. Since the mixing proportions π_i sum to unity, one of them is redundant. In defining $\boldsymbol{\Psi}$ as (1.21), we have arbitrarily omitted the gth mixing proportion π_g.

To demonstrate the notation above for defining a parametric mixture, we consider a mixture of univariate normal and Laplace components with a common mean μ, as considered in Kanji (1985) and Jones and McLachlan (1990a) in modeling the distribution of wind shears during aircraft landing. For this model, the mixture density of the measurement Y_j of wind shear can be represented as

$$f(y_j;\,\boldsymbol{\Psi}) = \pi_1\phi(y_j;\,\mu,\sigma^2) + \pi_2(2\kappa)^{-1}\exp(-\mid y_j - \mu \mid /\kappa),$$

where

$$\boldsymbol{\Psi} = (\pi_1,\,\boldsymbol{\xi}^T)^T$$

and

$$\boldsymbol{\xi} = (\mu,\,\sigma^2,\,\kappa)^T.$$

Often, the component densities are specified to belong to the same parametric family. The mixture density $f(\boldsymbol{y}_j;\,\boldsymbol{\Psi})$ will then have the form

$$f(\boldsymbol{y}_j;\,\boldsymbol{\Psi}) = \sum_{i=1}^{g} \pi_i f(\boldsymbol{y}_j;\,\boldsymbol{\theta}_i), \tag{1.22}$$

where $f(\cdot;\,\boldsymbol{\theta})$ denotes a generic member of the parametric family,

$$\{f(\boldsymbol{y}_j;\,\boldsymbol{\theta}) : \boldsymbol{\theta} \in \boldsymbol{\Theta}\} \tag{1.23}$$

and $\boldsymbol{\Theta}$ denotes the parameter space for $\boldsymbol{\theta}$. Throughout this book, we use f as a generic symbol for a density; for example, $f(\boldsymbol{y}_j;\ \boldsymbol{\Psi})$ denotes the mixture density and $f(\boldsymbol{y}_j;\ \boldsymbol{\theta}_i)$ denotes the ith component density under (1.23).

In practice, the components are often taken to belong to the normal family, leading to normal mixtures. In the case of multivariate normal components, we have that

$$f(\boldsymbol{y}_j;\ \boldsymbol{\theta}_i) = \phi(\boldsymbol{y}_j;\ \boldsymbol{\mu}_i,\ \boldsymbol{\Sigma}_i), \tag{1.24}$$

where

$$\phi(\boldsymbol{y}_j; \boldsymbol{\mu}_i, \boldsymbol{\Sigma}_i) = (2\pi)^{-\frac{p}{2}} \mid \boldsymbol{\Sigma}_i \mid^{-\frac{1}{2}} \exp\{-\tfrac{1}{2}(\boldsymbol{y}_j - \boldsymbol{\mu}_i)^T \boldsymbol{\Sigma}_i^{-1}(\boldsymbol{y}_j - \boldsymbol{\mu}_i)\}$$

denotes the multivariate normal density with mean (vector) $\boldsymbol{\mu}_i$ and covariance matrix $\boldsymbol{\Sigma}_i$ $(i = 1,\ \ldots,\ g)$. In this case, the vector $\boldsymbol{\Psi}$ of unknown parameters is given by

$$\boldsymbol{\Psi} = (\pi_1,\ \ldots,\ \pi_{g-1},\ \boldsymbol{\xi}^T)^T,$$

where $\boldsymbol{\xi}$ consists of the elements of the component means, $\boldsymbol{\mu}_1,\ \ldots,\ \boldsymbol{\mu}_g$, and the distinct elements of the component-covariance matrices, $\boldsymbol{\Sigma}_1,\ \ldots,\ \boldsymbol{\Sigma}_g$. In the case of normal homoscedastic components where the component-covariance matrices $\boldsymbol{\Sigma}_i$ are restricted to being equal,

$$\boldsymbol{\Sigma}_i = \boldsymbol{\Sigma} \qquad (i = 1,\ \ldots,\ g), \tag{1.25}$$

$\boldsymbol{\xi}$ consists of the elements of the component means $\boldsymbol{\mu}_1,\ \ldots,\ \boldsymbol{\mu}_g$ and the distinct elements of the common component-covariance matrix $\boldsymbol{\Sigma}$.

1.12 NONPARAMETRIC ML ESTIMATION OF A MIXING DISTRIBUTION

In the case of a finite mixture model defined by (1.22), each of $\boldsymbol{\theta}_1,\ \ldots,\ \boldsymbol{\theta}_g$ is an element of the same parameter space $\boldsymbol{\Theta}$. We can then think of

$$\boldsymbol{\pi} = (\pi_1,\ \ldots,\ \pi_g)^T$$

as defining a discrete probability distribution $H(\boldsymbol{\theta})$ over $\boldsymbol{\Theta}$, where

$$\begin{aligned} H(\boldsymbol{\theta}_i) &= \operatorname{pr}\{\boldsymbol{\theta} = \boldsymbol{\theta}_i\} \\ &= \pi_i \qquad\qquad (i = 1,\ \ldots,\ g). \end{aligned} \tag{1.26}$$

In this way, we can equate the set of unknown probabilities π_i and parameters $\boldsymbol{\theta}_i$ with the discrete probability measure H on the space $\boldsymbol{\Theta}$ for $\boldsymbol{\theta}$, with g points of support $\boldsymbol{\theta}_1, \ldots,\ \boldsymbol{\theta}_g$ and corresponding masses $\pi_1, \ldots,\ \pi_g$. The function H is called the *mixing distribution*, and with the discrete probability measure H on $\boldsymbol{\Theta}$ defined by (1.26), (1.22) may be formally rewritten as

$$f(\boldsymbol{y}_j;\ H) = \int f(\boldsymbol{y}_j;\ \boldsymbol{\theta})\, dH(\boldsymbol{\theta}). \tag{1.27}$$

If H is degenerate with mass one at an unknown point $\boldsymbol{\theta}_1$ (that is, $g = 1$), then (1.27) is an ordinary parametric model, referred to in the literature as the one-component, unmixed, or homogeneity model. If H is discrete, with g points of support, then $f(\boldsymbol{y}_j;\ H)$ is a g-component mixture model. In the latter case, Chen (1992) has shown that the optimal rate of convergence in estimating H is $n^{-1/4}$.

Although H is a (finite) discrete probability measure defined on $\boldsymbol{\Theta}$, the form (1.27) does suggest its generalization to more general mixture densities by taking H to be a more general probability measure. Lindsay (1983) considered the "nonparametric" ML estimation of the mixing distribution H, where the latter is not necessarily restricted to being a finite, discrete probability measure. He showed that finding the MLE involved a standard problem of convex optimization, that of maximizing a concave function over a convex set. Among the consequences is that, in spite of the relaxation of this finite restriction, as long as the likelihood is bounded, the MLE of H is concentrated on a support of cardinality at most that of the number of distinct data points in the sample. This is a very useful, if somewhat surprising, result because it means that a potentially difficult nonparametric estimation problem reduces to a simple one having finite dimensions, and hence algorithms can be constructed to find the solution. In particular, it is possible to determine by evaluation of a simple gradient function how close a candidate estimator H^* is to a ML solution $\hat{H}$. The gradient function can itself be the basis for an algorithm or it can be used in combination with the EM algorithm (Laird, 1978) or other general-purpose algorithms (Böhning, 1995).

The identifiability question for nonparametric ML estimation of a mixing distribution has been considered by Teicher (1963), Barndorff-Nielsen (1965), Chandra (1977), Jewell (1982), and Lindsay and Roeder (1992b), among others. The reader is referred to the recent monographs by Lindsay (1995) and Böhning (1999) for detailed accounts of the nonparametric ML estimation of a mixing distribution.

The mixture model (1.27) is also appropriate for empirical Bayes estimation (Robbins, 1964 and 1983; Laird, 1982). In this context, the mixing distribution H is an unknown prior distribution; the objective is to estimate the posterior distribution of $\boldsymbol{\theta}$ without assuming a functional form for the prior distribution.

If in (1.27), the density $f(\boldsymbol{y}_j;\ \boldsymbol{\theta})$ also depends on other parameters, say regression parameters, then (1.27) is referred to as a semiparametric mixture model. A common example of such a model is a GLM in which a random intercept term is introduced into the linear predictor to handle overdispersion; see Section 5.6. A general account of semiparametric mixture models may be found in the review paper by Lindsay and Lesperance (1995) and in the monograph of Lindsay (1995).

1.13 ESTIMATION OF MIXTURE DISTRIBUTIONS

Over the years, a variety of approaches have been used to estimate mixture distributions. They include graphical methods, method of moments, minimum-distance methods, maximum likelihood, and Bayesian approaches. As surmised by Titterington (1996), perhaps the main reason for the huge literature on estimation methodology for mixtures is the fact that explicit formulas for parameter estimates are typically not available. For example, the MLE for the mixing proportions and the component means and variances/covariances cannot be written down in closed form for normal mixtures. These MLEs have to be computed iteratively. However, as to be pursued further in Section 2.8, their computation is straightforward using the EM algorithm of Dempster et al. (1977).

As discussed in Section 1.1.2, it can be argued that the problem of fitting mixture models started in earnest with the work of Pearson (1894) on the estimation of the parameters of a mixture of two univariate normal heteroscedastic components by the method of moments. In recent times there has been renewed interest in this method of estimation for normal mixtures with the work by Lindsay and Basak (1993) and by Furman and Lindsay (1994a, 1994b); see also Withers (1996). In the special case of $g = 2$ normal components with common covariance matrix, Lindsay and Basak (1993) derived a system of moment equations whose unique solution gives a consistent estimator of $\boldsymbol{\Psi}$. Recently, Craigmile and Titterington (1998) have considered the method of moments as well as maximum likelihood for mixtures of uniform distributions, extending material in Gupta and Miyawaki (1978).

Another way of estimating the vector of parameters $\boldsymbol{\Psi}$ in a mixture model is by using the value of $\boldsymbol{\Psi}$ that minimizes

$$\delta(\hat{F}_n, F_{\boldsymbol{\Psi}}), \tag{1.28}$$

the distance between the mixture distribution $F_{\boldsymbol{\Psi}}$ and the empirical distribution function $\hat{F}_n$ that places mass one at each data point $\boldsymbol{y}_j$ $(j = 1, \ldots, n)$. Titterington et al. (1985) have provided a comprehensive account of the properties of minimum-distance estimators for mixtures, in particular, for the estimation of the mixing proportions. They described various distances δ that have been used, including those where densities rather than distribution functions were used in (1.28). This class includes the MLE if δ is taken to be the Kullback–Leibler (1951) distance between $F_{\boldsymbol{\Psi}}$ and $\hat{F}_n$,

$$I(\hat{F}_n, F_{\boldsymbol{\Psi}}) = \int \log\{d\hat{F}_n(\boldsymbol{w})/dF_{\boldsymbol{\Psi}}(\boldsymbol{w})\}\, d\hat{F}_n(\boldsymbol{w}).$$

When only the mixing proportions are unknown, some distances that have been considered include the Wolfowitz distance (Choi and Bulgren, 1968), the Levy distance (Yakowitz, 1969), the Cramér–von Mises distance (Macdonald, 1971), the squared L_2 norm (Clarke, 1989; Clarke and Heathcote, 1994), and the Hellinger distance (Woodward, Whitney, and Eslinger, 1995). For the more general problem in which all the parameters are unknown, the Wolfowitz distance was considered by Choi (1969), the Cramér–von Mises distance by Woodward et al. (1984) and Lindsay (1994), the squared L_2 norm by Clarke and Heathcote (1994), the Kolmogorov distance by Deely and Kruse (1968) and Blum and Susarla (1977), the Hellinger distance by Cutler and Cordiero-Braña (1996) and Karlis and Xekalaki (1998), and a distance using a kernel density estimate by Cao, Cuevas, and Fraiman (1995). Chen and Kalbfleisch (1996) considered a penalized minimum-distance approach. A penalized likelihood approach was discussed in Leroux (1992b) and Leroux and Puterman (1992).

Often in practice, where the primary interest in fitting a mixture model is to estimate the mixing proportions, the component densities either are specified or can be estimated separately from available classified data. This considerably simplifies the problem. McLachlan and Basford (1988, Chapter 4) have devoted a full chapter to this problem. As discussed there, other methods of estimation besides maximum likelihood, such as minimum distance as mentioned above, have been applied; see Ganesalingam and McLachlan (1981) and McLachlan (1982b) for some efficiency results of the discriminant analysis and method of moments estimators relative to the MLE.

In a recent development, DasGupta (1999) has presented an algorithm for the fitting of normal components with equal component-covariance matrices. His algorithm fits the component means to within the precision specified by the user with high probability. It runs in time only linear in the dimension p of the data and polynomial in the number of components g. This algorithm has three phases. The first phase involves projecting the data into a very small subspace without significantly increasing the overlap of the clusters. The dimension of this subspace is independent of the number of data points n and of p. After this projection, clusters that were elliptical become more spherical in shape, and hence more manageable. In the second phase, a clustering procedure is used to locate the modes of the data in this low-dimensional subspace. Finally, in the third phase, the low-dimensional modes are used to reconstruct the original centers.

In this book, the emphasis is on the fitting of mixture models by ML estimation via the EM algorithm. However, we shall also cover the recent developments with a Bayesian approach. The use of Bayesian methods for estimation of mixture distributions had been somewhat limited until the appearance of the paper by Gelfand and Smith (1990). They brought into focus the tremendous potential of the Gibbs sampler in a wide variety of statistical problems. In particular, they observed that almost any Bayesian computation could be carried out via the Gibbs sampler.

1.14 IDENTIFIABILITY OF MIXTURE DISTRIBUTIONS

The estimation of $\boldsymbol{\Psi}$ on the basis of the observations $\boldsymbol{y}_j$ is only meaningful if $\boldsymbol{\Psi}$ is identifiable. In general, a parametric family of densities $f(\boldsymbol{y}_j;\,\boldsymbol{\Psi})$ is identifiable if distinct values of the parameter $\boldsymbol{\Psi}$ determine distinct members of the family of densities

$$\{f(\boldsymbol{y}_j;\,\boldsymbol{\Psi}):\quad \boldsymbol{\Psi}\in\boldsymbol{\Omega}\},$$

where $\boldsymbol{\Omega}$ is the specified parameter space; that is,

$$f(\boldsymbol{y}_j;\,\boldsymbol{\Psi}) = f(\boldsymbol{y}_j;\,\boldsymbol{\Psi}^*), \tag{1.29}$$

if and only if

$$\boldsymbol{\Psi} = \boldsymbol{\Psi}^*. \tag{1.30}$$

Identifiability for mixture distributions is defined slightly different. To see why this is necessary, suppose that $f(\boldsymbol{y}_j;\, \boldsymbol{\Psi})$ has two component densities, say, $f_i(\boldsymbol{y};\, \boldsymbol{\theta}_i)$ and $f_h(\boldsymbol{y}_j;\, \boldsymbol{\theta}_h)$, that belong to the same parametric family. Then (1.29) will still hold when the component labels i and h are interchanged in $\boldsymbol{\Psi}$. That is, although this class of mixtures may be identifiable, $\boldsymbol{\Psi}$ is not. Indeed, if all the g component densities belong to the same parametric family, then $f(\boldsymbol{y}_j;\, \boldsymbol{\Psi})$ is invariant under the $g!$ permutations of the component labels in $\boldsymbol{\Psi}$.

Let

$$f(\boldsymbol{y}_j;\, \boldsymbol{\Psi}) = \sum_{i=1}^{g} \pi_i f_i(\boldsymbol{y}_j;\, \boldsymbol{\theta}_i)$$

and

$$f(\boldsymbol{y}_j;\, \boldsymbol{\Psi}^*) = \sum_{i=1}^{g^*} \pi_i^* f_i(\boldsymbol{y}_j;\, \boldsymbol{\theta}_i^*)$$

be any two members of a parametric family of mixture densities. This class of finite mixtures is said to be identifiable for $\boldsymbol{\Psi} \in \boldsymbol{\Omega}$ if

$$f(\boldsymbol{y}_j;\, \boldsymbol{\Psi}) \equiv f(\boldsymbol{y}_j;\, \boldsymbol{\Psi}^*)$$

if and only if $g = g^*$ and we can permute the component labels so that

$$\pi_i = \pi_i^* \quad \text{and} \quad f_i(\boldsymbol{y}_j;\, \boldsymbol{\theta}_i) = f_i(\boldsymbol{y}_j;\, \boldsymbol{\theta}_i^*) \quad (i = 1, \ldots, g). \tag{1.31}$$

Here $\equiv$ implies equality of the densities for almost all $\boldsymbol{y}_j$ relative to the underlying measure on $\mathbb{R}^p$ for $f(\boldsymbol{y}_j;\, \boldsymbol{\Psi})$.

The lack of identifiability of $\boldsymbol{\Psi}$ due to the interchanging of component labels is generally handled by the imposition of an appropriate constraint on $\boldsymbol{\Psi}$.

For example, the approach of Aitkin and Rubin (1985) is to impose the restriction that

$$\pi_1 \le \pi_2 \le \ldots \le \pi_g, \tag{1.32}$$

but to carry out the ML estimation without this restriction on the estimates of the mixing proportions $\pi_1, \ldots, \pi_g$. Alternatively, one may wish to impose some other restriction on the solution, for example,

$$(\boldsymbol{\mu}_1)_1 \le (\boldsymbol{\mu}_2)_1 \le \ldots \le (\boldsymbol{\mu}_g)_1. \tag{1.33}$$

Sometimes in practice, in particular with univariate data, there may be a natural ordering of the components according to the size of their means. In the work to be presented here, we do not explicitly impose any restriction on the π_i, but in any situation, we report the result for only one of the possible arrangements of the elements of $\boldsymbol{\Psi}$.

Thus this lack of identifiability is not of concern in the normal course of events in the fitting of mixture models by maximum likelihood, say, via the EM algorithm. However, it does cause major problems in a Bayesian framework where posterior simulation is used to make inferences from the mixture model. It is known as the label-switching problem and will be one of the issues addressed in Chapter 4 on the Bayesian analysis of mixture models.

Another approach to the identifiability problem is to use an identifying function (Kadane, 1974). This is essentially the same as Redner's (1981) approach of using the quotient topological space $\tilde{\boldsymbol{\Omega}}$ obtained by mapping equivalent values of $\boldsymbol{\Psi}$ into a single point.

As noted by Crawford (1994), among others, nonidentifiability due to overfitting (that is, fitting too many components in the model) is more problematic. For example, modeling a mixture of $g-1$ components incorrectly by a mixture of g components can be handled in two ways:

1. One of the mixing proportions in the g-component mixture can be set equal to zero.

2. Two component densities in the g-component mixture can be taken to be the same.

For example, suppose that the true density is $f(\boldsymbol{y}_j;\, \boldsymbol{\theta})$, but that a model with two component densities, $f(\boldsymbol{y}_j;\, \boldsymbol{\theta}_1)$ and $f(\boldsymbol{y}_j;\, \boldsymbol{\theta}_2)$, is used instead. Define

$$\boldsymbol{\Omega}_1 = \{\boldsymbol{\Psi} : \pi_1 = 1,\, \boldsymbol{\theta}_1 = \boldsymbol{\theta}\} \cup \{\boldsymbol{\Psi} : \pi_1 = 0,\, \boldsymbol{\theta}_2 = \boldsymbol{\theta}\}$$

and

$$\boldsymbol{\Omega}_2 = \{\boldsymbol{\Psi} : \pi_1 \in (0,1),\, \boldsymbol{\theta}_1 = \boldsymbol{\theta}_2 = \boldsymbol{\theta}\}.$$

Then for all $\boldsymbol{\Psi}$ belonging to

$$\boldsymbol{\Omega}_1 \cup \boldsymbol{\Omega}_2, \tag{1.34}$$

we have

$$f(\boldsymbol{y}_j;\, \boldsymbol{\Psi}) = f(\boldsymbol{y}_j;\, \boldsymbol{\theta}).$$

The extension to arbitrary g is straightforward.

As to be discussed later, convergence results can still be obtained in this case. For instance, Redner (1981) extended Wald's (1949) results on the consistency of the MLE of $\boldsymbol{\Psi}$ by using the quotient topological space $\tilde{\boldsymbol{\Omega}}$. In the above example, all $\boldsymbol{\Psi}$ in (1.34) correspond to the same point in $\tilde{\boldsymbol{\Omega}}$. Feng and McCulloch (1996) obtained the parallel result in Euclidean space, as to be discussed in Chapter 2.

Titterington et al. (1985, Section 3.1) have given a lucid account of the concept of identifiability for mixtures. They pointed out that most finite mixtures of continuous densities are identifiable; an exception is a mixture of uniform densities. Teicher (1960) showed that a finite mixture of Poisson distributions is identifiable, whereas mixtures of binomial distributions are not identifiable if

$$N < 2g - 1,$$

where N is the common number of trials in the component binomial distributions. Yakowitz and Spragins (1968) showed that finite mixtures of negative binomial component distributions are identifiable.

1.15 CLUSTERING OF DATA VIA MIXTURE MODELS

1.15.1 Mixture Likelihood Approach to Clustering

In some applications of mixture models, questions related to clustering may arise only after the mixture model has been fitted. For instance, suppose that in the first instance the reason for fitting a mixture model was to obtain a satisfactory model for the distribution of the data. If this were achieved by the fitting of, say, a three-component mixture model, then it may then be of interest to consider the problem further to see if the three components can be identified with three externally existing groups or subpopulations or if the clusters implied by the fitted mixture model reveal the existence of previously unrecognized or undefined subpopulations.

However, in other applications of mixture models, the clustering of the data at hand is the primary aim of the analysis. In this case, the mixture model is being used purely as a device for exposing any grouping that may underlie the data. McLachlan and Basford (1988) highlighted the usefulness of mixture models as a way of providing an effective clustering of various data sets under a variety of experimental designs.

With a mixture model-based approach to clustering, it is assumed that the data to be clustered are from a mixture of an initially specified number g of groups in various proportions. That is, each data point is taken to be a realization of the mixture density (1.20), where the g components correspond to the g groups. On specifying a parametric form for each component density $f_i(\boldsymbol{y}_j; \boldsymbol{\theta}_i)$, the vector $\boldsymbol{\Psi}$ can be estimated by maximum likelihood (or some other method). Once the mixture model has been fitted, a probabilistic clustering of the data into g clusters can be obtained in terms of the fitted posterior probabilities of component membership for the data. An outright assignment of the data into g clusters is achieved by assigning each data point to the component to which it has the highest estimated posterior probability of belonging. Although these estimated posterior probabilities may have limited reliability in small samples, they may well give a satisfactory outright assignment of the data. This point is considered further in the next section, where the theoretical basis for performing the clustering in terms of the posterior probabilities of component membership of the mixture is presented.

In the above, there is a one-to-one correspondence between the mixture components and the groups. In those cases where the underlying population consists of groups in which the feature vector is unable to be modeled by a single normal distribution but needs a normal mixture formulation, the components in the fitted g-component normal mixture model and the consequent clusters will correspond to g subgroups rather than to the smaller number of actual groups represented in the data.

It can be seen that this mixture likelihood-based approach to clustering is model based in that the form of each component density of an observation has to be specified in advance. Hawkins, Muller, and ten Krooden (1982, p. 353) commented that most writers on cluster analysis "lay more stress on algorithms and criteria in the belief that intuitively reasonable criteria should produce good results over a wide range of possible (and generally unstated) models." For example, the trace $\boldsymbol{W}$ criterion, where $\boldsymbol{W}$ is the pooled within-cluster sums of squares and products matrix, is predicated on normal groups with (equal) spherical covariance matrices; but as they pointed out, many users apply this criterion even in the face of evidence of nonspherical clusters or, equivalently, would use Euclidean distance as a metric. They strongly supported the increasing emphasis on a model-based approach to clustering. Indeed, as remarked by Aitkin, Anderson, and Hinde (1981) in the reply to the discussion of their paper, "when clustering samples from a population, no cluster method is, *a priori* believable without a statistical model." Concerning the use of mixture models to represent nonhomogeneous populations, they noted in their paper that "Clustering methods based on such mixture models allow estimation and hypothesis testing within the framework of standard statistical theory." Previously, Marriott (1974, p. 70) had noted that the mixture likelihood-based approach "is about the only clustering technique that is entirely satisfactory from the mathematical point of view. It assumes a well-defined mathematical model, investigates it by well-established statistical techniques, and provides a test of significance for the results."

The mixture likelihood-based approach to clustering can obviously play a major role in any exploratory data analysis in both searching for groupings in the data and testing the validity of any cluster structure discovered; that is, testing whether the apparent clusters are due to random fluctuations or whether they reflect a real separation of the data into distinct groups.

1.15.2 Decision-Theoretic Approach

Decision theory provides a convenient framework for the construction of discriminant rules in the situation where an allocation of an unclassified entity is required. In the context of a finite mixture model, the allocation is with respect to the components of the mixture. For this purpose we let $r(\boldsymbol{y}_j)$ denote an allocation rule for assigning the feature vector $\boldsymbol{y}_j$ to one of the components of the mixture model, where $r(\boldsymbol{y}_j) = i$ implies that the observation is assigned to the ith component $(i = 1, \ldots, g)$.

The optimal or Bayes rule $r_B(\boldsymbol{y}_j)$ for the allocation of $\boldsymbol{y}_j$ is defined by

$$r_B(\boldsymbol{y}_j) = i \quad \text{if} \quad \tau_i(\boldsymbol{y}_j) \geq \tau_h(\boldsymbol{y}_j) \qquad (h = 1, \ldots, g). \tag{1.35}$$

That is,

$$r_B(\boldsymbol{y}_j) = \arg\max_h \tau_h(\boldsymbol{y}_j).$$

The rule $r_B(\boldsymbol{y}_j)$ is not uniquely defined at $\boldsymbol{y}_j$ if the maximum of the posterior probabilities of component membership is achieved with respect to more than one component. In this case, the entity can be assigned arbitrarily to one of the components for which the corresponding posterior probabilities are equal to the maximum value. We are assuming here that the cost of a correct allocation is zero and all misallocations are taken to have the same cost; see McLachlan (1992, Chapter 1).

As the posterior probabilities of component membership $\tau_i(\boldsymbol{y}_j)$ have the same common denominator $f(\boldsymbol{y}_j)$, $r_B(\boldsymbol{y}_j)$ can be defined in terms of the relative sizes of the component densities weighted according to the mixing (prior) probabilities; that is,

$$r_B(\boldsymbol{y}_j) = i \quad \text{if} \quad \pi_i f_i(\boldsymbol{y}_j) \geq \pi_h f_h(\boldsymbol{y}_j) \qquad (h = 1, \ldots, g). \tag{1.36}$$

The Bayes rule can be estimated by the so-called plug-in rule, $r_B(\boldsymbol{y}_j;\, \hat{\boldsymbol{\Psi}})$, where $\hat{\boldsymbol{\Psi}}$ denotes the estimate of the unknown parameter vector $\boldsymbol{\Psi}$. This approach, where the component densities (that is, the group-conditional densities) are directly modeled for use in the formation of the posterior probabilities of group membership, is called the sampling approach by Dawid (1976). Another approach to the estimation of the Bayes rule is to model these posterior probabilities directly, as with the logistic model. Dawid (1976) calls this approach the diagnostic paradigm. With this approach, the interest is not on what the component densities of the feature vector $\boldsymbol{Y}_j$ look like, but on the distribution of the component-indicator vectors, $\boldsymbol{z}_1, \ldots, \boldsymbol{z}_n$, for the observed feature data $\boldsymbol{y}_1, \ldots, \boldsymbol{y}_n$ and similar values. This is the main approach with neural networks (Ripley, 1996, p. 7). However, the diagnostic paradigm is limited to the case where there are classified data available.

1.15.3 Clustering of I.I.D. Data

Suppose that the purpose of fitting the finite mixture model (1.20) is to cluster an observed random sample $\boldsymbol{y}_1, \ldots, \boldsymbol{y}_n$ into g components. In terms of the complete-data specification (1.15) of the mixture model, we wish to infer the associated component labels $\boldsymbol{z}_1, \ldots, \boldsymbol{z}_n$ of these feature data vectors. That is, we wish to infer the $\boldsymbol{z}_j$ on the basis of the feature data $\boldsymbol{y}_j$. After we fit the g-component mixture model to obtain the estimate $\hat{\boldsymbol{\Psi}}$ of the vector of unknown parameters in the mixture model, we can give a probabilistic clustering of the n feature observations $\boldsymbol{y}_1, \ldots, \boldsymbol{y}_n$ in terms of their fitted posterior probabilities of component membership. For each $\boldsymbol{y}_j$, the g probabilities $\tau_1(\boldsymbol{y}_j;\, \hat{\boldsymbol{\Psi}}), \ldots, \tau_g(\boldsymbol{y}_j;\, \hat{\boldsymbol{\Psi}})$ give the estimated posterior probabilities that this observation belongs to the first, second,..., and gth components, respectively, of the mixture $(j = 1, \ldots, n)$.

We can give an outright or hard clustering of these data by assigning each $\boldsymbol{y}_j$ to the component of the mixture to which it has the highest posterior probability of belonging. That is, we estimate the component-label vector $\boldsymbol{z}_j$ by $\hat{\boldsymbol{z}}_j$, where $\hat{z}_{ij} = (\hat{\boldsymbol{z}}_j)_i$ is defined by

$$\begin{aligned} \hat{z}_{ij} &= 1, \quad \text{if } i = \arg\max_h \tau_h(\boldsymbol{y}_j;\, \hat{\boldsymbol{\Psi}}), \\ &= 0, \quad \text{otherwise,} \end{aligned} \tag{1.37}$$

for $i = 1, \ldots, g;\; j = 1, \ldots, n$.

It follows from (1.35) that this use of the assignment criterion (1.37) corresponds to using the so-called plug-in sample version of the Bayes (optimal) rule, $r_B(\boldsymbol{y};\, \hat{\boldsymbol{\Psi}})$, whereby $\boldsymbol{\Psi}$ is replaced by $\hat{\boldsymbol{\Psi}}$ in $r_B(\boldsymbol{y};\, \boldsymbol{\Psi})$.

If the postulated component densities provide a good fit and the mixing proportions are able to be estimated with some precision, then the plug-in rule $r_B(\boldsymbol{y}_j;\, \hat{\boldsymbol{\Psi}})$ should be a good approximation to the Bayes rule $r_B(\boldsymbol{y}_j;\, \boldsymbol{\Psi})$ in the case where the components of the mixture can be identified with g externally existing groups $G_1, \ldots, G_g$ in which the ith group-conditional densities of $\boldsymbol{Y}_j$ can be modeled by $f_i(\boldsymbol{y}_j;\, \boldsymbol{\theta}_i)$ $(i = 1, \ldots, g)$.

However, even if $\hat{\boldsymbol{\Psi}}$ is a poor estimate of $\boldsymbol{\Psi}$, $r_B(\boldsymbol{y}_j;\, \hat{\boldsymbol{\Psi}})$ may still be a reasonable allocation rule. It can be seen from (1.36) that for $r_B(\boldsymbol{y}_j;\, \hat{\boldsymbol{\Psi}})$ to be a good approximation to $r_B(\boldsymbol{y}_j;\, \boldsymbol{\Psi})$, it is only necessary that the boundaries defining the allocation regions,

$$\{\boldsymbol{y}_j \;:\; \pi_i f_i(\boldsymbol{y}_j;\, \boldsymbol{\theta}_i) = \pi_h f_h(\boldsymbol{y}_j;\, \boldsymbol{\theta}_h), \qquad i < h = 2,\, \ldots,\, g\}, \tag{1.38}$$

be estimated precisely. This implies at least for well-separated groups that in consideration of the estimated group-conditional densities, it is the fit in the tails rather than in the main body of the distributions that is crucial. This is what one would expect. Any reasonable allocation rule should be able to allocate correctly an entity whose group of origin is obvious from its feature vector. Its accuracy is really determined by how well it can handle entities of doubtful origin. Their feature vectors tend to occur in the tails of the distributions. If reliable estimates of the posterior probabilities of component-membership $\tau_i(\boldsymbol{y}_j;\, \boldsymbol{\Psi})$ are sought in their own right and not just for the purposes of making an outright assignment, then the fit of the estimated group-conditional density ratios $f_i(\boldsymbol{y}_j;\, \boldsymbol{\theta}_i)/f_h(\boldsymbol{y}_j;\, \boldsymbol{\theta}_h)$ is important for all values of $\boldsymbol{y}_j$ and not just on the boundaries (1.38).

It can therefore be seen in clustering applications of mixture models that the estimates of the component densities are not of interest as an end in themselves, but rather how useful their ratios are in providing estimates of the posterior probabilities of component membership or at least an estimate of the Bayes rule. However, for convenience, the question of model fit in practice is usually approached by consideration of the individual fit of each estimated component density $f_i(\boldsymbol{y}_j;\, \boldsymbol{\theta}_i)$.

We have assumed here that the clustering is to be undertaken on i.i.d. data sets, which suffices for many applications in practice. However, the assumption of independence is not always tenable with some applications, in particular with those that occur in the field of image analysis.

1.15.4 Image Segmentation or Restoration

We suppose here that the observed feature data $\boldsymbol{y}_1,\, \ldots,\, \boldsymbol{y}_n$ are vectors of intensities measured on n pixels in some two-dimensional scene or voxels in a three-dimensional scene. Here the components of the postulated mixture model for the intensity vector correspond to the true colors of the pixels. The problem is to infer the totality of the component-label vectors

$$\boldsymbol{z} = (\boldsymbol{z}_1^T,\, \ldots,\, \boldsymbol{z}_n^T)^T$$

on the basis of the observed intensity vectors

$$\boldsymbol{y} = (\boldsymbol{y}_1^T,\, \ldots,\, \boldsymbol{y}_n^T)^T.$$

This clustering process is referred to as segmentation. It can also be referred to as image restoration.

An estimate $\hat{\boldsymbol{z}}$ can be produced by considering each $\boldsymbol{z}_j$ individually and allocating each $\boldsymbol{y}_j$ on the basis of an estimate of

$$\operatorname{pr}\{\boldsymbol{Z}_j = \boldsymbol{z}_j \mid \boldsymbol{y}_j\}. \tag{1.39}$$

For instance, the pixels can be individually allocated by choosing $\hat{z}_j$ to be the value of z_j that has maximum posterior probability given $Y_j = y_j$; that is, $\hat{z}_j$ maximizes (1.39). It follows from Section 1.15.2 that this approach of maximizing the posterior marginal probability for each pixel corresponds to maximizing the expected number of correctly assigned pixels in the scene. It is thus biased towards a low rate of mis-allocated pixels rather than overall appearance. With the widely used ICM algorithm of Besag (1986), segmentation is performed on the basis of

$$\text{pr}\{Z_j = z_j \mid y, z_{\partial j}\},$$

where $z_{\partial j}$ contains the label vectors of those pixels lying in a prescribed neighborhood ∂j of the jth pixel ($j = 1, \ldots, n$).

Alternatively, an estimate $\hat{z}$ of z can be produced by simultaneous estimation of its n subvectors z_j ($j = 1, \ldots, n$). An example of the former approach is taking $\hat{z}$ to be the value of z that maximizes

$$\text{pr}\{Z = z \mid y\}. \tag{1.40}$$

That is, $\hat{z}$ is taken to be the mode of the posterior distribution of Z. It is therefore referred to as the MAP (maximum *a posteriori*) estimate. From a decision-theoretic viewpoint, $\hat{z}$ corresponds to the adoption of a zero–one loss function according to whether the reconstructed image is perfect or not. The maximization of (1.40) would appear at first sight to be an ambitious task, given that there are g^n possible values of z. Geman and Geman (1984) approached this formidable problem using simulated annealing.

1.16 HIDDEN MARKOV MODELS

As to be considered in Chapter 13, one way of formally extending mixture models to the analysis of dependent data is to adopt a stationary Markovian model for the vector z containing the component-indicator labels, $z_1, \ldots, z_n$. For example, in the context of the segmentation of pixels as discussed in the previous section, we could introduce a Markov random field for the distribution of Z. It will be seen that the EM algorithm is unable to be implemented exactly for a Markov random field. However, an approximation can be obtained by first performing a restoration step, on which a current estimate of z is obtained by a segmentation, for example, by using the ICM or MAP estimates as mentioned above. The M-step can then be performed conditionally on this current estimate of z if the distribution of z is approximated by the pseudo likelihood (Besag, 1975).

In other work on mixture models for dependent data, Wong and Li (2000) recently generalized the Gaussian MTD (mixture transition distribution) model introduced by Le, Martin, and Raftery (1996) to the mixture autoregressive (MAR) model for the modeling of nonlinear time series. The MTD model was first introduced by Raftery (1985) in the discrete case as a model for high-order Markov chains; see also Raftery and Tavaré (1994). Le et al. (1996) illustrated the usefulness of their MAR model with two examples: the common stock closing prices series for International Business Machines (IBM) and the Canadian lynx data. They noted that many published nonlinear time series models can be shown to have multimodal marginal or conditional distributions. For example, the zeroth-order self-exciting threshold autoregressive model (Tong, 1990) can be shown to have a mixture of Gaussian distributions marginally (Jalali and Pemberton, 1995). In some other work on mixtures of time series, Shephard (1994) has considered Bayesian techniques appropriate for handling time series models whose noise is drawn from a Gaussian mixture. Harrison and Stevens (1976) call such a structure a multiprocess model. There is a huge literature on this model (Pẽna and Guttman, 1988; Kitagawa, 1989). Using simulation techniques like those developed by Shephard (1994), Billio, Montfort, and Robert (1999) have considered a Bayesian analysis of switching ARMA models.

1.17 TESTING FOR THE NUMBER OF COMPONENTS IN MIXTURE MODELS

In some applications of mixture models, there is sufficient *a priori* information for the number of components g in the mixture model to be specified with no uncertainty. For example, this would be the case where the components correspond to externally existing groups in which the feature vector is known to be normally distributed. However, on many occasions, the number of components has to be inferred from the data, along with the parameters in the component densities. If, say, a mixture model is being used to describe the distribution of some data, the number of components in the final version of the model may be of interest beyond matters of a technical or computational nature. For example, McLaren et al. (1991) used a two-component log normal mixture distribution to model the distribution of the volume of red blood cells in patients recovering from anemia. The red blood cell volume distribution of healthy individuals can be modeled adequately by a single log normal component. However, for patients not completely recovered from anemia, their red blood cell distribution, although unimodal in appearance toward the end of the iron therapy treatment, may still need to be modeled by a two-component log normal mixture due to the presence of a sufficient number of microcytic cells in relation to the normocytic cells. Thus the result of a statistical test on the number of components in the log normal mixture model for a specific patient can be used as an early guide to aid clinicians in making a decision when to suspend iron therapy treatment for the patient. A nonsignificant test result is consistent with the red blood cell distribution of the patient having returned to a healthy state (McLaren, 1996; McLachlan, McLaren, and Matthews, 1995).

Another example concerns the modeling of the distribution of *in vivo* insulin action in Pima Indians by Bogardus et al. (1989). If a single gene produced insulin resistance, with environmental effects creating some additional variance, insulin action might be distributed as a mixture of two normal distributions if the gene is dominant or recessive or as a mixture of three normal distributions if the gene is codominant. Bogardus et al. (1989) concluded that three components were needed in their normal mixture model, which they noted was consistent with the hypothesis that among Pima Indians, insulin resistance is determined by a single gene with a codominant mode of inheritance.

In applications of mixture models in cluster and latent analyses, the problem of assessing the number of components in a particular mixture model arises with the question of how many clusters or latent classes there are. An obvious way of approaching the problem is to use the likelihood ratio statistic to test for the smallest value of g compatible with the data. Unfortunately with mixture models, regularity conditions do not hold for $-2 \log \lambda$ to have its usual distribution of chi-squared with degrees of freedom equal to the difference between the number of parameters under the null and alternative hypotheses.

The problem of testing for the number of components in a mixture model is clearly of much theoretical and practical importance, and so has attracted considerable attention in many studies over the years. These studies are to be reviewed in Chapter 6.

1.18 BRIEF HISTORY OF FINITE MIXTURE MODELS

We have seen in Section 1.1 that the history of finite mixture models goes back to over a century ago with the classic paper of Pearson (1894) on his moments-based fitting of a mixture of two univariate normal components to some crab measurements provided by his colleague Weldon (1892, 1893). The possibility of resolving a normal mixture into its constituent components was, of course, implicit in Quetelet's (1846, 1852) work and was mentioned explicitly by Galton in 1869; see Stigler (1986, Chapter 10) for an absorbing account of this early work on mixtures. Another early reference on mixtures is Holmes (1892), who brought in the concept of mixtures of populations in his suggestion that an average alone was inadequate in consideration of wealth disparity; see Billard (1997).

Given the amount of algebra involved with the approach of Pearson (1894), various attempts were made in the early part of the twentieth century to simplify the method, including by Charlier (1906). During the next 30 years, work continued on the use of the method of moments for this mixture problem. It was extended to the case of bivariate normal components by Charlier and Wicksell (1924) and to the case of more than two univariate normal components by Doetsch (1928). Strömgren (1934) also considered the use of cumulants, while Rao (1948) considered the use of k-statistics. In more recent work, Cohen (1967) subsequently showed how the solving of Pearson's nonic could be circumvented through an iterative process which involves solving a cubic equation for a unique negative root. This approach was suggested by the solution in the case of equal variances where the estimates depend uniquely on the negative root of a cubic equation constructed from the first four moments; see Charlier and Wicksell (1924) and Rao (1952, Section 8b.6). However, Tan and Chang (1972) and Fryer and Robertson (1972), among others, showed that the method of moments was inferior to ML estimation for this problem. As noted earlier, there has been renewed interest in this method of estimation for normal mixtures with the work by Lindsay and Basak (1993), among others.

As speculated by Fowlkes (1979) in his study of diagnostic plotting procedures for the detection of univariate mixtures, it was probably because of the intractability of the moment estimators and the absence of modern computer technology that attention was focused on graphical techniques for mixtures during the early and mid-1900s. Pioneering work on these techniques was undertaken by Harding (1948), Preston (1953), and Cassie (1954) and was continued by Bhattacharya (1967) and Wilk and Gnanadesikan (1968), among others. Tarter and Silvers (1975) presented a graphical procedure based on the properties of the bivariate Gaussian density function, while Chhikara and Register (1979) developed a numerical classification technique based on computer-aided methods for the display of the data. More recently, Tarter and Lock (1993, Chapter 5) described a curve-estimation approach, called the λ method, which is mainly a graphical approach to the decomposition of mixtures. Loosely speaking, a λ curve is derived from the original density by reducing the standard deviations of any constituent components without affecting any other characteristics of the distribution. Thus, if a distribution has more than one component, the λ method enhances the differences between the components making them easier to see. Medgyessy (1961) brought the λ methodology, first proposed by the mathematician Doetsch (1928, 1936), to the attention of the applied statistical community; see Tarter and Lock (1993, Chapter 5).

With the advent of high-speed computers, attention was turned to ML estimation of the parameters in a mixture distribution. The first use of this method for a mixture model has been attributed to Rao (1948), who used Fisher's method of scoring for a mixture of two univariate distributions with equal variances. However, Butler (1986) pointed out that Newcomb (1886), predating even Pearson's (1894) early attempt on mixture models with the method of moments, suggested an iterative reweighting scheme which can be interpreted as an application of the EM algorithm of Dempster et al. (1977) to compute the MLE of the common mean of a mixture in known proportions of a finite number of univariate normal populations with known variances. Also, Butler (1986) noted that Jeffreys (1932) used essentially the EM algorithm in iteratively computing the estimates of the means of two univariate normal populations which had known variances and which were mixed in known proportions. Following Rao's (1948) paper, ML estimation appears not to have been pursued further until Hasselblad (1966, 1969) addressed the problem, initially for a mixture of g univariate normal distributions with equal variances, and then for mixtures of distributions from the exponential family. The former case was also considered briefly by Behboodian (1970), and its multivariate analogue was studied by Wolfe (1965, 1967, 1970, 1971) and Day (1969) in major papers. Day (1969) concentrated on the solution for two normal populations with the same covariance matrix, while Wolfe (1965, 1967, 1970, 1971) dealt with an arbitrary number of normal heteroscedastic populations as well as mixtures of multivariate Bernoulli distributions for use in latent class analysis. As with Hasselblad (1966, 1969), their solutions were presented in an iterative form corresponding to particular applications of the EM algorithm of Dempster et al. (1977). Other works on ML estimation of mixtures with the computation of the estimates expressed in this iterative form are by Peters and Coberly (1976), Duda and Hart (1973), and Hosmer (1973a, 1973b).

However, it was not until Dempster et al. (1977) had formalized this iterative scheme in a general context through their EM algorithm that the convergence properties of the ML solution for the mixture problem were established on a theoretical basis. This paper of Dempster et al. (1977) proved to be a timely catalyst for further research into the applications of finite mixture models. This is witnessed by the subsequent stream of papers on finite mixtures in the literature, commencing with, for example, Ganesalingam and McLachlan (1978, 1979a, 1979b, 1980a), O'Neill (1978), and Aitkin (1980).

By now, there is quite an extensive literature on finite mixture models. Hence it is not our intention to provide an exhaustive bibliography here. We have attempted to reference the main results, with the emphasis on the more recent developments and applications. Earlier references on mixture models may be found in the comprehensive bibliographies in the previous books on finite mixture distributions by Everitt and Hand (1981), Titterington et al. (1985), McLachlan and Basford (1988), Lindsay (1995), and Böhning (1999). In addition, there are the review articles of Holgersson and Jorner (1978), Gupta and Huang (1981), Redner and Walker (1984), and Titterington (1990), as well as the encyclopedia entries by Blischke (1978) and Everitt (1985). The latter entry has been updated by Titterington (1996). Previously, McLachlan (1982a) had reviewed the fitting of mixture models, concentrating on their a role as a device for clustering. In more recent papers, Flury, Airoldi, and Biber (1992) and McLachlan (1994) have provided nontechnical reviews of some basic concepts of mixture model-based analysis. Additional references on mixture models in a medical context may be found in the issue of the journal *Statistical Methods in Medical Research* on Finite Mixture Models (1995, **5**, 107–211). Applications of mixture models in marketing are given in Wedel and Kamakura (1998, Chapters 6 and 7).

In some applications, the component densities of a mixture are specified to be different. A special case of this, called a *nonstandard mixture*, is for a $g = 2$ component mixture with one of the components degenerate, having mass one placed at a single point. Nonstandard mixtures are considered in some depth in the report by a panel of the Committee on Applied and Theoretical Statistics of the Board on Mathematical Sciences of the National Research Council, chaired by Professor D. Guthrie (Panel on Nonstandard Mixtures of Distributions, 1989).

1.19 NOTATION

We now define the notation that is used consistently throughout the book. Less frequently used notation will be defined later when it is first introduced.

All vectors and matrices are in boldface. The superscript T denotes the transpose of a vector or matrix. The trace of a matrix $\boldsymbol{A}$ is denoted by $tr(\boldsymbol{A})$, while the determinant of $\boldsymbol{A}$ is denoted by $|\boldsymbol{A}|$. The null vector is denoted by $\mathbf{0}$. The notation diag$(a_1, \ldots, a_n)$ is used for a matrix with diagonal elements $a_1, \ldots, a_n$ and all off-diagonal elements zero.

We let $\boldsymbol{Y}_1, \ldots, \boldsymbol{Y}_n$ denote a random sample of size n where $\boldsymbol{Y}_j$ is a p-dimensional random vector with probability density function $f(\boldsymbol{y}_j)$ on $\mathbb{R}^p$. In practice, $\boldsymbol{Y}_j$ contains the random variables corresponding to p measurements made on the jth recording of some features on the phenomenon under study. We let $\boldsymbol{Y} = (\boldsymbol{Y}_1^T, \ldots, \boldsymbol{Y}_n^T)^T$. Note that we are using $\boldsymbol{Y}$ to represent the entire sample; that is, $\boldsymbol{Y}$ is an n-tuple of points in $\mathbb{R}^p$. Where possible, a realization of a random vector is denoted by the corresponding lower letter. For example, $\boldsymbol{y} = (\boldsymbol{y}_1^T, \ldots, \boldsymbol{y}_n^T)^T$ denotes an observed random sample, where $\boldsymbol{y}_j$ is the observed value of the random vector $\boldsymbol{Y}_j$. In order to avoid any confusion with the use of $\boldsymbol{Y}$ for the entire random sample, we shall always use $\boldsymbol{Y}_j$ for an individual observation even when only one observation is being considered.

The probability density function of the random vector $\boldsymbol{Y}_j$ under a g-component mixture model is written in parametric form as

$$f(\boldsymbol{y}_j; \boldsymbol{\Psi}) = \sum_{i=1}^{g} \pi_i f_i(\boldsymbol{y}_j; \boldsymbol{\theta}_i), \tag{1.41}$$

where the vector $\boldsymbol{\Psi}$ containing all the unknown parameters in the mixture model is written as

$$\boldsymbol{\Psi} = (\pi_1, \ldots, \pi_{g-1}, \boldsymbol{\xi}^T)^T, \tag{1.42}$$

and $\boldsymbol{\xi}$ is the vector containing all the parameters in $\boldsymbol{\theta}_1, \ldots, \boldsymbol{\theta}_g$ known *a priori* to be distinct. We let $\boldsymbol{\Omega}$ denote the specified parameter space for $\boldsymbol{\Psi}$, and we let

$$\boldsymbol{\pi} = (\pi_1, \ldots, \pi_g)^T$$

be the vector of mixing proportions. Since the mixing proportions π_i sum to unity, one of them is redundant. In defining $\boldsymbol{\Psi}$, we have arbitrarily omitted the gth mixing proportion π_g.

The ML fitting of the mixture density $f(\boldsymbol{y}_j; \boldsymbol{\Psi})$ to an observed random sample

$$\boldsymbol{y} = (\boldsymbol{y}_1^T, \ldots, \boldsymbol{y}_n^T)^T$$

is undertaken using the EM algorithm. In the EM framework, this problem is viewed as being incomplete due to the unavailability of the associated component-label vectors $\boldsymbol{z}_1, \ldots, \boldsymbol{z}_n$, where $z_{ij} = (\boldsymbol{z}_j)_i$ is defined to be one or zero, according to whether $\boldsymbol{y}_j$ is viewed or not viewed as having arisen from the ith component of the mixture model being fitted. The complete-data vector $\boldsymbol{y}_c$ is given by

$$\boldsymbol{y}_c = (\boldsymbol{y}^T, \boldsymbol{z}^T)^T, \tag{1.43}$$

where

$$\boldsymbol{z} = (\boldsymbol{z}_1^T, \ldots, \boldsymbol{z}_n^T)^T. \tag{1.44}$$

In this EM framework, the posterior probability that the entity belongs to the ith component with $\boldsymbol{y}_j$ having been observed on it is given by

$$\begin{aligned} \tau_i(\boldsymbol{y}_j) &= \text{pr}\{\text{ entity} \in i\text{th component } \mid \boldsymbol{y}_j\} \\ &= \text{pr}\{Z_{ij} = 1 \mid \boldsymbol{y}_j\} \\ &= \pi_i f_i(\boldsymbol{y}_j)/f(\boldsymbol{y}_j) \quad (i = 1, \ldots, g;\ j = 1, \ldots, n). \end{aligned} \tag{1.45}$$

The likelihood function for $\boldsymbol{\Psi}$ formed from the observed data $\boldsymbol{y}$ is denoted by $L(\boldsymbol{\Psi})$, while $L_c(\boldsymbol{\Psi})$ denotes the complete-data likelihood function for $\boldsymbol{\Psi}$ that could be formed from the complete-data vector $\boldsymbol{y}_c$ if it were completely observable.

The (incomplete-data) score statistic is given by

$$S(\boldsymbol{y};\ \boldsymbol{\Psi}) = \partial \log L(\boldsymbol{\Psi})/\partial \boldsymbol{\Psi}, \tag{1.46}$$

while

$$S_c(\boldsymbol{y}_c;\ \boldsymbol{\Psi}) = \partial \log L_c(\boldsymbol{\Psi})/\partial \boldsymbol{\Psi} \tag{1.47}$$

denotes the corresponding complete-data score statistic.

A given sequence of EM iterates is denoted by $\{\boldsymbol{\Psi}^{(k)}\}$, where $\boldsymbol{\Psi}^{(0)}$ denotes the starting value of $\boldsymbol{\Psi}$ and $\boldsymbol{\Psi}^{(k)}$ the value of $\boldsymbol{\Psi}$ after the kth subsequent iteration of the EM algorithm. The MLE of $\boldsymbol{\Psi}$ is denoted by $\hat{\boldsymbol{\Psi}}$; see Section 2.2 for the definition of the MLE.

The observed information matrix is denoted by $\boldsymbol{I}(\hat{\boldsymbol{\Psi}};\ \boldsymbol{y})$, where

$$\boldsymbol{I}(\boldsymbol{\Psi};\ \boldsymbol{y}) = -\partial^2 \log L(\boldsymbol{\Psi})/\partial \boldsymbol{\Psi} \partial \boldsymbol{\Psi}^T. \tag{1.48}$$

The (incomplete-data) expected information matrix is denoted by $\boldsymbol{\mathcal{I}}(\boldsymbol{\Psi})$, where under regularity conditions we obtain

$$\begin{aligned} \boldsymbol{\mathcal{I}}(\boldsymbol{\Psi}) &= E_{\boldsymbol{\Psi}}\{\boldsymbol{S}(\boldsymbol{Y};\ \boldsymbol{\Psi})\boldsymbol{S}^T(\boldsymbol{Y};\ \boldsymbol{\Psi})\} \\ &= E_{\boldsymbol{\Psi}}\{\boldsymbol{I}(\boldsymbol{\Psi};\ \boldsymbol{Y})\}. \end{aligned}$$

Here and elsewhere in this book, the operator $E_{\boldsymbol{\Psi}}$ denotes expectation using the parameter vector $\boldsymbol{\Psi}$.

For the complete data, we let

$$\boldsymbol{I}_c(\boldsymbol{\Psi};\ \boldsymbol{y}_c) = -\partial^2 \log L_c(\boldsymbol{\Psi})/\partial \boldsymbol{\Psi} \partial \boldsymbol{\Psi}^T,$$

while its conditional expectation given the observed data $\boldsymbol{y}$ is denoted by

$$\boldsymbol{\mathcal{I}}_c(\boldsymbol{\Psi};\ \boldsymbol{y}) = E_{\boldsymbol{\Psi}}\{\boldsymbol{I}_c(\boldsymbol{\Psi};\ \boldsymbol{Y}_c) \mid \boldsymbol{y}\}. \tag{1.49}$$

The expected information matrix corresponding to the complete data is given by

$$\boldsymbol{\mathcal{I}}_c(\boldsymbol{\Psi}) = E_{\boldsymbol{\Psi}}\{\boldsymbol{I}_c(\boldsymbol{\Psi};\ \boldsymbol{Y}_c)\}.$$

In other notations involving I, the symbol $\boldsymbol{I}_d$ is used to denote the $d \times d$ identity matrix, while $I_A(x)$ denotes the indicator function that is 1 if x belongs to the set A and is zero otherwise.

The density of a random vector $\boldsymbol{W}$ having a p-dimensional multivariate normal distribution with mean $\boldsymbol{\mu}$ and covariance $\boldsymbol{\Sigma}$ is denoted by $\phi(\boldsymbol{w};\ \boldsymbol{\mu},\ \boldsymbol{\Sigma})$, where

$$\phi(\boldsymbol{w};\ \boldsymbol{\mu},\ \boldsymbol{\Sigma}) = (2\pi)^{-\frac{p}{2}} \mid \boldsymbol{\Sigma} \mid^{-\frac{1}{2}} \exp\{-\tfrac{1}{2}(\boldsymbol{w}-\boldsymbol{\mu})^T \boldsymbol{\Sigma}^{-1}(\boldsymbol{w}-\boldsymbol{\mu})\}.$$

The notation $\phi(w;\ \mu,\ \sigma^2)$ is used to denote the density of a random variable having a univariate normal distribution with mean μ and variance σ^2.

2

ML Fitting of Mixture Models

2.1 INTRODUCTION

As noted earlier, since the advent of the EM algorithm, maximum likelihood (ML) has been by far the most commonly used approach to the fitting of mixture distributions. Hence the focus in this chapter is on the ML fitting of finite mixture models via the EM algorithm. We consider the general case of arbitrary component distributions. These results will be specialized in later chapters for particular components such as the normal and t distributions.

Before proceeding to consider ML estimation for finite mixture models, we first briefly define ML estimation in general and introduce some associated notation.

2.2 ML ESTIMATION

With the ML approach to the estimation of a d-dimensional parameter vector $\boldsymbol{\Psi}$ in a postulated density $f(\boldsymbol{y}_j;\, \boldsymbol{\Psi})$ for the random vector $\boldsymbol{Y}_j$ associated with the jth recording on the phenomenon under study, an estimate $\hat{\boldsymbol{\Psi}}$ is provided in regular situations by an appropriate solution of the likelihood equation,

$$\partial L(\boldsymbol{\Psi})/\partial \boldsymbol{\Psi} = \mathbf{0},$$

or, equivalently,

$$\partial \log L(\boldsymbol{\Psi})/\partial \boldsymbol{\Psi} = \mathbf{0}, \tag{2.1}$$

where

$$L(\boldsymbol{\Psi}) = \prod_{j=1}^{n} f(\boldsymbol{y}_j; \boldsymbol{\Psi}) \tag{2.2}$$

denotes the likelihood function for $\boldsymbol{\Psi}$ formed under the assumption of independent data $\boldsymbol{y}_1, \ldots, \boldsymbol{y}_n$.

Briefly, the aim of ML estimation (Lehmann, 1980, 1983) is to determine an estimate for each n ($\hat{\boldsymbol{\Psi}}$ in the present context), so that it defines a sequence of roots of the likelihood equation that is consistent and asymptotically efficient. Such a sequence is known to exist under suitable regularity conditions (Cramér, 1946). With probability tending to one, these roots correspond to local maxima in the interior of the parameter space. This consistent sequence of roots is essentially unique. The reader is referred to Huzurbazar (1948) and Perlman (1983) for a precise statement of the uniqueness of a consistent sequence of roots of the likelihood equation. For estimation models in general, the likelihood usually has a global maximum in the interior of the parameter space. Then typically a sequence of roots of the likelihood equation with the desired asymptotic properties is provided by taking $\hat{\boldsymbol{\Psi}}$ for each n to be the root that globally maximizes the likelihood function $L(\boldsymbol{\Psi})$. That is, $\hat{\boldsymbol{\Psi}}$ is the global maximizer of the likelihood and is called the maximum likelihood estimator (MLE).

We shall henceforth refer to $\hat{\boldsymbol{\Psi}}$ as the MLE even in situations where it may not globally maximize the likelihood. Indeed, in some of the examples on mixture models to be presented, the likelihood is unbounded. However, for these models there may still exist, under the usual regularity conditions, a sequence of roots of the likelihood equation corresponding to local maxima with the properties of consistency, efficiency, and asymptotic normality; see Section 2.5. The reader is referred to Cheng and Traylor (1995) for an account of nonregular maximum likelihood problems for estimation in general.

Given that a statistical model is at best an approximation to reality, it is worth considering the behavior of the MLE $\hat{\boldsymbol{\Psi}}$ if the structure of the postulated model (2.2) is not valid. It follows under mild regularity conditions that the MLE $\hat{\boldsymbol{\Psi}}$ under the invalid model (2.2) is still a meaningful estimator, in that it is a consistent estimator of $\boldsymbol{\Psi}_o$, the value of $\boldsymbol{\Psi}$ that minimizes the Kullback–Leibler distance between the actual density of $\boldsymbol{Y}_j$ and the postulated parametric family,

$$\{f(\boldsymbol{y}_j; \boldsymbol{\Psi}) : \quad \boldsymbol{\Psi} \in \boldsymbol{\Omega}\}; \tag{2.3}$$

see, for example, Hjort (1986).

2.3 INFORMATION MATRICES

The Fisher expected information matrix about the parameter vector $\boldsymbol{\Psi}$ is defined as

$$\mathcal{I}(\boldsymbol{\Psi}) = E_{\boldsymbol{\Psi}}\{\boldsymbol{S}(\boldsymbol{Y}; \boldsymbol{\Psi})\boldsymbol{S}^T(\boldsymbol{Y}; \boldsymbol{\Psi})\}, \tag{2.4}$$

where

$$S(y;\, \Psi) = \partial \log L(\Psi)/\partial \Psi$$

is the gradient vector of the log likelihood function (the score statistic) and $y = (y_1^T, \ldots, y_n^T)^T$ contains the observed data. Under regularity conditions, $\mathcal{I}(\Psi)$ can be expressed as

$$\mathcal{I}(\Psi) = E_{\Psi}\{I(\Psi;\, Y)\}, \tag{2.5}$$

where

$$I(\Psi;\, y) = -\partial^2 \log L(\Psi)/\partial \Psi \partial \Psi^T \tag{2.6}$$

is the negative of the Hessian of the log likelihood function. The so-called observed information matrix is defined to be $I(\hat{\Psi};\, y)$.

2.4 ASYMPTOTIC COVARIANCE MATRIX OF MLE

The asymptotic covariance matrix of the MLE $\hat{\Psi}$ is equal to the inverse of the expected information matrix $\mathcal{I}(\Psi)$, which can be approximated by $\mathcal{I}(\hat{\Psi})$; that is, the standard error of $\hat{\Psi}_r = (\hat{\Psi})_r$ is given by

$$\text{SE}(\hat{\Psi}_r) \approx (\mathcal{I}^{-1}(\hat{\Psi}))_{rr}^{1/2} \qquad (r = 1, \ldots, d), \tag{2.7}$$

where the standard notation $(A)_{rs}$ is used for the element in the rth row and sth column of a matrix A.

It is common in practice to estimate the inverse of the covariance matrix of the MLE by the observed information matrix $I(\hat{\Psi};\, y)$, rather than the expected information matrix $\mathcal{I}(\Psi)$ evaluated at $\Psi = \hat{\Psi}$. This approach gives the approximation

$$\text{SE}(\hat{\Psi}_r) \approx (I^{-1}(\hat{\Psi};\, y))_{rr}^{1/2} \quad (r = 1, \ldots, d). \tag{2.8}$$

Efron and Hinkley (1978) have provided a frequentist justification of (2.8) over (2.7) in the case of one-parameter ($d = 1$) families. Also, the observed information matrix is usually more convenient to use than the expected information matrix, as it does not require an expectation to be taken.

The provision of standard errors of MLEs in the context of mixture models is to be discussed in Section 2.16. It should be noted that the sample size n has to be very large before the asymptotic theory applies to mixture models.

2.5 PROPERTIES OF MLEs FOR MIXTURE MODELS

If the MLE of Ψ exists as a global maximizer of the likelihood for a mixture distribution with compact parameter space, then the conditions of Wald (1949) ensure that it is strongly consistent; see Redner (1981). This is assuming that Ψ is identifiable. It has been seen in Section 1.14 that, although a class of mixture densities may be identifiable, $\boldsymbol{\Psi}$ may not. Let $\boldsymbol{\Psi}_t$ denote the true value of $\boldsymbol{\Psi}$ and let $\boldsymbol{\Omega}_t$ denote the subset of the parameter space for which

$$f(\boldsymbol{y}_j;\, \boldsymbol{\Psi}) = f(\boldsymbol{y}_j;\, \boldsymbol{\Psi}_t)$$

for almost all $\boldsymbol{y}_j$ in $\mathbb{R}^p$. This set will contain more than the single point $\boldsymbol{\Psi}_t$ in the case of component densities belonging to the same parametric family. For then, as discussed in Section 1.14, a permutation of the component labels in $\boldsymbol{\Psi}$ does not alter the value of $f(\boldsymbol{y}_j;\, \boldsymbol{\Psi})$. This particular identifiability problem can be avoided by the imposition of a constraint on $\boldsymbol{\Psi}$ like (1.32).

Redner's (1981) work, however, was specifically aimed at families of distributions which are not identifiable. His results imply for mixture families with compact parameter space that, under the conditions of Wald (1949) except for the identifiability of $\boldsymbol{\Psi}$, $L(\boldsymbol{\Psi})$ is almost surely maximized in a neighborhood of $\boldsymbol{\Omega}_t$. More precisely, Redner (1981) referred to it as convergence of the MLE in the topology of the quotient space obtained by collapsing $\boldsymbol{\Omega}_t$ into a single point; see Ghosh and Sen (1985), Li and Sedransk (1988), and Redner and Walker (1984) for further discussion.

More recently, Feng and McCulloch (1996) obtained the parallel result in Euclidean space. Their result is also valid when the true value $\boldsymbol{\Psi}_t$ of $\boldsymbol{\Psi}$ lies on the boundary of the parameter space $\boldsymbol{\Omega}$ and is in a nonidentifiable subset $\boldsymbol{\Omega}_t$. They showed that the MLE of $\boldsymbol{\Psi}$ converges to the subset $\boldsymbol{\Omega}_t$. More precisely, they showed that there exists a $\boldsymbol{\Psi}_t^*(\hat{\boldsymbol{\Psi}}) \in \boldsymbol{\Omega}_t$ which depends on $\hat{\boldsymbol{\Psi}}$ such that $\hat{\boldsymbol{\Psi}} - \boldsymbol{\Psi}_t^*(\hat{\boldsymbol{\Psi}})$ converges to zero with probability one. The regularity conditions assumed by Feng and McCulloch (1996) are essentially those of Cramér (1946) but modified to allow for the fact that $\boldsymbol{\Psi}_t$ lies in a nonidentifiable subset of the parameter space and may also be on the boundary. In other work, Self and Liang (1987) proved the consistency of the MLE when the parameter vector is identifiable but on the boundary of the parameter space. In the identifiable case, Feng and McCulloch (1992) proposed an unrestricted MLE which is consistent and is asymptotically normally distributed. Thus the result of Feng and McCulloch (1996) is an extension of those of Redner (1981), Self and Liang (1987), and Feng and McCulloch (1992). The work of Feng and McCulloch (1996) is to be discussed further in Section 6.4.2 on the null distribution of the likelihood ratio test statistic for the number of components in a mixture model.

In the sequel, unless explicitly stated otherwise, it is implicitly assumed that $\boldsymbol{\Omega}_t$ is a singleton set; that is, $\boldsymbol{\Psi}$ is identifiable or, in the terminology of Ghosh and Sen (1985), the class of mixture densities is strongly identifiable.

As noted earlier, in some instances the likelihood $L(\boldsymbol{\Psi})$ under the mixture model (1.20) is unbounded, and so then the MLE of $\boldsymbol{\Psi}$ does not exist, at least as a global maximizer of the likelihood function. However, it may still exist as a local maximizer. Peters and Walker (1978) and Redner and Walker (1984) have given, for the class of identifiable mixtures, the regularity conditions that must be satisfied in order for there to exist a sequence of roots of the likelihood equation, $\partial L(\boldsymbol{\Psi})/\partial \boldsymbol{\Psi} = \mathbf{0}$, with the properties of consistency, efficiency and asymptotic normality. The form of the conditions, which are essentially multivariate generalizations of Cramér's (1946) results for the corresponding properties in a general context, suggests that they should hold for many parametric families. Available results for the normal parametric family are to be reviewed in Section 3.8.

2.6 CHOICE OF ROOT

With many applications of mixture models, the likelihood equation will have multiple roots corresponding to local maxima, and so there is the problem of identifying the desired root to define the MLE $\hat{\boldsymbol{\Psi}}$. Since the global maximizer of the likelihood function is typically a root of the likelihood equation, the problem of identifying $\hat{\boldsymbol{\Psi}}$ is solved in principle for those mixture models that satisfy Wald's (1949) conditions for the consistency of the global maximizer. For then if all the roots of the likelihood equation have been located, one can take $\hat{\boldsymbol{\Psi}}$ to be the root that gives the largest local maximum. However, in practice, the problem is not really solved since a search for all roots corresponding to local maximizers of the likelihood function may take considerable time and there will generally be no guarantee that all local maximizers will have been found. Moreover, as mentioned above, the likelihood function for mixture models may be unbounded and so then the MLE will correspond to a local maximizer (assuming regularity conditions).

If there is available some consistent estimator of $\boldsymbol{\Psi}$, then the estimator defined by the root of the likelihood equation closest to it is also consistent, and hence efficient (Lehmann, 1983, pp. 421), providing regularity conditions hold for the existence of a sequence of consistent and asymptotically efficient solutions of the likelihood equation. Where computationally feasible, the method of moments provides a way of constructing a $\sqrt{n}$-consistent estimator of $\boldsymbol{\Psi}$. The usefulness of a $\sqrt{n}$-consistent estimator of $\boldsymbol{\Psi}$, say $\tilde{\boldsymbol{\Psi}}$, is that under the same regularity conditions referred to above, a consistent and asymptotically efficient estimator of $\boldsymbol{\Psi}$ is obtained by a one-step approximation to the solution of the likelihood equation according to a Newton iterative scheme started from $\tilde{\boldsymbol{\Psi}}$. This approach does not require the determination of the closest root of the likelihood equation to $\tilde{\boldsymbol{\Psi}}$, but its implementation does require the inversion of a symmetric matrix whose dimension is equal to the total of unknown parameters. Hence it may not be as convenient computationally as the EM algorithm applied from $\tilde{\boldsymbol{\Psi}}$.

In the above, it is not suggested that inference or a clustering should be based solely on a single root of the likelihood equation. For example, in a clustering application, the clusters implied by the various local maximizers may have a number of points in common, which may be useful collectively in confirming that some or all of them are distant from the main body of points.

2.7 TEST FOR A CONSISTENT ROOT

2.7.1 Basis of Test

Recently, Gan and Jiang (1999) investigated a test of the null hypothesis that a given root of the likelihood equation should be adopted as the MLE of $\boldsymbol{\Psi}$; that is, that a given root is consistent and asymptotically efficient. It is based on the result (2.5) that under regularity conditions the expected information matrix is equal to the negative expectation of the Hessian of the log likelihood. Hence if $\hat{\boldsymbol{\Psi}}$ is the MLE, then

$$\mathcal{I}(\boldsymbol{\Psi}) = E_{\boldsymbol{\Psi}}\{\boldsymbol{I}(\boldsymbol{\Psi};\ \boldsymbol{Y})\} \tag{2.9}$$

should hold approximately at $\boldsymbol{\Psi} = \hat{\boldsymbol{\Psi}}$.

2.7.2 Example 2.1: Likelihood Function with Two Maximizers

To illustrate this point, Gan and Jiang (1999) generated a random sample of size $n = 5000$ from a mixture of two univariate normal components in proportions $\pi_1 = 0.4$ and $\pi_2 = 0.6$ with means $\mu_1 = -3$ and $\mu_2 = 6$, and variances $\sigma_1^2 = 1$ and $\sigma_2^2 = 16$. They chose this model where $\Psi = \mu_1$ is the single unknown parameter, as the likelihood equation usually has two roots that correspond to a global maximizer (Ψ_1) and a local maximizer (Ψ_2). For the generated sample, they calculated $\text{SE}_1(\Psi) = \{\mathcal{I}(\Psi)\}^{-1/2}$ and $\text{SE}_2(\Psi) = \{I(\Psi)\}^{-1/2}$ for the two roots, and the results are displayed in Table 2.1.

It is clear from Table 2.1 that the $\text{SE}_1(\Psi)$ and $\text{SE}_2(\Psi)$ are virtually the same at $\Psi = \Psi_1$ (the global maximizer), but not so at $\Psi = \Psi_2$ (the local maximizer).

Table 2.1 Local Maximizers of Normal Mixture Likelihood

	Global Maximizer ($i = 1$)	Local Maximizer ($i = 2$)
$\Psi^{(0)}$	-3.0	6.0
Ψ_i	-2.9931	6.4778
$\log L(\Psi_i)$	-9901	-17443
$\text{SE}_1(\Psi_i)$	0.0234	0.0492
$\text{SE}_2(\Psi_i)$	0.0237	0.2807

Source: Adapted from Gan and Jiang (1999).

2.7.3 Formulation of Test Statistic

To form an unbiased estimator of the difference

$$\mathcal{I}(\boldsymbol{\Psi}) - E_{\boldsymbol{\Psi}}\{\boldsymbol{I}(\boldsymbol{\Psi};\ \boldsymbol{Y})\}, \tag{2.10}$$

we note that the expected information matrix $\mathcal{I}(\boldsymbol{\Psi})$ is estimated unbiasedly by

$$\hat{\mathcal{I}}(\boldsymbol{\Psi}) = \sum_{j=1}^{n} \boldsymbol{s}(\boldsymbol{y}_j;\ \boldsymbol{\Psi})\boldsymbol{s}^T(\boldsymbol{y}_j;\ \boldsymbol{\Psi}), \tag{2.11}$$

where

$$\boldsymbol{s}(\boldsymbol{y}_j;\ \boldsymbol{\Psi}) = \partial \log f(\boldsymbol{y}_j;\ \boldsymbol{\Psi})/\partial \boldsymbol{\Psi} \qquad (j = 1, \ldots, n). \tag{2.12}$$

Thus

$$\boldsymbol{T}(\boldsymbol{y};\, \boldsymbol{\Psi}) = \hat{\mathcal{I}}(\boldsymbol{\Psi}) - \boldsymbol{I}(\boldsymbol{\Psi};\, \boldsymbol{y}) \tag{2.13}$$

is an unbiased estimator of the difference (2.10).

If $\hat{\boldsymbol{\Psi}}$ is a consistent estimator of $\boldsymbol{\Psi}$, then it follows that

$$\boldsymbol{T}(\boldsymbol{y};\, \hat{\boldsymbol{\Psi}}) \approx \mathbf{0}. \tag{2.14}$$

Gan and Jiang (1999) proposed using

$$\boldsymbol{T}(\boldsymbol{y};\, \hat{\boldsymbol{\Psi}}) \tag{2.15}$$

as the basis of a large sample test of the null hypothesis that $\hat{\boldsymbol{\Psi}}$ is a consistent root of the likelihood equation; that is, $\hat{\boldsymbol{\Psi}}$ is the MLE of $\boldsymbol{\Psi}$. They investigated the case where there is only one unknown parameter so that Ψ, and hence $T(\boldsymbol{y};\, \hat{\Psi})$, is real-valued (a scalar). In this simplified case, they derived a normalized version of $T(\boldsymbol{y};\, \hat{\Psi})$ that is asymptotically standard normal under the null hypothesis. Their simulations suggest that the test performs quite well when the sample size is large, but may suffer the problem of over-rejection with relatively small samples.

For convenience, Gan and Jiang (1999) suggested

$$\{T(\boldsymbol{y};\, \hat{\Psi})/n\}/[\text{var}\{T(\boldsymbol{Y};\, \hat{\Psi})/n\}]^{1/2} \tag{2.16}$$

as an alternative test statistic, where the variance in the denominator of (2.16) is approximated by Monte Carlo methods, for example, the bootstrap. The asymptotic null distribution of the normalized version (2.16) of $T(\boldsymbol{y};\, \hat{\Psi})$ is also standard normal.

The test of Gan and Jiang (1999) is based on the work of White (1982), who used the (normalized) statistic (2.15) to test the null hypothesis that the model was not misspecified. The latter is assumed to hold in the test of Gan and Jiang (1999). Thus if their test rejects the null hypothesis, then one can conclude that there is either a problem of misspecification or a problem of $\hat{\Psi}$ being an inconsistent root of the likelihood equation. One cannot distinguish between the two, but there are other ways of assessing the validity of a model (Section 3.5).

Given the questionable small sample performance of this test and its present limitation to the case of a single unknown parameter, in practice we will usually have to rely on other ways of deciding which root to choose as the MLE of $\boldsymbol{\Psi}$. This is to be pursued further in Section 3.10 in the context of normal components, where we shall see that there are good reasons to choose the root of the likelihood equation corresponding to the largest of the local maxima located, after eliminating any so-called spurious local maximizers. This is in the absence of any other information on the parameters and their estimates, and it also assumes that we are reasonably confident about the suitability of the parametric forms adopted for the component densities.

We now discuss how roots of the likelihood equation corresponding to local maxima can be obtained for mixture models by application of the EM algorithm.

2.8 APPLICATION OF EM ALGORITHM FOR MIXTURE MODELS

2.8.1 Direct Approach

We describe here the application of the EM algorithm for the ML fitting of the parametric mixture model,

$$f(\boldsymbol{y}_j;\, \boldsymbol{\Psi}) = \sum_{i=1}^{g} \pi_i\, f_i(\boldsymbol{y}_j;\, \boldsymbol{\theta}_i), \tag{2.17}$$

to an observed random sample, $\boldsymbol{y} = (\boldsymbol{y}_1^T, \ldots, \boldsymbol{y}_n^T)^T$. In (2.17),

$$\boldsymbol{\Psi} = (\pi_1, \ldots, \pi_{g-1}, \boldsymbol{\xi}^T)^T \tag{2.18}$$

is the vector containing all the unknown parameters in this mixture model and $\boldsymbol{\xi}$ is the vector containing all the parameters in $\boldsymbol{\theta}_1, \ldots, \boldsymbol{\theta}_g$ known *a priori* to be distinct.

The log likelihood for $\boldsymbol{\Psi}$ that can be formed from the observed data is given by

$$\begin{aligned} \log L(\boldsymbol{\Psi}) &= \sum_{j=1}^{n} \log f(\boldsymbol{y}_j;\, \boldsymbol{\Psi}) \\ &= \sum_{j=1}^{n} \log\{\sum_{i=1}^{g} \pi_i f_i(\boldsymbol{y}_j;\, \boldsymbol{\theta}_i)\}. \end{aligned} \tag{2.19}$$

Computation of the MLE of $\boldsymbol{\Psi}$ by direct consideration of this log likelihood function requires solving the likelihood equation,

$$\partial \log L(\boldsymbol{\Psi})/\partial \boldsymbol{\Psi} = \boldsymbol{0}. \tag{2.20}$$

It can be manipulated so that the MLE of $\boldsymbol{\Psi}$, $\hat{\boldsymbol{\Psi}}$, satisfies

$$\hat{\pi}_i = \sum_{j=1}^{n} \tau_i(\boldsymbol{y}_j;\, \hat{\boldsymbol{\Psi}})/n \qquad (i = 1, \ldots, g) \tag{2.21}$$

and

$$\sum_{i=1}^{g} \sum_{j=1}^{n} \tau_i(\boldsymbol{y}_j;\, \hat{\boldsymbol{\Psi}}) \partial \log f_i(\boldsymbol{y}_j;\, \hat{\boldsymbol{\theta}}_i)/\partial \boldsymbol{\xi} = \boldsymbol{0}, \tag{2.22}$$

where

$$\tau_i(\boldsymbol{y}_j;\, \boldsymbol{\Psi}) = \pi_i\, f_i(\boldsymbol{y}_j;\, \boldsymbol{\theta}_i) / \sum_{h=1}^{g} \pi_h\, f_h(\boldsymbol{y}_j;\, \boldsymbol{\theta}_h) \tag{2.23}$$

is the posterior probability that $\boldsymbol{y}_j$ belongs to the ith component of the mixture; see McLachlan and Krishnan (1997, Section 1.4) for further details.

These manipulations were carried out by various researchers in the past in their efforts to solve the likelihood equation for mixture models with specific component densities as, for example, by Hasselblad (1966, 1969), Wolfe (1965, 1967, 1970), and Day (1969). They observed in their special cases that the equations (2.21) and (2.22) suggest an iterative computation of the solution whereby for an initial value $\boldsymbol{\Psi}^{(0)}$ of $\boldsymbol{\Psi}$ in the right-hand side of these equations, a new estimate $\boldsymbol{\Psi}^{(1)}$ can be computed for $\boldsymbol{\Psi}$, which in turn can be substituted into the right-hand side of these equations to produce a new update $\boldsymbol{\Psi}^{(2)}$, and so on, until convergence. Now this iterative method of solution of the likelihood equation can be identified with the direct application of the EM algorithm of Dempster et al. (1977) for finding solutions of the likelihood equation. The application of the EM algorithm to the mixture problem automatically reveals the iterative scheme to be followed for the computation of the MLE. Furthermore, it ensures that the likelihood values increase monotonically. Prior to the appearance of the paper by Dempster et al. (1977), various researchers did note the monotone convergence of the likelihood sequences produced in their particular applications, but were only able to speculate on this monotonicity holding in general.

2.8.2 Formulation as an Incomplete-Data Problem

As already discussed in Section 1.9, within the formulation of the mixture problem in the EM framework, the observed-data vector

$$\boldsymbol{y} = (\boldsymbol{y}_1^T, \ldots, \boldsymbol{y}_n^T)^T$$

is viewed as being incomplete, as the associated component-label vectors, $\boldsymbol{z}_1, \ldots, \boldsymbol{z}_n$, are not available. In this framework, where each $\boldsymbol{y}_j$ is conceptualized as having arisen from one of the components of the mixture model (2.17) being fitted, $\boldsymbol{z}_j$ is a g-dimensional vector with $z_{ij} = (\boldsymbol{z}_j)_i = 1$ or 0, according to whether $\boldsymbol{y}_j$ did or did not arise from the ith component of the mixture $(i = 1, \ldots, g;\ j = 1, \ldots, n)$. The complete-data vector is therefore declared to be

$$\boldsymbol{y}_c = (\boldsymbol{y}^T, \boldsymbol{z}^T)^T, \tag{2.24}$$

where

$$\boldsymbol{z} = (\boldsymbol{z}_1^T, \ldots, \boldsymbol{z}_n^T)^T \tag{2.25}$$

The component-label vectors $\boldsymbol{z}_1, \ldots, \boldsymbol{z}_n$ are taken to be the realized values of the random vectors $\boldsymbol{Z}_1, \ldots, \boldsymbol{Z}_n$, where, for independent feature data, it is appropriate to assume that they are distributed unconditionally according to the multinomial distribution (1.18). This assumption means that the distribution of the complete-data vector $\boldsymbol{Y}_c$ implies the appropriate distribution for the incomplete-data vector $\boldsymbol{Y}$. The complete-data log likelihood for $\boldsymbol{\Psi}$, $\log L_c(\boldsymbol{\Psi})$, is given by

$$\log L_c(\boldsymbol{\Psi}) = \sum_{i=1}^{g} \sum_{j=1}^{n} z_{ij} \{\log \pi_i + \log f_i(\boldsymbol{y}_j;\ \boldsymbol{\theta}_i)\}. \tag{2.26}$$

2.8.3 E-Step

The EM algorithm is applied to this problem by treating the $\boldsymbol{z}_{ij}$ as missing data. It proceeds iteratively in two steps, E (for expectation) and M (for maximization).

The addition of the unobservable data to the problem (here the $\boldsymbol{z}_j$) is handled by the E-step, which takes the conditional expectation of the complete-data log likelihood, $\log L_c(\boldsymbol{\Psi})$, given the observed data $\boldsymbol{y}$, using the current fit for $\boldsymbol{\Psi}$. Let $\boldsymbol{\Psi}^{(0)}$ be the value specified initially for $\boldsymbol{\Psi}$. Then on the first iteration of the EM algorithm, the E-step requires the computation of the conditional expectation of $\log L_c(\boldsymbol{\Psi})$ given $\boldsymbol{y}$, using $\boldsymbol{\Psi}^{(0)}$ for $\boldsymbol{\Psi}$, which can be written as

$$Q(\boldsymbol{\Psi}; \boldsymbol{\Psi}^{(0)}) = E_{\boldsymbol{\Psi}^{(0)}}\{\log L_c(\boldsymbol{\Psi}) \mid \boldsymbol{y}\}. \tag{2.27}$$

The expectation operator E has the subscript $\boldsymbol{\Psi}^{(0)}$ to explicitly convey that this expectation is being effected using $\boldsymbol{\Psi}^{(0)}$ for $\boldsymbol{\Psi}$. It follows that on the $(k+1)$th iteration, the E-step requires the calculation of $Q(\boldsymbol{\Psi}; \boldsymbol{\Psi}^{(k)})$, where $\boldsymbol{\Psi}^{(k)}$ is the value of $\boldsymbol{\Psi}$ after the kth EM iteration. As the complete-data log likelihood, $\log L_c(\boldsymbol{\Psi})$, is linear in the unobservable data z_{ij}, the E-step (on the $(k+1)$th iteration) simply requires the calculation of the current conditional expectation of Z_{ij} given the observed data $\boldsymbol{y}$, where Z_{ij} is the random variable corresponding to z_{ij}. Now

$$\begin{aligned} E_{\boldsymbol{\Psi}^{(k)}}(Z_{ij} \mid \boldsymbol{y}) &= \operatorname{pr}_{\boldsymbol{\Psi}^{(k)}}\{Z_{ij} = 1 \mid \boldsymbol{y}\} \\ &= \tau_i(\boldsymbol{y}_j; \boldsymbol{\Psi}^{(k)}), \end{aligned} \tag{2.28}$$

where, corresponding to (2.23),

$$\begin{aligned} \tau_i(\boldsymbol{y}_j; \boldsymbol{\Psi}^{(k)}) &= \pi_i^{(k)} f_i(\boldsymbol{y}_j; \boldsymbol{\theta}_i^{(k)}) / f(\boldsymbol{y}_j; \boldsymbol{\Psi}^{(k)}) \\ &= \pi_i^{(k)} f_i(\boldsymbol{y}_j; \boldsymbol{\theta}_i^{(k)}) / \sum_{h=1}^{g} \pi_h^{(k)} f_h(\boldsymbol{y}_j; \boldsymbol{\theta}_h^{(k)}) \end{aligned} \tag{2.29}$$

for $i = 1, \ldots, g$; $j = 1, \ldots, n$. The quantity $\tau_i(\boldsymbol{y}_j; \boldsymbol{\Psi}^{(k)})$ is the posterior probability that the jth member of the sample with observed value $\boldsymbol{y}_j$ belongs to the ith component of the mixture. Using (2.28), we have on taking the conditional expectation of (2.26) given $\boldsymbol{y}$ that

$$Q(\boldsymbol{\Psi}; \boldsymbol{\Psi}^{(k)}) = \sum_{i=1}^{g} \sum_{j=1}^{n} \tau_i(\boldsymbol{y}_j; \boldsymbol{\Psi}^{(k)}) \{\log \pi_i + \log f_i(\boldsymbol{y}_j; \boldsymbol{\theta}_i)\}. \tag{2.30}$$

2.8.4 M-Step

The M-step on the $(k+1)$th iteration requires the global maximization of $Q(\boldsymbol{\Psi}; \boldsymbol{\Psi}^{(k)})$ with respect to $\boldsymbol{\Psi}$ over the parameter space $\boldsymbol{\Omega}$ to give the updated estimate $\boldsymbol{\Psi}^{(k+1)}$. For the finite mixture model, the updated estimates $\pi_i^{(k+1)}$ of the mixing proportions π_i are calculated independently of the updated estimate $\boldsymbol{\xi}^{(k+1)}$ of the parameter vector $\boldsymbol{\xi}$ containing the unknown parameters in the component densities.

If the z_{ij} were observable, then the complete-data MLE of π_i would be given simply by

$$\hat{\pi}_i = \sum_{j=1}^{n} z_{ij}/n \qquad (i = 1, \ldots, g). \tag{2.31}$$

As the E-step simply involves replacing each z_{ij} with its current conditional expectation $\tau_i(\boldsymbol{y}_j;\, \boldsymbol{\Psi}^{(k)})$ in the complete-data log likelihood, the updated estimate of π_i is given by replacing each z_{ij} in (2.31) by $\tau_i(\boldsymbol{y}_j;\, \boldsymbol{\Psi}^{(k)})$ to give

$$\pi_i^{(k+1)} = \sum_{j=1}^{n} \tau_i(\boldsymbol{y}_j;\, \boldsymbol{\Psi}^{(k)})/n \qquad (i = 1,\, \ldots,\, g). \tag{2.32}$$

Thus in forming the estimate of π_i on the $(k+1)$th iteration, there is a contribution from each observation $\boldsymbol{y}_j$ equal to its (currently assessed) posterior probability of membership of the ith component of the mixture model. This EM solution therefore has an intuitively appealing interpretation.

Concerning the updating of $\boldsymbol{\xi}$ on the M-step of the $(k+1)$th iteration, it can be seen from (2.30) that $\boldsymbol{\xi}^{(k+1)}$ is obtained as an appropriate root of

$$\sum_{i=1}^{g}\sum_{j=1}^{n} \tau_i(\boldsymbol{y}_j;\, \boldsymbol{\Psi}^{(k)})\partial \log f_i(\boldsymbol{y}_j;\, \boldsymbol{\theta}_i)/\partial\boldsymbol{\xi} = \boldsymbol{0}. \tag{2.33}$$

One nice feature of the EM algorithm is that the solution of (2.33) often exists in closed form, as is to be demonstrated for the normal mixture model in Section 3.2.

The E- and M-steps are alternated repeatedly until the difference

$$L(\boldsymbol{\Psi}^{(k+1)}) - L(\boldsymbol{\Psi}^{(k)})$$

changes by an arbitrarily small amount in the case of convergence of the sequence of likelihood values $\{L(\boldsymbol{\Psi}^{(k)})\}$. Dempster et al. (1977) showed that the (incomplete-data) likelihood function $L(\boldsymbol{\Psi})$ is not decreased after an EM iteration; that is,

$$L(\boldsymbol{\Psi}^{(k+1)}) \geq L(\boldsymbol{\Psi}^{(k)}) \tag{2.34}$$

for $k = 0, 1, 2, \ldots$. Hence convergence must be obtained with a sequence of likelihood values that are bounded above. Dempster et al. (1977) also showed that if the very weak condition that $Q(\boldsymbol{\Psi};\, \boldsymbol{\phi})$ is continuous in both $\boldsymbol{\Psi}$ and $\boldsymbol{\phi}$ holds, then L^* will be a local maximum of $L(\boldsymbol{\Psi})$, provided that the sequence is not trapped at some saddle point. A detailed account of the convergence properties of the EM algorithm in a general setting has been given by Wu (1983), who addressed, in particular, the problem that the convergence of $L(\boldsymbol{\Psi}^{(k)})$ to L^* does not automatically imply the convergence of $\boldsymbol{\Psi}^{(k)}$ to a point $\boldsymbol{\Psi}^*$. Further details may be found in the monograph of McLachlan and Krishnan (1997) on the EM algorithm.

Titterington (1984a) developed a recursive algorithm for data that are available sequentially. More recently for the same situation, Jorgensen (1999) has considered a recursive algorithm for the estimation of the mixing proportions when the component densities are specified.

2.8.5 Assessing the Implied Error Rates

Ganesalingam and McLachlan (1980b) and Basford and McLachlan (1985a) considered a measure of strength of the clustering implied by the MLE $\hat{\boldsymbol{\Psi}}$ in terms of the fitted posterior probabilities of component membership $\tau_i(\boldsymbol{y}_j;\, \hat{\boldsymbol{\Psi}})$. For example, if the maximum of $\tau_i(\boldsymbol{y}_j;\, \hat{\boldsymbol{\Psi}})$ is near to one for most of the observations $\boldsymbol{y}_j$, then it suggests that the clusters are well separated. They proposed estimates of the error rates of the outright clustering obtained by assigning the observed data point $\boldsymbol{y}_j$ to the component to which it has the greatest (estimated) posterior probability of belonging. As noted in Section 1.14 on the identifiability of mixture models, there are $g!$ permutations of the component labels for a mixture of g component densities belonging to the same parametric family. Therefore in speaking about the error rates of the clustering of the mixture model-based approach, it is implicitly assumed that the clusters correspond appropriately to the g external groups assumed to exist for the purpose of defining the error rates. The overall error rate can be assessed by the average of the maximum of the component-posterior probabilities over the data, and the bootstrap can be used to correct for bias; see McLachlan and Basford (1988, Chapter 5) for details.

2.9 FITTING MIXTURES OF MIXTURES

In the past, discriminant analysis was frequently carried out for continuous data by taking the group-conditional densities to be normal. However, these days the flexibility of normal mixture models is being increasingly exploited by modeling the group-conditional densities by normal mixture densities (McLachlan and Gordon, 1989). More recently, Hastie and Tibshirani (1996) have exploited the flexibility of normal mixtures by proposing a discriminant rule where the group-conditional densities are modeled as finite mixtures of normal homoscedastic components. Also, Streit and Luginbuhl (1994) have considered normal mixtures for the group-conditional distributions in the training of probabilistic neural networks by maximum likelihood; see also Yiu, Mak, and Li (1999).

In a typical application of discriminant analysis, there are classified data from each group (so-called training data) from which to estimate the group-conditional densities. If the latter are modeled as mixtures, then the estimation of a given group-conditional density can be undertaken by fitting a mixture model to the data from that group, using the EM algorithm as described above.

However, in some applications, the component densities for a group may have some parameters in common with the component parameters for another group. As mentioned above, in the approach of Hastie and Tibshirani(1996), the normal component densities for the groups all have the same covariance matrix. In some applications, the component densities may have more than common scale parameters. For example, in the case study of Belbin and Rubin (1995), the density for the log reaction time Y_j for a schizophrenic subject is modeled by the normal mixture,

$$f(y_j;\, \boldsymbol{\Psi}_1) = \pi_{11}\, \phi(y_j;\, \mu_{11},\, \sigma^2) + \pi_{12}\, \phi(y_j;\, \mu_{12},\, \sigma^2),$$

where μ_{11} is taken to be the mean of the log reaction time for a nonschizophrenic subject (assumed to have a normal distribution with variance σ^2).

In such situations, we can implement the E-step (on the $(k+1)$th iteration) by still carrying out separately the E-step for the fitting of a mixture model to the data from each group to give the Q-function specific to each group, $Q_i(\boldsymbol{\Psi}_i;\ \boldsymbol{\Psi}_i^{(k)})$, where now $\boldsymbol{\Psi}_i$ denotes the vector of parameters in the ith mixture model for the ith group $(i = 1, \ldots, M)$, and M denotes the number of groups.

To perform the M-step, we consider the maximization of a combined Q-function given by

$$Q(\boldsymbol{\Psi};\ \boldsymbol{\Psi}^{(k)}) = \sum_{i=1}^{M} Q_i(\boldsymbol{\Psi}_i;\ \boldsymbol{\Psi}_i^{(k)}),$$

where $\boldsymbol{\Psi}$ is the vector containing the elements of $\boldsymbol{\Psi}_1, \ldots, \boldsymbol{\Psi}_M$ known *a priori* to be distinct.

2.10 MAXIMUM *A POSTERIORI* ESTIMATION

The EM algorithm is easily modified to produce the maximum *a posteriori* (MAP) estimate in a Bayesian framework for some prior density $p(\boldsymbol{\Psi})$ for $\boldsymbol{\Psi}$. The E-step is effectively the same as for the computation of the MLE of $\boldsymbol{\Psi}$ in a frequentist framework, requiring the calculation of the Q-function, $Q(\boldsymbol{\Psi};\ \boldsymbol{\Psi}^{(k)})$. The M-step differs in that the objective function for the maximization process is equal to $Q(\boldsymbol{\Psi};\ \boldsymbol{\Psi}^{(k)})$ augmented by the log prior density, $\log p(\boldsymbol{\Psi})$.

The MAP estimate of $\boldsymbol{\Psi}$ can be used as an initial value for $\boldsymbol{\Psi}$ in posterior simulations for $\boldsymbol{\Psi}$ (Gelman and Rubin, 1996). Also, Rubin and Wu (1997) have argued that the use of the EM algorithm or its variants is a wise first step for solid computing involving the simulation of posterior distributions. Firstly, it gives a rough picture of the posterior distribution at a lower cost than the Gibbs sampler. Secondly, the monotone iterates of the (incomplete-data) likelihood are effective detectors for seemingly inevitable programming errors, and thus help to ensure clean code for the EM algorithm, which can then be translated into the code for the Gibbs sampler in a straightforward manner. Thirdly, the results from the EM algorithm and the Gibbs sampler can be checked with each other to assess the normality of the posterior distribution and also to detect possible programming errors.

The relationship between the EM algorithm and the Gibbs sampler has been examined recently in Sahu and Roberts (1999).

2.11 AN AITKEN ACCELERATION-BASED STOPPING CRITERION

The stopping criterion usually adopted with the EM algorithm is in terms of either the size of the relative change in the parameter estimates or the log likelihood, $\log L(\boldsymbol{\Psi})$. As Lindstrom and Bates (1988) emphasize, this is a measure of lack of progress but not of actual convergence. Böhning et al. (1994) have exploited Aitken's acceleration procedure in its application to the sequence of log likelihood values to provide a useful estimate of the limiting value. It is applicable in the case where the sequence of log likelihood values $\{l^{(k)}\}$ is linearly convergent to some value l^*, where here for brevity of notation

$$l^{(k)} = \log L(\boldsymbol{\Psi}^{(k)}).$$

Under this assumption,

$$l^{(k+1)} - l^* \approx a\,(l^{(k)} - l^*), \tag{2.35}$$

for all k and some a $(0 < a < 1)$. Equation (2.35) can be rearranged to give

$$l^{(k+1)} - l^{(k)} \approx (1-a)(l^* - l^{(k)}) \tag{2.36}$$

for all k. It can be seen from (2.36) that, if a is very close to one, a small increment in the log likelihood, $l^{(k+1)} - l^{(k)}$, does not necessarily mean that $l^{(k)}$ is very close to l^*.

From (2.36), we have that

$$l^{(k+1)} - l^{(k)} \approx a\,(l^{(k)} - l^{(k-1)}) \tag{2.37}$$

for all k. Just as Aitken's acceleration procedure can be applied to accelerate the sequence of EM iterates $\{\boldsymbol{\Psi}^{(k)}\}$ (McLachlan and Krishnan, 1997, Section 4.8), it can be applied to obtain the corresponding result for the limit l^* of the sequence of log likelihood values $\{l^{(k)}\}$, namely,

$$l^* = l^{(k)} + \frac{1}{(1-a)}(l^{(k+1)} - l^{(k)}). \tag{2.38}$$

Since a is unknown, it has to be estimated in (2.38), for example, by the ratio of successive increments,

$$a^{(k)} = (l^{(k+1)} - l^{(k)})/(l^{(k)} - l^{(k-1)}).$$

This leads to the Aitken accelerated estimate of l^*,

$$l_A^{(k+1)} = l^{(k)} + \frac{1}{(1-a^{(k)})}(l^{(k+1)} - l^{(k)}). \tag{2.39}$$

In applications where the primary interest is on the sequence of log likelihood values rather than the sequence of parameter estimates, Böhning et al. (1994) suggest the EM algorithm can be stopped if

$$\mid l_A^{(k+1)} - l_A^{(k)} \mid < \text{tol},$$

where tol is the desired tolerance. An example concerns the resampling approach (McLachlan, 1987) to the problem of assessing the null distribution of the likelihood ratio test statistic for the number of components in a mixture model. The criterion (2.39) is applicable for any log likelihood sequence that is linearly convergent.

2.12 STARTING VALUES FOR EM ALGORITHM

2.12.1 Specification of an Initial Parameter Value

As explained in Section 2.8, the EM algorithm is started from some initial value of $\boldsymbol{\Psi}$, $\boldsymbol{\Psi}^{(0)}$. Hence in practice we have to specify a value for $\boldsymbol{\Psi}^{(0)}$. Recently, Seidel, Mosler, and Alker (2000a) have demonstrated how different starting strategies and stopping rules can lead to quite different estimates in the context of fitting mixtures of exponential components via the EM algorithm; see also Seidel, Mosler, and Alker (2000b).

As convergence with the EM algorithm is slow, the situation will be exacerbated by a poor choice of $\boldsymbol{\Psi}^{(0)}$. Indeed, in some cases where the likelihood is unbounded on the edge of the parameter space, the sequence of estimates $\{\boldsymbol{\Psi}^{(k)}\}$ generated by the EM algorithm may diverge if $\boldsymbol{\Psi}^{(0)}$ is chosen too close to the boundary. This matter is to be discussed further in Section 3.8 in the case of normal mixture models. Another problem with mixture models is that the likelihood equation will usually have multiple roots corresponding to local maxima, and so the EM algorithm should be applied from a wide choice of starting values in any search for all local maxima.

In the absence of the observed value of any known consistent estimator of $\boldsymbol{\Psi}$ or any other information, an obvious choice for the root of the likelihood equation is the one corresponding to the largest of the local maxima located (excluding so-called spurious local maximizers), although it does not necessarily follow that this choice defines the sequence of roots of the likelihood equation that is consistent and asymptotically efficient (Lehmann, 1980, p. 234). This point and the identification of spurious local maximizers is to be considered further in Section 3.10 in the context of normal mixture models with unrestricted covariance matrices.

For independent data in the case of mixture models, the effect of the E-step is to update the posterior probabilities of component membership. Hence an alternative approach is to perform the first E-step by specifying a value $\boldsymbol{\tau}_j^{(0)}$ for $\boldsymbol{\tau}(\boldsymbol{y}_j;\, \boldsymbol{\Psi})$ for each j $(j = 1, \ldots, n)$, where

$$\boldsymbol{\tau}(\boldsymbol{y}_j;\, \boldsymbol{\Psi}) = (\tau_1(\boldsymbol{y}_j;\, \boldsymbol{\Psi}), \ldots, \tau_g(\boldsymbol{y}_j;\, \boldsymbol{\Psi}))^T$$

is the vector containing the g posterior probabilities of component membership for $\boldsymbol{y}_j$. The latter is usually undertaken by setting $\boldsymbol{\tau}_j^{(0)} = \boldsymbol{z}_j^{(0)}$ for $j = 1, \ldots, n$, where

$$\boldsymbol{z}^{(0)} = (\boldsymbol{z}_1^{(0)^T}, \ldots, \boldsymbol{z}_n^{(0)^T})^T$$

defines an initial partition of the data into g groups. For example, an *ad hoc* way of initially partitioning the data in the case of, say, a mixture of $g = 2$ normal components with the same covariance matrices would be to plot the data for selections of two of the p variables, and then draw a line that divides the bivariate data into two groups that have a scatter that appears normal.

For higher-dimensional data, an initial value $\boldsymbol{z}^{(0)}$ for $\boldsymbol{z}$ might be obtained through the use of some clustering algorithm, such as k-means or, say, an hierarchical procedure if n is not too large.

2.12.2 Random Starting Values

Another way of specifying an initial partition $\boldsymbol{z}^{(0)}$ of the data is to randomly divide the data into g groups corresponding to the g components of the mixture model. That is, for each observation $\boldsymbol{y}_j$, we randomly generate an integer between 1 and g, both inclusive. If this random integer is equal to h, then we set the ith element of $\boldsymbol{z}_j^{(0)}$ equal to one for $i = h$ and equal to zero for $i \neq h$ $(i = 1, \ldots, g)$.

Usually, the EM algorithm would be applied from a number of random starts. With random starts, the effect of the central limit theorem tends to have the component parameters initially being similar at least in large samples. One way to reduce this effect is to first select a small random subsample from the data, which is then randomly assigned to the g components. The first M-step is then performed on the basis of the subsample. The subsample has to be sufficiently large to ensure that the first M-step is able to produce a nondegenerate estimate of the parameter vector $\boldsymbol{\Psi}$. For example, in the fitting of a mixture of p-dimensional normal components with unrestricted covariance matrices, there needs to be at least $(p+1)$ observations assigned to each component to ensure nonsingular estimates of the component-covariance matrices on the first M-step.

An alternative method of specifying a random start, at least in the context of g normal components with means $\boldsymbol{\mu}_i$ and covariance matrices $\boldsymbol{\Sigma}_i$, is to randomly generate the $\boldsymbol{\mu}_i^{(0)}$ independently as

$$\boldsymbol{\mu}_1^{(0)}, \ldots, \boldsymbol{\mu}_g^{(0)} \overset{\text{i.i.d.}}{\sim} N(\overline{\boldsymbol{y}}, \boldsymbol{V}), \tag{2.40}$$

where $\overline{\boldsymbol{y}}$ is the sample mean and

$$\boldsymbol{V} = \sum_{j=1}^{n} (\boldsymbol{y}_j - \overline{\boldsymbol{y}})(\boldsymbol{y}_j - \overline{\boldsymbol{y}})^T / n \tag{2.41}$$

is the sample covariance matrix of the observed data. With this method, there is more variation between the initial values $\boldsymbol{\mu}_i^{(0)}$ for the component means $\boldsymbol{\mu}_i$ than with a random partition of the data into g groups, and it is also computationally less demanding.

The component-covariance matrices $\boldsymbol{\Sigma}_i$ and the mixing proportions π_i can be specified as

$$\boldsymbol{\Sigma}_i^{(0)} = \boldsymbol{V} \text{ and } \pi_i^{(0)} = 1/g \qquad (i = 1, \ldots, g). \tag{2.42}$$

As illustrated in McLachlan and Basford (1988, Section 3.2), a key factor in the fitting of a mixture model is the accuracy of the estimate of the vector of the mixing proportions. For univariate mixtures Fowlkes (1979) suggested the determination of the point of inflection in quantile–quantile (Q–Q) plots to estimate first the mixing proportions of the underlying populations. The remaining parameters can then be estimated from the sample partitioned into groups in accordance with the estimates of the mixing proportions.

2.12.3 Example 2.2: Synthetic Data Set 1

To illustrate the use of random starting values, we consider the fitting of six normal components with unrestricted covariance matrices to an artificial data set (Synthetic Data Set 1), as described in McLachlan and Peel (1996). It consists of 204 bivariate observations. In Figure 2.1 we have plotted the (asymptotic) 95% ellipsoids

$$(\boldsymbol{y}_j - \boldsymbol{\mu}_i^{(1)})^T \boldsymbol{\Sigma}_i^{(1)^{-1}} (\boldsymbol{y}_j - \boldsymbol{\mu}_i^{(1)})^T \leq \chi^2_{p;.95}$$

for each component i after the first M-step, corresponding to a random assignment of all the data to the g components. Here $\chi^2_{p;.95}$ denotes the 95th percentile of the chi-squared distribution with p degrees of freedom. It can be seen that there is a high degree of overlap between these first iterates of the estimated component densities.

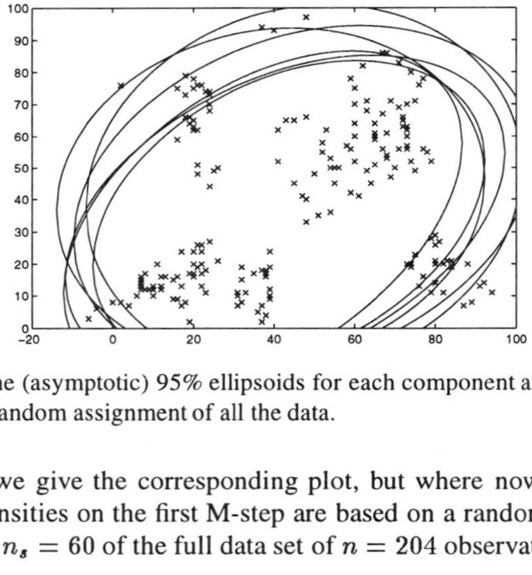

Fig. 2.1 Plot of the (asymptotic) 95% ellipsoids for each component after the first M-step, corresponding to a random assignment of all the data.

In Figure 2.2 we give the corresponding plot, but where now the estimates of the component densities on the first M-step are based on a random assignment of a subsample of size $n_s = 60$ of the full data set of $n = 204$ observations.

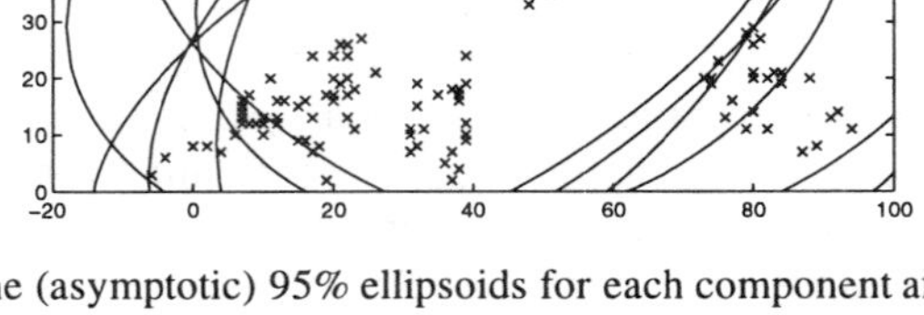

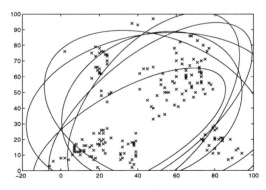

Fig. 2.2 Plot of the (asymptotic) 95% ellipsoids for each component after the first M-step, corresponding to a random assignment of a subsample of the data set.

To further illustrate the differences between using all the data and subsamples, we fitted a mixture of two normal components with unrestricted covariance matrices to the *Iris virginica* data set, consisting of $n = 50$ four-dimensional observations. In Figure 2.3 we give the frequencies for the different local maxima found, using 1,000 random starts based on a random assignment into $g = 2$ groups, firstly of all the data and, secondly, on subsamples of size $n_s = 25$. It can be seen that the latter approach found a wider range of local maxima although, as to be expected, the median and mode are similar for both approaches. Of course, some of the additional local maxima located correspond to spurious local maximizers.

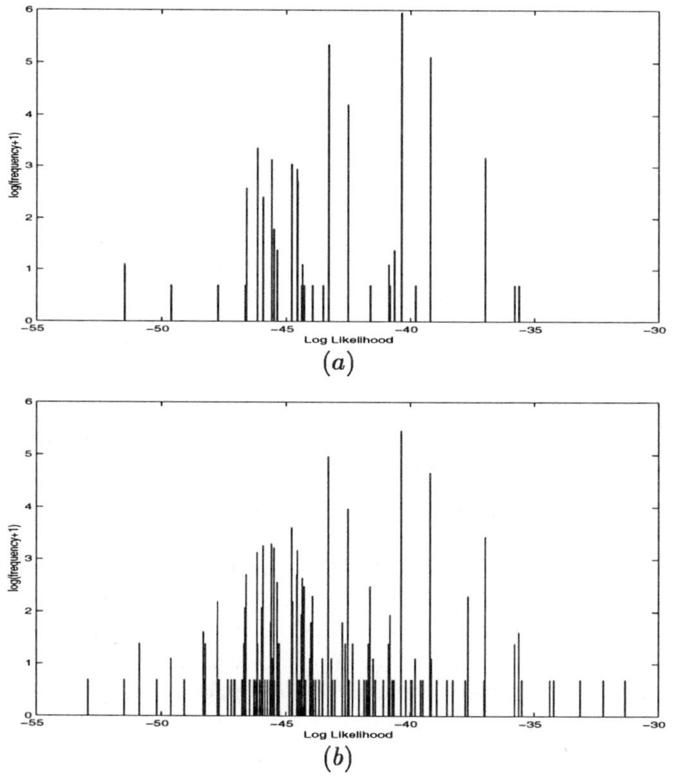

Fig. 2.3 Frequencies of the different local maxima found based on a random assignment into g groups using (a) all the data and (b) a subsample of the data.

2.12.4 Deterministic Annealing EM Algorithm

Recently, Ueda and Nakano (1998) considered a deterministic annealing EM (DAEM) algorithm in order for the EM iterative process to be able to recover from a poor choice of starting value. They proposed using the principle of maximum entropy and the statistical mechanics analogy, whereby a parameter (β) is introduced with $1/\beta$ corresponding to the "temperature" in an annealing sense; see also Yuille, Stolorz, and Utans (1994). With their DAEM algorithm, the E-step is effected by averaging the complete-data log likelihood over the distribution taken to be proportional to that of the current estimate of the conditional density of the complete data (given the observed data) raised to the power β. In the present context of the fitting of finite mixtures to independent data, it means that the current estimate of the posterior probability of the ith component of the mixture for the jth observation $\boldsymbol{y}_j$, $\tau_i(\boldsymbol{y}_j;\, \boldsymbol{\Psi}^{(k)})$, is replaced by

$$\{\tau_i(\boldsymbol{y}_j;\, \boldsymbol{\Psi}^{(k)})\}^{\beta} \Big/ \sum_{h=1}^{g} \{\tau_h(\boldsymbol{y}_j;\, \boldsymbol{\Psi}^{(k)})\}^{\beta} \qquad (i = 1,\, \ldots,\, g).$$

They suggest starting with a value of β close to zero $(0 \leq \beta \leq 1)$, and then increasing it after each EM iteration in the manner

$$\beta^{(k+1)} = c\,\beta^{(k)},$$

where c is in the range 1 to 1.5, until $\beta^{(k+1)} = 1$. For β near zero, the effect of each $\boldsymbol{y}_j$ is almost uniform, so that the initial estimates of the posterior probabilities of component membership of the mixture are close to $1/g$, producing estimates of the component densities that overlap considerably. As β is increased toward one, the effect of each $\boldsymbol{y}_j$ is gradually localized. In this way, Ueda and Nakano (1998) argue that the DAEM algorithm is able to recover from a choice of starting value that may be far from the true value of the parameter vector. Thus the DAEM algorithm guards against a poor choice of starting value essentially by letting the estimates of the component densities on the first few iterations overlap considerably. But a similar effect can be achieved through the use of random starting values, as described in the previous section.

To illustrate this now, we consider the simulated example in Ueda and Nakano (1998). They simulated a sample from a mixture of $g = 3$ bivariate normal components with parameters as given in Table 2.2 and (asymptotic) component ellipsoids as shown in Figure 2.4. They considered a poor choice of starting value (see Table 2.2) from which the EM algorithm was unable to recover, converging to an inferior local maximizer, as can be seen in Figure 2.5. On the other hand, the DAEM algorithm converged to a local maximizer close to the true value of $\boldsymbol{\Psi}$. However, we found that if we applied the EM algorithm from only a few random starts to our simulated sample (we did not have access to the actual data set of Ueda and Nakano (1998)), it converged like the DAEM algorithm to a local maximizer that was close to the true value of $\boldsymbol{\Psi}$, as can be seen from Table 2.2 and Figure 2.6. In more recent work on this problem, Ueda et al. (2000) have proposed a split and merge version of the EM algorithm.

Table 2.2 Initial Values and Estimates for EM and DAEM Algorithms

$\boldsymbol{\Psi}$	True Values	Initial Values	Estimates by DAEM	Estimates by EM* (Random)
π_1	0.333	0.333	0.344	0.294
π_2	0.333	0.333	0.342	0.337
π_3	0.333	0.333	0.341	0.370
$\boldsymbol{\mu}_1^T$	$(0\ -2)$	$(-1\ 0)$	$(-0.050\ -1.984)$	$(-0.154\ -1.961)$
$\boldsymbol{\mu}_2^T$	$(0\ 0)$	$(0\ 0)$	$(0.057\ 0.032)$	$(0.360\ 0.115)$
$\boldsymbol{\mu}_3^T$	$(0\ 2)$	$(1\ 0)$	$(0.050\ 1.994)$	$(-0.004\ 2.027)$
$\boldsymbol{\Sigma}_1$	$\begin{pmatrix} 2 & 0 \\ 0 & 0.2 \end{pmatrix}$	$\begin{pmatrix} 1 & 0 \\ 0 & 1 \end{pmatrix}$	$\begin{pmatrix} 2.192 & -0.078 \\ -0.078 & 0.214 \end{pmatrix}$	$\begin{pmatrix} 1.961 & -0.016 \\ -0.016 & 0.218 \end{pmatrix}$
$\boldsymbol{\Sigma}_2$	$\begin{pmatrix} 2 & 0 \\ 0 & 0.2 \end{pmatrix}$	$\begin{pmatrix} 1 & 0 \\ 0 & 1 \end{pmatrix}$	$\begin{pmatrix} 1.984 & -0.088 \\ -0.088 & 0.162 \end{pmatrix}$	$\begin{pmatrix} 2.346 & -0.553 \\ -0.553 & 0.150 \end{pmatrix}$
$\boldsymbol{\Sigma}_3$	$\begin{pmatrix} 2 & 0 \\ 0 & 0.2 \end{pmatrix}$	$\begin{pmatrix} 1 & 0 \\ 0 & 1 \end{pmatrix}$	$\begin{pmatrix} 1.987 & -0.042 \\ -0.042 & 0.215 \end{pmatrix}$	$\begin{pmatrix} 2.339 & 0.042 \\ 0.042 & 0.206 \end{pmatrix}$

*Applied to a different sample (but generated from the same mixture) as that for the DAEM algorithm.

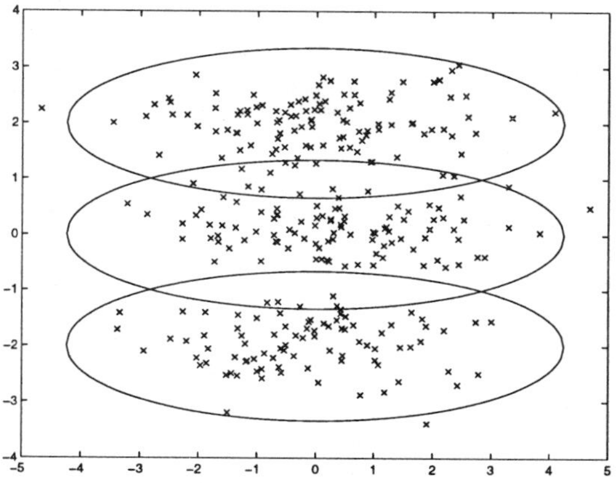

Fig. 2.4 Plot of the (asymptotic) 95% ellipsoids for each component based on the true parameters specified in Ueda and Nakano (1998).

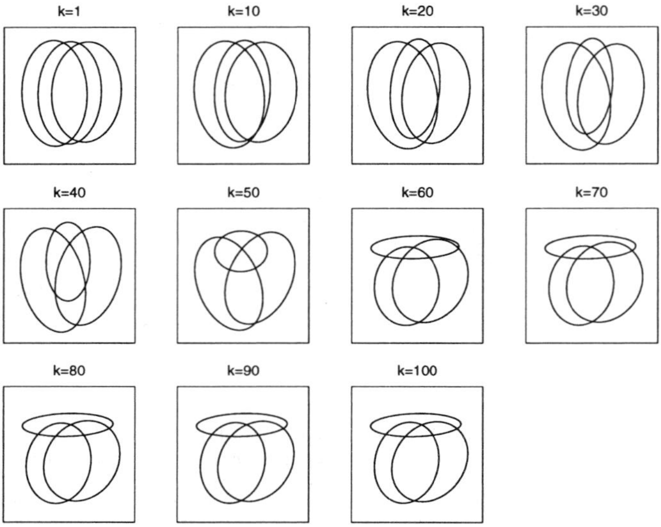

Fig. 2.5 Plot of the (asymptotic) 95% ellipsoids for each component based on the parameters produced by the kth M-step for various k for the EM algorithm started from parameter values specified in Ueda and Nakano (1998).

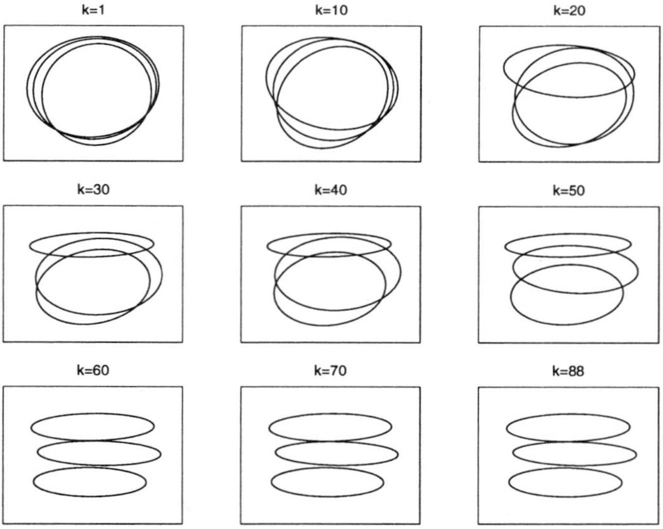

Fig. 2.6 Plot of the (asymptotic) 95% ellipsoids for each component based on the parameters produced by the kth M-step for various k for the EM algorithm using random starts.

2.13 STOCHASTIC EM ALGORITHM

Broniatowski, Celeux, and Diebolt (1983), Celeux and Diebolt (1985, 1986a, 1986b), and Celeux, Chauveau, and Diebolt (1996) considered a modified version of the EM algorithm in the context of computing the MLE for finite mixture models. They called it the Stochastic EM algorithm. It is the same as the Monte Carlo EM algorithm with a single replication; see McLachlan and Krishnan (1997, Section 6.3).

With the Stochastic EM algorithm, the current posterior probabilities are used in a Stochastic E-step, wherein a single draw is made from the current conditional distribution of $\boldsymbol{z}$ given the observed data $\boldsymbol{y}$. Because of the assumption of independence of the complete-data observations, this is effected by conducting a draw for each $j\ (j = 1, \ldots, n)$. That is, a draw $\boldsymbol{z}_j^{(k)}$ is made from the multinomial distribution with g categories having probabilities specified by

$$\boldsymbol{\tau}(\boldsymbol{y}_j;\, \boldsymbol{\Psi}^{(k)}) = (\tau_1(\boldsymbol{y}_j;\, \boldsymbol{\Psi}^{(k)}),\, \ldots,\, \tau_g(\boldsymbol{y}_j;\, \boldsymbol{\Psi}^{(k)}))^T. \tag{2.43}$$

This effectively assigns each observation $\boldsymbol{y}_j$ outright to one of the g components of the mixture rather than weighting it fractionally with respect to the g components using the current posterior probabilities of component membership given by (2.43).

The Stochastic EM algorithm thus gives the iterative process a chance of escaping from a current path of convergence to a local maximizer to other paths. This is desirable if the algorithm has been started from a poor value of the parameter vector. On the other hand, it would not be desirable if the process is near to convergence to a suitable local maximizer. Thus it has been suggested that consideration be given to using the ordinary version of the EM algorithm in the latter stages of the iterative process. Also, we have found that the Stochastic EM algorithm has the tendency to find unwanted spurious local maximizers; see Section 3.10.

2.14 RATE OF CONVERGENCE OF THE EM ALGORITHM

We now discuss briefly some properties of the EM algorithm that are used in attempts to speed up its convergence, to provide estimates of standard errors, and to assess the number of components in the mixture model. For a detailed account of these properties, the reader is referred to McLachlan and Krishnan (1997).

2.14.1 Rate Matrix for Linear Convergence

The EM algorithm implicitly defines a mapping $\boldsymbol{\Psi} \rightarrow M(\boldsymbol{\Psi})$, from the parameter space of $\boldsymbol{\Psi}$, $\boldsymbol{\Omega}$, to itself such that each iteration $\boldsymbol{\Psi}^{(k)} \rightarrow \boldsymbol{\Psi}^{(k+1)}$ is defined by

$$\boldsymbol{\Psi}^{(k+1)} = M(\boldsymbol{\Psi}^{(k)}) \qquad (k = 0, 1, 2, \ldots).$$

If $\boldsymbol{\Psi}^{(k)}$ converges to some point $\boldsymbol{\Psi}^*$ and $M(\boldsymbol{\Psi})$ is continuous, then $\boldsymbol{\Psi}^*$ is a fixed point of the algorithm; that is, $\boldsymbol{\Psi}^*$ must satisfy

$$\boldsymbol{\Psi}^* = M(\boldsymbol{\Psi}^*). \tag{2.44}$$

By a Taylor series expansion of $\boldsymbol{\Psi}^{(k+1)} = \boldsymbol{M}(\boldsymbol{\Psi}^{(k)})$ about the point $\boldsymbol{\Psi}^{(k)} = \boldsymbol{\Psi}^*$, we have in a neighborhood of $\boldsymbol{\Psi}^*$ that

$$\boldsymbol{\Psi}^{(k+1)} - \boldsymbol{\Psi}^* \approx \boldsymbol{J}(\boldsymbol{\Psi}^*)(\boldsymbol{\Psi}^{(k)} - \boldsymbol{\Psi}^*), \tag{2.45}$$

where $\boldsymbol{J}(\boldsymbol{\Psi})$ is the $d \times d$ Jacobian matrix for $\boldsymbol{M}(\boldsymbol{\Psi}) = (M_1(\boldsymbol{\Psi}), \ldots, M_d(\boldsymbol{\Psi}))^T$, having (r, s)th element $J_{rs}(\boldsymbol{\Psi})$ equal to

$$J_{rs}(\boldsymbol{\Psi}) = \partial M_r(\boldsymbol{\Psi})/\partial \Psi_s,$$

where $\Psi_s = (\boldsymbol{\Psi})_s$; see Meng and Rubin (1991) and McLachlan and Krishnan (1997, Chapter 3).

Thus, in a neighborhood of $\boldsymbol{\Psi}^*$, the EM algorithm is essentially a linear iteration with rate matrix $\boldsymbol{J}(\boldsymbol{\Psi}^*)$, since $\boldsymbol{J}(\boldsymbol{\Psi}^*)$ is typically nonzero. For this reason, $\boldsymbol{J}(\boldsymbol{\Psi}^*)$ is often referred to as the matrix rate of convergence, or simply the rate of convergence.

For vector $\boldsymbol{\Psi}$, a measure of the actual observed convergence rate is the global rate of convergence, which is defined as

$$r = \lim_{k \to \infty} ||\boldsymbol{\Psi}^{(k+1)} - \boldsymbol{\Psi}^*|| / ||\boldsymbol{\Psi}^{(k)} - \boldsymbol{\Psi}^*||,$$

where $||\cdot||$ is any norm on d-dimensional Euclidean space $\mathbb{R}^d$. It is well known that under certain regularity conditions,

$$r = \lambda_{\max} \equiv \text{ the largest eigenvalue of } \boldsymbol{J}(\boldsymbol{\Psi}^*).$$

In practice, r is typically assessed as

$$r = \lim_{k \to \infty} ||\boldsymbol{\Psi}^{(k+1)} - \boldsymbol{\Psi}^{(k)}|| / ||\boldsymbol{\Psi}^{(k)} - \boldsymbol{\Psi}^{(k-1)}||. \tag{2.46}$$

2.14.2 Rate Matrix in Terms of Information Matrices

Suppose that $\{\boldsymbol{\Psi}^{(k)}\}$ is an EM sequence for which

$$\partial Q(\boldsymbol{\Psi}; \boldsymbol{\Psi}^{(k)})/\partial \boldsymbol{\Psi} = \mathbf{0} \tag{2.47}$$

is satisfied by $\boldsymbol{\Psi} = \boldsymbol{\Psi}^{(k+1)}$, which will be the case with standard complete-data estimation. Then Dempster et al. (1977) showed that if $\boldsymbol{\Psi}^{(k)}$ converges to a point $\boldsymbol{\Psi}^*$, then

$$\boldsymbol{J}(\boldsymbol{\Psi}^*) = \mathcal{I}_c^{-1}(\boldsymbol{\Psi}^*; \boldsymbol{y})\mathcal{I}_m(\boldsymbol{\Psi}^*; \boldsymbol{y}), \tag{2.48}$$

where $\mathcal{I}_c(\boldsymbol{\Psi}; \boldsymbol{y})$ is the conditional expected complete-data matrix defined by (1.49) and where

$$\mathcal{I}_m(\boldsymbol{\Psi}; \boldsymbol{y}) = -E_{\boldsymbol{\Psi}}\{\partial^2 \log k(\boldsymbol{Y}_c \mid \boldsymbol{y}; \boldsymbol{\Psi})/\partial \boldsymbol{\Psi} \partial \boldsymbol{\Psi}^T \mid \boldsymbol{y}\} \tag{2.49}$$

is the expected information matrix for $\boldsymbol{\Psi}$ based on $\boldsymbol{y}_c$ (or equivalently, the unobservable data) when conditioned on $\boldsymbol{y}$, and $k(\boldsymbol{y}_c \mid \boldsymbol{y}; \boldsymbol{\Psi})$ is the conditional density of $\boldsymbol{Y}_c$ given $\boldsymbol{y}$. This result was obtained also by Sundberg (1974).

Thus the rate of convergence of the EM algorithm is given by the largest eigenvalue of the information ratio matrix $\mathcal{I}_c^{-1}(\boldsymbol{\Psi}^*;\, \boldsymbol{y})\mathcal{I}_m(\boldsymbol{\Psi}^*;\, \boldsymbol{y})$, which measures the proportion of information about $\boldsymbol{\Psi}$ that is missing by not also observing the "missing" data (the component-label vectors $\boldsymbol{z}_1, \ldots, \boldsymbol{z}_n$ in the present context of mixture models), in addition to $\boldsymbol{y}$. The greater the proportion of missing information, the slower the rate of convergence.

The rate of convergence of the EM algorithm can be expressed equivalently in terms of the smallest eigenvalue of

$$\mathcal{I}_c^{-1}(\boldsymbol{\Psi}^*;\, \boldsymbol{y})\boldsymbol{I}(\boldsymbol{\Psi}^*;\, \boldsymbol{y}).$$

This is because we can express $\boldsymbol{J}(\boldsymbol{\Psi}^*)$ also in the form

$$\boldsymbol{J}(\boldsymbol{\Psi}^*) = \boldsymbol{I}_d - \mathcal{I}_c^{-1}(\boldsymbol{\Psi}^*;\, \boldsymbol{y})\boldsymbol{I}(\boldsymbol{\Psi}^*;\, \boldsymbol{y}), \tag{2.50}$$

where $\boldsymbol{I}_d$ denotes the $d \times d$ identity matrix.

As to be discussed in Section 6.8.5, Windham and Cutler (1992) based a test for the number of components in a mixture model on the smallest eigenvalue of $\mathcal{I}_c^{-1}(\hat{\boldsymbol{\Psi}};\, \boldsymbol{y})\boldsymbol{I}(\hat{\boldsymbol{\Psi}};\, \boldsymbol{y})$. Their motivation is that, heuristically, a large value of this smallest eigenvalue is suggestive of a good clustering of the data, whereas a small value is not.

2.15 INFORMATION MATRIX FOR MIXTURE MODELS

One initial criticism of the EM algorithm was that it does not automatically provide an estimate of the covariance matrix of the MLE, as do some other procedures, such as Newton-type methods. A number of methods have since been proposed for assessing the covariance matrix of the MLE $\hat{\boldsymbol{\Psi}}$ of the parameter vector $\boldsymbol{\Psi}$, obtained via the EM algorithm. Most of these methods are based on the observed information matrix $\boldsymbol{I}(\hat{\boldsymbol{\Psi}};\, \boldsymbol{y})$, which is defined by (1.48). For independent data, $\boldsymbol{I}(\hat{\boldsymbol{\Psi}};\, \boldsymbol{y})$ can be approximated without additional work beyond the calculations used to compute the MLE in the first instance.

2.15.1 Direct Evaluation of Observed Information Matrix

As explained in Section 2.4, it is common in practice to estimate the inverse of the covariance matrix of the MLE $\hat{\boldsymbol{\Psi}}$ by the observed information matrix $\boldsymbol{I}(\hat{\boldsymbol{\Psi}};\, \boldsymbol{y})$. Hence we shall consider in the following subsections a number of ways for calculating or approximating $\boldsymbol{I}(\hat{\boldsymbol{\Psi}};\, \boldsymbol{y})$. One way to proceed is to directly evaluate $\boldsymbol{I}(\hat{\boldsymbol{\Psi}};\, \boldsymbol{y})$ after the computation of the MLE $\hat{\boldsymbol{\Psi}}$. However, analytical evaluation of the second-order derivatives of the log likelihood, $\log L(\boldsymbol{\Psi})$, may be difficult, or at least tedious, for most mixture models, in particular for multivariate data. Indeed, often it is for reasons of this nature that the EM algorithm is used to compute the MLE in the first instance.

2.15.2 Extraction of Observed Information Matrix in Terms of the Complete-Data Log Likelihood

Louis (1982) showed that $\boldsymbol{I}(\boldsymbol{\Psi};\ \boldsymbol{y})$, the negative of the Hessian of the incomplete-data log likelihood, can be expressed in the form

$$\begin{aligned} \boldsymbol{I}(\boldsymbol{\Psi};\ \boldsymbol{y}) &= \mathcal{I}_c(\boldsymbol{\Psi};\ \boldsymbol{y}) - \text{cov}_{\boldsymbol{\Psi}}\{\boldsymbol{S}_c(\boldsymbol{Y}_c;\ \boldsymbol{\Psi}) \mid \boldsymbol{y}\} && (2.51) \\ &= \mathcal{I}_c(\boldsymbol{\Psi};\ \boldsymbol{y}) \\ &\quad -E_{\boldsymbol{\Psi}}\{\boldsymbol{S}_c(\boldsymbol{Y}_c;\ \boldsymbol{\Psi})\boldsymbol{S}_c^T(\boldsymbol{Y}_c;\ \boldsymbol{\Psi}) \mid \boldsymbol{y}\} \\ &\quad +\boldsymbol{S}(\boldsymbol{y};\ \boldsymbol{\Psi})\boldsymbol{S}^T(\boldsymbol{y};\ \boldsymbol{\Psi}). && (2.52) \end{aligned}$$

In (2.52), $\boldsymbol{S}(\boldsymbol{y};\ \boldsymbol{\Psi})$ and $\boldsymbol{S}_c(\boldsymbol{Y}_c;\ \boldsymbol{\Psi})$ denote the incomplete-data and complete-data score statistics, as defined by (1.46) and (1.47), respectively. It can be shown that

$$\boldsymbol{S}(\boldsymbol{y};\ \boldsymbol{\Psi}) = E_{\boldsymbol{\Psi}}\{\boldsymbol{S}_c(\boldsymbol{Y}_c;\ \boldsymbol{\Psi}) \mid \boldsymbol{y}\}. \tag{2.53}$$

From (2.53), the observed information matrix $\boldsymbol{I}(\hat{\boldsymbol{\Psi}})$ can be computed as

$$\boldsymbol{I}(\hat{\boldsymbol{\Psi}};\ \boldsymbol{y}) = \mathcal{I}_c(\hat{\boldsymbol{\Psi}};\ \boldsymbol{y}) - [E_{\boldsymbol{\Psi}}\{\boldsymbol{S}_c(\boldsymbol{Y}_c;\ \boldsymbol{\Psi})\boldsymbol{S}_c^T(\boldsymbol{Y}_c;\ \boldsymbol{\Psi}) \mid \boldsymbol{y}\}]_{\boldsymbol{\Psi}=\hat{\boldsymbol{\Psi}}}, \tag{2.54}$$

since the last term on the right-hand side of (2.52) is zero as $\hat{\boldsymbol{\Psi}}$ satisfies

$$\boldsymbol{S}(\boldsymbol{y};\ \boldsymbol{\Psi}) = \boldsymbol{0}.$$

Hence the observed information matrix for the original (incomplete-data) problem can be computed in terms of the conditional moments of the gradient and curvature of the complete-data log likelihood function introduced within the EM framework.

The calculation of $\mathcal{I}_c(\hat{\boldsymbol{\Psi}};\ \boldsymbol{y})$ is readily facilitated by standard complete-data computations if the complete-data density belongs to the regular exponential family, since then

$$\mathcal{I}_c(\hat{\boldsymbol{\Psi}};\ \boldsymbol{y}) = \mathcal{I}_c(\hat{\boldsymbol{\Psi}}). \tag{2.55}$$

2.15.3 Approximations to Observed Information Matrix: I.I.D. Case

It can be seen from expression (2.54) for the observed information matrix $\boldsymbol{I}(\hat{\boldsymbol{\Psi}};\ \boldsymbol{y})$ that it requires, in addition to the code for the E- and M-steps, the calculation of the conditional (on the observed data $\boldsymbol{y}$) expectation of the complete-data information matrix $\boldsymbol{I}_c(\boldsymbol{\Psi};\ \boldsymbol{Y}_c)$ and of the complete-data score statistic $\boldsymbol{S}_c(\boldsymbol{Y}_c;\ \boldsymbol{\Psi})$ times its transpose. An illustration of the use of the result (2.54) to calculate the observed information matrix is given in McLachlan and Krishnan (1997, Section 4.25) for a mixture of two univariate normal densities with known common variance. But for more complicated mixture models, the calculation of the observed information matrix via (2.54) would be algebraically tedious. Hence we now consider some practical methods for approximating the observed information matrix.

In the present case of i.i.d. data, an approximation to the observed information matrix is readily available without any additional analyses having to be performed.

The log likelihood, $\log L(\boldsymbol{\Psi})$, can be expressed then in the form

$$\log L(\boldsymbol{\Psi}) = \sum_{j=1}^{n} \log L_j(\boldsymbol{\Psi}),$$

where

$$L_j(\boldsymbol{\Psi}) = f(\boldsymbol{y}_j;\ \boldsymbol{\Psi})$$

is the likelihood function for $\boldsymbol{\Psi}$ formed from the single observation $\boldsymbol{y}_j$ $(j = 1, \ldots, n)$.

We can now write the score vector $\boldsymbol{S}(\boldsymbol{y};\ \boldsymbol{\Psi})$ as

$$\boldsymbol{S}(\boldsymbol{y};\ \boldsymbol{\Psi}) = \sum_{j=1}^{n} \boldsymbol{s}(\boldsymbol{y}_j;\ \boldsymbol{\Psi}),$$

where

$$\boldsymbol{s}(\boldsymbol{y}_j;\ \boldsymbol{\Psi}) = \partial \log L_j(\boldsymbol{\Psi}) / \partial \boldsymbol{\Psi}.$$

The expected information matrix $\mathcal{I}(\boldsymbol{\Psi})$ can be written as

$$\mathcal{I}(\boldsymbol{\Psi}) = n\boldsymbol{i}(\boldsymbol{\Psi}), \tag{2.56}$$

where

$$\begin{aligned} \boldsymbol{i}(\boldsymbol{\Psi}) &= E_{\boldsymbol{\Psi}}\{\boldsymbol{s}(\boldsymbol{Y};\ \boldsymbol{\Psi})\boldsymbol{s}^T(\boldsymbol{Y};\ \boldsymbol{\Psi})\} \\ &= \mathrm{cov}_{\boldsymbol{\Psi}}\{\boldsymbol{s}(\boldsymbol{Y};\ \boldsymbol{\Psi})\} \end{aligned} \tag{2.57}$$

is the information contained in a single observation. Corresponding to (2.57), the empirical information matrix (in a single observation) can be defined to be

$$\begin{aligned} \bar{\boldsymbol{i}}(\boldsymbol{\Psi}) &= n^{-1} \sum_{j=1}^{n} \boldsymbol{s}(\boldsymbol{y}_j;\ \boldsymbol{\Psi})\boldsymbol{s}^T(\boldsymbol{y}_j;\ \boldsymbol{\Psi}) - \bar{\boldsymbol{s}}\,\bar{\boldsymbol{s}}^T \\ &= n^{-1} \sum_{j=1}^{n} \boldsymbol{s}(\boldsymbol{y}_j;\ \boldsymbol{\Psi})\boldsymbol{s}^T(\boldsymbol{y}_j;\ \boldsymbol{\Psi}) \\ &\quad -n^{-2}\boldsymbol{S}(\boldsymbol{y};\ \boldsymbol{\Psi})\boldsymbol{S}^T(\boldsymbol{y};\ \boldsymbol{\Psi}), \end{aligned} \tag{2.58}$$

where

$$\bar{\boldsymbol{s}} = n^{-1} \sum_{j=1}^{n} \boldsymbol{s}(\boldsymbol{y}_j;\ \boldsymbol{\Psi}).$$

Corresponding to this empirical form (2.58) for $\boldsymbol{i}(\boldsymbol{\Psi})$, $\mathcal{I}(\boldsymbol{\Psi})$ is estimated by

$$\begin{aligned} \boldsymbol{I}_{\mathrm{e}}(\boldsymbol{\Psi};\ \boldsymbol{y}) &= n\bar{\boldsymbol{i}}(\boldsymbol{\Psi}) \\ &= \sum_{j=1}^{n} \boldsymbol{s}(\boldsymbol{y}_j;\ \boldsymbol{\Psi})\boldsymbol{s}^T(\boldsymbol{y}_j;\ \boldsymbol{\Psi}) \\ &\quad -n^{-1}\boldsymbol{S}(\boldsymbol{y};\ \boldsymbol{\Psi})\boldsymbol{S}^T(\boldsymbol{y};\ \boldsymbol{\Psi}). \end{aligned} \tag{2.59}$$

On evaluation at $\boldsymbol{\Psi} = \hat{\boldsymbol{\Psi}}$, $\boldsymbol{I}_e(\hat{\boldsymbol{\Psi}};\, \boldsymbol{y})$ reduces to

$$\boldsymbol{I}_e(\hat{\boldsymbol{\Psi}};\, \boldsymbol{y}) = \sum_{j=1}^{n} \boldsymbol{s}(\boldsymbol{y}_j;\, \hat{\boldsymbol{\Psi}})\boldsymbol{s}^T(\boldsymbol{y}_j;\, \hat{\boldsymbol{\Psi}}), \tag{2.60}$$

since $\boldsymbol{S}(\boldsymbol{y};\, \hat{\boldsymbol{\Psi}}) = \boldsymbol{0}$.

Meilijson (1989) termed $\boldsymbol{I}_e(\hat{\boldsymbol{\Psi}};\, \boldsymbol{y})$ the empirical observed information matrix. It is used commonly in practice to approximate the observed information matrix $\boldsymbol{I}(\hat{\boldsymbol{\Psi}};\, \boldsymbol{y})$. Further justification of the use of the empirical information matrix may be found in McLachlan and Krishnan (1997, Section 4.3). It follows from the result (2.53) that

$$\boldsymbol{s}(\boldsymbol{y}_j;\, \boldsymbol{\Psi}) = E_{\boldsymbol{\Psi}}\{\partial \log L_{cj}(\boldsymbol{\Psi})/\partial \boldsymbol{\Psi} \mid \boldsymbol{y}\},$$

where $L_{cj}(\boldsymbol{\Psi})$ is the complete-data likelihood formed from the single observation $\boldsymbol{y}_j$ $(j = 1, \ldots, n)$. Thus the approximation $\boldsymbol{I}_e(\hat{\boldsymbol{\Psi}};\, \boldsymbol{y})$ to the observed information matrix $\boldsymbol{I}(\hat{\boldsymbol{\Psi}};\, \boldsymbol{y})$ can be expressed in terms of the conditional expectation of the gradient vector of the complete-data log likelihood function evaluated at the MLE $\hat{\boldsymbol{\Psi}}$. It thus avoids the computation of second-order partial derivatives of the complete-data log likelihood.

For finite mixture models,

$$\log L_{cj} = \sum_{i=1}^{g} z_{ij}\{\log \pi_i + \log f_i(\boldsymbol{y}_j;\, \boldsymbol{\theta}_i)\} \quad (j = 1, \ldots, n), \tag{2.61}$$

so that

$$\boldsymbol{s}(\boldsymbol{y}_j;\, \boldsymbol{\Psi}) = \sum_{i=1}^{g} \tau_i(\boldsymbol{y}_j;\, \boldsymbol{\Psi})\partial\{\log \pi_i + \log f_i(\boldsymbol{y}_j;\, \boldsymbol{\theta}_i)\}/\partial \boldsymbol{\Psi} \quad (j = 1, \ldots, n). \tag{2.62}$$

Thus with the use of the empirical information matrix for the fitting of mixture models, the observed information matrix is approximated solely in terms of the gradient of the complete-data log likelihood, where the unobservable component-label variables z_{ij} are replaced by their fitted conditional expectations, that is, by the posterior probabilities of component membership of the mixture evaluated at the MLE for $\boldsymbol{\Psi}$. If only the mixing proportions are unknown, then the observed information matrix is exactly equal to the empirical information matrix.

2.15.4 Supplemented EM Algorithm

The methods presented in the previous section are applicable only to i.i.d. data. For the general case, Meng and Rubin (1989, 1991) define a procedure that obtains a numerically stable estimate of the asymptotic covariance matrix of the EM-computed estimate, using only the code for computing the complete-data covariance matrix, the code for the EM algorithm itself, and the code for standard matrix operations.

In particular, neither likelihoods, nor partial derivatives of log likelihoods need to be evaluated.

The basic idea is to use the fact that the rate of convergence is governed by the fraction of the missing information to find the increased variability due to missing information to add to the assessed complete-data covariance matrix. Meng and Rubin (1991) refer to the EM algorithm with their modification for the provision of the asymptotic covariance matrix as the Supplemented EM algorithm; see McLachlan and Krishnan (1997, Section 4.5) for further details.

2.15.5 Conditional Bootstrap Approach

We shall shortly describe the full bootstrap approach to the approximation of the covariance matrix of the MLE $\hat{\boldsymbol{\Psi}}$, but we first mention the conditional bootstrap approach of Diebolt and Ip (1996) for approximating the observed information matrix $\boldsymbol{I}(\hat{\boldsymbol{\Psi}};\, \boldsymbol{y})$. As they warn, Efron (1994) has pointed out that the supplemented EM method of Meng and Rubin (1991) and the empirical information-based method of Meilijson (1989) are basically delta methods, and so will often underestimate the standard errors. Their method may also be subject to the same kind of warning.

From (1.49) and (2.54), the observed information matrix $\boldsymbol{I}(\hat{\boldsymbol{\Psi}};\, \boldsymbol{y})$ can be expressed as

$$\begin{aligned} \boldsymbol{I}(\hat{\boldsymbol{\Psi}};\, \boldsymbol{y}) &= [E_{\boldsymbol{\Psi}}\{\boldsymbol{I}_c(\boldsymbol{\Psi};\, \boldsymbol{Y}_c) \mid \boldsymbol{y}\}]_{\boldsymbol{\Psi}=\hat{\boldsymbol{\Psi}}} \\ &\quad -[E_{\boldsymbol{\Psi}}\{\boldsymbol{S}_c(\boldsymbol{Y}_c;\, \boldsymbol{\Psi})\boldsymbol{S}_c^T(\boldsymbol{Y}_c;\, \boldsymbol{\Psi}) \mid \boldsymbol{y}\}]_{\boldsymbol{\Psi}=\hat{\boldsymbol{\Psi}}}, \end{aligned} \tag{2.63}$$

which requires computing the conditional expectations of two quantities over the conditional distribution of the component-indicator vectors $\boldsymbol{z}_1, \ldots, \boldsymbol{z}_n$, given the observed data $\boldsymbol{y} = (\boldsymbol{y}_1^T, \ldots, \boldsymbol{y}_n^T)^T$.

Diebolt and Ip (1996) proposed approximating the conditional expectations of these two quantities by averaging them over B independent bootstrap samples,

$$(\boldsymbol{z}_{1b}^{*T}, \ldots, \boldsymbol{z}_{nb}^{*T})^T \quad (b = 1, \ldots, B)$$

where $\boldsymbol{z}_{jb}^*$ is distributed according to a multinomial distribution consisting of one draw with g categories having probabilities

$$\hat{\boldsymbol{\tau}}_j = (\tau_1(\boldsymbol{y}_j;\, \hat{\boldsymbol{\Psi}}), \ldots, \tau_g(\boldsymbol{y}_j;\, \hat{\boldsymbol{\Psi}}))^T \quad (j = 1, \ldots, n).$$

In their empirical studies involving the EM algorithm and its stochastic version, they found that their method (with $B = 100$ bootstrap replications) gave similar results for the standard errors as obtained by direct calculation of the observed information matrix.

2.16 PROVISION OF STANDARD ERRORS

2.16.1 Information-Based Methods

One way of obtaining standard errors of the estimates of the parameters in a mixture model is to approximate the covariance matrix of $\hat{\boldsymbol{\Psi}}$ by the inverse of the observed information matrix, which can be computed as discussed in the previous section. Another information-type method is to adopt the approach of Dietz and Böhning (1996), which is based on the result that in large samples from regular models for which the log likelihood is quadratic in the parameters, the likelihood ratio and Wald's tests for the significance of an individual parameter are equivalent, so that the deviance change (that is, twice the change in the log likelihood) on omitting that variable is equal to the square of the t-statistic (that is, is equal to the square of the ratio of the parameter estimate to its standard error). Thus the standard error can be calculated as the absolute value of the parameter estimate divided by the square root of the deviance change. This requires the fitting of a set of reduced models in which each variable is omitted from the final version of the model.

Liu (1998) recently considered a related approximation to the observed information and hence to the asymptotic variance of the MLE of a parameter. He extended his result to the multiparameter case using the fact that the covariance matrix of a normal distribution can be obtained from its one-dimensional conditional distributions whose sample spaces span the sample space of the joint distribution.

It is important to emphasize that estimates of the covariance matrix of the MLE based on the expected or observed information matrices are guaranteed to be valid inferentially only asymptotically. In particular for mixture models, it is well known that the sample size n has to be very large before the asymptotic theory of maximum likelihood applies. Hence we shall now consider a resampling approach, the bootstrap, to this problem. Basford et al. (1997a) and Peel (1998) compared the bootstrap and information-based approaches for some normal mixture models. They found that unless the sample size was very large, the standard errors found by an information-based approach were too unstable to be recommended. In such situations the bootstrap approach is recommended.

2.16.2 Bootstrap Approach to Standard Error Approximation

The bootstrap was introduced by Efron (1979), who has investigated it further in a series of articles; see Efron (1982), Efron and Tibshirani (1993), Davison and Hinkley (1997), Chernick (1999), and the references therein. Over the past twenty years, the bootstrap has become one of the most popular developments in statistics. Hence there now exists an extensive literature on it.

The bootstrap is a powerful technique that permits the variability in a random quantity to be assessed using just the data at hand. An estimate $\hat{F}$ of the underlying distribution is formed from the observed data $\boldsymbol{y}$. Conditional on the latter, the sampling distribution of the random quantity of interest with F replaced by $\hat{F}$ defines its so-called bootstrap distribution, which provides an approximation to its true distribution.

It is assumed that $\hat{F}$ has been so formed that the stochastic structure of the model has been preserved. Usually, it is impossible to express the bootstrap distribution in simple form, and it must be approximated by Monte Carlo methods whereby pseudo-random samples (bootstrap samples) are drawn from $\hat{F}$. In recent times there have been a number of papers written on improving the efficiency of the bootstrap computations with the latter approach. If a parametric form is adopted for the distribution function of $\boldsymbol{Y}$, where $\boldsymbol{\Psi}$ denotes the vector of unknown parameters, then the parametric bootstrap uses an estimate $\hat{\boldsymbol{\Psi}}$ formed from $\boldsymbol{y}$ in place of $\boldsymbol{\Psi}$. That is, if we write F as $F_{\boldsymbol{\Psi}}$ to signify its dependence on $\boldsymbol{\Psi}$, then the bootstrap data are generated from $\hat{F} = F_{\hat{\boldsymbol{\Psi}}}$.

Standard error estimation of $\hat{\boldsymbol{\Psi}}$ may be implemented according to the bootstrap as follows.

Step 1. A new set of data, $\boldsymbol{y}^*$, called the bootstrap sample, is generated according to $\hat{F}$, an estimate of the distribution function of $\boldsymbol{Y}$ formed from the original observed data $\boldsymbol{y}$. That is, in the case where $\boldsymbol{y}$ contains the observed values of a random sample of size n, $\boldsymbol{y}^*$ consists of the observed values of the random sample

$$\boldsymbol{Y}_1^*, \ldots, \boldsymbol{Y}_n^* \overset{\text{i.i.d.}}{\sim} \hat{F}, \tag{2.64}$$

where the estimate $\hat{F}$ (now denoting the distribution function of a single observation $\boldsymbol{Y}_j$) is held fixed at its observed value.

Step 2. The EM algorithm is applied to the bootstrap observed data $\boldsymbol{y}^*$ to compute the MLE for this data set, $\hat{\boldsymbol{\Psi}}^*$.

Step 3. The bootstrap covariance matrix of $\hat{\boldsymbol{\Psi}}^*$ is given by

$$\text{cov}^*(\hat{\boldsymbol{\Psi}}^*) = E^*[\{\hat{\boldsymbol{\Psi}} - E^*(\hat{\boldsymbol{\Psi}})\}\{\hat{\boldsymbol{\Psi}}^* - E^*(\hat{\boldsymbol{\Psi}}^*)\}^T], \tag{2.65}$$

where E^* denotes expectation over the bootstrap distribution specified by $\hat{F}$.

The bootstrap covariance matrix can be approximated by Monte Carlo methods. Steps (1) and (2) are repeated independently a number of times (say, B) to give B independent realizations of $\hat{\boldsymbol{\Psi}}^*$, denoted by $\hat{\boldsymbol{\Psi}}_1^*, \ldots, \hat{\boldsymbol{\Psi}}_B^*$. Then (2.65) can be approximated by the sample covariance matrix of these B bootstrap replications to give

$$\text{cov}^*(\hat{\boldsymbol{\Psi}}^*) \approx \sum_{b=1}^{B} (\hat{\boldsymbol{\Psi}}_b^* - \overline{\hat{\boldsymbol{\Psi}}}^*)(\hat{\boldsymbol{\Psi}}_b^* - \overline{\hat{\boldsymbol{\Psi}}}^*)^T/(B-1), \tag{2.66}$$

where

$$\overline{\hat{\boldsymbol{\Psi}}}^* = \sum_{b=1}^{B} \hat{\boldsymbol{\Psi}}^*/B. \tag{2.67}$$

The standard error of the ith element of $\hat{\boldsymbol{\Psi}}$ can be estimated by the positive square root of the ith diagonal element of (2.66). It has been shown that 50 to 100 bootstrap replications are generally sufficient for standard error estimation (Efron and Tibshirani, 1993). The above results on the number of replications to be used are based on the unconditional coefficient of variation of the Monte Carlo approximation. Recently, Booth and Sarkar (1998) argued that the conditional coefficient of variation should be used, which indicates that approximately 800 replications are required for this purpose. They have given a simple formula for determining a lower bound on the number of bootstrap replications required to approximate a d-dimensional covariance matrix.

In Step 1 of the above algorithm, the nonparametric version of the bootstrap would take $\hat{F}$ to be the empirical distribution function formed from the observed data $\boldsymbol{y}$. Given that we are concerned here with ML estimation in the context of a parametric model for the mixture distribution of $\boldsymbol{Y}_j$, we would tend to use the parametric version of the bootstrap instead of the nonparametric version. Situations where we may still wish to use the latter include problems where the observed data are censored or are missing in the conventional sense. In these cases the use of the nonparametric bootstrap avoids having to postulate a suitable model for the underlying mechanism that controls the censorship or the absence of the data.

The EMMIX software (McLachlan et al., 1999) has the provision for the bootstrap to be implemented parametrically or nonparametrically. Also, the weighted bootstrap (Newton and Raftery, 1994) can be used. It is a generalization of the nonparametric version of the bootstrap. A bootstrap sample is given by

$$w_1\boldsymbol{y}_1, \ldots, w_n\boldsymbol{y}_n, \tag{2.68}$$

where the weights w_j are nonnegative and sum to n. For sampling with replacement with the nonparametric bootstrap, the w_j are nonnegative integers that sum to n, as w_j is a count of the number of times that the original point $\boldsymbol{y}_j$ occurs in the bootstrap sample. That is, $w_1, \ldots, w_n$ have a multinomial distribution consisting of n draws on n categories with equal probabilities $1/n$.

As discussed in Section 1.14 on the identifiability of a mixture model, if the component densities of the mixture belong to the same parametric family, then the likelihood does not change under a permutation of the component labels in the parameter $\boldsymbol{\Psi}$ and hence its MLE $\hat{\boldsymbol{\Psi}}$. This raises the question of whether the so-called label switching problem occurs in the generation of the bootstrap replications of the MLE, as in Monte Carlo Markov chain computations involving mixture models. However, in our experience it has not arisen, as we always take the MLE $\hat{\boldsymbol{\Psi}}$ calculated from the original data to be the initial value of the parameter in applying the EM algorithm to each bootstrap sample. In the parametric application of the bootstrap, $\hat{\boldsymbol{\Psi}}$ corresponds to the true value of the parameter, and so it should be a reasonable starting value.

2.17 SPEEDING UP CONVERGENCE

2.17.1 Introduction

The other common criticism that has been leveled at the EM algorithm is that its convergence can be quite slow. We shall therefore consider some methods that have been proposed for accelerating the EM algorithm in the context of mixture models. However, methods to accelerate the EM algorithm do tend to sacrifice the simplicity it usually enjoys.

As remarked by Lange (1995b), it is likely that no acceleration method can match the stability and simplicity of the unadorned EM algorithm. This has essentially been the case in general up to now. However, recently there have been some promising developments on modifications to the EM algorithm that speed up its performance without sacrificing the simplicity and stability that it enjoys. These new versions of the EM algorithm are called the Incremental EM (IEM), the Sparse EM (SPEM), and the Lazy EM. We shall consider these algorithms in Chapter 12, where we also present the approach of Moore (1999), using multiresolution kd-trees to speed up the conventional EM algorithm, and the work of Bradley, Fayyad, and Reina (1999) on scaling the EM algorithm to large databases.

The methods discussed below are applicable for a given specification of the complete data. There are also methods that approach the problem of speeding up convergence in terms of the choice of the missing data in the specification of the complete-data problem in the EM framework. These methods include the expectation–conditional maximization either (ECME) algorithm of Liu and Rubin (1994), the alternating ECM (AECM) algorithm of Meng and van Dyk (1997), and the parameter-expanded EM (PX-EM) algorithm of Liu, Rubin, and Wu (1998). In later work, we shall make use of the ECME and AECM algorithms in the fitting of mixtures of t components and mixtures of factor analyzers. We shall delay our discussion of them until where they are first used.

2.17.2 Louis' Method

The most commonly used method for EM acceleration is the multivariate version of Aitken's acceleration method. Suppose that $\boldsymbol{\Psi}^{(k)} \rightarrow \boldsymbol{\Psi}^*$, as $k \rightarrow \infty$. Then as demonstrated in McLachlan and Krishnan (1997, Section 4.7), we can express $\boldsymbol{\Psi}^*$ as

$$\begin{aligned}\boldsymbol{\Psi}^* &\approx \boldsymbol{\Psi}^{(k)} + \sum_{h=0}^{\infty} \{\boldsymbol{J}(\boldsymbol{\Psi}^*)\}^h (\boldsymbol{\Psi}^{(k+1)} - \boldsymbol{\Psi}^{(k)}) \\ &= \boldsymbol{\Psi}^{(k)} + \{\boldsymbol{I}_d - \boldsymbol{J}(\boldsymbol{\Psi}^*)\}^{-1} (\boldsymbol{\Psi}^{(k+1)} - \boldsymbol{\Psi}^{(k)}), \end{aligned} \tag{2.69}$$

as the power series

$$\sum_{h=0}^{\infty} \{\boldsymbol{J}(\boldsymbol{\Psi}^*)\}^h$$

converges to $\{\boldsymbol{I}_d - \boldsymbol{J}(\boldsymbol{\Psi}^*)\}^{-1}$ if $\boldsymbol{J}(\boldsymbol{\Psi}^*)$ has all its eigenvalues between 0 and 1.

The multivariate version of Aitken's acceleration method suggests trying the sequence of iterates $\{\boldsymbol{\Psi}_A^{(k)}\}$, where $\boldsymbol{\Psi}_A^{(k+1)}$ is defined by

$$\boldsymbol{\Psi}_A^{(k+1)} = \boldsymbol{\Psi}_A^{(k)} + \{\boldsymbol{I}_d - \boldsymbol{J}(\boldsymbol{\Psi}_A^{(k)})\}^{-1} (\boldsymbol{\Psi}_{EMA}^{(k+1)} - \boldsymbol{\Psi}_A^{(k)}), \tag{2.70}$$

where $\boldsymbol{\Psi}_{EMA}^{(k+1)}$ is the EM iterate produced using $\boldsymbol{\Psi}_A^{(k)}$ as the current fit for $\boldsymbol{\Psi}$.

Hence this method proceeds on the $(k+1)$th iteration by first producing $\boldsymbol{\Psi}_{EMA}^{(k+1)}$ using an EM iteration with $\boldsymbol{\Psi}_A^{(k)}$ as the current fit for $\boldsymbol{\Psi}$. One then uses the EM iterate $\boldsymbol{\Psi}_{EMA}^{(k+1)}$ in Aitken's acceleration procedure (2.70) to yield the final iterate $\boldsymbol{\Psi}_A^{(k+1)}$ on the $(k+1)$th iteration. This is the method proposed by Louis (1982) for speeding up the convergence of the EM algorithm.

Louis (1982) suggests making use of the relationship (2.50) to estimate $\boldsymbol{J}(\boldsymbol{\Psi}_A^{(k)})$ in (2.70). Using (2.50) in (2.70) gives

$$\boldsymbol{\Psi}_A^{(k+1)} = \boldsymbol{\Psi}_A^{(k)} + \boldsymbol{I}^{-1}(\boldsymbol{\Psi}_A^{(k)}; \boldsymbol{y})\boldsymbol{\mathcal{I}}_c(\boldsymbol{\Psi}_A^{(k)}; \boldsymbol{y})(\boldsymbol{\Psi}_{EMA}^{(k+1)} - \boldsymbol{\Psi}_A^{(k)}). \tag{2.71}$$

As cautioned by Louis (1982), the relationship (2.71) is an approximation useful only local to the MLE, and so it should not be used until some EM iterations have been performed. As noted by Meilijson (1989), the use of (2.71) is approximately equivalent to using the Newton-Raphson algorithm to find a zero of the (incomplete-data) score statistic $\boldsymbol{S}(\boldsymbol{y}; \boldsymbol{\Psi})$.

2.17.3 Quasi-Newton Methods

Meilijson (1989) and Jamshidian and Jennrich (1993) note that the accelerated sequence (2.70) as proposed by Louis (1982) is precisely the same as that obtained by applying the Newton-Raphson method to find a zero of the difference

$$\boldsymbol{\delta}(\boldsymbol{\Psi}) = \boldsymbol{M}(\boldsymbol{\Psi}) - \boldsymbol{\Psi},$$

where $\boldsymbol{M}$ is the map defined by the EM sequence; see McLachlan and Krishnan (1997, Section 4.7). On further ways to approximate $\{\boldsymbol{I}_d - \boldsymbol{J}(\boldsymbol{\Psi}_A^{(k)})\}$ for use in (2.70), Meilijson (1989) suggests using symmetric quasi-Newton updates. However, as cautioned by Jamshidian and Jennrich (1993), these will not work, as $\{\boldsymbol{I}_d - \boldsymbol{J}(\boldsymbol{\Psi})\}$ is in general not symmetric.

Lange (1995b) used the EM gradient algorithm that he proposed in Lange (1995a) to form the basis of a quasi-Newton approach to accelerate convergence of the EM algorithm. Recently, Jamshidian and Jennrich (1997) proposed two new methods for accelerating the EM algorithm, both based on quasi-Newton methods. They demonstrated their methods in some examples that included the ML fitting of Poisson and normal mixtures. They listed the necessary derivatives needed to implement their method for these two mixture models, although the formulas for the normal mixture model are for the special case of $g = 2$ normal components with a common covariance matrix.

2.17.4 Hybrid Methods

Various authors, including Redner and Walker (1984), propose a hybrid approach to the computation of the MLE that switches from the EM algorithm after a few iterations to the Newton-Raphson or some quasi-Newton method. The idea is to use the EM algorithm initially to take advantage of its good global convergence properties and to then exploit the rapid local convergence of Newton-type methods by switching to such a method. It can be seen that the method of Louis (1982) is a hybrid algorithm of this nature, for in effect after a few initial EM iterations, it uses the Newton-Raphson method to accelerate convergence after performing each subsequent EM iteration. Of course, there is no guarantee that these hybrid algorithms increase the likelihood $L(\boldsymbol{\Psi})$ monotonically. Hybrid algorithms have been considered also by Atkinson(1992), Heckman and Singer (1984), Atwood et al. (1996), Jones and McLachlan (1992a), and Aitkin and Aitkin (1996).

Aitkin and Aitkin (1996) have considered a hybrid method that combines the EM algorithm with a modified Newton-Raphson method whereby the information matrix is replaced by the empirical information matrix. In the context of fitting finite normal mixture models, they constructed a hybrid algorithm that starts with five EM iterations before switching to the modified Newton-Raphson method until convergence or until the log likelihood decreases. In the case of the latter, Aitkin and Aitkin (1996) proposed halving the step size up to five times. As further step-halvings would generally leave the step size smaller than that of an EM step, if the log likelihood decreases after five step-halves, the algorithm of Aitkin and Aitkin (1996) returns to the previous EM iterate and runs the EM algorithm for a further five iterations, before switching back again to the modified Newton-Raphson method. Their choice of performing five EM iterations initially is based on the work of Redner and Walker (1984), who report that, in their experience, 95 percent of the change in the log likelihood from its initial value to its maximum generally occurs in five iterations. Obviously, the choice of the starting point can have a major influence on such guidelines. For example, if the EM algorithm is started from a random starting point for which the component means are close together, then the EM algorithm may require a few iterations to recover from such a start before making being able to make good progress.

Aitkin and Aitkin (1996) replicated part of the study by Everitt (1988a) on the fitting of a mixture of two normal densities with means μ_1 and μ_2 and variances σ_1^2 and σ_2^2 in proportions π_1 and π_2. Their stopping criterion was a difference in successive values of the log likelihood of 10^{-5}. They found that their hybrid algorithm required 70 percent of the time required for the EM algorithm to converge, consistently over all starting values of $\boldsymbol{\Psi}$. They noted that the EM algorithm was impressively stable. Their hybrid algorithm almost always decreased the log likelihood when the switch to the modified Newton-Raphson was first applied, and sometimes required a large number of EM controlling steps (after full step-halving) before finally increasing the log likelihood, and then usually converging rapidly to the same maximizer as with the EM algorithm. As is well known, mixture likelihoods for small sample sizes are badly behaved with multiple maxima. Aitkin and Aitkin (1996) liken the maximization of the normal mixture log likelihood in their simulation studies as to the progress of a "traveler following the narrow EM path up a hazardous mountain with chasms on all sides. When in sight of the summit, the modified Newton-Raphson method path leapt to the top, but when followed earlier, it caused repeated falls into the chasms, from which the traveler had to be pulled back onto the EM track."

2.18 OUTLIER DETECTION FROM A MIXTURE

2.18.1 Introduction

We consider here the problem of testing whether a new observation $\boldsymbol{y}_{n+1}$ is an outlier of a multivariate mixture density

$$f(\boldsymbol{y}_j;\, \boldsymbol{\Psi}) = \sum_{i=1}^{g} \pi_i\, f_i(\boldsymbol{y}_j;\, \boldsymbol{\theta}_i) \quad (j = 1, \ldots, n). \tag{2.72}$$

This problem was considered recently by Wang et al. (1997) and Sain et al. (1999) in the context of determining whether an observed seismic event may be a nuclear explosion or earthquake, or some other seismic disturbance. It arose from multinational efforts to monitor the Comprehensive Test Ban Treaty, which was approved by the United Nations (with India dissenting) in 1996 and which mandates a worldwide ban on nuclear testing. As explained by Wang et al. (1997), the distinguishing characteristics of small nuclear explosions are regional in nature, and so the features that characterize such events are not transportable from region to region around the world. As a consequence, no historical data on nuclear explosions are available in most locations around the world. Were such historical nuclear available, the problem would fall in the field of discriminant analysis; see, for example, McLachlan (1992, Chapter 6).

One way to approach this problem in the absence of classified data would be to compare the new observation $\boldsymbol{y}_{n+1}$ to each of the fitted components of the mixture in turn, forming an atypicality measure (such as the Mahalanobis squared distance) with respect to each component mean. If this measure is sufficiently large for all the components, then $\boldsymbol{y}_{n+1}$ can be declared an outlier. This approach is described in McLachlan and Basford (1988, Section 2.7). As Wang et al. (1997) note, with this approach there is poor control over the overall significance level, except through conservative techniques such as use of Bonferroni bounds.

2.18.2 Modified Likelihood Ratio Test

Wang et al. (1997) considered the likelihood ratio test for this problem, which we shall now describe, assuming there are no classified data available. They considered the calculation of the likelihood ratio statistic λ for testing the null hypothesis H_0 that $\boldsymbol{Y}_{n+1}$ is from the postulated mixture model for the first n observations versus the alternative hypothesis H_1 that it is not. Let $h(\boldsymbol{y}_{n+1};\, \boldsymbol{\beta})$ denote the density of $\boldsymbol{Y}_{n+1}$ under H_1. Unless $\boldsymbol{\beta}$ is functionally related to $\boldsymbol{\Psi}$, there is only the one observation $\boldsymbol{y}_{n+1}$ from which to estimate $\boldsymbol{\beta}$, even if the parametric form h is specified. Accordingly, Wang et al. (1997) took $h(\boldsymbol{y}_{n+1};\, \boldsymbol{\beta})$ to be constant over its practical (finite) support. This allowed them to drop the constant form for the density of $\boldsymbol{Y}_{n+1}$ from the likelihood under H_1, thereby giving a modified version of λ given by

$$\tilde{\lambda} = L(\hat{\boldsymbol{\Psi}}_{n+1})/L(\hat{\boldsymbol{\Psi}}), \tag{2.73}$$

where $\hat{\boldsymbol{\Psi}}$ denotes the MLE of $\boldsymbol{\Psi}$ formed from the mixture likelihood for the n observations in $\boldsymbol{y} = (\boldsymbol{y}_1^T, \ldots, \boldsymbol{y}_n^T)^T$, and $\hat{\boldsymbol{\Psi}}_{n+1}$ denotes the MLE of $\boldsymbol{\Psi}$ using $\boldsymbol{y}$ and $\boldsymbol{y}_{n+1}$. Not surprisingly, they showed asymptotically that the use of $\tilde{\lambda}$ is equivalent to the test that rejects H_0 if $f(\boldsymbol{y}_{n+1}; \hat{\boldsymbol{\Psi}})$ is sufficiently small.

In order to assess the null distribution of their modified test statistic $\tilde{\lambda}$, Wang et al. (1997) suggested using the bootstrap with either parametric or nonparametric resampling. They also suggested a computationally less burdensome method, which they called the bootstrap-one method. With this approximation, only the $(n+1)$th observation is resampled each time, and so the bootstrap replications of $\tilde{\lambda}$ are formed using the same (original) values for the first n observations. Their empirical studies demonstrated that the effect of this conditioning was minimal, provided that the sample size was sufficiently large.

2.19 PARTIAL CLASSIFICATION

We now consider the situation where the observed data contain some observations whose component of origin is known. That is, the observed data, $\boldsymbol{y}_1, \ldots, \boldsymbol{y}_n$, contain some data that are classified with respect to the components of the mixture model. We suppose that the data have been labeled so that $\boldsymbol{y}_j$ $(j = 1, \ldots, m)$ denote the $m\,(m < n)$ classified observations; that is, for these $\boldsymbol{y}_j$, the associated component-indicator vectors $\boldsymbol{z}_j$ are known. Hosmer (1973b) and Hosmer and Dick (1977) have discussed various models under which the classified data may have been obtained.

A situation where this may arise in practice is where the components correspond to externally existing groups, and some of the observed data have been classified (without error) with respect to these groups. This covers the updating problem in discriminant analysis where a discriminant rule has been formed from the classified data and the intent is to use the subsequent unclassified data to improve the performance of the rule by forming it on the basis of the combined classified and unclassified data; see McLachlan (1992, Section 2.7).

Another example concerns improving the accuracy of learned text classifiers by augmenting a small number of classified documents with a large pool of unclassified documents (Nigam et al., 2000). This is significant because in many important text classification problems, obtaining the true classification of documents is expensive, whereas large quantities of unclassified documents are readily available.

Whatever the reason for wishing to carry out the estimation on the basis of both classified and unclassified data, it can be undertaken in a straightforward manner by maximum likelihood via the EM algorithm. The equation (2.33) for the update $\boldsymbol{\xi}^{(k+1)}$ containing the parameters in the component densities of the mixture model still applies in the presence of some classified data, except that we use the known value of the component-indicator variable z_{ij} instead of its currently assessed expectation $\tau_i(\boldsymbol{y}_j; \boldsymbol{\Psi}^{(k)})$ $(i = 1, \ldots, g)$. And we make the same modification in the equation (2.32) for the update of the ith mixing proportion π_i, assuming that the classified data have been obtained by sampling from the mixture. If the classified data do not provide any information on the π_i as, for example, where the classified data have been obtained by sampling separately from each of the groups corresponding to the components of the mixture model being fitted, then the updated estimate of π_i is given by

$$\pi_i^{(k+1)} = \sum_{j=m+1}^{n} \tau_i(\boldsymbol{y}_j;\, \boldsymbol{\Psi}^{(k)})/(n-m). \tag{2.74}$$

Maximum likelihood estimation is facilitated by the presence of data of known origin with respect to each component of the mixture. For example, there may be singularities in the likelihood on the edge of the parameter space in the case of component densities that are multivariate normal with unequal covariance matrices. However, no singularities will occur if there are more than p classified observations available from each component. Also, there can be difficulties with the choice of suitable starting values for the EM algorithm. But in the presence of classified data, an obvious choice of a starting point is the MLE of $\boldsymbol{\Psi}$ based solely on the classified data, assuming there are enough classified entities from each group for this purpose.

Recently, Cooley and MacEachern (1999) have considered the behavior of the Bayes rule as the number of unclassified observations grows without bound; see McLachlan (1975), O'Neill (1978), Lavine and West (1992), and Castelli and Cover (1996) for some previous work on this problem.

2.20 PARTIAL NONRANDOM CLASSIFICATION

2.20.1 Introduction

In some situations in practice with a partially classified training set, the classified data may not represent an observed random sample from the sample space of the feature vector. For example, in medical screening, patients are often initially diagnosed on the basis of some simple rule, for instance whether one feature variable is above or below a certain threshold, corresponding to a positive or negative test. Patients with a positive test are investigated further, from which their true condition may be ascertained. However, patients with a negative test may be regarded as apparently healthy and so may not be investigated further. This would be the case if a true diagnosis can be made, say, only by an invasive technique whose application would not be ethical in apparently healthy patients. In these circumstances if only the data of known origin were used in the estimation of the unknown parameters, then it would generally bias the results, unless appropriate steps were taken, such as fitting truncated densities or using logistic regression (McLachlan, 1992, Chapter 8).

Another approach that avoids this bias problem is to perform the estimation on the basis of all the data collected, including the data of unknown origin, by fitting a mixture model. This approach was adopted by McLachlan and Gordon (1989) in their development of a probabilistic allocation rule as an aid in the diagnosis of renal artery stenosis (RAS), which is potentially curable by surgery. Another example involves the modeling of consumer credit ratings, where the only information available on whether a loan was satisfactorily serviced or not is for a client who achieved an acceptable rating in the first instance. That is, there is no information on the subsequent loan patterns for those applicants who failed to get a loan.

2.20.2 A Nonrandom Model

We now define the partial nonrandom classification model as considered in McLachlan and Gordon (1989). It is supposed that a random sample is drawn from a mixture of the two groups, G_1 and G_2, in proportions π_1 and π_2. If a member of this sample has feature vector falling in the region $\boldsymbol{R}$, then its true classification is determined. Thus if the feature vector falls in the region $\boldsymbol{R}^c$, the complement of $\boldsymbol{R}$ in the feature space, it remains unclassified. We let $\boldsymbol{y}_1, \ldots, \boldsymbol{y}_m$ denote the feature vectors that are classified (that is, fall in $\boldsymbol{R}$), and $\boldsymbol{y}_j$ $(j = m+1, \ldots, n)$ the $n - m$ unclassified feature vectors that fall in $\boldsymbol{R}^c$.

McLachlan and Gordon (1989) showed for this nonrandom partial classification model that the MLE of $\boldsymbol{\Psi}$ is obtained by consideration of the likelihood function for $\boldsymbol{\Psi}$ that is applicable if the classified data had been obtained by sampling randomly from a mixture of the groups. This likelihood function $L_{pc}(\boldsymbol{\Psi})$ is given by

$$\begin{aligned}\log L_{pc}(\boldsymbol{\Psi}) &= \sum_{i=1}^{g}\sum_{j=1}^{m} z_{ij} \log\{\pi_i f_i(\boldsymbol{y}_j;\, \boldsymbol{\theta}_i)\} \\ &+ \sum_{j=m+1}^{n} \log \sum_{i=1}^{g} \pi_i f_i(\boldsymbol{y}_j;\, \boldsymbol{\Psi}),\end{aligned} \tag{2.75}$$

where the component-indicator vectors $\boldsymbol{z}_1, \ldots, \boldsymbol{z}_m$ corresponding to the m classified feature observations are known.

This result can also be established on noting that the missing-data mechanism is *ignorable* in the terminology introduced by Little and Rubin (1987). Here the missing-data mechanism is ignorable if the "missing" component-indicator $\boldsymbol{z}_j$ for an unclassified feature point $\boldsymbol{y}_j$ has the same conditional distribution given $\boldsymbol{y}_j$, irrespective of whether $\boldsymbol{y}_j$ is classified or unclassified. Obviously, this is true for the partial classification model being considered here.

2.20.3 Asymptotic Relative Efficiencies

We let $\hat{\boldsymbol{\Psi}}_{pc}$ denote the MLE obtained by consideration of the likelihood $L_{pc}(\boldsymbol{\Psi})$, and $\hat{\boldsymbol{\Psi}}$ and $\hat{\boldsymbol{\Psi}}_c$ be the MLEs in the totally unclassified case $(m = 0)$ and the totally classified case $(m = n)$, respectively. Also, we let $r_B(y_o;\, \boldsymbol{\Psi})$ be the Bayes rule for the allocation of a new observation y_o in the case of two groups with univariate normal distributions having a common variance. That is, this rule will be based on the linear discriminant function.

In this univariate case, we let the region $R = (K, \infty)$ be the interval in which a feature variable must fall for it to be subsequently classified in the partial nonrandom classification model introduced in the previous section. Under this model, McLachlan and Scot (1995) derived the asymptotic relative efficiencies ARE($\hat{\boldsymbol{\Psi}}_{\text{pc}}$; $\hat{\boldsymbol{\Psi}}_{\text{c}}$) and ARE($\hat{\boldsymbol{\Psi}}$; $\hat{\boldsymbol{\Psi}}_{\text{pc}}$), where the former is the ARE of the sample rule $r_B(y_o;\ \hat{\boldsymbol{\Psi}}_{\text{pc}})$ based on a partially classified sample relative to $r_B(y_o;\ \hat{\boldsymbol{\Psi}}_{\text{c}})$ based on a totally classified sample, and the latter is the ARE of the sample rule $r_B(y_o;\ \hat{\boldsymbol{\Psi}})$ based on a totally unclassified sample relative to $r_B(y_o;\ \hat{\boldsymbol{\Psi}}_{\text{pc}})$ based on a partially classified sample. The asymptotic relative efficiency in each case was defined in terms of the ratio of the increases in the overall error rate over the Bayes error. The results are presented here in Tables 2.3 and 2.4 for various levels of the Mahalanobis distance Δ and the threshold K in the case of equally likely groups ($\pi_1 = \pi_2 = 0.5$).

The entries for ARE($\hat{\boldsymbol{\Psi}}_{\text{pc}}$; $\hat{\boldsymbol{\Psi}}_{\text{c}}$) at $K = \infty$ give the values of the asymptotic relative efficiency of the rule formed from a completely unclassified training set. They agree with the results reported for this case in Ganesalingam and McLachlan (1978) and O'Neill (1978). For a given Δ, ARE($\hat{\boldsymbol{\Psi}}_{\text{pc}}$; $\hat{\boldsymbol{\Psi}}_{\text{c}}$) increases monotonically to 1 as K tends to $-\infty$. The value of ARE($\hat{\boldsymbol{\Psi}}_{\text{pc}}$; $\hat{\boldsymbol{\Psi}}_{\text{c}}$) is, not surprisingly, quite low for $\Delta = 1$, representing the case of two groups with small Mahalanobis distance between them.

Table 2.3 Asymptotic (Percentage) Efficiency of Rule $r_B(y_o;\ \hat{\boldsymbol{\Psi}}_{\text{pc}})$ Relative to $r_B(y_o;\ \hat{\boldsymbol{\Psi}}_{\text{c}})$ for Various Levels of Δ and K

				K			
Δ	0.5	1.0	1.5	2.0	2.5	3.0	∞
1.0	20.83	8.39	5.00	3.10	1.57	0.80	0.51
1.5	38.18	18.47	11.55	7.50	4.79	3.63	3.24
2.0	64.91	37.37	23.85	16.37	12.20	10.59	10.08
2.5	70.44	62.57	41.81	30.14	24.21	22.06	21.40
3.0	95.64	83.84	63.20	47.43	39.53	36.74	35.89

Source: Adapted from McLachlan and Scot (1995).

Table 2.4 Asymptotic (Percentage) Efficiency of Rule $r_B(y_o;\ \hat{\boldsymbol{\Psi}})$ Relative to $r_B(y_o;\ \hat{\boldsymbol{\Psi}}_{\text{pc}})$ for Various Levels of Δ and K

				K			
Δ	$-\infty$	0.5	1.0	1.5	2.0	2.5	3.0
1.0	0.51	2.43	6.04	10.14	16.37	32.28	63.63
1.5	3.24	8.48	17.54	28.06	43.20	67.59	89.25
2.0	10.08	15.53	26.97	42.27	61.59	82.61	95.16
2.5	21.40	30.38	34.20	51.18	71.01	88.38	96.99
3.0	35.89	37.53	42.81	56.79	75.68	90.80	97.69

Source: Adapted from McLachlan and Scot (1995).

In the case of two groups that are moderately separated ($\Delta = 2$), it can be seen from Table 2.3 that ARE($\hat{\boldsymbol{\Psi}}_{\text{pc}}$; $\hat{\boldsymbol{\Psi}}_{\text{c}}$) is 64.91% at $K = 0.5$, falling to 16.37% at $K = 2$. That is, suppose that G_2 (which has mean $\mu_2 = 0$) corresponds to a healthy group of patients and that G_1 (with $\mu_1 = 1$) corresponds to an unhealthy group. Then if a definitive diagnosis is made when the feature observation y_o is more than one-half of a standard deviation greater than the mean ($\mu_2 = 0$) of the healthy group, the asymptotic relative efficiency of the rule formed under this partial (nonrandom) classification of the training feature data is moderately high. But it drops to approximately 16% if a definitive classification is made only when the feature observation is more than two standard deviations greater than the mean of the healthy group.

Concerning the asymptotic efficiency of the rule $r_B(y_o;\, \hat{\boldsymbol{\Psi}})$ formed from a totally unclassified sample relative to the rule $r_B(y_o;\, \hat{\boldsymbol{\Psi}}_{\text{pc}})$ formed under partial nonrandom classification of the training set, ARE($\hat{\boldsymbol{\Psi}}$; $\hat{\boldsymbol{\Psi}}_{\text{pc}}$) increases monotonically with K for a given Δ, tending to one as $K \to \infty$. At $K = -\infty$, ARE($\hat{\boldsymbol{\Psi}}$; $\hat{\boldsymbol{\Psi}}_{\text{pc}}$) gives the asymptotic efficiency of $r_B(y_o;\, \hat{\boldsymbol{\Psi}})$ relative to $r_B(y_o;\, \hat{\boldsymbol{\Psi}}_{\text{c}})$, as represented in Table 2.3 by ARE($\hat{\boldsymbol{\Psi}}_{\text{pc}}$; $\hat{\boldsymbol{\Psi}}_{\text{c}}$) at $K = \infty$.

It can be seen from Table 2.4 that in the case of two groups not widely separated, partial nonrandom classification of the training set can considerably improve the performance of the sample linear discriminant rule over its version formed in the absence of any classification.

2.21 CLASSIFICATION ML APPROACH

Another likelihood-based approach to clustering besides the mixture likelihood approach is what is sometimes called the classification likelihood approach. With this approach, $\boldsymbol{\Psi}$ and the unknown component-indicator vectors $\boldsymbol{z}_1, \ldots, \boldsymbol{z}_n$ of the observed feature data $\boldsymbol{y}_1, \ldots, \boldsymbol{y}_n$ are chosen to maximize $L_c(\boldsymbol{\Psi})$, the likelihood for $\boldsymbol{\Psi}$ formed on the basis of the so-called complete-data as introduced within the EM framework for the ML fitting of the mixture likelihood. That is, the unknown vector $\boldsymbol{z} = (\boldsymbol{z}_1^T, \ldots, \boldsymbol{z}_n^T)^T$ containing the component-indicators is treated as a parameter to be estimated along with $\boldsymbol{\Psi}$. Accordingly, the maximization of $L_c(\boldsymbol{\Psi})$ is over the set of zero-one values of the elements of the unknown $\boldsymbol{z}_j$, corresponding to all possible assignments of the n entities to the g components, as well as over all values of $\boldsymbol{\Psi}$. In principle, the maximization process for the classification likelihood approach can be carried out for arbitrary n, since it is just a matter of computing the maximum value of $L_c(\boldsymbol{\Psi})$ over all possible partitions of the n observations to the g components. In some situations, for example with multivariate normal component densities with unequal covariance matrices, the restriction that at least $p+1$ observations belong to each component is needed to avoid the degenerate case of infinite likelihood. Unless n is small, however, searching over all possible partitions is prohibitive. If $\hat{\boldsymbol{z}}_j$ ($j = 1, \ldots, n$) denotes the optimal partition of the n observations, then $\hat{z}_{ij} = 1$ or 0, according to whether

$$\hat{\pi}_i f_i(\boldsymbol{y}_j;\, \hat{\boldsymbol{\theta}}_i) \geq \hat{\pi}_h f_h(\boldsymbol{y}_j;\, \hat{\boldsymbol{\theta}}_h) \quad (h = 1, \ldots, g;\, h \neq i)$$

holds or not, where $\hat{\boldsymbol{\theta}}_i$ and $\hat{\pi}_i$ are the MLEs of $\boldsymbol{\theta}_i$ and π_i, respectively, for the n observations partitioned according to $\hat{\boldsymbol{z}}_1, \ldots, \hat{\boldsymbol{z}}_n$. Hence, as noted by McLachlan (1982a), a solution corresponding to a local maximum can be computed iteratively by alternating a modified version of the E-step with the same M-step, as described in Section 2.8 for the application of the EM algorithm in fitting the mixture model (2.17). In the E-step on the $(k+1)$th iteration, z_{ij} is replaced not by the current estimate of the posterior probability that the jth entity belongs to the ith component, but by one or zero according to whether

$$\pi_i^{(k)} f_i(\boldsymbol{y}_j;\, \boldsymbol{\theta}_i^{(k)}) \geq \pi_h^{(k)} f_h(\boldsymbol{y}_j;\, \boldsymbol{\theta}_h^{(k)}) \quad (h = 1, \ldots, g;\; h \neq i)$$

holds or not $(i = 1, \ldots, g;\; j = 1, \ldots, n)$; see also McLachlan (1975) for a related approach with a partially classified sample.

A more detailed account of this approach may be found in McLachlan and Basford (1988, Section 1.12); see also Ganesalingam and McLachlan (1980a). More recent references on the classification approach include Celeux and Govaert (1991, 1993, 1995), Banfield and Raftery (1993), and Govaert and Nadif (1996). Although the classification ML approach may provide a slightly better clustering in some situations (small samples in comparable proportions from normal groups), this apparent superiority is more than offset by its poor performance for groups that are not widely separated or are in disparate proportions. Further, from an estimation point of view, it yields inconsistent estimates of the parameters; see Bryant and Williamson (1978, 1986), Bryant (1991), and Titterington (1984b).

The classification ML approach can be shown to be equivalent to some commonly used clustering criteria under the assumption of normal groups with various constraints on their covariance matrices, as noted originally by Scott and Symons (1971). For example, if the mixing proportions are taken to be equal or, equivalently, a separate sampling scheme is assumed for the data, then the classification ML approach with the constraint of equal covariance matrices leads to the $\mid \boldsymbol{W} \mid$ criterion, as originally suggested by Friedman and Rubin (1967). If the covariance matrices are further assumed to be diagonal, then it yields the trace $\boldsymbol{W}$ criterion or, equivalently, the k-means procedure. More recently, Celeux and Govaert (1995) have considered the equivalence of the classification ML approach to other clustering criteria under varying assumptions on the group densities.

3

Multivariate Normal Mixtures

3.1 INTRODUCTION

In this chapter we focus on mixture models with normal components. A common assumption in practice is to take the component densities to be (multivariate) normal. For example, in a clustering context, it is typical to proceed on the basis that any nonnormal features in the data are due to some underlying group structure. Further, often clusters in the data are essentially elliptical, so that it is reasonable to consider fitting mixtures of elliptically symmetric component densities. Within this class of component densities, the multivariate normal density is a convenient choice given its computational tractability. In Chapter 7 we consider a robust extension of the normal mixture model through the use of t component distributions. The family of t distributions provides a heavy-tailed alternative to the normal family.

3.2 HETEROSCEDASTIC COMPONENTS

In Section 2.8 we have described the application of the EM algorithm for the ML fitting of a mixture of arbitrary component distributions to an observed random sample, $\boldsymbol{y}_1, \ldots, \boldsymbol{y}_n$. We now specialize these results to the case of a mixture of normal components,

$$f(\boldsymbol{y}_j; \boldsymbol{\Psi}) = \sum_{i=1}^{g} \pi_i \phi(\boldsymbol{y}_j; \boldsymbol{\mu}_i, \boldsymbol{\Sigma}_i). \tag{3.1}$$

Here $\boldsymbol{\Psi} = (\pi_1, \ldots, \pi_{g-1}, \boldsymbol{\xi}^T)^T$, where $\boldsymbol{\xi}$ contains the elements of the component means $\boldsymbol{\mu}_i$ and the distinct elements of the component-covariance matrices $\boldsymbol{\Sigma}_i$ $(i = 1, \ldots, g)$.

We first consider the unrestricted (heteroscedastic) case where the component-covariance matrices $\boldsymbol{\Sigma}_i$ are unequal; that is, there are no restrictions placed on them. Concerning the E-step on the $(k+1)$th iteration, we have seen in Section 2.8.3 that it effectively replaces the unknown zero–one component-label variables z_{ij} by their current conditional expectations given by the posterior probabilities of component membership of the observed data $\boldsymbol{y}_j$, $\tau_i(\boldsymbol{y}_j; \boldsymbol{\Psi}^{(k)})$, where now

$$\tau_i(\boldsymbol{y}_j; \boldsymbol{\Psi}) = \pi_i \phi(\boldsymbol{y}_j; \boldsymbol{\mu}_i, \boldsymbol{\Sigma}_i) / \sum_{h=1}^{g} \pi_h \phi(\boldsymbol{y}_j; \boldsymbol{\mu}_h, \boldsymbol{\Sigma}_h) \tag{3.2}$$

for $i = 1, \ldots, g$; $j = 1, \ldots, n$.

The M-step for normal components exists in closed form. The updates of the component means $\boldsymbol{\mu}_i$ and component-covariance matrices $\boldsymbol{\Sigma}_i$ are given simply by

$$\boldsymbol{\mu}_i^{(k+1)} = \sum_{j=1}^{n} \tau_{ij}^{(k)} \boldsymbol{y}_j / \sum_{j=1}^{n} \tau_{ij}^{(k)} \tag{3.3}$$

and

$$\boldsymbol{\Sigma}_i^{(k+1)} = \sum_{j=1}^{n} \tau_{ij}^{(k)} (\boldsymbol{y}_j - \boldsymbol{\mu}_i^{(k+1)})(\boldsymbol{y}_j - \boldsymbol{\mu}_i^{(k+1)})^T / \tau_{ij}^{(k)} \tag{3.4}$$

for $i = 1, \ldots, g$, where

$$\tau_{ij}^{(k)} = \tau_i(\boldsymbol{y}_j; \boldsymbol{\Psi}^{(k)}) \qquad (i = 1, \ldots, g;\ j = 1, \ldots, n).$$

The updated estimate of the ith mixing proportion π_i is as given by (2.32).

Computationalwise, it is advantageous to express the update (3.4) of $\boldsymbol{\Sigma}_i$ directly in terms of the current conditional expectations of the sufficient statistics T_{i1}, $\boldsymbol{T}_{i2}$, and $\boldsymbol{T}_{i3}$ for $\boldsymbol{\Psi}$ in the complete-data framework, given by

$$T_{i1}^{(k)} = \sum_{j=1}^{n} \tau_{ij}^{(k)}, \tag{3.5}$$

$$\boldsymbol{T}_{i2}^{(k)} = \sum_{j=1}^{n} \tau_{ij}^{(k)} \boldsymbol{y}_j, \tag{3.6}$$

and

$$\boldsymbol{T}_{i3}^{(k)} = \sum_{j=1}^{n} \tau_{ij}^{(k)} \boldsymbol{y}_j \boldsymbol{y}_j^T. \tag{3.7}$$

We have then that

$$\boldsymbol{\Sigma}_i^{(k+1)} = \{\boldsymbol{T}_{i3}^{(k)} - T_{i1}^{(k)^{-1}} \boldsymbol{T}_{i2}^{(k)} \boldsymbol{T}_{i2}^{(k)^T}\} / T_{i1}^{(k)} \quad (i = 1, \ldots, g). \tag{3.8}$$

Using (3.8) instead of (3.4) to update the estimate of the ith component-covariance matrix gives a reduction in CPU time of around 50%.

Liu and Sun (1997) have considered the use of the ECME algorithm to speed up convergence of the M-step. With their approach, the component means and covariance matrices are updated as above. However, the mixing proportions are updated by a conditional maximization of the incomplete-data log likelihood, $\log L(\boldsymbol{\Psi})$, with the component means and covariance matrices fixed at $\boldsymbol{\mu}_i^{(k+1)}$ and $\boldsymbol{\Sigma}_i^{(k+1)}$ $(i = 1, \ldots, g)$. The updates $\pi_i^{(k+1)}$ do not exist in closed form and have to be obtained iteratively, for example, by Newton-Raphson. Liu and Sun (1997) report that typically only one Newton-Raphson iteration is required.

3.3 HOMOSCEDASTIC COMPONENTS

Often in practice, the component-covariance matrices $\boldsymbol{\Sigma}_i$ are restricted to being the same,

$$\boldsymbol{\Sigma}_i = \boldsymbol{\Sigma} \qquad (i = 1, \ldots, g), \tag{3.9}$$

where $\boldsymbol{\Sigma}$ is unspecified. In this case of homoscedastic normal components, the updated estimate of the common component-covariance matrix $\boldsymbol{\Sigma}$ is given by

$$\boldsymbol{\Sigma}^{(k+1)} = \sum_{i=1}^{g} T_{i1}^{(k)} \boldsymbol{\Sigma}_i^{(k+1)} / n, \tag{3.10}$$

where $\boldsymbol{\Sigma}_i^{(k+1)}$ is given by (3.8), and the updates of π_i and $\boldsymbol{\mu}_i$ are as above in the heteroscedastic case.

In the isotropic case, the component-covariance matrices are further restricted by taking the common component-covariance matrix $\boldsymbol{\Sigma}$ to be diagonal with equal diagonal elements σ^2; that is,

$$\boldsymbol{\Sigma} = \sigma^2 \boldsymbol{I}_p. \tag{3.11}$$

3.4 STANDARD ERRORS

The provision of standard errors for the parameters in a general finite mixture model has been discussed in Chapter 2, including the use of the bootstrap. With the information-based approach that uses the observed information matrix $\boldsymbol{I}(\hat{\boldsymbol{\Psi}}; \boldsymbol{y})$, we need to calculate the Hessian matrix for the log likelihood. This has been done in Roberts et al. (1998) for the normal mixture model, initially in the simplified case of diagonal component-covariance matrices. They then performed a transformation to convert their results to the general case of unrestricted component-covariance matrices. However, this part of their calculations would appear to be incorrect.

We have seen in Section 2.15.3 that the observed information matrix can be approximated by the empirical matrix $\boldsymbol{I}_e(\hat{\boldsymbol{\Psi}}; \boldsymbol{y})$, which only requires the gradient of the log likelihood function to be evaluated. Moreover, it can be expressed in terms of the gradient of the complete-data log likelihood function. We have from (2.60) that

$$I_e(\hat{\boldsymbol{\Psi}};\, \boldsymbol{y}) = \sum_{j=1}^{n} \boldsymbol{s}(\boldsymbol{y}_j;\, \hat{\boldsymbol{\Psi}})\boldsymbol{s}^T(\boldsymbol{y}_j;\, \hat{\boldsymbol{\Psi}}) \tag{3.12}$$

where, from (2.62) for finite mixture models,

$$\boldsymbol{s}(\boldsymbol{y}_j;\, \boldsymbol{\Psi}) = \sum_{i=1}^{g} \tau_i(\boldsymbol{y}_j;\, \boldsymbol{\Psi})\partial\{\log \pi_i + \log f_i(\boldsymbol{y}_j;\, \boldsymbol{\theta}_i)\}/\partial\boldsymbol{\Psi} \quad (j = 1,\, \ldots,\, n). \tag{3.13}$$

We now give expressions for the elements of $\boldsymbol{s}(\boldsymbol{y}_j;\, \hat{\boldsymbol{\Psi}})$ in the case of normal components with no restrictions on the component-covariance matrices $\boldsymbol{\Sigma}_i$.

Let $\boldsymbol{\pi}_{g-1} = (\pi_1,\, \ldots,\, \pi_{g-1})^T$ and let $\boldsymbol{\omega}_i$ contain the $\frac{1}{2}p(p+1)$ elements of $\boldsymbol{\Sigma}_i$ $(i = 1,\, \ldots,\, g)$. Then on partitioning the vector $\boldsymbol{\Psi}$ of unknown parameters as

$$\boldsymbol{\Psi} = (\boldsymbol{\pi}_{g-1}^T,\, \boldsymbol{\mu}_1^T,\, \ldots,\, \boldsymbol{\mu}_g^T,\, \boldsymbol{\omega}_1^T,\, \ldots,\, \boldsymbol{\omega}_g^T)^T,$$

we let

$$(\boldsymbol{d}_{\pi_{g-1},j}^T,\, \boldsymbol{d}_{\mu_{1j}}^T,\, \ldots,\, \boldsymbol{d}_{\mu_{gj}}^T,\, \boldsymbol{d}_{\omega_{1j}}^T,\, \ldots,\, \boldsymbol{d}_{\omega_{gj}}^T)^T$$

be the corresponding partition of $\partial\boldsymbol{s}(\boldsymbol{y}_j;\, \hat{\boldsymbol{\Psi}})/\partial\boldsymbol{\Psi}$.

We have from McLachlan and Basford (1988) that for $j = 1,\, \ldots,\, n$,

$$(\boldsymbol{d}_{\pi_{g-1},j})_i = \hat{\tau}_{ij}/\pi_i - \hat{\tau}_{gj}/\pi_g \quad (i = 1,\, \ldots,\, g-1), \tag{3.14}$$

$$\boldsymbol{d}_{\mu_{ij}} = \hat{\tau}_{ij}\hat{\boldsymbol{\Sigma}}_i^{-1}(\boldsymbol{y}_j - \hat{\boldsymbol{\mu}}_i) \quad (i = 1,\, \ldots,\, g), \tag{3.15}$$

and

$$(\boldsymbol{d}_{\omega_{ij}})_h = \tfrac{1}{2}\hat{\tau}_{ij}(2 - \delta_{rs})[-(\hat{\boldsymbol{\Sigma}}_i^{-1})_{rs} + (\boldsymbol{y}_j - \hat{\boldsymbol{\mu}}_i)^T\hat{\boldsymbol{\sigma}}_i^{(r)}(\boldsymbol{y}_j - \hat{\boldsymbol{\mu}}_i)^T\hat{\boldsymbol{\sigma}}_i^{(s)}], \tag{3.16}$$

where the hth element of $\boldsymbol{d}_{\omega_{ij}}$ corresponds to the $(r,\, s)$th element of $\boldsymbol{\Sigma}_i$ and where $\hat{\boldsymbol{\sigma}}_i^{(r)}$ is the rth column of $\hat{\boldsymbol{\Sigma}}_i^{-1}$ $(i = 1,\, \ldots,\, g)$.

3.5 ASSESSMENT OF MODEL FIT

The problem of assessing model fit is not straightforward for mixture models at least for multivariate data. For univariate data, we can compare the fitted mixture density with the data in histogram form; see, for example, Figure 1.1. Alternatively, in the univariate case, we can compare the fitted mixture distribution function with the empirical distribution function, as advocated by Aitkin (1997), among others.

In the case of normal components, one way to proceed with the assessment of normality and homoscedasticity is to follow the approach of Hawkins et al. (1982) and apply the test of Hawkins' (1981) to the clusters implied by the MLE. As cautioned by McLachlan and Basford (1988), it is self-serving to cluster the data under the assumption of normality when one is subsequently to test for it, but it is brought about by the limitations on such matters as hypothesis testing in the absence of classified data. The reader is referred to McLachlan and Basford (1988, Section 3.2) and McLachlan (1992, Chapter 6) on the application of Hawkins' (1981) test in the case of a partially or totally unclassified sample.

3.6 EXAMPLES OF UNIVARIATE NORMAL MIXTURES

Here we briefly describe the important role of univariate normal mixtures in the field of genetics where they provide the basic model. Also, as mentioned in Section 1.1.1, mixture models have proven to be very useful in assessing the error rates (sensitivity and specificity) of diagnostic and screening procedures in the absence of a gold standard (Qu, Tan, and Kutner, 1996). We illustrate this in two examples. The first is on screening for hemochromatosis; the second is on a diagnostic test for diabetes.

3.6.1 Basic Model in Genetics

One area in which mixture modeling has played a very important role is genetics, following Karl Pearson's pioneering work in the area in the late 1800s. An example of mixture modeling in genetics may be found in Shoukri and McLachlan (1994), who fitted a mixture distribution to nuclear family data. The reader is referred to Schork et al. (1996) for a comprehensive review of the theory and application of mixture models in human genetics.

A basic mixture model can be used to assess the impact of possible underlying genotypes associated with a particular locus on phenotypes that display continuous or quantitative variation in the population (for example, blood pressure, height, weight). Consider a locus with two alleles, denoted A and a. There are three possible genotypes an individual can possess at this locus: AA, Aa (equivalent to aA), and aa. Suppose the phenotype or trait values associated with individuals possessing the AA genotypes, the heterozygote genotypes Aa, and the homozygote genotypes aa are distributed $N(\mu_1, \sigma_1^2)$, $N(\mu_2, \sigma_2^2)$, and $N(\mu_3, \sigma_3^2)$, respectively. The proportions of these genotypes in the population at large are denoted by p_{AA}, p_{Aa}, and p_{aa}.

If y_j denotes the phenotype of interest, then it has the normal mixture distribution given by

$$f(y_j) = \sum_{i=1}^{3} \pi_i \, \phi(y_j;\, \mu_i,\, \sigma_i^2), \tag{3.17}$$

where $\pi_1 = p_{AA}$, $\pi_2 = p_{Aa}$, and $\pi_3 = p_{aa}$. The genotype frequencies, π_1, π_2, and π_3, are frequently written in terms of the allele frequencies p_A and p_a, using the Hardy–Weinberg equilibrium (HWE) structure,

$$\pi_1 = p_A^2, \pi_2 = 2p_A p_a, \pi_3 = p_a^2. \tag{3.18}$$

Under the additive model, we have $\mu_3 - \mu_2 = \mu_2 - \mu_1$.

An alternative genetics model to the model above is the simple dominance model in which genotypes AA and Aa have the same phenotype with frequency $\pi_1 = p_A^2 + 2p_A p_a$. This leads to a two-component normal mixture model with components corresponding to the phenotypes AA and Aa and to aa in proportions $\pi_1 = p_A^2 + 2p_A p_a$ and $\pi_2 = p_a^2$, respectively.

A recent focus of mixture models in genetics is the construction of gene marker maps for percentage testing, disease resistance diagnosis, and *quantitative trait loci* (QTL) detection; see, for example, Doerge, Zeng, and Weir (1997), Kao and Zeng (1997), and the references therein.

3.6.2 Example 3.1: PTC Sensitivity Data

We report in Table 3.1 the results of Jones and McLachlan (1991), who fitted a mixture of three normal components to data on phenylthiocarbamide (PTC) sensitivity for three groups of people. These data sets had been considered by Kalmus and Maynard Smith (1965). They are in interval (binned) form, as PTC sensitivity was recorded as falling between the second highest and highest detectable dilutions, equally spaced on the $\log_2$ scale. Jones and McLachlan (1991) fitted normal mixture models to these data, using the methodology of McLachlan and Jones (1988) for binned data as to be discussed in Chapter 9.

Of interest is the likelihood ratio test of the Hardy–Weinberg equilibrium (HWE) structure (3.18) versus unconstrained mixing proportions (apart from requiring that they sum to one). It can be seen from Table 3.1 that it is unlikely that the HWE structure of the mixing proportions would be rejected. Indeed, $-2 \log \lambda$ for this test is practically zero. Jones and McLachlan (1991) suggested bootstrapping this statistic if a P-value were required.

Table 3.1 Fit of Mixture Model to Three Data Sets (Standard Errors in Parentheses)

Parameter	Data Set 1	Data Set 2	Data Set 3
p_A	0.572 (.027)	0.626 (.025)	0.520 (.026)
μ_1	2.49 (.15)	1.62 (.14)	1.49 (.09)
μ_2	9.09 (.18)	8.09 (.15)	7.47 (.47)
μ_3	10.37 (.28)	8.63 (.50)	9.08 (.08)
σ_1^2	1.34 (.29)	1.44 (.28)	0.34 (.09)
σ_2^2	2.07 (.39)	1.19 (.22)	6.23 (2.06)
σ_3^2	0.57 (.33)	0.10 (.18)	0.48 (.10)
Test statistic:			
$-2 \log \lambda(\sigma_2^2 = \sigma_3^2)$	3.60	6.87	58.36
$-2 \log \lambda$ (HWE)	0.00	3.76	1.06

Source: Adapted from Jones and McLachlan (1991).

Another test of interest was $\sigma_2^2 = \sigma_3^2$, as assumed by Kalmus and Maynard Smith (1965) in their analysis. Jones and McLachlan (1991) noted from their estimates of these two variances relative to their standard errors (based on the inverse of the empirical information matrix) that this assumption did not appear to be justified. Furthermore, the value of $-2 \log \lambda$ for $\sigma_2^2 = \sigma_3^2$ versus $\sigma_2^2 \neq \sigma_3^2$ is extremely large for the third data set (given that its asymptotic null distribution is chi-squared with one degree of freedom).

3.6.3 Example 3.2: Screening for Hemochromatosis

We consider the case study of McLaren et al. (1998) on the screening for hemochromatosis. There is incontrovertible evidence that early diagnosis of hereditary hemochromatosis prevents virtually all manifestations of the disease and results in normal life expectancy. In contrast, when unrecognized and untreated, the disease leads to cirrhosis, hepatocellular carcinoma, and other lethal complications (McLaren et al., 1998). The identification of heterozygotes for hereditary hemochromatosis is important too as, if undetected, they may go on to develop an iron load sufficient to cause overt organ damage. Also, the identification of a heterozygote provides the opportunity to conduct studies of family members in order to identify individuals who are homozygous for the disease. Thus early identification of heterozygotes and homozygotes for hemochromatosis is an important clinical challenge. An elevated level of transferrin saturation remains the most useful noninvasive screening test for affected individuals, but there is some debate as to the appropriate screening level.

Studies have suggested that mean transferrin saturation values for heterozygotes are higher than among unaffected subjects, but lower than homozygotes. Since the distribution of transferrin saturation is known to be well approximated by a single normal distribution in unaffected subjects, the physiologic models used in the study of McLaren et al. (1998) were a single normal component and a mixture of two normal components. Using this approach, they confirmed that hemochromatosis heterozygotes form a distinct subpopulation with respect to transferrin saturation, with mean transferrin saturation greater than that for unaffected individuals. This can be seen from Figure 3.1, which gives the plots of the densities of the mixture of two (univariate) normal heteroscedastic components fitted to some transferrin values on asymptomatic Australians (stratified by sex), which were known to be either unaffected or heterozygotes with respect to hereditary hemochromatosis. The estimated component means are given in Table 3.2, where one component corresponds to unaffected individuals and the other to heterozygotes. This table also contains the means from a classified sample on 485 subjects known to be homozygous or heterozygous for hemochromatosis. With a fasting transferrin saturation of $\geq$ 45%, virtually all affected homozygous subjects would be identified in the second sample. From the fitted mixture distributions, it was found for this threshold that none of the postulated unaffected individuals would undergo further testing unnecessarily and 19.1% of the heterozygotes would be identified.

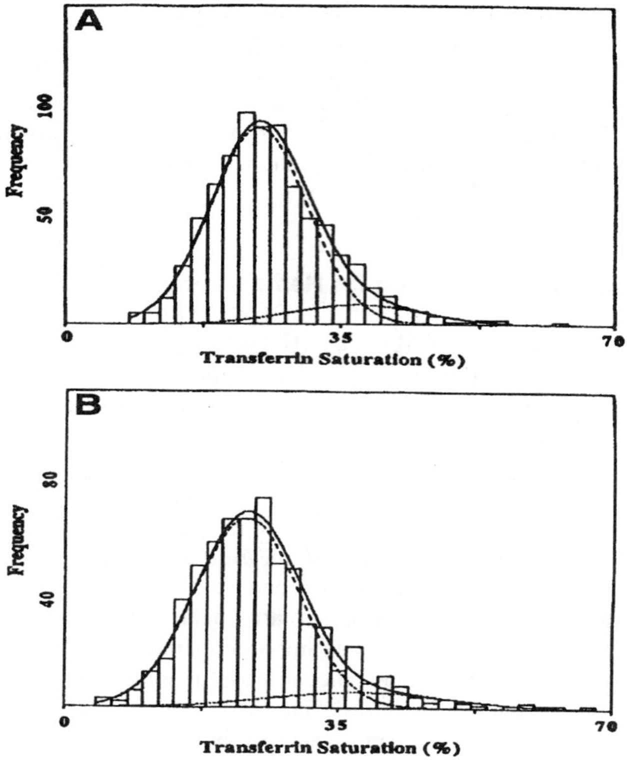

Fig. 3.1 Plots of the densities of the mixture of two (univariate) normal heteroscedastic components fitted to some transferrin values on asymptomatic Australians (stratified by sex: (A) 796 men and (B) 669 women), which were known to be either unaffected or heterozygotes with respect to hereditary hemochromatosis. From McLaren et al. (1998).

Table 3.2 Transferrin Saturation Results Expressed as Mean Percentage ± SD

	Asymptomatic Individuals		Individuals Identified by Pedigree Analysis	
Sex	Postulated Unaffected	Postulated Heterozygotes	Known Heterozygotes	Known Homozygotes
Male	24.1 ± 6.0	37.3 ± 7.7	37.1 ± 17.0	82.7 ± 14.4
Female	22.5 ± 6.4	37.6 ± 10.4	32.5 ± 15.3	75.3 ± 19.3

Source: McLaren et al. (1998).

3.6.4 Example 3.3: Diagnostic Criteria for Diabetes

In this example, a normal mixture model is used to estimate the prevalence of diabetes and to assess the sensitivity and specificity of a diagnostic test for diabetes as a function of some covariates. This example also serves as an introductory example to the case where there is direct inclusion of covariates in the mixture model by allowing the component densities and also the mixing proportions to depend on some available covariates. In Chapter 5 we shall consider the problem of fitting mixtures of GLMs in proportions that may also depend on covariates.

The data analyzed by Thompson, Smith, and Boyle (1998) consisted of $n = 919$ observations on each of $p = 6$ variables. The response variable y_j is the 2-h plasma glucose measurement after an oral glucose load. The vector $\boldsymbol{x}_j$ of associated covariates contained five variables, age (36–65 years), sex, residence (urban or rural), physical activity (active or sedentary), and obesity (no or yes). These data were from a population-based survey conducted in Cairo and surrounding rural villages between 1991 and 1994. Thompson et al. (1998) fitted a two-component normal mixture model to the logarithm of y_j initially with all five covariates in both component densities and in the logistic model for the mixing proportions π_i. Consideration was given to two-way interactions and squared terms in age, using the BIC criterion in a forward selection fitting strategy. The fit for their final model is displayed in Table 3.3, where the variable age has been coded to be $(age - 50)/10$. Table 3.3 also contains standard errors based on 2000 bootstrap replications. In this final fit, no covariates were included in the second component density. The logistic regression model for the mixing proportion π_1 included three covariates, age, age^2, and urban. The covariates included in the first component were urban, obese, and two variables (corresponding to active females and active males) for an effect of sex nested within physical activity.

Table 3.3 Generalized Linear Finite Mixture Model

Distribution	Parameter	Estimate	SE	Z-Value	P-Value
Mixing	Intercept	1.142	0.1939	5.89	
	Age	−0.511	0.0993	−5.15	<0.001
	(Age)^2	0.370	0.1215	3.04	0.002
	Urban	−1.368	0.1914	−7.15	<0.001
First component	Intercept	2.024	0.0109	185.16	
	Urban	−0.052	0.0122	−4.24	<0.001
	Obese	0.071	0.0123	5.80	<0.001
	Active female	0.054	0.0149	3.66	<0.001
	Active male	−0.050	0.0211	−2.35	0.019
	σ_1	0.117	0.0048	24.59	
Second component	Intercept	2.510	0.0085	296.34	
	σ_2	0.132	0.0074	17.67	

Source: Adapted from Thompson et al. (1998).

As there is no definitive standard for diagnosing diabetes, the current World Health Organization (WHO) standard was developed via consensus. The WHO cutoff point for the diagnosis of diabetes is 200 mg dl^{-1}. Thompson et al. (1998) used the fitted mixture model to investigate the choice of a cutoff point C above which an individual is diagnosed as having diabetes. It was taken to be the point that minimizes the overall misclassification rate as determined from the fitted mixture model. In Table 3.4 we report the results so obtained by Thompson et al. (1998) for the cutoff point as a function of age and place of residence with the other covariates replaced by their mean values.

Table 3.4 Cutoff Points, Sensitivities, Specificities, and Misclassification Rates by Residence and Age

Cutoff Point Criteria	Residence	Age (years)	Cutoff Point (mg dl^{-1})	Sensitivity (%)	Specificity (%)	Error Rate (%)
Min. Error	Rural	40	223	89.1 (2.63)	99.2 (0.80)	2.0 (0.62)
		60	206	93.3 (2.05)	98.3 (0.66)	3.1 (0.58)
	Urban	40	192	95.9 (1.55)	98.4 (0.66)	2.5 (0.48)
		60	178	97.6 (1.08)	96.8 (1.06)	2.7 (0.53)
WHO	Rural	40	200	94.4 (1.83)	97.8 (0.80)	2.6 (0.62)
		60	200	94.4 (1.83)	97.8 (0.80)	3.1 (0.56)
	Urban	40	200	94.4 (1.83)	98.9 (0.50)	2.6 (0.59)
		60	200	94.4 (1.83)	98.9 (0.50)	3.7 (1.07)

Source: Thompson et al. (1998).

3.7 EXAMPLES OF MULTIVARIATE NORMAL MIXTURES

3.7.1 Example 3.4: Crab Data

To illustrate the use of the normal mixture model in the clustering of some multivariate data, we consider in this example the clustering of the crab data set of Campbell and Mahon (1974) on the genus *Leptograpsus*, which has been analyzed further in Ripley (1996) and McLachlan and Peel (1998a). Attention is focused on the sample of $n = 100$ blue crabs, there being $n_1 = 50$ males and $n_2 = 50$ females corresponding to groups G_1 and G_2 respectively. Each specimen has measurements (in mm) on the width of the frontal lip *FL*, the rear width *RW*, the length along the midline *CL* and the maximum width *CW* of the carapace, and the body depth *BD*. In Figure 3.2 we give the scatter plot of the second and third variates with their group of origin noted. Hawkins' (1981) simultaneous test for multivariate normality and equal covariance matrices (homoscedasticity) suggests that it is reasonable to assume that the group-conditional distributions are normal with a common covariance matrix. Consistent with this, it was found that the sample linear discriminant function (formed using the known classification of the data) misallocates only two more observations than the quadratic version formed without the restriction of equal group-covariance matrices. There are 6 misallocations overall with the latter (4 from G_1 and 2 from G_2).

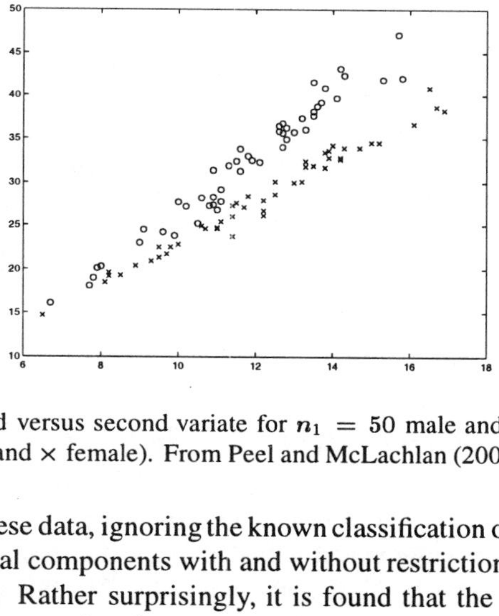

Fig. 3.2 Plot of third versus second variate for $n_1 = 50$ male and $n_2 = 50$ female blue crabs (o denotes male and × female). From Peel and McLachlan (2000).

We now cluster these data, ignoring the known classification of the data, by fitting a mixture of two normal components with and without restrictions on the component-covariance matrices. Rather surprisingly, it is found that the assumption of equal component-covariance matrices has a marked impact on the implied clustering of the data. For without any restrictions on the component-covariance matrices, the normal mixture model-based clustering results in one cluster containing 39 observations from G_1 and another containing all 50 observations from G_2, along with the remaining 11 observations from G_1. Identifying the larger-sized cluster with G_2 (and the smaller-sized cluster with G_1), the overall misallocation rate is 11. However, if we impose the restriction of equal component-covariance matrices, then a further 8 observations from G_1 are misallocated, as these 8 observations are moved from the smaller-sized cluster to the larger-sized cluster. This is an interesting example as it is a situation where, although it is reasonable to assume that the true group-structure is homoscedastic, the assumption of homoscedasticity in fitting the normal mixture model leads to a much inferior clustering of the data. This is brought about by there being several observations from G_1 (male crabs) that are near the boundary of the feature space on which the group-conditional densities weighted by their prior probabilities are apparently equal, and when knowledge of their true classification is not used, these points have an adverse effect on the group-density estimates (and the consequent clustering) produced under the assumption of homoscedasticity.

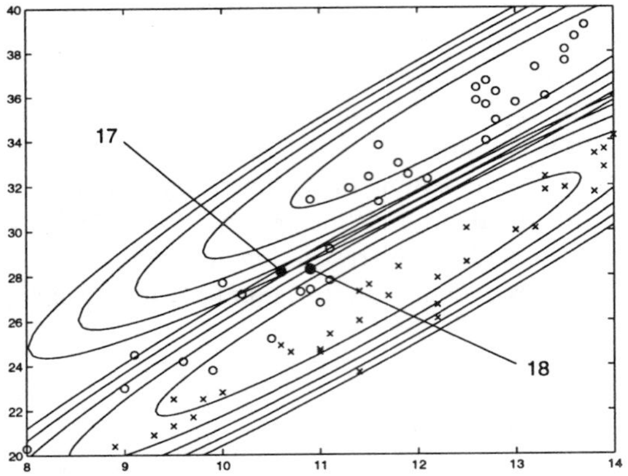

Fig. 3.3 Contours of the fitted component densities on the second and third variates in the region of points 17 and 18 for the $n_1 = 50$ male and $n_2 = 50$ female blue crabs (o denotes male and × female). From Peel and McLachlan (2000).

To illustrate just how sensitive the process is to the location of some of these observations, we considered the implied clustering of the 17th and 18th observations from G_1,

$$y_{17} = (13.1, 10.6, 28.2, 32.3, 11.0)^T$$

and

$$y_{18} = (13.1, 10.9, 28.3, 32.4, 11.2)^T.$$

Although these two points are very close in terms of Euclidean and also Mahalanobis distance, their estimated posterior probabilities of belonging to G_1 are quite different under the normal mixture model with equal covariance matrices, being equal to 0.564 and 0.170, respectively. To see why there should be this somewhat unexpected difference between these estimated posterior probabilities, we have plotted in Figure 3.3 the contours of the estimated component densities for the 2nd and 3rd feature variables. It can be seen that the 18th observation lies in the feature space where the contours of $f_2(y_j;\, \hat{\Psi})$ are very steep, so that even slight changes in the values of the feature variables can result in relatively large changes in the values of the component densities. Without the restriction of equality on the component-covariance matrices, the estimates of these two posterior probabilities are very similar, being 1.000 and 0.996, and now observation 18 is correctly allocated to G_1.

3.7.2 Example 3.5: Hemophilia Data

We consider here the hemophilia data set, as analyzed by Basford and McLachlan (1985d), McLachlan and Basford (1988) and, more recently, Celeux et al. (1996). It consists of only two bivariate components, and its analysis illustrates the caution that needs to be exercised in practice with the fitting of mixture models. The data set was originally taken from Habbema, Hermans, and van den Broek (1974), where, in the context of genetic counseling, the question of discriminating between normal women and hemophilia A carriers was considered on the basis of two feature variates, $(\boldsymbol{y}_j)_1 = \log_{10}$(AHF activity) and $(\boldsymbol{y}_j)_2 = \log_{10}$(AHF-like antigen). Both variates are scaled up by 100 here to simplify the presentation of the results. We let G_1 and G_2 be the groups in the population corresponding to the noncarriers and the carriers, respectively. The classification of the $n = 75$ data points $\boldsymbol{y}_j$ are known, with there being $n_1 = 30$ and $n_2 = 45$ observations from G_1 and G_2, respectively. We let $\boldsymbol{y}_j$ $(j = 1, \ldots, 30)$ denote the $n_1 = 30$ observations from G_1 and $\boldsymbol{y}_j$ $(j = 31, \ldots, 75)$ denote the $n_2 = 45$ observations from G_2. These points are plotted in Figure 3.4.

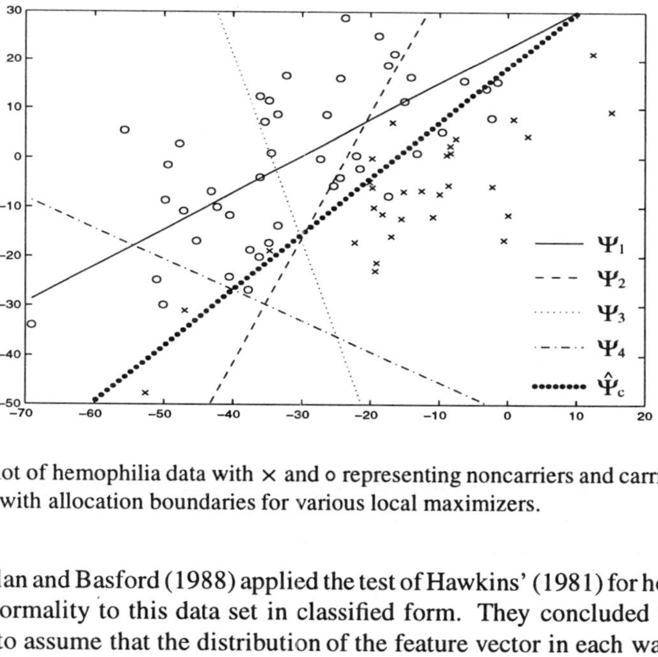

Fig. 3.4 Plot of hemophilia data with × and o representing noncarriers and carriers, respectively, along with allocation boundaries for various local maximizers.

McLachlan and Basford (1988) applied the test of Hawkins' (1981) for homoscedasticity and normality to this data set in classified form. They concluded that it was reasonable to assume that the distribution of the feature vector in each was bivariate normal. The test indicated some presence of heteroscedasticity, which an inspection of the fitted covariance matrices revealed was limited to the second variate of the feature vector $\boldsymbol{Y}_j$.

We first cluster these data, ignoring their known classification, by fitting a two-component normal mixture model with equal component-covariance matrices. Four local maxima were located, and the corresponding estimates of the parameter vector $\boldsymbol{\Psi}$ are displayed in Table 3.5. The linear allocation boundary, $\tau_i(\boldsymbol{y}_j;\, \boldsymbol{\Psi}) = 0.5$ $(i = 1, 2)$, is plotted in Figure 3.4 for the various local maximizers, along with the boundary corresponding to $\hat{\boldsymbol{\Psi}}_c$, which is the MLE of $\boldsymbol{\Psi}$ obtained by maximizing the complete-data likelihood; that is, $\boldsymbol{\Psi}_c$ was obtained using the known values of the component-label vectors of the data points.

As $\boldsymbol{\Psi}_1$ corresponds to the largest of the local maxima located, it is apparently the global maximizer and hence the MLE under homoscedasticity. It can be seen from Table 3.5 that $\boldsymbol{\Psi}_1$ gives a poor assessment of the proportions in which G_1 and G_2 are represented in the sample and, as a consequence, it gives a worse allocation of the sample than $\boldsymbol{\Psi}_2$, with a total of 25 members, all from G_2, misallocated. This compares with 3 and 13 members from G_1 and G_2, respectively, being misallocated according to the solution $\boldsymbol{\Psi}_2$.

Table 3.5 Local Maximizers of $L(\boldsymbol{\Psi})$ under Homoscedasticity

	$\boldsymbol{\Psi}_1$	$\boldsymbol{\Psi}_2$	$\boldsymbol{\Psi}_3$	$\boldsymbol{\Psi}_4$	$\hat{\boldsymbol{\Psi}}_c$
π_1	0.72	0.53	0.68	0.89	0.4
$(\boldsymbol{\mu}_1)_1$	−20.6	−12.1	−15.2	−21.2	−13.5
$(\boldsymbol{\mu}_1)_2$	−8.0	−1.9	1.2	−0.9	−7.8
$(\boldsymbol{\mu}_2)_1$	−32.1	−37.0	−42.0	−45.4	−30.8
$(\boldsymbol{\mu}_2)_2$	7.9	−5.2	−13.5	−24.7	−0.6
$(\boldsymbol{\Sigma})_{11}$	265	137	138	235	226
$(\boldsymbol{\Sigma})_{22}$	171	220	175	167	216

It is not difficult to understand why the local maximizer $\boldsymbol{\Psi}_2$ is not the MLE under the condition of homoscedasticity. This is because this root of the likelihood equation gives a probabilistic clustering in which the variation of $(\boldsymbol{Y}_j)_2$ is substantially smaller in the cluster associated with G_1 than in the other associated with G_2, which is not consistent with homoscedasticity. In particular, this root provides estimates of π_1 and π_2 much closer to the actual proportions in which G_1 and G_2 occur in the sample than the MLE under the homoscedastic model; hence the apparent reason why this root leads to a better allocation of the sample than the MLE. This root gives almost the same clustering as obtained using the MLE under the heteroscedastic model without restrictions on the component-covariance matrices. The latter misallocates one less observation from G_2.

3.8 PROPERTIES OF MLE FOR NORMAL COMPONENTS

3.8.1 Heteroscedastic Components

As noted in Section 2.5, the log likelihood for a mixture of normal distributions with unequal covariance matrices is unbounded, and so the global maximizer does not exist. Each observation $\boldsymbol{y}_j$ gives rise to a singularity on the edge of the parameter space. For instance, if we set $\mu_1 = y_1$ and let $\sigma_1^2 \rightarrow 0$ in a mixture of two univariate normal densities with means μ_1 and μ_2 and variances σ_1^2 and σ_2^2, then the likelihood will tend to infinity. This was first noticed by Kiefer and Wolfowitz (1956). However, as remarked in Section 2.2, the nonexistence of a global maximizer of the likelihood function estimate does not place a caveat on proceedings, as the essential aim of likelihood estimation is to find a sequence of roots of the likelihood equation that is consistent, and hence efficient if the usual regularity conditions hold. However, with mixture models the likelihood equation will generally have multiple roots. Thus, even if it were known that there exists a sequence of roots of the likelihood equation with the desired asymptotic properties, there is the problem of identifying this sequence, as discussed in Section 2.2 for ML estimation in general.

For mixtures of normal distributions, at least in the univariate case, there is a sequence of roots corresponding to local maxima that is consistent and asymptotically normal and efficient. The last two results that the asymptotic distribution of this sequence is normal with covariance matrix equal to the inverse of the expected Fisher information matrix $\mathcal{I}(\boldsymbol{\Psi})$ require the additional conditions that

$$\pi_i \neq 0 \qquad (i = 1, \ldots, g) \tag{3.19}$$

and

$$(\mu_h, \sigma_h^2) \neq (\mu_i, \sigma_i^2) \qquad (h \neq i = 1, \ldots, g), \tag{3.20}$$

on writing Σ_i as σ_i^2 in the univariate case. In the subsequent discussion of these two results, these conditions are implicitly assumed to hold.

In the case where (3.19) and (3.20) or their multivariate analogues do not both hold, the true value $\boldsymbol{\Psi}_t$ of $\boldsymbol{\Psi}$ will lie in a nonidentifiable subset $\boldsymbol{\Omega}_t$ of the parameter space $\boldsymbol{\Omega}$ for which $f(\boldsymbol{y}_j;\, \boldsymbol{\Psi}) = f(\boldsymbol{y}_j;\, \boldsymbol{\Psi}_t)$ for almost all $\boldsymbol{y}_j$ in $\mathbb{R}^p$, and it may also lie on the boundary. In this case, Feng and McCulloch (1996) showed that the MLE of $\boldsymbol{\Psi}$ converges to the nonidentifiable subset $\boldsymbol{\Omega}_t$ to which $\boldsymbol{\Psi}_t$ belongs, as explained in Section 2.5.

Kiefer (1978) verified for a mixture of univariate normal distributions in the more general case of the switching regression model that there is a sequence of roots of the likelihood equation which is consistent and asymptotically efficient and normally distributed. With probability tending to one, these roots correspond to local maxima in the interior of the parameter space; see also Peters and Walker (1978). Kiefer's (1978) verification was for $g = 2$ components, but his result will hold for $g > 2$, as noted in Hathaway (1985) who reported some univariate results on the maximization of $L(\boldsymbol{\Psi})$ over the constrained parameter space

$$\boldsymbol{\Omega}_C = \{\boldsymbol{\Psi} \in \boldsymbol{\Omega} : \; \sigma_h^2/\sigma_i^2 \geq C > 0, \quad 1 \leq h \neq i \leq g\}, \tag{3.21}$$

where $\boldsymbol{\Omega}$ denotes the unconstrained parameter space. For any $C \in (0,\, 1]$, Hathaway (1985) showed that the global maximizer $\hat{\boldsymbol{\Psi}}_C$ of $L(\boldsymbol{\Psi})$ over $\boldsymbol{\Omega}_C$ exists (assuming that the sample contains at least $g + 1$ distinct points) and that, provided that the true value of $\boldsymbol{\Psi}$ belongs to $\boldsymbol{\Omega}_C$, $\hat{\boldsymbol{\Psi}}_C$ is then strongly consistent for $\boldsymbol{\Psi}$. The constraint (3.21) is imposed to avoid the singularities in $L(\boldsymbol{\Psi})$ that occur when the mean of one component, say μ_i, is set equal to any observed value and σ_i^2 tends to zero.

As pointed out by Hathaway (1985), the global constrained maximizer shares all the good asymptotic properties of the consistent local maximizer of the likelihood function $L(\boldsymbol{\Psi})$. Thus the constrained formulation is statistically well posed. Problems with singularities do not exist, and those associated with spurious local maximizers are at least lessened. The only problem in practice is to choose a value of C for which the true value of $\boldsymbol{\Psi}$ satisfies (3.21).

For mixtures of univariate normal distributions, Hathaway (1983, 1986b) and Bezdek, Hathaway, and Huggins (1985) have investigated a constrained version of the EM algorithm which incorporates the constraint (3.21). Concerning mixtures of multivariate normal distributions, Redner and Walker (1984) have given the necessary regularity conditions for there to exist a consistent solution of the likelihood equation (see Section 2.5), but to the authors' knowledge, these conditions have not been verified, although they should hold as they are fairly weak. Also, Hathaway (1985) has indicated how a constrained (global) maximum likelihood formulation can be given in the multivariate case by constraining all the eigenvalues of $\boldsymbol{\Sigma}_h \boldsymbol{\Sigma}_i^{-1}$ $(1 \leq h \neq i \leq g)$ to be greater than or equal to some minimum value $C > 0$ (satisfied by the true value of $\boldsymbol{\Psi}$).

As noted by Hathaway (1985), several questions remain concerning solutions of the likelihood for normal components with unrestricted component-covariance matrices. For example, is it possible to let C decrease to zero as the sample size n increases to infinity while maintaining consistency? If the answer is yes, then at what rate can C be decreased to zero? For fixed C, do all the spurious maximizers of $L(\boldsymbol{\Psi})$ in $\boldsymbol{\Omega}_C$ disappear as n tends to infinity? According to Hathaway (1985), empirical studies indicate that the answer to the last question could be yes.

3.8.2 Homoscedastic Components

For normal components with covariance matrices $\boldsymbol{\Sigma}_i$ restricted to being equal to $\boldsymbol{\Sigma}$,

$$\boldsymbol{Y} \sim N(\boldsymbol{\mu}_i, \boldsymbol{\Sigma}) \text{ with prob. } \pi_i \quad (i = 1, \ldots, g), \tag{3.22}$$

the MLE of $\boldsymbol{\Psi} = (\pi_1, \ldots, \pi_{g-1}, \boldsymbol{\xi}^T)^T$ exists as the global maximizer and is strongly consistent; $\boldsymbol{\xi}$ denotes now the elements of the $\boldsymbol{\mu}_i$ and the distinct elements of the common component-covariance matrix $\boldsymbol{\Sigma}$. As remarked in Section 2.5, Redner (1981) noted that under the conditions of Wald (1949), the MLE is strongly consistent for finite mixture models where attention is restricted to a compact subset of the parameter space. But the strong consistency of the MLE under (3.22), even for the unrestricted (noncompact) parameter space, follows from the device of Kiefer and Wolfowitz (1956). Their conditions require that the mixture density converges, though not to zero, as $\boldsymbol{\Psi}$ tends to the boundary of the parameter space. To overcome the unboundedness of the density here, it is necessary to apply the device in Section 6 of Kiefer and Wolfowitz (1956), as discussed in Perlman (1972). Using this device, one works with the joint density of $g + p$ observations instead of the density corresponding to a single observation. This device was employed by Hathaway (1985) in his proof of the strong consistency of his constrained MLE for univariate normal mixtures.

The choice of root of the likelihood equation is therefore straightforward under homoscedasticity, in the sense that the MLE exists as the global maximizer, which is known to be consistent. For normal component distributions that are widely separated or for small values of p, there should be little difficulty in locating the global maximizer, but as p increases, the problem becomes more difficult.

3.9 OPTIONS

3.9.1 Choice of Local Maximizer

As noted in the previous section, the choice of root of the likelihood equation in the case of homoscedastic components is straightforward in the sense that the MLE exists as the global maximizer of the likelihood function. The situation is less straightforward in the case of heteroscedastic components as the likelihood function is unbounded. But assuming the univariate result of Hathaway (1985) extends to the case of multivariate normal components, then the constrained global maximizer is consistent provided the true value of the parameter vector $\boldsymbol{\Psi}$ belongs to the parameter space constrained so that the component generalized variances are not too disparate; for example,

$$|\boldsymbol{\Sigma}_h| / |\boldsymbol{\Sigma}_i| \geq C > 0 \quad (1 \leq h \neq i \leq g). \tag{3.23}$$

If we wish to proceed in the heteroscedastic case by the prior imposition of a constraint of the form (3.23), then there is the problem of how small the lower bound C must be to ensure that the constrained parameter space contains the true value of the parameter vector $\boldsymbol{\Psi}$. We shall see shortly that with this approach there is the possibility that some legitimate solutions or solutions worthy of consideration from, say, a data-analytic aspect, may be missed.

Therefore to avoid having to specify a value for C beforehand, we prefer where possible to fit the normal mixture without any constraints on the component covariances $\boldsymbol{\Sigma}_i$. It thus means we have to be careful to check that the EM algorithm has actually converged and is not on its way to a singularity which exists since the likelihood is unbounded for unequal component-covariance matrices. Even if we can be sure that the EM algorithm has converged to a local maximizer, we have to be sure that it is not a spurious solution that deserves to be discarded. After these checks, we can take the MLE of $\boldsymbol{\Psi}$ to be the root of the likelihood equation corresponding to the largest of the remaining local maxima located.

We shall look at this issue of spurious solutions more closely in Section 3.10. Broadly speaking, we tend to be suspicious of any solution for which the component-covariance matrices are dissimilar. For example, if the components of the mixture model are supposed to be representing some externally existing groups, then it is not unreasonable to expect that the feature variables would not have greatly differing component-covariance matrices.

3.9.2 Choice of Model for Component-Covariance Matrices

A normal mixture model without restrictions on the component-covariance matrices may be viewed as too general for many situations in practice. At the same time, though, we are reluctant to impose the homoscedastic condition (3.22), as we have noted in our analyses that the imposition of the constraint of equal component-covariance matrices can have a marked effect on the resulting estimates and the implied clustering. This was illustrated in Examples 3.4 and 3.5 in Section 3.7.

In a discriminant analysis, the observed data (the training data) are of known origin. It has been found that the normal homoscedastic solution-based classifier (equivalent to Fisher's linear discriminant function) is fairly robust against mild departures from homoscedasticity (McLachlan, 1992, Chapter 5). However, it appears that the normal mixture-based solution formed without the knowledge of the component labels can have its performance reduced with the imposition of homoscedasticity even though it may be a reasonable restriction. A possible explanation is that with the fitting of homoscedastic components, the estimates of the component means and, in particular, the mixing proportions, suffer at the expense of the need for the estimates of the component-covariance matrices to be approximately equal. Hence the flexibility afforded by a general model for the component-covariance matrices is preferable, provided that it does not take the solution too far away from what we would expect to be a realistic situation for the comparability of the component-covariance matrices.

Another way of proceeding is to adopt some model for the component-covariance matrices that is intermediate between homoscedasticity and the general model, as in the approach of Banfield and Raftery (1993) to be introduced in Section 3.12. A further approach, particularly when the data are of very high dimension relative to the sample size n, is to adopt a mixture of factor analyzers model. This model is to be covered in Chapter 8. And there is also the Bayesian approach to be considered in the next chapter.

Another factor in the consideration of restrictions on the component-covariance matrices is the purpose for which the mixture model is being fitted. In the above, it has been implicitly assumed that the mixture model is being used detect or represent the group structure, if any, in the data. But another purpose may be to provide an adequate estimate of the density of the data. In this case, we may wish to avoid the issues associated with the use of heteroscedastic components and to restrict the component-covariance matrices to be equal. An example concerns the use of mixture models to represent individual group densities for use in a discriminant analysis.

3.9.3 Starting Points

In Section 2.12 we have discussed the possible choices for the starting value $\boldsymbol{\Psi}^{(0)}$ for $\boldsymbol{\Psi}$ in the iterative fitting of a mixture model via the EM algorithm. There we mentioned that in addition to random starts provided by random partitions of the data into g groups corresponding to the g components of the mixture model, we can use (2.40) to generate random starting values for the component means $\boldsymbol{\mu}_i$. The EMMIX program of McLachlan et al. (1999) can be used to fit a normal mixture model from a user-specified starting value or from starting values provided automatically according to the user's specifications. The latter include random starts obtained by either random partitions or randomly generated initial component means according to (2.40). The user also can specify a variety of other starting points obtained by clustering procedures such as k-means or commonly used hierarchical ones.

3.10 SPURIOUS LOCAL MAXIMIZERS

3.10.1 Introduction

In practice, consideration has to be given to the problem of relatively large local maxima that occur as a consequence of a fitted component having a very small (but nonzero) variance for univariate data or generalized variance (the determinant of the covariance matrix) for multivariate data. Such a component corresponds to a cluster containing a few data points either relatively close together or almost lying in a lower-dimensional subspace in the case of multivariate data. There is thus a need to monitor the relative size of the fitted mixing proportions and of the component variances for univariate observations, or of the generalized component variances for multivariate data, in an attempt to identify these spurious local maximizers. The possibility that for a given starting point the EM algorithm may converge to a spurious local maximizer or may not converge at all is not a failing of this algorithm. Rather it is a consequence of the properties of the likelihood function for the normal mixture model with unrestricted component-covariance matrices in the case of unbinned data.

In the absence of any information on the normal components in the mixture model to be fitted to some independent data, a common strategy is to run the EM algorithm from a variety of starting values as discussed in the previous section. The intent is to choose as the MLE of the parameter vector $\boldsymbol{\Psi}$ the local maximizer corresponding to the largest of the local maxima located. As noted in the previous sections, some care needs to be exercised in the case of unrestricted component-covariance matrices as the likelihood is then unbounded. The detection of these singularities is not a problem in practice, as they can be detected by one or more of the component-covariance matrices becoming singular in the EM iterative process.

However, there often exist other solutions which may be regarded as spurious, lying very close to the edge of the parameter space, but with component-covariance matrices that are not actually singular, although they may be close to singular for some components. These solutions often have a high likelihood, but are of little practical use or real-world interpretation. It often seems in these cases that the model is fitting a small localized random pattern in the data rather then any underlying group structure. These spurious solutions frequently have very few points in one cluster, which has little variation in one or more of the cluster's axes (small eigenvalues of the component-covariance matrix) compared to the other clusters.

One situation where an apparent spurious solution would be of practical interest is where one (or more) of the fitted components correspond to a small number of points that are distant from the rest of the sample; see Aitkin and Tunnicliffe Wilson (1980) and Jorgensen (1990).

Before we discuss how the presence of these spurious solutions can be detected in the automatic fitting of normal mixture models with unrestricted component-covariance matrices, we shall give some examples.

3.10.2 Example 3.6: Synthetic Data Set 2

To demonstrate the occurrence of what is meant by a spurious local maximizer due to a few points lying close together, we generated a random sample of size $n = 50$ points from a mixture of two univariate normal densities in equal proportions with means 0 and 2, and common unit variance. We fitted a mixture of $g = 2$ univariate normal densities with unspecified mixing proportions π_1 and π_2, means μ_1 and μ_2, and unrestricted variances, σ_1^2 and σ_2^2,

$$f(y_j;\, \boldsymbol{\Psi}) = \sum_{i=1}^{2} \pi_i \phi(y_j;\, \mu_i,\, \sigma_i^2). \tag{3.24}$$

Two local maximizers, $\boldsymbol{\Psi}_1$ and $\boldsymbol{\Psi}_2$, were located, with the latter being found using the Stochastic EM algorithm as described in Section 2.13. The results are summarized in Table 3.6.

Table 3.6 Local Maximizers for Synthetic Data Set 2

Local Maximizer	$\log L$	π_1	μ_1	μ_2	σ_1^2	σ_2^2	σ_1^2/σ_2^2
$\boldsymbol{\Psi}_1$	-170.56	0.157	-0.764	1.359	0.752	1.602	0.4696
$\boldsymbol{\Psi}_2$	-165.94	0.020	-2.161	1.088	5.22×10^{-9}	2.626	1.97×10^{-9}
$\boldsymbol{\Psi}_3$ (binned)	-187.63	0.205	-0.598	1.400	0.399	1.612	0.2475

The likelihood is greater at $\boldsymbol{\Psi} = \boldsymbol{\Psi}_2$ than at $\boldsymbol{\Psi} = \boldsymbol{\Psi}_1$, but the latter is obviously a spurious solution. This is because its first fitted component corresponds to a cluster of the two points closest to each other in the generated sample, with the second fitted component corresponding to the rest of the sample; see Figure 3.5 in which we have fitted the normal mixture density for this fit $\boldsymbol{\Psi}_2$. The corresponding fit for $\boldsymbol{\Psi}_1$, which we would conclude to be the MLE of $\boldsymbol{\Psi}$, is plotted in Figure 3.6.

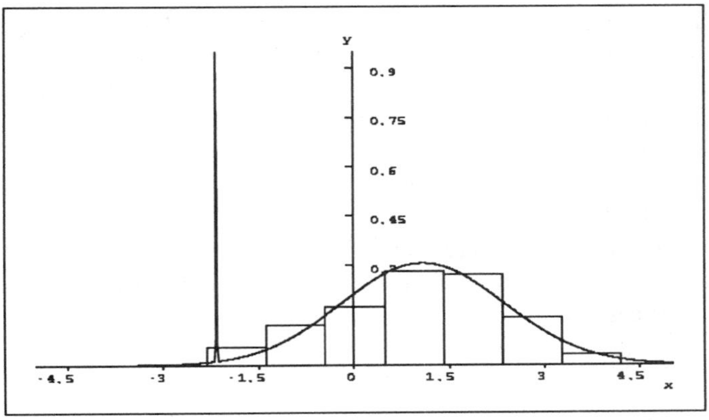

Fig. 3.5 Histogram of Synthetic Data Set 2 for fit $\boldsymbol{\Psi}_2$ of the normal mixture density.

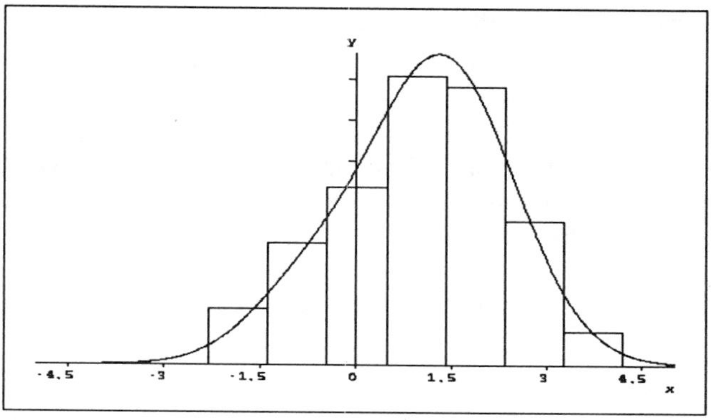

Fig. 3.6 Histogram of Synthetic Data Set 2 for fit $\boldsymbol{\Psi}_1$ of the normal mixture density.

The occurrence of spurious solutions of this nature are due to the fact that the data are continuous. If we were to bin the data into intervals and then fit the mixture model to the binned data, they do not occur. To illustrate this, we have reported in Table 3.6 the summary of the fit $\boldsymbol{\Psi}_3$ after first binning the data into seven intervals of equal width. The plot of the fitted normal mixture for $\boldsymbol{\Psi}_3$ is displayed in Figure 3.7. It can be seen that it is fairly similar to that for $\boldsymbol{\Psi}_1$, confirming that $\boldsymbol{\Psi}_1$ is the MLE for the data in unbinned form. In this example, the tell-tale sign that $\boldsymbol{\Psi}_2$ is a spurious local maximizer is the very small value for the mixing proportion of the first component and the huge imbalance between the estimated component variances.

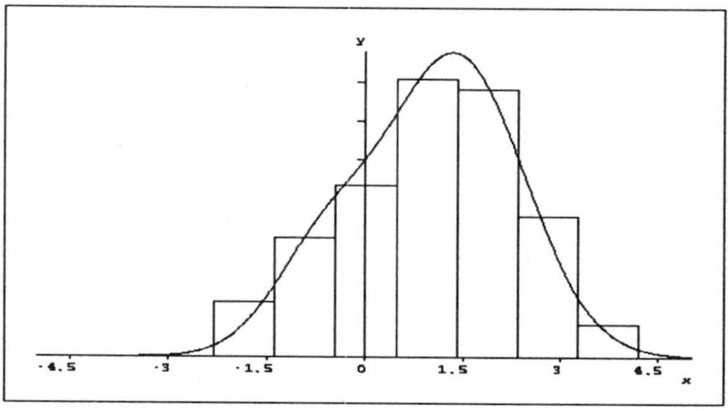

Fig. 3.7 Histogram of Synthetic Data Set 2 for fit $\boldsymbol{\Psi}_3$ of the normal mixture density.

3.10.3 Example 3.7: Synthetic Data Set 3

This data set consists of 100 observations randomly generated from a mixture of two bivariate normal components in equal proportions. In Figure 3.8 we give the clustering implied by what we concluded to be the ML solution under a bivariate normal mixture model without any restrictions on the component-covariance matrices. The log likelihood at this solution was equal to -336.021. In Figure 3.9 we give the implied clustering from another solution found corresponding to a higher value (-333.321) of the likelihood function. This solution would qualify as being spurious, as the points in the cluster corresponding to one of the two components are almost collinear.

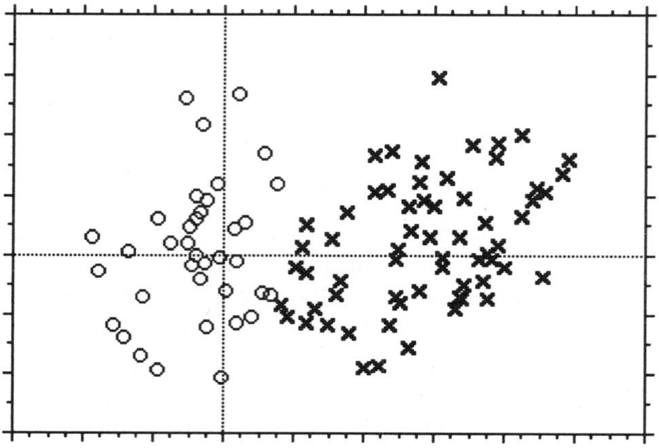

Fig. 3.8 Plot of two-group clustering of Synthetic Data Set 3 implied by MLE.

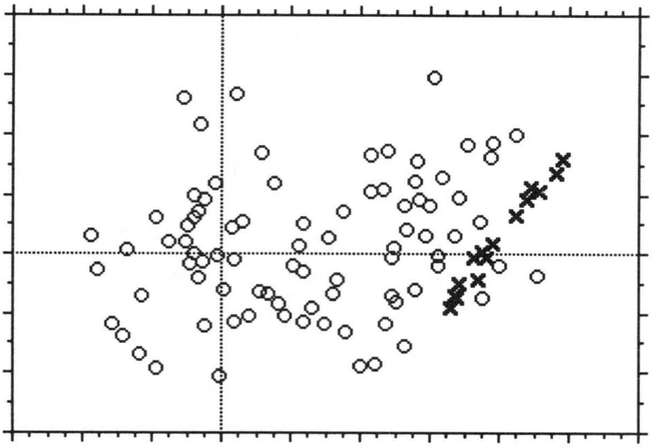

Fig. 3.9 Plot of two-group clustering of Synthetic Data Set 3 implied by solution corresponding to a higher local maximum than MLE.

3.10.4 Example 3.8: Modeling Hemophilia Data under Heteroscedasticity

We return here to the hemophilia data set introduced in Example 3.5, but now we fit a two-component normal mixture model with unrestricted component-covariance matrices. Five local maximizers were located. The first three were obtained in the analysis of McLachlan and Basford (1988), while the fourth and fifth were obtained in the analyses of this data set by Celeux et al. (1996) and Peel (1998), respectively. The values of the component-generalized variances for the five solutions are listed in Table 3.7, along with the values of the eigenvalues of the component-covariance matrices. It is clear from the imbalances between the component-generalized variances and the eigenvalues for the fourth and fifth solutions that they are spurious. This can be confirmed from a plot of the asymptotic 95% confidence ellipsoidal region for the mean of the second component in these two solutions.

Table 3.7 Local Maximizers for Hemophilia Data Set under Heteroscedasticity

Sol.	$\mid \boldsymbol{\Sigma}_1 \mid$	λ_{11}	λ_{12}	$\mid \boldsymbol{\Sigma}_2 \mid$	λ_{21}	λ_{22}	$\log L$
1	9428	51	182	28860	70	410	−613.73
2	28895	420	68	6315	204	31	−613.97
3	52962	395	134	38	13	3	−612.09
4	54336.9	402.7	134.9	16.0	0.0963	166.5	−608.262
5	54215.7	400.8	135.3	3.4	0.0237	144.1	−607.944

3.10.5 Detection of Spurious Local Maximizers

As noted in the previous section, spurious solutions typically have a small number of points in at least one cluster which has a relatively small generalized variance. Hence the ratio of the fitted generalized component variances can be a useful guide, or warning, that a spurious solution has been found. A more informative approach is to examine the actual eigenvalues of the covariance matrix in question, rather than the determinant (which is the product of the eigenvalues). The individual eigenvalues offer a much better reflection of the shape of the clusters, with each eigenvalue corresponding to the variance along the elliptical axis (eigenvector) of the cluster. In this way the user can discern between small compact clusters and long thin clusters.

There is also a need to monitor the distances between the fitted component means where there appear to be spurious local maximizers. The Euclidean distances between apparent spurious and nonspurious component means could be calculated, but may be unreliable if the feature variables are measured on disparate scales. In such cases, one may want to consider the Mahalanobis distances between the apparent spurious and nonspurious component means, using as covariance matrix the estimated covariance matrix for the relevant nonspurious component. Even then, small intercomponent mean distances need not reflect spurious clusters, as one can have a situation where two clusters have similar means but are quite different in shape due to their covariance matrices being quite disparate; see Figure 3.10. Hence there is really a need to monitor the distances between points in an apparent spurious cluster and the points in nearby nonspurious clusters.

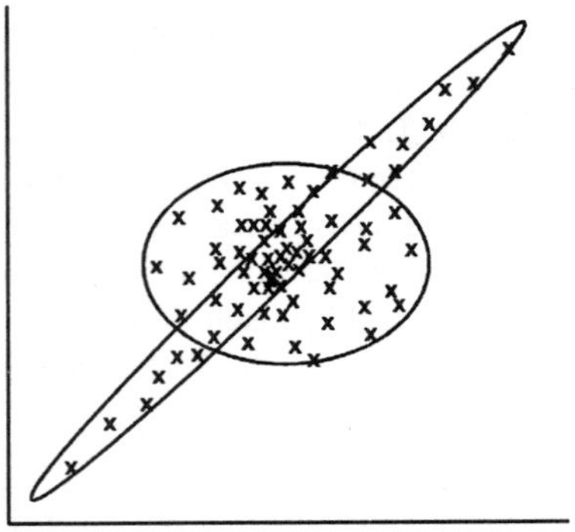

Fig. 3.10 Representation of a legitimate solution that would seem spurious when only inter-component mean distances are considered.

3.10.6 Example 3.9: Galaxy Data Set

To illustrate the point that a relatively small component variance does not necessarily imply a spurious solution, we consider the galaxy data set analyzed in Roeder (1990). This set contains measurements of the velocities of 82 galaxies diverging away from our own galaxy. In Figure 3.11 we give the plot of the six-component solution corresponding to the largest of the local maximizers found, along with the data in histogram form. The estimated component variances are given in Table 3.8.

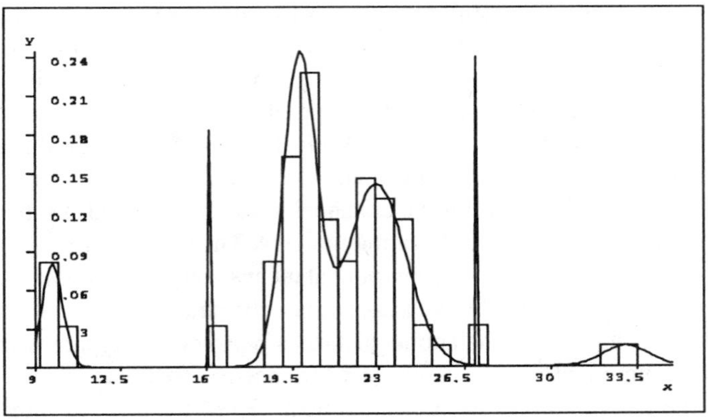

Fig. 3.11 Plot of fitted six-component normal mixture density for galaxy data set.

Table 3.8 A Six-Component Normal Mixture Solution for the Galaxy Data Set

Component i	π_i	μ_i	σ_i^2
1	0.085	9.710	0.178515
2	0.024	16.127	0.001849
3	0.037	33.044	0.849564
4	0.425	22.920	1.444820
5	0.024	26.978	0.00030
6	0.404	19.790	0.454717

From Figure 3.11 we have a seemingly spurious solution; the two small clusters centered around 16 and 26.98, corresponding to components 2 and 5, have relatively very small variances. However, on closer examination, it seems reasonable that the clusters in question may be legitimate, with their points not belonging to the main two clusters in the center of the data set. This can only be confirmed if more observations become available. For this data set, Richardson and Green (1997) concluded from their approach that the number of components ranged from 5 to 7, while McLachlan and Peel (1997b) provided support for $g = 6$ components.

Another property of spurious solutions is that the EM algorithm converges to them very rarely and often only when a particular starting point is given. This reflects a very localized peak in the likelihood function. This is a tell-tale sign that a spurious solution has been found. This idea is very useful since potentially legitimate solutions such as the one seen in the galaxy data set example above will not have this property. However, for extremely large data sets, especially where not enough clusters are fitted, the repeatability aspect might not be as useful.

3.11 EXAMPLE 3.10: PREVALENCE OF LOCAL MAXIMIZERS

We consider now the well-known set of *Iris* data as originally collected by Anderson (1935) and first analyzed by Fisher (1936). It consists of measurements of the length and width of both sepals and petals of 50 plants for each of three types of *Iris* species *setosa, versicolor*, and *virginica*. As pointed out by Wilson (1982), the *Iris* data were collected originally by Anderson (1935) with the view to seeing whether there was "evidence of continuing evolution in any group of plants." The aural approach of Wilson (1982) to data analysis suggested that both the *versicolor* and *virginica* species should each be split into two subspecies.

Hence we focus on the clustering of the $n = 50$ observations in the *Iris virginica* set, which are plotted in Figure 3.12 in the space of the first two principal components using the sample correlation matrix of the data. We considered a clustering of this data set into two clusters C_1 and C_2 by fitting a mixture of two heteroscedastic normal components. The membership of the smaller-sized cluster (C_1) is reported in Table 3.9 for the clustering implied by each of fifteen solutions of the likelihood equation. Also listed in Table 3.9 for each of these local maximizers is the value of the determinant of each of the two component-covariance matrices, $|\boldsymbol{\Sigma}_1|$ and $|\boldsymbol{\Sigma}_2|$, and the value of the log likelihood. The clustering implied by the first solution S_1 listed in Table 3.9, which had been obtained previously by Wilson (1982), has nine observations in the first cluster. This solution can be found by running the EM algorithm from an initial partition of the data given by either Ward's or the farthest neighbor hierarchical clustering procedures in standardized or unstandardized forms. It can also be found using random starts, but our results suggest that a very large number of random starts is needed if it is to be found with a high degree of probability in a given run. It was not found when running the EM algorithm from the initial partition provided by the k-means algorithm, which was carried out 1,000 times, using different initial seeds each time. On each of these 1,000 runs, the EM algorithm so initialized converged to the same local maximum, which was less than the local maximum with S_1.

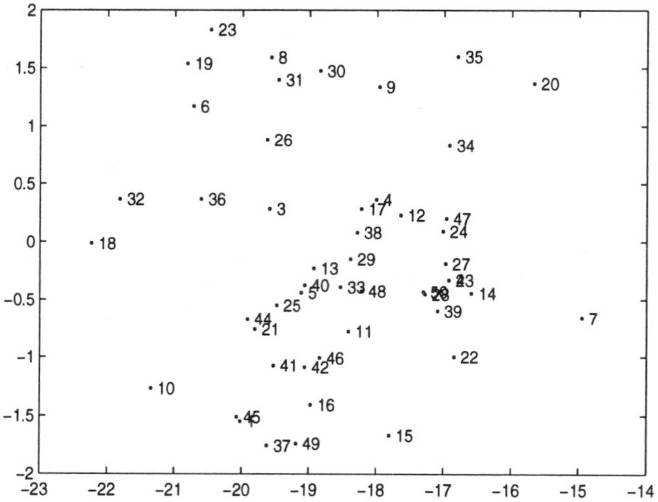

Fig. 3.12 Plot of the first two principal components of $n = 50$ four-dimensional observations on the *Iris virginica* species, using the sample correlation matrix.

It can be seen from Figure 3.12 that the nine points in the smaller-sized cluster C_1 of solution S_1 lie in an extreme portion of the scatter plot of the first two principal components using the sample correlation matrix of the data, and they can be separated almost from the other points by a straight line. However, on the question of whether there are signs of continuing evolution in the *virginica* species, this two-group structure would not be considered significant at a conventional level. The value of $-2 \log \lambda$ for the test of $g = 1$ versus $g = 2$ is 43.2. The assessed P-value, as obtained by resampling on the basis of 99 replications, is 40%. Since $g = 1$ under the null hypothesis, the bootstrap replications of $-2 \log \lambda$ are actual applications of this test statistic.

In order to demonstrate the occurrence of multiple maxima, some of which may be spurious, in data sets without a strong group structure, we ran the stochastic version of the EM algorithm (Celeux and Diebolt, 1985) from 1,000 random starting points, limiting attention to solutions at which the likelihood had a greater value than that for the first solution S_1. The stochastic version allows the EM algorithm to have a chance to escape from the current EM sequence. But evidently in this example, such escapes often led to convergence in the end to what we have concluded to be spurious local maximizers. In Table 3.9 we list fourteen such solutions found, labeled S_2 to S_{15}, in order of increasing value of the corresponding local maxima. They are a selection from the 51 distinct local maxima found that were greater than that for S_1 and they include the six largest found. Given the imbalance in the estimates of the component generalized variances for these fourteen solutions, they would appear to be spurious. However, it is not suggested that the clustering of a data set should be based solely on a single solution of the likelihood equation, but rather on the various solutions considered collectively. The smaller-sized clusters implied by eight of these fourteen solutions have only five members. Hence given that the data are of dimension $p = 4$, it is not surprising that the fitted covariance matrix for the first component of the mixture is nearly singular for each of them, with a generalized variance equal to only 7.6×10^{-8} or smaller. In this sense, these eight solutions would be regarded as spurious. Not withstanding that, the solutions S_2, S_4, S_6, and S_{13} provide some support to the clustering implied by S_1 in that at least four of the five members of the first cluster C_1 implied by these four solutions (actually all five members except in the case of S_6) are a subset of C_1 as implied by S_1. The first clusters C_1 implied by the solutions S_9 to S_{12} have no members in common with that of S_1. Further, it can be confirmed from scatter plots and the Mahalanobis distances between the fitted component means that solutions S_3, S_5, S_7 to S_{12}, S_{14}, and S_{15} do not provide as much separation between the means of the implied clusters. Thus these solutions would appear to be more spurious in nature rather than representing a genuine grouping. In Table 3.10 we have listed for each solution the Mahalanobis squared distance between the two fitted component means and also the average Mahalanobis squared distance between the points in the smaller- and larger-sized clusters. In each instance, the covariance matrix used in these calculations was the estimated component-covariance matrix corresponding to the larger-sized cluster.

Another way of proceeding to reduce the prevalence of solutions corresponding to artificially small values of the generalized fitted variances is to restrict the covariance matrices to be the same. Under homoscedasticity, the first cluster implied by the ML solution (assuming it is the global maximizer) contains the union of all members of the first clusters implied by the heteroscedastic solutions S_1, S_2, S_4, S_6, and S_{13}, along with observations 9 and 36. The solution corresponding to the largest maximum located under the less restrictive assumption of equal correlation matrices gives almost the same clustering as S_4, but with the observation numbered 31 moved to the larger-sized cluster. The solution corresponding to the second largest of the local maxima located under the assumption of equal correlation matrices gives the same clustering as S_1.

Table 3.9 Results of Fitting a Mixture of $g = 2$ Heteroscedastic Normal Components to *Iris Virginica* Species

Solution No.	Cluster C_1	log L	$\|\boldsymbol{\Sigma}_1\|$	$\|\boldsymbol{\Sigma}_2\|$
1	6,8,18,19,23, 26,30,31,32	−36.994	1.4×10^{-6}	3.7×10^{-5}
2	6,18,19,23,32	−36.987	7.6×10^{-8}	5.2×10^{-5}
3	10,13,17,21,26 30,32,40,41,42 44,46,47	−35.622	2.9×10^{-7}	7.3×10^{-5}
4	6,18,19,23,31	−35.406	6.8×10^{-9}	6.3×10^{-5}
5	5,18,21,32,35, 40,41,42,44,46	−34.427	6.2×10^{-8}	7.2×10^{-5}
6	6,18,19,32,35	−34.063	1.5×10^{-8}	5.5×10^{-5}
7	2,14,17,20,30, 32,36,43	−33.690	3.6×10^{-9}	9.7×10^{-5}
8	2,7,13,20,30 32,36,43	−32.862	3.5×10^{-10}	1.4×10^{-4}
9	1,16,41,42,45 46,49	−32.225	4.1×10^{-9}	8.8×10^{-5}
10	1,37,41,42,49	−30.374	2.9×10^{-11}	9.3×10^{-5}
11	20,35,42,46,47	−29.756	8.0×10^{-11}	8.0×10^{-5}
12	2,7,17,40,42, 43,48	−28.581	7.6×10^{-11}	1.2×10^{-4}
13	8,19,23,30,31	−27.899	8.0×10^{-11}	7.4×10^{-5}
14	8,19,23,28,39	−25.071	1.3×10^{-11}	8.0×10^{-5}
15	3,6,17,39,40	−23.536	8.9×10^{-12}	1.4×10^{-4}

Table 3.10 Intercomponent Mean and Point Mahalanobis Squared Distances

Solution	$\log L(\boldsymbol{\Psi})$	Intercomponent Mean Distance	Intercomponent Point Distance
1	−36.994	11.12	16.78
2	−36.987	8.77	16.48
3	−35.622	1.78	5.40
4	−35.406	9.96	14.52
5	−34.427	0.40	7.21
6	−34.063	5.62	14.27
7	−33.690	0.49	6.09
8	−32.862	0.80	2.03
9	−32.225	3.96	7.45
10	−30.374	1.52	6.31
11	−29.756	2.29	9.80
12	−28.581	0.79	4.53
13	−27.899	8.47	12.05
14	−25.071	2.05	11.04
15	−23.536	0.96	3.92

3.12 ALTERNATIVE MODELS FOR COMPONENT-COVARIANCE MATRICES

3.12.1 Spectral Representation

We have seen that a g-component normal mixture model with unrestricted component-covariance matrices is a highly parameterized model with $\frac{1}{2}p(p+1)$ parameters for each component-covariance matrix $\boldsymbol{\Sigma}_i$ $(i = 1, \ldots, g)$. Banfield and Raftery (1993) introduced a parameterization of the component-covariance matrix $\boldsymbol{\Sigma}_i$ based on a variant of the standard spectral decomposition of $\boldsymbol{\Sigma}_i$,

$$\boldsymbol{\Sigma}_i = \sum_{v=1}^{p} \lambda_{iv} \boldsymbol{a}_{iv} \boldsymbol{a}_{iv}^T, \tag{3.25}$$

where $\boldsymbol{a}_{i1}, \ldots, \boldsymbol{a}_{ip}$ denote the eigenvectors corresponding to the eigenvalues $\lambda_{i1} \geq \lambda_{i2} \geq \cdots \lambda_{ip} > 0$ of $\boldsymbol{\Sigma}_i$ $(i = 1, \ldots, g)$. They expressed $\boldsymbol{\Sigma}_i$ further as

$$\boldsymbol{\Sigma}_i = \lambda_i \boldsymbol{A}_i \boldsymbol{\Lambda}_i \boldsymbol{A}_i^T, \tag{3.26}$$

where $\boldsymbol{A}_i = (\boldsymbol{a}_{i1}, \ldots, \boldsymbol{a}_{ip})$ is the (orthogonal) matrix of the eigenvectors of $\boldsymbol{\Sigma}_i$. Conventions for normalizing λ_i and $\boldsymbol{\Lambda}_i$ include taking $\lambda_i = \lambda_{i1}$ (the largest eigenvalue of $\boldsymbol{\Sigma}_i$) for which then

$$\boldsymbol{\Lambda}_i = \operatorname{diag}(1, \lambda_{i2}/\lambda_{i1}, \ldots, \lambda_{ip}/\lambda_{i1}). \tag{3.27}$$

Another requires $\mid \boldsymbol{\Lambda}_i \mid = 1$ for which $\lambda_i = \mid \boldsymbol{\Sigma}_i \mid^{1/p}$ and

$$\boldsymbol{\Lambda}_i = \operatorname{diag}(\lambda_{i1}/\lambda_i, \ldots, \lambda_{ip}/\lambda_i).$$

The parameter λ_i controls the volume of the cluster corresponding to the ith component, $\boldsymbol{\Lambda}_i$ its shape, and $\boldsymbol{A}_i$ its orientation. A reduction in the number of parameters is achieved by imposing various constraints on the $\boldsymbol{A}_i$, $\boldsymbol{\Lambda}_i$, and the λ_i. For example, the constraint $\boldsymbol{A}_i = \boldsymbol{A}$ $(i = 1, \ldots, g)$ imposes the same orientation on the g clusters.

Applications of mixture models under the model (3.26) for the component-covariance matrices have been considered since by Bensmail and Celeux (1996), Bensmail et al. (1997), Campbell et al. (1997), Dasgupta and Raftery (1998), and Fraley and Raftery (1998).

3.12.2 Example 3.11: Minefield Data Set

We consider the simulated minefield data as analyzed by Dasgupta and Raftery (1998) and Fraley and Raftery (1998); see Figure 3.13. The data arose from the processing of a series of images taken by a reconnaissance aircraft in which a large number of points are identified as representing possible mines, but many of these are in fact false positives (noise). The assumption is that the imaged area does not lie completely within a minefield, and that if there is a minefield, it will occur in an area where there is a higher density of identified points. The aims were to determine whether the image contains one or more minefields, and to give the location of any minefields that may be present. The denoising of the data in Figure 3.13 was undertaken using the nearest-neighbor-based procedure of Byers and Raftery (1998).

Dasgupta and Raftery (1998) fitted a g-component normal mixture for various g plus a uniform component (with unknown mixing proportion),

$$f_0(\boldsymbol{y}_j) = 1/v, \tag{3.28}$$

corresponding to the background noise (clutter), such as other metal objects and rocks, which was assumed to be distributed as a homogeneous spatial point Poisson process. In (3.28), v is the area of the image. In Dasgupta and Raftery (1998), v was

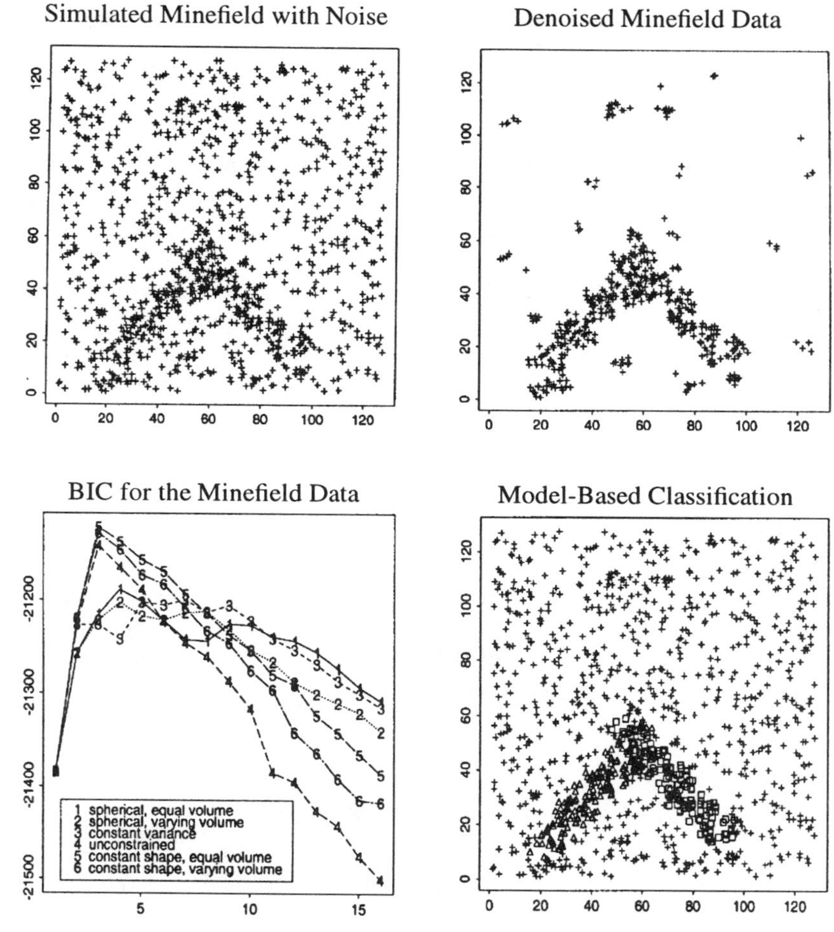

Fig. 3.13 Simulated minefield with noise and mixture model-based clustering after a nearest neighbor denoising. From Fraley and Raftery (1998).

computed as the volume of the smallest hyperrectangle with sides parallel to the axes that contains all the data points. However, as noted by Fraley and Raftery (1998), this could overestimate v in many cases. Other possibilities include taking v to be the volume of the smallest hyperrectangle with sides parallel to the principal component axes that contains all the data points or taking v to be the volume of the convex hull of the data.

To specify the component-covariance matrices, Dasgupta and Raftery (1998) used the parameterized forms (3.26) and (3.27) for the $\boldsymbol{\Sigma}_i$. Of particular interest in this application for $p = 2$ dimensions is the model in which the shapes of the clusters implied by (3.26) are the same, but the volumes and orientations are different. That is,

$$\boldsymbol{A}_i = \text{diag}(1, \alpha) \quad (i = 1, \ldots, g),$$

where $\alpha < 1$ is the common value of the ratio of the second to first eigenvalue, $\lambda_{i2}/\lambda_{i1}$. When α is near one, the clusters will tend to be almost circular, but when α is much smaller, they will be tend to be long and narrow (highly linear), as in the case of minefields.

From the plot of the BIC criterion in Figure 3.13, it can be seen that three clusters ($g = 2$ normal components plus the noise component) and component-covariance matrices of the same shape ($\alpha_i = \alpha$) and equal volume ($\lambda_{i1} = \lambda_1$) for $i = 1, 2$ are favored. Note that the BIC criterion in Figure 3.13 is written as the negative of its usual form (see Section 6.9.3).

3.13 SOME OTHER MODELS

In this section we consider normal mixture models for some experimental designs that involve more than a simple random sample having been observed on the phenomenon under study. There is also the work of Wedel, ter Hofstede, and Steenkamp (1998), who have investigated the effects of a complex sampling design on the estimation of mixture models.

3.13.1 Clustering of Treatment Means in ANOVA

A common problem in practice is the analysis of experiments designed to compare treatments. In the situation where an analysis of variance (ANOVA) leads to a significant F-test for the difference between treatment means, there is the problem of isolating those treatments that do not appear to be different. This problem continues to attract attention in the literature; see, for example, Bautista, Smith, and Steiner (1997) and a comment on this paper by McLachlan (2000).

One approach to this problem is to cluster the treatment means. The possibility of using cluster analysis in place of multiple comparison procedures had been suggested by O'Neill and Wetherill (1971). In the discussion of this paper, Nelder (1971) commented that one of the patterns to look for in the sample means of the treatments was whether "the means divide into two or more groups within which they look like samples from normal distributions." The normal mixture model-based approach to this problem, as developed in Aitkin (1980) and Basford and McLachlan (1985b), is in keeping with the spirit of this comment.

With this mixture approach, the n treatment means, denoted here by $\bar{y}_1, \ldots, \bar{y}_n$, are assumed to be distributed (independently) about their means $\mu_1, \ldots, \mu_n$, with variances equal to $\sigma^2/r_1, \ldots, \sigma^2/r_n$, respectively, where r_j denotes the number of observations made on the jth treatment. In addition, s^2 denotes an independent estimate of σ^2, distributed as

$$\nu s^2/\sigma^2 \sim \chi^2_\nu. \tag{3.29}$$

This covers various experimental designs, including a completely randomized design with unequal replicates and, for equal r_j, more involved designs with equireplicated treatments; see McLachlan and Basford (1988, Chapter 6).

Under a normal mixture model with g components, it is assumed further that $\bar{Y}_j$ is distributed as

$$\bar{Y}_j \sim N(\mu_i, \sigma^2/r_j) \quad \text{in } G_i \text{ with prob. } \pi_i \quad (i = 1, \ldots, g). \tag{3.30}$$

The log likelihood for the vector of unknown parameters formed on the basis of $\bar{y}_1, \ldots, \bar{y}_n$ under the mixture model (3.30) and also s^2 under (3.29) is given, up to terms not involving the unknown parameters, by

$$\log L(\boldsymbol{\Psi}) = \sum_{j=1}^{n} \log \sum_{i=1}^{g} \pi_i \phi(y_j;\, \mu_i,\, \sigma^2/r_j) - \tfrac{1}{2}\nu \log \sigma^2 - \tfrac{1}{2}\nu s^2/\sigma^2. \tag{3.31}$$

This mixture model can be fitted to the treatments using the EM algorithm. The FORTRAN listing of a program for this purpose is given in the appendix of McLachlan and Basford (1988). A probabilistic clustering of the treatments is obtained in terms of their fitted posterior probabilities of component membership. An outright clustering into distinct groups is obtained by assigning each treatment mean to the group to which it has the highest posterior probability of belonging.

McLachlan (2000) applied this approach to the six treatments in the example considered by Bautista et al. (1997). These treatments, designated as 3DOk1, 3DOk5, 3DOk7, COMP, 3DOk4, and 3DOk13, have means equal to 28.82, 23.98, 19.92, 18.70, 14.74, and 13.26, respectively. For $g = 2$, it results in the first four treatments being clustered into one group and the remaining two in another group. For $g = 3$, the first two treatments are grouped together in one cluster, the next three in another, while the last treatment forms a group on its own. These partitions of the treatment means agree with those of Bautista et al. (1997). On the question of whether there should be $g = 2$ or 3 groups, the value of $-2\log\lambda$ (that is, twice the increase in the log likelihood for $g = 3$ over $g = 2$) is 4.702. An assessment of the P-value can be obtained by using a resampling approach as in McLachlan (1987). Its application here with $B = 19$ bootstrap replications leads to the null hypothesis of $g = 2$ groups being retained at any conventional level of significance.

3.13.2 Three-Way Models

Normal mixture models have been applied successfully to so-called three-way data by Basford and McLachlan (1985c) and, more recently, by Hunt and Basford (1999) in the case where there are also categorical data present. Specifically, in the terminology of Basford and McLachlan (1985c), suppose we wish to cluster genotypes on the basis of genotype by attribute by environment data. We let the observation

$$\boldsymbol{y}_j = (\boldsymbol{y}_{j1}^T, \ldots, \boldsymbol{y}_{jr}^T)^T$$

contain the multiattribute responses of the jth genotype in all r environments, where $\boldsymbol{y}_{jm}$ is a vector of dimension p giving the response of the jth genotype in the mth environment ($m = 1, \ldots, r$). The vectors $\boldsymbol{y}_{jm}$ ($j = 1, \ldots, n$; $m = 1, \ldots, r$) are taken to be independently distributed. Under the mixture model proposed by Basford and McLachlan (1985d), it is assumed that each genotype belongs to one of g possible groups $G_1, \ldots, G_g$ in proportions $\pi_1, \ldots, \pi_g$, so that in a given environment m,

$$\boldsymbol{Y}_j \sim N(\boldsymbol{\mu}_{im}, \boldsymbol{\Sigma}_i) \quad \text{in } G_i \text{ with prob. } \pi_i \quad (i = 1, \ldots, g). \tag{3.32}$$

The group-conditional covariance matrix $\boldsymbol{\Sigma}_i$ is taken not to depend on the environment. This model covers the general situation where there may be some interpretation between genotypes and environments.

3.13.3 Example 3.12: Consumer Data on Cat Food

To illustrate the use of a normal mixture model applied to so-called three-way data, we consider here the consumer study on a cat food trial, as analyzed in Jones and McLachlan (1992b) using mixtures of normal linear regression models. In other work on mixtures of normal regression models, Quandt and Ramsey (1978) and Kiefer (1978) studied the case of data comprising two regression equations, including the use of moment generating function techniques for parameter estimation. De Veaux (1989) developed an EM approach to fitting the two regression situation. Similar estimating equations were used by DeSarbo and Cron (1988) for the case of relating an attribute scored by individuals to auxiliary information on the individuals and where the number of groups in the data were unknown. Mixtures of generalized linear models are to be covered in some depth in Chapter 5.

Jones and McLachlan (1992b) considered the assessment of $r = 10$ different cat food items by $n = 86$ individuals who rated the products with regard to $p = 4$ attributes: appearance, texture, aroma and overall acceptability. So this is a three-way data set, consisting of data on individuals by attributes by items. For each of these attributes, the consumers marked a survey sheet on a scale that ranged between 0 and 100. These measurements were scaled to lie between -1 and 1 before the following analyses. In this example, an auxiliary measurement was taken. It consisted of a variable derived from scores of the products made by an expert sensory panel. These scores were composed from a variety of individual descriptors of meat quality and also took values between 0 and 100. These scores were reduced to a single dimension using principal component analysis and this dimension was interpreted as measuring the "chunkiness" of the products. Its values over the ten products ranged between −2 and 2.

Jones and McLachlan (1992b) proposed a normal mixture model with g components corresponding to the existence of g groups $G_1, \ldots, G_g$ among the individuals. Under their model, the vector $\boldsymbol{y}_{jm}$ containing the scores by the jth individual for the $p = 4$ attributes for the mth item is assumed to be normally distributed with mean in each group related by a linear regression model to the auxiliary measurement x_m; that is,

$$Y_{jm} \sim N(\boldsymbol{\beta}_i x_m, \boldsymbol{\Sigma}_i) \text{ in } G_i = (i = 1, \ldots, g), \tag{3.33}$$

where $\boldsymbol{\beta}_i$ is a $p \times 1$ vector of parameters for group i and x_m is the auxiliary measurement for the mth item. The log likelihood was formed under the assumption of independence of the $\boldsymbol{y}_{jm}$.

In Table 3.11 we report the estimates obtained by Jones and McLachlan (1992b) for the regression coefficients and the attribute variances for $g = 1$ to 3 groups. From the coefficients in Table 3.11, Jones and McLachlan (1992b) inferred that the group with the largest proportion of consumers places the least weight on the "chunkiness" of the product, and the other two groups place increasing weight on this aspect of the product. This was of interest at the time to product developers as "chunkiness" was one of the strong selling points for this type of product in advertising campaigns. The third attribute, aroma, had (as one would expect) the weakest link with "chunkiness" of the four variables over all three groups. This attribute also had the largest variance within each of the three groups.

Table 3.11 Attribute Regression Coefficients for Three-Group Clustering of Eighty-Six Subjects in a Cat Food Trial

Group No.		Regression Coefficients Attribute				Variances Attribute			
i	π_i	1	2	3	4	1	2	3	4
1	.41	.055	.132	.019	.098	.097	.084	.100	.091
	.09	.021	.018	.021	.031	.010	.007	.008	.007
2	.31	.126	.161	.120	.137	.086	.077	.087	.083
	.05	.027	.016	.041	.018	.007	.006	.007	.007
3	.27	.237	.277	.071	.213	.049	.031	.095	.059
	.07	.020	.022	.017	.017	.007	.008	.009	.007

Source: Adapted from Jones and McLachlan (1992b).

3.13.4 Errors-In-Variables Model

To further illustrate the usefulness of normal mixture models in providing a flexible framework for parametric modeling, we briefly consider the errors-in-variables problem studied by Carroll, Roeder, and Wasserman (1999). In a typical errors-in-variables model, the response variable Y_j is functionally related to a predictor variable U_j that is unobserved because either it is impractical or impossible to measure it directly. To compensate, a surrogate variable X_j is observed instead of the latent predictor variable U_j. This surrogate variable is assumed to be related to the truth by a simple measurement process. The simplest example of such a model is the basic measurement error model,

$$\begin{aligned} Y_j &= \beta_0 + \beta_1 U_j + \epsilon_j, \\ X_j &= U_j + e_j, \end{aligned}$$

where ϵ_j and e_j are independent of U_j. The random variable Y_j has mean μ_y and variance σ_y^2, e_j has mean zero and variance σ_e^2, and ϵ_j has mean zero and variance σ_ϵ^2; all these random variables are uncorrelated. Carroll et al. (1999) considered a version of this model with missing surrogate data and a change-point Berkson model. Their model was motivated by a dietary assessment study where Y_j corresponds to the Food Frequency Questionnaire (FFQ) and X_j to the multiple-day food record (FR). The unobservable U_j corresponds to the true long-term usual daily diet. Data on FFQs and FRs were obtained in two stages. At the first of the two stages, FFQ is observed for a number (M) of individuals. At the second stage, a smaller number of these individuals are selected by stratified sampling on the basis of their FFQ scores. On each individual so selected, the FR is observed for a number of days, as it is an imprecise measurement of the true usual intake.

For this missing-data version of the errors-in-variables problem, the method of moments is still consistent and asymptotically efficient. But the other commonly used approach, maximum likelihood based on all random variables being normally distributed, is not. By modeling the distribution of U_j by a mixture of g normal distributions, Carroll et al. (1999) were able to provide a parametric model that has the efficiencies of a parametric model but also shares the robustness properties of the method of moments estimator. The mixture model can be fitted in either a frequentist framework or a Bayesian framework. Carroll et al. (1999) adopted the latter approach, using the posterior simulation methods to be described in the next chapter. They were able also to develop a similar model for the change-point problem by modeling the measurement error by a normal mixture.

4

Bayesian Approach to Mixture Analysis

4.1 INTRODUCTION

We have seen that estimation for mixture models is straightforward using the EM algorithm. Estimation in a Bayesian framework is now feasible using posterior simulation via recently developed Markov chain Monte Carlo (MCMC) methods. Bayes estimators for mixture models are well defined so long as the prior distributions are proper. However, it has only been with the development of MCMC methods, as proposed by Tanner and Wong (1987) and Gelfand and Smith (1990), that the implementation of the Bayesian approach for mixtures has been made practical. For a perspective on MCMC methods, the reader is referred to Tierney (1994) and to the books by Gilks, Richardson, and Spiegelhalter (1996) and Robert and Casella (1999), as well as to the references therein.

Among the first papers on Bayesian estimation for mixture models via posterior simulation are those by Gilks, Oldfield, and Rutherford (1989), Diebolt and Robert (1990, 1994), Gelman and King (1990), Verdinelli and Wasserman (1991), Evans, Guttman, and Olkin (1992), and Lavine and West (1992). It can be seen that it took almost 100 years from Pearson's (1894) classic mixtures paper for a truly Bayesian solution to the mixture problem to emerge. An account of approximation techniques for Bayesian mixture analysis before the use of MCMC methods may be found in Titterington et al. (1985, Chapter 6). Key initial papers on the Bayesian analysis of mixtures following MCMC methods include Diebolt and Robert (1990, 1994) and Escobar and West (1995). Also, the mixture problem is treated in the review of MCMC methods by Smith and Roberts (1993). Further advances are covered in West, Müller, and Escobar (1994), Carlin and Chib (1995), Cao and West (1996), Mengersen and

Robert (1996), Phillips and Smith (1996), Raftery (1996), Robert (1996), Bensmail et al. (1997), Richardson and Green (1997), Roeder and Wasserman (1997), Robert and Mengersen (1999), and Yu and Tanner (1999). Additional applications of Bayesian mixture estimation via posterior simulation are given in Gelman et al. (1995, Chapter 16) and the references therein. Other recent case studies involving the Bayesian analysis of mixture distributions include those by Dellaportas (1998) and Vounatsou, Smith, and Smith (1998).

Provided that suitable (conjugate) priors are used, the posterior density will be proper, thereby allowing the application of MCMC methods such as the Gibbs sampler to provide an accurate approximation to the Bayes solution. Although the application of MCMC methods is now routine, there are some difficulties that have to be addressed with the Bayesian approach in the context of mixture models. One main hindrance is that improper priors yield improper posterior distributions. An alternative is to use a "partially proper prior." By this it is meant a prior that does not require subjective input for the component parameters, yet the posterior is proper (Roeder and Wasserman, 1997). Another hindrance is that when the number of components g is unknown, the parameter space is simultaneously ill-defined and of infinite dimension. This prevents the use of classical testing procedures and priors. The usual approach therefore is to fit the mixture model for fixed g and then to consider the choice of g according to some so-called information criterion that typically penalizes the log likelihood for the complexity of the adopted model, possibly adjusted for the sample size. Recently, Phillips and Smith (1996) and Richardson and Green (1997) presented a fully Bayesian approach with g taken to be an unknown parameter. Their MCMC methods allow *jumps* to be made for variable dimension parameters and thus can handle g being unspecified.

A further hindrance is the effect of label switching, which arises when there is no real prior information that allows one to discriminate between the components of a mixture model belonging to the same parametric family. Since the likelihood is invariant then under a permutation of the component labels in $\boldsymbol{\Psi}$, it effectively has $g!$ modes. Hence the posterior also will have this property for a prior distribution that is symmetric in the components. A label switching occurs when some of the labels of the mixture components permute. The effect of label switching is very important when the solution is being calculated iteratively and there is the possibility that the labels of the components may be switched on different iterations. This switching of component labels is not a problem in the normal course of events in the iterative computation of the MLE via the EM algorithm, but it is a serious issue in the simulation of (approximate) realizations of the parameter vector $\boldsymbol{\Psi}$ from its posterior distribution.

In this chapter we shall briefly review the main approaches to Bayesian mixture analysis and discuss some of the methods that have been proposed for overcoming the aforementioned difficulties. We shall first start with the case of a proper prior distribution before considering some extensions that have been proposed for noninformative settings.

4.2 ESTIMATION FOR PROPER PRIORS

We consider here the case of a proper prior density $p(\boldsymbol{\Psi})$ for the parameter vector $\boldsymbol{\Psi}$. In the sequel, we shall use $p(\cdot)$ as a generic notation for a density function. We can write the posterior density of $\boldsymbol{\Psi}$ as

$$\begin{aligned} p(\boldsymbol{\Psi} \mid \boldsymbol{y}) &= C^{-1}L(\boldsymbol{\Psi})p(\boldsymbol{\Psi}) \\ &= C^{-1}\sum_{\boldsymbol{z}} L_c(\boldsymbol{\Psi})p(\boldsymbol{z} \mid \boldsymbol{\Psi})p(\boldsymbol{\Psi}), \end{aligned} \tag{4.1}$$

where $p(\boldsymbol{z} \mid \boldsymbol{\Psi})$ denotes the conditional density of $\boldsymbol{Z}$ given $\boldsymbol{\Psi}$. The normalizing constant C in (4.1) is given by

$$C = \int \sum_{\boldsymbol{z}} L_c(\boldsymbol{\Psi})p(\boldsymbol{z} \mid \boldsymbol{\Psi})p(\boldsymbol{\Psi})\, d\boldsymbol{\Psi}. \tag{4.2}$$

As in the previous chapters, $L_c(\boldsymbol{\Psi})$ denotes the complete-data log likelihood formed on the basis of the feature data $\boldsymbol{y} = (\boldsymbol{y}_1^T, \ldots, \boldsymbol{y}_n^T)^T$ and also their component-indicator vectors given by $\boldsymbol{z} = (\boldsymbol{z}_1^T, \ldots, \boldsymbol{z}_n^T)^T$. In (4.1), the sum is over all possible values of $\boldsymbol{z}$ defining the component membership of $\boldsymbol{y}_j$ $(j = 1, \ldots, n)$.

This approach may be viewed as hierarchical with, on top, the parameters in $\boldsymbol{\Psi}$ for the mixture, then the unobservable component-indicator vectors in $\boldsymbol{z}$, whose distribution depends on $\boldsymbol{\Psi}$ and, at the bottom, the observed data $\boldsymbol{y}$ whose distribution depends on $\boldsymbol{z}$ and $\boldsymbol{\Psi}$.

If a conjugate prior is specified, then the posterior expectation of $\boldsymbol{\Psi}$ can be written in closed form. However, its direct use is only feasible with small sample sizes. We shall demonstrate this for component densities belonging to a general exponential family.

4.3 CONJUGATE PRIORS

We assume now with little loss of generality that the component densities $f_i(\boldsymbol{y}_j; \boldsymbol{\theta}_i)$ belong to the same exponential family, so that $f_i(\boldsymbol{y}_j; \boldsymbol{\theta}_i) = f(\boldsymbol{y}_j; \boldsymbol{\theta}_i)$, where

$$f(\boldsymbol{y}_j; \boldsymbol{\theta}_i) = \exp\{\boldsymbol{\theta}_i^T \boldsymbol{y}_j - b(\boldsymbol{\theta}_i) + c(\boldsymbol{y}_j)\}. \tag{4.3}$$

This family allows a conjugate prior for $\boldsymbol{\theta}_i$, which is taken to be distinct for each component i (that is, if the component-indicator $z_{ij} = 1$), having the form

$$p(\boldsymbol{\theta}_i; \boldsymbol{\omega}_i, \gamma_i) \propto \exp\{\boldsymbol{\theta}_i^T \boldsymbol{\omega}_i - \gamma_i b(\boldsymbol{\theta}_i)\}, \tag{4.4}$$

where $\boldsymbol{\omega}_i$ is a real-valued vector of constants and γ_i is a scalar constant $(i = 1, \ldots, g)$. A conjugate prior for the vector $\boldsymbol{\pi} = (\pi_1, \ldots, \pi_g)^T$ of mixing proportions is the Dirichlet distribution $\mathcal{D}(\alpha_1, \ldots, \alpha_g)$, which has density

$$p_D(\boldsymbol{\pi}) = \Gamma(\sum_{i=1}^{g} \alpha_i - g) \prod_{i=1}^{g} \pi_i^{\alpha_i - 1}/\Gamma(\alpha_i). \tag{4.5}$$

The Dirichlet distribution is a generalization of the beta distribution $Be(\alpha_1, \alpha_2)$.

For this prior for $\boldsymbol{\Psi}$, the posterior density $p(\boldsymbol{\Psi} \mid \boldsymbol{y})$ is proportional to

$$p(\boldsymbol{\Psi} \mid \boldsymbol{y}) \propto \sum_{\boldsymbol{z}} p_D(\alpha_1 + n_1, \ldots, \alpha_g + n_g) \prod_{i=1}^{g} \{p(\boldsymbol{\theta}_i;\ \omega_i + n_i \overline{\boldsymbol{y}}_i,\ \gamma_i + n_i)\}, \quad (4.6)$$

where $n_i = \sum_{j=1}^{n} z_{ij}$ and $\overline{\boldsymbol{y}}_i = \sum_{j=1}^{n} z_{ij} \boldsymbol{y}_j / n_i$. While the posterior expectation of $\boldsymbol{\Psi}$ can be written in closed form from (4.6), the time required to calculate (4.6) is far too high for the Bayesian approach to be applied in practice, even for moderate sample sizes.

In the special case where only the mixing proportions are unknown, sequential approximations have been provided by Smith and Makov (1978) and Bernardo and Girón (1988).

4.4 MARKOV CHAIN MONTE CARLO

4.4.1 Posterior Simulation

We can approximate posterior quantities of interest through the use of MCMC methods. Such methods allow the construction of an ergodic Markov chain with stationary distribution equal to the posterior distribution of the parameter of interest, here $\boldsymbol{\Psi}$ containing the parameters in the mixture model. Gibbs sampling achieves this by simulating directly from the conditional distribution of a subvector of $\boldsymbol{\Psi}$ given all the other parameters in $\boldsymbol{\Psi}$ (and the observed data $\boldsymbol{y}$). This conditional is called the complete conditional. We then cycle through all the parameters iteratively, each time drawing from each parameter's complete conditional, until we have N draws, $\boldsymbol{\Psi}^{(1)}, \ldots, \boldsymbol{\Psi}^{(N)}$, from the Markov chain. After a sufficiently long *burn-in* of, say, N_1 draws, the points $\boldsymbol{\Psi}^{(k)}$ $(k = N_1 + 1, \ldots, N)$ will be dependent draws approximately from the posterior distribution of $\boldsymbol{\Psi}$. In practice, consideration has to be given to the choice of suitable starting values for $\boldsymbol{\Psi}$, and choice of values for N_1 and N that are sufficiently large to ensure the consequent approximation is accurate. Reviews of convergence diagnostics have been given by Cowles and Carlin (1996), Brooks (1998), Brooks and Roberts (1998), and Mengersen, Robert, and Guihenneuc-Jouyaux (1999); see also Robert (1998). Among other sampling methods, there is the Metropolis–Hastings algorithm, which, in contrast to the Gibbs sampler, simulates from a convenient proposal distribution and then accepts the proposed value with some defined probability.

In the implementation of MCMC methods for mixture models, the unobservable component-indicator vector $\boldsymbol{z}$ is introduced and $\boldsymbol{\Psi}$ is augmented by $\boldsymbol{z}$ during the Gibbs sampling. Thus samples for the missing-data vector $\boldsymbol{z}$ and the parameter vector $\boldsymbol{\Psi}$ are alternately generated, producing a missing-data chain and a parameter chain. The proof of Diebolt and Robert (1994) of the convergence of the algorithm to the true posterior distribution of the parameter $\boldsymbol{\Psi}$ is based on a *duality principle*. The finite-state structure of the missing-data chain allows many convergence results to be easily established for it and transferred automatically to the parameter chain. They include geometric convergence, ϕ-mixing, and a central limit theorem; see Robert (1996). Diebolt and Robert (1994) established their results in the context of univariate normal component densities, and Bensmail et al. (1997) subsequently confirmed their applicability for multivariate normal component densities. The validity of the central limit theorem for MCMC algorithms is now well documented; see Meyn and Tweedie (1993) and Roberts and Rosenthal (1998).

The simulated sample, $\boldsymbol{\Psi}^{(N_1+1)}, \ldots, \boldsymbol{\Psi}^{(N)}$, can be used to approximate any well-defined posterior quantity, such as some function of $\boldsymbol{\Psi}$, $E\{a(\boldsymbol{\Psi}) \mid \boldsymbol{y}\}$, by the ergodic average,

$$E\{a(\boldsymbol{\Psi}) \mid \boldsymbol{y}\} \approx \sum_{k=N_1+1}^{N} \frac{a(\boldsymbol{\Psi}^{(k)})}{(N - N_1)}, \tag{4.7}$$

where the first N_1 burn-in samples have been discarded in forming (4.7). This provides a point estimate of $a(\boldsymbol{\Psi})$. A 95% interval estimate for $a(\boldsymbol{\Psi})$ is obtained by ordering the sample values $a(\boldsymbol{\Psi}^{(k)})$ for $k = N_1 + 1, \ldots, N$ and finding the 0.025 and 0.975 sample quantiles.

4.4.2 Perfect Sampling

Propp and Wilson (1996) showed that it is possible to use a MCMC sampler to simulate exactly from the distribution of interest by the technique of *coupling from the past*. Hobert, Robert, and Titterington (1999) have examined this approach for two- and three-component mixture models with known components. Casella et al. (2000) have proposed a perfect sampler for mixtures of distributions, in the spirit of Mira and Roberts (1999), building on the paper of Hobert et al. (1999); see also Casella, Robert, and Wells (1999). Their results apply to an arbitrary number of continuous components where their parameters and the mixing proportions need not be specified. Thus the authors claim that their method is the first general i.i.d. sampling method for mixture posterior distributions.

4.5 EXPONENTIAL FAMILY COMPONENTS

We now describe the steps to effect posterior simulation by the Gibbs sampler in the case of the conjugate priors (4.4) and (4.5) for component densities belonging to a general exponential family.

Step 1. Simulate

$$\boldsymbol{\pi} \sim \mathcal{D}(\alpha_1 + n_1, \ldots, \alpha_g + n_g) \tag{4.8}$$

and

$$\boldsymbol{\theta}_i \sim p(\boldsymbol{\theta}; \omega_i + n_i \overline{\boldsymbol{y}}_i, \gamma_i + n_i) \quad (i = 1, \ldots, g). \tag{4.9}$$

Step 2. Simulate

$$\boldsymbol{Z}_j \sim \text{Mult}_g(1,\, \boldsymbol{\tau}_j) \quad (j = 1,\, \ldots,\, n), \tag{4.10}$$

a multinomial distribution consisting of one draw on g categories with probabilities

$$\boldsymbol{\tau}_j = (\tau_1(\boldsymbol{y}_j;\, \boldsymbol{\Psi}),\, \ldots,\, \tau_g(\boldsymbol{y}_j;\, \boldsymbol{\Psi}))^T, \tag{4.11}$$

where

$$\tau_i(\boldsymbol{y}_j;\, \boldsymbol{\Psi}) = \pi_i\, f(\boldsymbol{y}_j;\, \boldsymbol{\theta}_i) / \sum_{h=1}^{g} \pi_h\, f(\boldsymbol{y}_j;\, \boldsymbol{\theta}_h).$$

Step 3. Update n_i and $\overline{\boldsymbol{y}}_i$ $(i = 1,\, \ldots,\, g)$.

The Gibbs sampler is run by simulating successively from the distributions in (4.8) to (4.10) and replacing the conditioning parameters. Robert (1996) reported that in his posterior simulations for univariate and bivariate mixture models, 5,000 iterations were "enough" in the sense that a substantial increase in the number of iterations did not usually perturb values of ergodic averages.

4.6 NORMAL COMPONENTS

4.6.1 Conjugate Priors

Conjugate priors for normal densities have been considered by Diebolt and Robert (1994) and Bensmail et al. (1997). They are given by

$$\boldsymbol{\pi} \sim D(\alpha_1,\, \ldots,\, \alpha_g), \tag{4.12}$$

$$\boldsymbol{\mu}_i \sim N(\boldsymbol{\omega}_i,\, \boldsymbol{\Sigma}_i/\kappa_i) \quad (i = 1,\, \ldots,\, g), \tag{4.13}$$

and

$$\boldsymbol{\Sigma}_i^{-1} \sim W(r_i,\, \boldsymbol{C}_i) \quad (i = 1,\, \ldots,\, g), \tag{4.14}$$

where $W(r,\, \boldsymbol{C})$ denotes a Wishart distribution, which has density

$$\frac{|\, \boldsymbol{U}\,|^{\frac{1}{2}(r-p-1)} \exp\{-\frac{1}{2}\text{tr}(\boldsymbol{U}\boldsymbol{C}^{-1})\}}{2^{\frac{1}{2}rp}\pi^{p(p-1)/4}\,|\,\boldsymbol{C}\,|^{\frac{1}{2}r} \prod_{v=1}^{p} \Gamma(\frac{1}{2}(r+1-v))}. \tag{4.15}$$

The Wishart distribution reduces to the gamma $(\frac{1}{2}r, \frac{1}{2}C^{-1})$ distribution. The gamma $(\alpha,\, \beta)$ distribution has density

$$\{\beta^\alpha u^{\alpha-1}/\Gamma(\alpha)\} \exp(-\beta u) I_{[0,\infty)}(u) \qquad (\alpha,\, \beta > 0). \tag{4.16}$$

4.6.2 Gibbs Sampler

The Gibbs sampler is implemented by simulating from the following conditional distributions:

Step 1. Simulate

$$\boldsymbol{\pi} \sim \mathcal{D}(\alpha_1 + n_1, \ldots, \alpha_g + n_g), \tag{4.17}$$

$$\boldsymbol{\mu}_i \sim N(\boldsymbol{\omega}_i^*, (n_i + \kappa_i)^{-1}\boldsymbol{\Sigma}_i), \tag{4.18}$$

$$\boldsymbol{\Sigma}_i^{-1} \sim W(n_i + r_i, \boldsymbol{C}_i^*), \tag{4.19}$$

where

$$\boldsymbol{\omega}_i^* = (n_i \overline{\boldsymbol{y}}_i + \kappa_i \boldsymbol{\omega}_i)/(n_i + \kappa_i),$$

$$\boldsymbol{C}_i^* = \{\boldsymbol{C}_i^{-1} + n_i \boldsymbol{V}_i + \frac{n_i r_i}{n_i + r_i}(\overline{\boldsymbol{y}}_i - \boldsymbol{\omega}_i)(\overline{\boldsymbol{y}}_i - \boldsymbol{\omega}_i)^T\}^{-1},$$

and

$$\boldsymbol{V}_i = \sum_{j=1}^{n} z_{ij}(\boldsymbol{y}_j - \overline{\boldsymbol{y}}_i)(\boldsymbol{y}_j - \overline{\boldsymbol{y}}_i)^T / n_i.$$

Step 2. Simulate

$$\boldsymbol{Z}_j \sim \text{Mult}_g(1, \boldsymbol{\tau}_j),$$

where

$$\boldsymbol{\tau}_j = (\tau_1(\boldsymbol{y}_j; \boldsymbol{\Psi}), \ldots, \tau_g(\boldsymbol{y}_j; \boldsymbol{\Psi}))^T.$$

Step 3. Update $n_i, \overline{\boldsymbol{y}}_i$, and $\boldsymbol{V}_i$ $(i = 1, \ldots, g)$.

In the above, the component-covariance matrices are unrestricted. Bensmail et al. (1997) give the conjugate priors and the corresponding conditional distributions for implementing the Gibbs sampler under the reduced models in Section 3.12 for the component-covariance matrices.

The approach of Bensmail et al. (1997) in choosing their conjugate priors is to have them fairly flat in the region where the likelihood is substantial and not much greater elsewhere. This is to ensure that the Bayesian estimates are relatively insensitive to reasonable changes in the prior. This was confirmed empirically by Bensmail et al. (1997) in their examples in which they took

$$\kappa_i = 1, \; \boldsymbol{\omega}_i = \overline{\boldsymbol{y}}_i, \; r_i = 5, \; \boldsymbol{C}_i^{-1} = \boldsymbol{V}_i \quad (i = 1, \ldots, g).$$

They noted that the amount of information in this prior is similar to that contained in a typical single observation.

4.7 PRIOR ON NUMBER OF COMPONENTS

In the Bayesian work considered up to now, the approach has been to sample from the posterior of $\boldsymbol{\Psi}$ for a fixed number of components in the mixture model under consideration. With the approach of Phillips and Smith (1996) and Richardson and Green (1997), the number of components g is treated as an unknown parameter with a prior distribution $p(g)$. In the former paper, it is taken to be a Poisson distribution truncated at the origin, namely,

$$p(g) = \lambda^g / \{(\exp(\lambda) - 1)g!\}, \quad g = 1, 2, \ldots.$$

Richardson and Green (1997) suggested also the use of a Poisson prior for g but, for convenience of presentation and interpretation, took a prior uniform distribution between 1 and some prespecified integer g_u.

With both approaches, a joint posterior distribution is formed for g and $\boldsymbol{\Psi}$, containing the component parameters and mixing proportions. Inference is then undertaken by simulating realizations from the resulting posterior distribution. The methodology developed by Grenader and Miller (1994) is exploited to enable discrete transitions or *jumps* to be made between the mixture models with different number of components g. Phillips and Smith (1996) use an iterative jump-diffusion sampling algorithm, while Richardson and Green (1997) use a reversible-jump Metropolis–Hastings algorithm (Green, 1994 and 1995). The process to split or combine components is rather complicated and the reader is referred to their papers for the details.

Phillips and Smith (1996) and Richardson and Green (1997) implemented their respective methods for mixtures of univariate normal densities with means μ_i and unrestricted variances σ_i^2. Within a model for fixed g, simulated approximation to the posterior distribution is carried out as in the posterior simulations described in the previous sections of this chapter. The priors of Richardson and Green (1997) for the π_i, μ_i, and σ_i^2 are all drawn independently with the Dirichlet prior for $\boldsymbol{\pi}$ and normal and inverse gamma priors for μ_i and σ_i^2 of the form

$$\mu_i \sim N(\omega, \kappa^2),$$

$$\sigma_i^{-2} \sim \text{gamma}(\beta_1, \beta_2).$$

The $N(\omega, \kappa^2)$ distribution is taken to be flat over an interval of variation of the data, for example, by setting ω to be the midpoint of this interval and setting κ^2 equal to a small multiple of R^2, where R is the length of the interval.

Concerning the σ_i^{-2}, which are taken to be gamma(β_1, β_2), Richardson and Green (1997) introduced an additional hierarchical level by allowing β_2 to follow a gamma (a, b) distribution. In their hierarchical model, Richardson and Green (1997) noted there was substantial change in the posterior distribution of g as the prior mean of the precision σ_i^{-2}, β_1/β_2, was varied between $(R/5)^{-2}$, $(R/10)^{-2}$, and $(R/20)^{-2}$. Hence in the standard model with fixed β_1 and β_2, the values of the latter will crucially influence the posterior distribution of g, and so it is difficult to be weakly informative. Richardson and Green (1997) showed that with their hierarchical model with fixed

β_1, but random β_2 that allows weak information to be put in at a higher level, the posterior distribution of g is not so affected. They presented some results to support their choice of an hierarchical model with $\beta_1 = 2, a = 0.2$, and $b = 100(a/\beta_1)/R^2$, as their default option.

Phillips and Smith (1996) adopted priors similar to those in Richardson and Green (1997), with the mixing proportions following a Dirichlet distribution with all parameters α_i set equal to one, that is, a uniform prior over the region $\pi_1+\pi_2+\cdots+\pi_g = 1$. They also used normal-inverse gamma priors for the means μ_i and variances σ_i^2, as in Richardson and Green (1997), but with all the parameters in the prior distributions of these parameters fixed; that is, there were no unspecified hyperparameters.

In more recent work on reversible jump techniques in the case of an unknown number of components g, Gruet, Philippe, and Robert (1999) have implemented a reversible jump MCMC technique for exponential mixture estimation. On extensions of the mixture model, Robert, Rydén, and Titterington (2000) have shown how reversible jump MCMC techniques can be used in the case of a normal hidden Markov chain model, which will be considered in Chapter 13.

4.8 NONINFORMATIVE SETTINGS

With the specification of the proper (conjugate) priors in the above, the intent is not to use strong prior information on the mixture parameters. For example, Richardson and Green (1997) use weakly informative priors, which may or may not be data-dependent, a line taken by Raftery (1996) and Noble (1994). We shall show that it is not possible to have fully noninformative priors and obtain proper posterior distributions. This is because there is always the possibility that no observations will be allocated to one or more components, and so the data are uninformative about them. Thus standard choices of improper noninformative priors cannot be used.

We shall discuss some attempts to circumvent this problem through partially proper prior distributions.

4.8.1 Improper Priors

If an improper prior is adopted for the component parameters then the posterior distribution will be improper. To see this, consider a mixture of $g = 2$ univariate normal densities in proportions π_1 and π_2 with means μ_1 and μ_2 and variances σ_1^2 and σ_2^2. Then the term in the sum of the numerator of (4.1) with $\boldsymbol{z}_j = (1,\ 0)^T$ for all j contains no information on μ_2 and σ_2^2. Hence if an improper prior is adopted, then the integral of this term in the denominator of (4.1) will diverge.

To look at this more closely, we assume further, as in Wasserman (1999), that $\pi_1 = \pi_2 = 0.5, \sigma_1^2 = \sigma_2^2 = 1$, and $\mu_1 = 0$, so that the only unknown parameter is

$\Psi = \mu_2$. Then the likelihood $L(\Psi)$ is given by

$$L(\Psi) = \prod_{j=1}^{n} \{0.5\phi(y;\, 0, 1) + 0.5\phi(y;\, \mu_2,\, 1)\}. \tag{4.20}$$

To see the impropriety of the posterior for a flat prior $p(\Psi)$, we note that

$$\begin{aligned} \int L(\Psi)\, p(\Psi)\, d\Psi &= \int \prod_{j=1}^{n} \{0.5\phi(y_j;\, 0, 1) + 0.5\phi(y_j;\, \mu_2,\, 1)\}\, p(\Psi)\, d\Psi \\ &\geq \prod_{j=1}^{n} \{0.5\phi(y_j;\, 0, 1)\} \int p(\Psi)\, d\Psi \\ &= \infty. \end{aligned} \tag{4.21}$$

Wasserman (1999) also showed that the posterior distribution is improper here for the Jeffreys (1961) prior for which

$$\begin{aligned} p(\Psi) &= \{\mathcal{I}(\Psi)\}^{1/2} \\ &= [E\{\partial \log f(Y_j;\, \Psi)/\partial\Psi)\}^2]^{1/2}. \end{aligned}$$

4.8.2 Data-Dependent Priors

One approach to avoiding improper posterior distributions in mixture models as above is to adopt a subjective prior. However, as argued by Wasserman (1999), there are good reasons to favor partially proper priors. Firstly, it provides an objective method of choosing noninformative priors. Secondly, this approach may be viewed as a convenient way of doing valid frequentist inference. Thirdly, as established by Wasserman (1999), the only priors that produce intervals with second-order correct coverage are data-dependent.

Concerning the specification of partially proper priors, the approach used by Diebolt and Robert (1994) is to adopt conjugate priors and then to discard part of the likelihood corresponding to the troublesome assignment that assigns all the observations to one component.

Wasserman (1999) shows that the *ad hoc* fix used by Diebolt and Robert (1994) that directly modifies the posterior to force propriety has a formal justification as a data-dependent prior. The idea is to follow Diebolt and Robert (1994) and simply throw away the part of the likelihood corresponding to the troublesome allocations. Wasserman (1999) begins with the Jeffreys (1961) prior, whereas Diebolt and Robert (1994) begin with a conjugate prior.

4.8.3 Markov Prior on Component Means

For the univariate normal mixture model, Roeder and Wasserman (1997) adopt a partially proper prior that imposes a Markov prior on the component means. This is a partially proper prior in the sense that the marginal distribution of each μ_i is flat, but the conditional distribution of μ_i, given the other μ_h's, is proper. The former property avoids having to specify subjective prior information on the location. Roeder and Wasserman (1997) construct a partially proper prior for the component standard deviations by introducing a common scale parameter β. Conditional on β, the joint prior is a product of scaled inverse chi-squared distributions with common scale parameter β and η degrees of freedom. The prior for β is chosen to be the usual reference prior for a scale parameter. They call this a partially proper prior, because the marginal distribution of each σ_i has the usual reference prior σ_i^{-1}.

4.8.4 Reparameterization for Univariate Normal Components

A partially proper prior can be obtained through a reparameterization of a global location-scale parameter. It was first proposed by Robert and Mengersen (1995) in the case of a mixture of $g = 2$ univariate normal densities,

$$f(y_j;\, \boldsymbol{\Psi}) = \pi_1 \phi(y_j;\, \mu_1,\, \sigma_1^2) + \pi_2 \phi(y_j;\, \mu_2,\, \sigma_2^2). \tag{4.22}$$

Under their proposed reparameterization, the component means and variances, μ_i and σ_i^2, in (4.22) are expressed in terms of the ω_i and κ_i, where

$$\begin{aligned}
\mu_1 &= \omega_1, \\
\mu_2 &= \omega_1 + \kappa_1 \omega_2, \\
\sigma_1^2 &= \kappa_1^2, \\
\sigma_2^2 &= \kappa_1^2 \kappa_2^2.
\end{aligned}$$

For identifiability purposes, the global location-scale parameter corresponds to the first component and $\kappa_2 < 1$. Their prior does make it difficult to handle cases where the two normal components have the same variance. On the other hand, the prior of Roeder and Wasserman (1997) makes it difficult to handle cases where the two component normals have the same means.

Robert and Mengersen (1995) showed how it could be extended to an arbitrary number of components for estimation purposes. Starting with the two-component normal model, the second component is replaced by a two-component mixture

$$q_2 \phi(y_j;\, \mu_2,\, \sigma_2^2) + (1 - q_2)\phi(y_j;\, \mu_2 + \sigma_2 \omega_3,\, \sigma_2^2 \kappa_3^2).$$

This leads to the three-component mixture,

$$f(y_j;\, \boldsymbol{\Psi}) = \sum_{i=1}^{3} \pi_i \phi(y_j;\, \mu_i,\, \sigma_i^2), \tag{4.23}$$

where

$$\pi_1 = q_1,$$

$$\begin{aligned}
\pi_2 &= (1-q_1)q_2 \\
\pi_3 &= (1-q_1)(1-q_2), \\
\mu_1 &= \omega_1 \\
\mu_2 &= \omega_1 + \kappa_1\omega_2, \\
\mu_3 &= \omega_1 + \kappa_1\omega_2 + \kappa_1\kappa_2\omega_3, \\
\sigma_1^2 &= \kappa_1^2 \\
\sigma_2^2 &= \kappa_1^2\kappa_2^2 \\
\sigma_3^2 &= \kappa_1^2\kappa_2^2\kappa_3^2.
\end{aligned} \tag{4.24}$$

This reparameterization can be extended in an obvious way from (4.24) to the case of a g-component normal mixture model for arbitrary g.

As shown in Robert and Mengersen (1995), the link between the different components created by the reparameterization dispenses with using a proper prior for the μ_i and the σ_i^2 and allows for the following prior distribution:

$$p(\omega_1, \kappa_1) = 1/\kappa_1,\ \kappa_i \sim U(0,1), \omega_i \sim N(0, \zeta^2)\ (i = 2, \ldots, g),$$

where the hyperparameter ζ is to be specified, although it has little bearing on the result.

Robert and Mengersen (1999) claimed that a direct Gibbs sampling implementation is not possible for the estimates, because of the presence of the scale factors $\kappa_1, \kappa_2, \ldots, \kappa_{g-1}$ in the means of the components. Robert and Mengersen (1999) consequently used a hybrid algorithm with the Gibbs sampler in conjunction with a Metropolis step. Contrary to this approach, Robert and Titterington (1998), in the more general setting of a hidden Markov model, implemented a full Gibbs algorithm for the simulation of the posterior distribution. They noted that the Jacobian of the transformation from the ω_i and the κ_i $(i = 1, \ldots, g)$ to the μ_i and the σ_i is

$$\prod_{i=1}^{g-1} \sigma_i^{-2}.$$

The prior $p(\boldsymbol{\xi})$ for $\boldsymbol{\xi} = (\mu_1, \ldots, \mu_g, \sigma_1, \ldots, \sigma_g)^T$ can be expressed as

$$p(\boldsymbol{\xi}) = \sigma_1^{-3}\sigma_g^2 \prod_{i=2}^{g} \sigma_i^{-2} \exp\{-\tfrac{1}{2}(\mu_i - \mu_{i-1})^2/(\zeta^2\sigma_{i-1}^2)\}$$

for $\sigma_g^2 \leq \sigma_{g-1}^2 \leq \cdots \leq \sigma_1^2$. The posterior density for $\boldsymbol{\Psi}$, conditional on $\boldsymbol{z}$, is

$$p(\boldsymbol{\xi}) \prod_{i=1}^{g} \pi_i^{n_i}\sigma_i^{-n_i} \prod_{i=1}^{g} \exp\{-\tfrac{1}{2}[n_i(\bar{y}_i - \mu_i)^2 + n_i v_i^2]/\sigma_i^2\}, \tag{4.25}$$

where

$$v_i^2 = \sum_{j=1}^{n} z_{ij}(y_j - \bar{y}_i)^2/n_i \quad (i = 1, \ldots, g).$$

Robert and Titterington (1998) give the full conditional distributions that can be obtained from (4.25) for implementing the Gibbs sampler.

4.9 LABEL-SWITCHING PROBLEM

As mentioned in Section 1.1, the so-called label-switching problem is an issue that has to be addressed in posterior simulations for mixtures with components belonging to the same parametric family. If the prior distribution is symmetric in the components of the mixture, then the posterior distribution will be invariant under a permutation of the component labels. This lack of identifiability of $\boldsymbol{\Psi}$ due to the interchanging of component labels is generally handled by a constraint on the mixing proportions of the form

$$\pi_1 \leq \pi_2 \leq \cdots \leq \pi_g, \tag{4.26}$$

or in the case of univariate component densities, say, normal, by the constraint

$$\mu_1 \leq \mu_2 \leq \cdots \leq \mu_g \tag{4.27}$$

on the component means, or a similar one on the component variances.

However, as recently stressed by Celeux, Hurn, and Robert (2000), this does not always work. Indeed, they go so far to say that, "Although somewhat presumptuous, we consider that almost the entirety of Markov chain Monte Carlo samplers implemented for mixture models has failed to converge!"

Suppose we are fitting a mixture of $g = 2$ components with both mixing proportions close to 0.5. Then if we simulate under the ordering restriction (1.27) on the mixing proportions (with a weak prior on the mixing proportions), we will get a biased sample of values for π_1 and π_2, since they are "pushed apart" under this ordering restriction. Similarly, if their means are close together, and we carry out posterior simulation under the restriction $\mu_1 < \mu_2$ (under a weakly informative prior), we will get a biased sample for these means, as they get "pushed apart" under the ordering restriction on them. With biased samples, ergodic averages will give misleading estimates.

To further demonstrate this label switching, we consider the illustration given by Richardson and Green (1997). They simulated a data set of $n = 250$ points from a mixture of $g = 2$ univariate normal densities in equal proportions $\pi_1 = \pi_2 = 0.5$ with means $\mu_1 = 0.0$ and $\mu_2 = 1.0$ and variances $\sigma_1^2 = 2.25$ and $\sigma_2^2 = 1.0$. We report in Figures 1.1 and 1.2 their results for the posterior distribution of the mixture parameters obtained by simulation for $g = 2$ components. Figure 1.1 gives the results obtained under an ordering of their means, while Figure 1.2 gives the corresponding results for an ordering of their variances. With the latter restriction, the simulated densities so obtained for the component means are bimodal, indicating that label switching took place on about half the runs. The estimated posterior densities under the ordering of the component means are much clearer, as they are unimodal for all the parameters, although there is still some evidence of label switching. If there is no obvious choice of labeling in practice, Richardson and Green (1997) suggest post-processing of the posterior simulation runs according to different choices of labels to obtain the clearest picture of the component parameters.

Celeux (1999) recommends performing the simulations without any constraints on the parameters and then, at the end of the simulations, applying a clustering-like

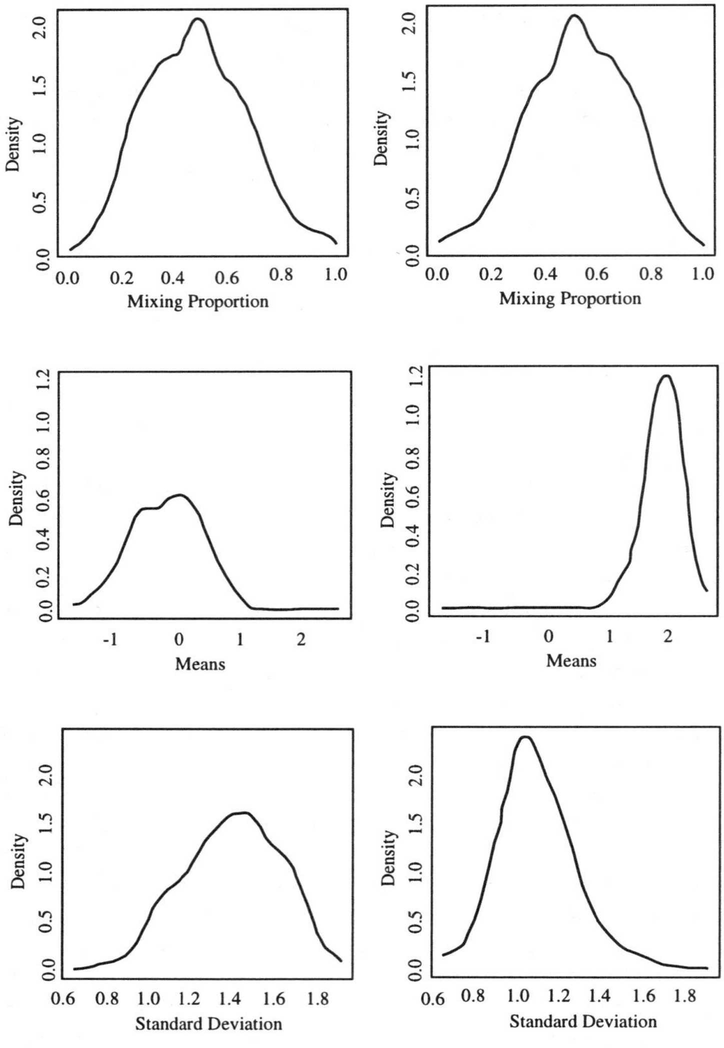

Fig. 4.1 Posterior densities of parameter estimates labeled according to ordering of component means. From Richardson and Green (1997).

method to change where appropriate the component labels of the simulated values for $\boldsymbol{\Psi}$. Stephens (1999) proposes a similar way of handling the problem.

West (1999) also recommends simulating without restrictions on the parameters, noting that in his experience, imposing an identifying ordering on parameters hinders convergence of the resulting MCMC on the constrained parameter space. He points out that an additional gain by operating the MCMC on the unrestricted parameter space is an aid in assessing convergence of the simulation. As the posterior distribution of the parameters is symmetric in the component labels under a symmetric prior,

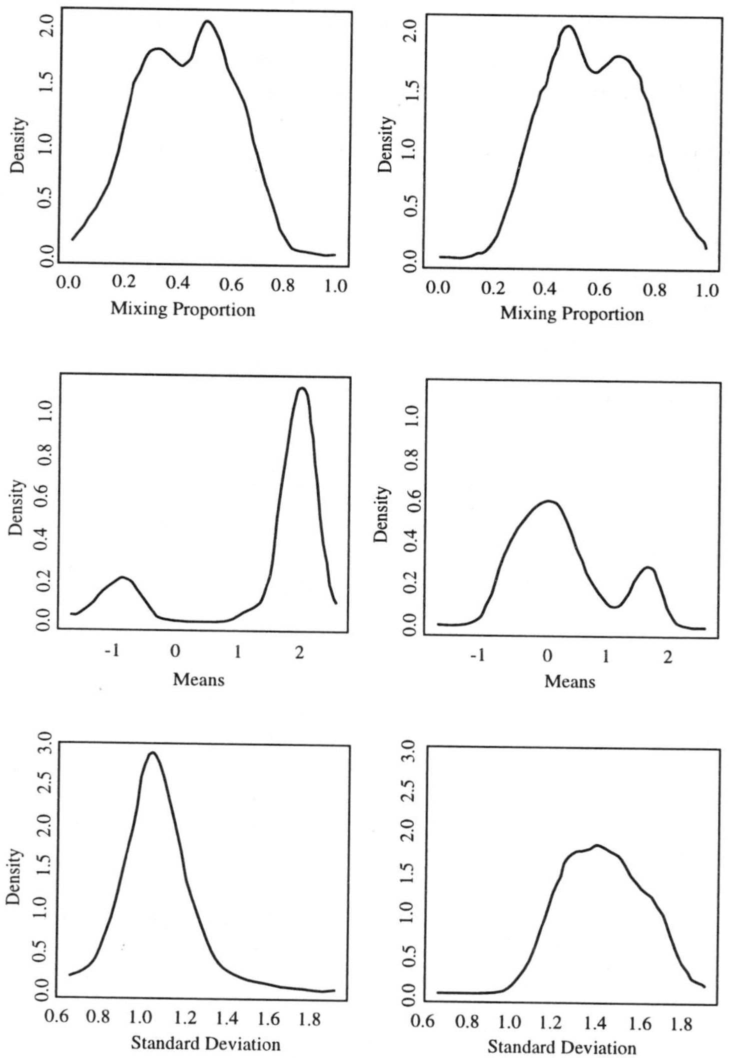

Fig. 4.2 Posterior densities of parameter estimates labeled according to ordering of component variances. From Richardson and Green (1997).

symmetry should be evident in the marginal distributions of the parameters. Thus the MCMC simulations on the unrestricted parameter space should exhibit switching effects, as parameter draws jump between the models representing the identification issue. However, as warned by West (1999), convergence will be slow. One reason is that the sampler may be unable to leave the vicinity of one mode (or a subset of modes) to move to another mode (or subset of modes) of equal importance, because of its inability to step over valleys of low probability.

4.10 PRIOR FEEDBACK APPROACH TO ML ESTIMATION

As discussed in Robert and Titterington (1998), several approaches have exploited Gibbs sampling to come up with new approximation techniques for ML estimation, thus circumventing the EM algorithm approach; see, for example, Geyer and Thompson (1992). The method considered by Robert and Titterington (1998) is based on a remark, recurrent in the literature (Pincus, 1968; Rubinstein, 1981; Aitkin, 1991), that the effect of the prior density fades away when the likelihood function is taken to a high enough power. As they argue, when Bayesian posterior expectation is straightforward, it is reasonable to consider this alternative approach to the computation of the MLE (Robert, 1993; Robert and Soubiran, 1993).

The prior feedback method is implemented by running the Gibbs sampler with an increasing number of replications of the original sample $\boldsymbol{y} = (\boldsymbol{y}_1^T, \ldots, \boldsymbol{y}_n^T)^T$, until the Bayesian posterior expectations stabilize. More precisely, let $\boldsymbol{y}^{(r)}$ denote the rth copy (replication) of $\boldsymbol{y}$ $(r = 1, \ldots, m)$. The Gibbs sampler is run on each copy, firstly generating the vector $\boldsymbol{z}^{(r)}$ of component-indicator variables and then proceeding conditionally on $\boldsymbol{z}^{(r)}$ and generating the subvectors of $\boldsymbol{\Psi}$.

The theoretical justification of the prior feedback method is that the posterior density of $\boldsymbol{\Psi}$ associated with m replications of the sample,

$$p_m(\boldsymbol{\Psi}) \propto p(\boldsymbol{\Psi})\, L(\boldsymbol{\Psi})^m, \tag{4.28}$$

converges to a Dirac mass at the global maximizer of $L(\boldsymbol{\Psi})$, as m goes to infinity. The index size m can be interpreted as the temperature factor in simulated annealing.

As noted by Robert and Titterington (1998), for mixtures of normal components with unequal covariance matrices, the likelihood is unbounded. They therefore recommend choosing the prior so that the posterior density $p_m(\boldsymbol{\Psi})$ is bounded. In this case, the posterior expectation should converge to the global maximizer of the likelihood function within some truncated version of the support. However, they note that due to multiple local maxima in the likelihood function, coupled with the difficulty of the Gibbs sampler in leaving the neighborhood of a strongly attractive mode, that the prior feedback estimate can converge to a local maximizer of the likelihood function. They demonstrate this for a hidden Markov chain. In Chapter 13 we report an example of Robert and Titterington (1998) in which the prior feedback method is applied to some data simulated from a hidden Markov chain.

4.11 VARIATIONAL APPROACH TO BAYESIAN ESTIMATION

Recently, Attias (1999a) described the variational framework for providing an approximation to the Bayesian solution. This approach approximates the full posterior distribution for the unknown parameter vector $\boldsymbol{\Psi}$ and any hidden (latent) variables such as $\boldsymbol{z}$ in the case of the fitting of mixture distributions. It draws together variational ideas from intractable hidden variables models (Saul, Jaakkola, and Jordan, 1996; Ghahramani and Jordan, 1997; Ghahramani and Beal, 2000) and from Bayesian inference (Jaakkola and Jordan, 2000), which, in turn, draw on the work of Neal and Hinton (1998).

The variational approach considers a lower bound on the log integrated density, $\log p(\boldsymbol{y})$, of the form

$$\log p(\boldsymbol{y}) \geq \int \sum_{\boldsymbol{z}} q(\boldsymbol{\Psi}, \boldsymbol{z} \mid \boldsymbol{y}) \log \frac{p(\boldsymbol{y}, \boldsymbol{z}, \boldsymbol{\Psi})}{q(\boldsymbol{\Psi}, \boldsymbol{z} \mid \boldsymbol{y})} d\boldsymbol{\Psi} \tag{4.29}$$

for a given model, where $q(\boldsymbol{\Psi}, \boldsymbol{z} \mid \boldsymbol{y})$ is any arbitrary density. It proceeds to optimize this lower bound over a restricted class of functions q, noting that the solution in the unrestricted case is the posterior density, for which equality is achieved in (4.29). This approach restricts the class of allowed functions q to densities that imply $\boldsymbol{\Psi}$ and $\boldsymbol{z}$ are conditionally independent given the observed data $\boldsymbol{y}$ and the model. The optimization of the variational posterior is carried out over the parameter space via an iterative EM-like algorithm whose convergence is guaranteed. Attias (1999a) demonstrated the variational approach to the fitting of mixture models to some toy and real data sets.

4.12 MINIMUM MESSAGE LENGTH

The mixture modeling program, Snob, uses the Minimum Message Length (MML) principle to fit a mixture model. More details on this program are given in the Appendix on mixture software. For a review of MML inductive inference, the reader is referred to Wallace and Boulton (1968), Wallace and Freeman (1987), and Wallace and Dowe (2000).

Wallace and Freeman (1987) argue that statistical estimation can be performed as a coding process. They stated the basic philosophy of the minimum encoding methods as follows (Wallace and Freeman, 1987; Rissanen, 1986):

> "We may first estimate the parameters and then the data under the assumption that these are the true values. The encoded string must now, however, contain a specification of the estimated values. Any model is, therefore, only worth considering if the shortening of the encoded data string achieved by adopting it more than compensates for the lengthening caused by having to quote estimated parameter values. We thus naturally arrive at a very simple trade-off between the complexity of a model and its goodness of fit."

As explained in Oliver, Baxter, and Wallace (1996) and Baxter and Oliver (2000) in the context of mixture models, the message consists of two parts. The first part states g, followed by the parameters for each component. The second part of the message states the observed data $\boldsymbol{y}$ under the assumption that the estimates of the parameters are the true values.

As reported in Baxter and Oliver (2000), Wallace and Freeman (1987) estimate the expected message length as

$$-\log L(\boldsymbol{\Psi}) - \log p(\boldsymbol{\Psi}) + \tfrac{1}{2}\log \mid \mathcal{I}(\boldsymbol{\Psi}) \mid + \tfrac{1}{2}d + \tfrac{1}{2}d \log \kappa_d, \tag{4.30}$$

where $p(\boldsymbol{\Psi})$ is the prior density for $\boldsymbol{\Psi}$, d is the number of unknown parameters, and κ_d is the d-dimensional optimal quantizing lattice constant. For example, $\kappa_1 = 1/12$ and $\kappa_2 = 5/(36\sqrt{3})$; see Table 2.3 in Conway and Sloane (1988).

Thus using (4.30) for a given number of components g, the MML estimate of $\boldsymbol{\Psi}$ is the value of $\boldsymbol{\Psi}$ that maximizes the log posterior density penalized by the term $\frac{1}{2} \mid \mathcal{I}(\boldsymbol{\Psi}) \mid$.

The approximated expected message length (4.30) is very similar to approximations to the negative of the log integrated likelihood, $\log p(\boldsymbol{y})$, obtained by Laplace's method; see (6.47) and (6.48). The integrated likelihood $p(\boldsymbol{y})$ is given by

$$p(\boldsymbol{y}) = \int p(\boldsymbol{\Psi}, \boldsymbol{y})\, d\boldsymbol{\Psi}. \tag{4.31}$$

For example, from (6.48), we have

$$-\log p(\boldsymbol{y}) \approx -\log L(\hat{\boldsymbol{\Psi}}) - \log p(\hat{\boldsymbol{\Psi}}) + \tfrac{1}{2}\log \mid \boldsymbol{I}(\hat{\boldsymbol{\Psi}}; \boldsymbol{y}) \mid - \tfrac{1}{2}d \log(2\pi), \tag{4.32}$$

where $\boldsymbol{I}(\hat{\boldsymbol{\Psi}}; \boldsymbol{y})$ is the observed information matrix.

The expression (4.30) is unable to be used directly in practice for, as noted in Chapter 2, the expected information matrix $\mathcal{I}(\boldsymbol{\Psi})$ is very difficult to calculate for mixture models. Even its commonly used approximation, the observed information $\boldsymbol{I}(\hat{\boldsymbol{\Psi}}; \boldsymbol{y})$, is difficult to calculate for mixture models. It would seem that in applications of the MML approach, the expected information matrix is usually approximated by the complete-data expected information matrix, $\mathcal{I}_c(\boldsymbol{\Psi})$ (Baxter and Oliver, 2000). We note that the determinants of the complete-data expected information matrix $\mathcal{I}_c(\boldsymbol{\Psi})$ and (the incomplete-data) expected information matrix $\mathcal{I}(\boldsymbol{\Psi})$ can be quite different. To see this, we have from (2.50), for the EM algorithm applied to compute the MLE $\hat{\boldsymbol{\Psi}}$, that the rate of convergence depends on the smallest eigenvalue of

$$\mathcal{I}_c^{-1}(\hat{\boldsymbol{\Psi}}; \boldsymbol{y})\boldsymbol{I}(\hat{\boldsymbol{\Psi}}; \boldsymbol{y}).$$

The EM algorithm can converge very slowly, in particular when the components of the mixture are not well separated, which indicates that these two matrices, $\mathcal{I}_c(\hat{\boldsymbol{\Psi}}; \boldsymbol{y})$ and $\boldsymbol{I}(\hat{\boldsymbol{\Psi}}; \boldsymbol{y})$, can be quite different. Now the observed information matrix $\boldsymbol{I}(\hat{\boldsymbol{\Psi}})$ should be similar to the estimated expected information matrix $\mathcal{I}(\hat{\boldsymbol{\Psi}})$; and for component densities belonging to the regular exponential family, $\mathcal{I}_c(\hat{\boldsymbol{\Psi}}; \boldsymbol{y})$ is equal to $\mathcal{I}_c(\hat{\boldsymbol{\Psi}})$. Thus it suggests that the determinants of $\mathcal{I}_c(\boldsymbol{\Psi})$ and $\mathcal{I}(\boldsymbol{\Psi})$ can be quite different.

Baxter and Oliver (2000) considered an improved estimate of the expected information matrix $\mathcal{I}(\boldsymbol{\Psi})$ for use in the MML criterion (4.30) by adopting $\boldsymbol{I}(\boldsymbol{\Psi}; \boldsymbol{y})$ as approximated by the empirical information matrix (2.60). They also considered a computationally less demanding approximation that forms the determinant of the empirical information matrix by ignoring its off-diagonal elements.

5

Mixtures with Nonnormal Components

5.1 INTRODUCTION

In this chapter we consider the fitting of mixture models with nonnormal component densities. We first consider the case of mixed feature variables, where some are continuous and some are categorical. We shall outline the use of the location model for the component densities, as in Jorgensen and Hunt (1996), Lawrence and Krzanowski (1996), and Hunt and Jorgensen (1999).

The ML fitting of commonly used components, such as the binomial and Poisson, can be undertaken within the framework of a mixture of generalized linear models (GLMs). This mixture model also has the capacity to handle the regression case, where the random variable $\boldsymbol{Y}_j$ for the jth entity is allowed to depend on the value $\boldsymbol{x}_j$ of a vector $\boldsymbol{x}$ of covariates. If the first element of $\boldsymbol{x}$ is taken to be one, then we can specialize this model to the nonregression situation by setting all but the first element in the vector of regression coefficients to zero.

One common use of mixture models with discrete data is to handle overdispersion in count data. For example, in medical research, data are often collected in the form of counts, corresponding to the number of times that a particular event of interest occurs. Because of their simplicity, one-parameter distributions for which the variance is determined by the mean are often used, at least in the first instance to model such data. Familiar examples are the Poisson and binomial distributions, which are members of the one-parameter exponential family. However, there are many situations where these models are inappropriate, in the sense that the mean–variance relationship implied by the one-parameter distribution being fitted is not valid. In most of these situations, the data are observed to be overdispersed; that is, the observed sample variance is larger than that predicted by inserting the sample mean into the mean–variance relationship. This phenomenon is called overdispersion. There are occasions in data analysis where the data are underdispersed (Faddy, 1994); that is, the sample variance is smaller than that implied by the mean–variance relationship. These phenomena are also observed with the fitting of regression models, where the mean (say, of the Poisson or the binomial distribution) is modeled as a function of some covariates. If this dispersion is not taken into account, then using these models may lead to biased estimates of the parameters and consequently incorrect inferences about the parameters (Wang, 1994; Wang et al., 1996). In this chapter, we focus on the more common case of overdispersion.

Concerning mixtures for multivariate discrete data, a common application arises in latent class analyses, in which the feature variables (or response variables in a regression context) are taken to be independent in the component distributions. This latter assumption allows mixture models in the context of a latent class analysis to be fitted within the above framework of mixtures of GLMs.

In work outside this chapter on mixtures of nonnormal components, we consider in Chapter 10 mixture models for inference from nonnormal univariate data arising in the context of survival or reliability analysis. In Chapter 7 we present some recent results on the fitting of mixtures of multivariate t components, while in Chapter 11 we shall give a case study using mixtures of Kent distributions for the modeling of multivariate directional data.

5.2 MIXED CONTINUOUS AND CATEGORICAL VARIABLES

We consider now the problem of fitting a mixture model

$$f(\boldsymbol{y}_j;\, \boldsymbol{\Psi}) = \sum_{i=1}^{g} f_i(\boldsymbol{y}_j;\, \boldsymbol{\theta}_i) \tag{5.1}$$

to some data, $\boldsymbol{y} = (\boldsymbol{y}_1^T, \ldots, \boldsymbol{y}_n^T)^T$, where some of the feature variables are categorical. The simplest way to model the component densities of these mixed feature variables is to proceed on the basis that the categorical variables are independent of each other and of the continuous feature variables, which are taken to have, say, a multivariate normal distribution. Although this seems a crude way in which to proceed, it often does well in practice as a way of clustering mixed feature data. In the case where there are data of known origin available, this procedure is known as the naive Bayes classifier.

We can refine this approach by adopting the location model, introduced by Olkin and Tate (1961) and used in discriminant analysis initially by Chang and Afifi (1974) and Krzanowski (1975). This model has come into prominence more recently because of its appearance in the graphical modeling of mixed variables, where it is known as the conditional Gaussian distribution model (Whittaker, 1990; Cox and Wermuth, 1992). The location model has since been used by Jorgensen and Hunt (1996), Lawrence and Krzanowski (1996), and Hunt and Jorgensen (1999) for the fitting of mixture models to mixed categorical and continuous variables. The identifiability of this model for mixtures has been investigated by Willse and Boik (1999).

Previously, Everitt (1988b) proposed, and Everitt and Merette (1990) studied, a model in which the categorical variables were assumed to have arisen through thresholding of unobservable continuous variables. The thresholds that define the categories are treated as extra parameters. As the log likelihood contains multivariate normal integrals over as many variables as are thresholded with limits of integration depending on these thresholds, the procedure is only a practical proposition if the number of categorical variables is relatively small (Lawrence and Krzanowski, 1996).

5.2.1 Location Model-Based Approach

Suppose that p_1 of the p feature variables in $\boldsymbol{Y}_j$ are categorical, where the qth categorical variable takes on m_q distinct values ($q = 1, \ldots, p_1$). Then there are $m = \prod_{q=1}^{p_1} m_q$ distinct patterns of these p_1 categorical variables. With the location model, the p_1 categorical variables are replaced by a single multinomial random variable $\boldsymbol{Y}_j^{(1)}$ with m cells; that is, $(\boldsymbol{Y}_j^{(1)})_s = 1$ if the realizations of the p_1 categorical variables in $\boldsymbol{Y}_j$ correspond to the sth pattern. Any associations between the original categorical variables are converted into relationships among the resulting multinomial cell probabilities. The location model assumes further that conditional on $(\boldsymbol{y}_j^{(1)})_s = 1$ and membership of the ith component of the mixture model, the distribution of the $p - p_1$ continuous feature variables is normal with mean $\boldsymbol{\mu}_{is}$ and covariance matrix $\boldsymbol{\Sigma}_i$, which is the same for all cells s (that is, for all possible combinations of the realizations of the p_1 categorical variables). This assumption of conditional normality means that ML estimation is straightforward to implement via the EM algorithm. The expressions for the updated estimates on the M-step of the multinomial cell probabilities and of the component means and component-covariance matrices can be given in closed form as follows.

Let p_{is} be the conditional probability that $(\boldsymbol{Y}_j^{(1)})_s = 1$ given its membership of the ith component of the mixture ($s = 1, \ldots, m$; $i = 1, \ldots, g$). Also, let δ_{js} be one or zero, according to whether $(\boldsymbol{y}_j^{(1)})_s$ equals one or zero, and let $\boldsymbol{y}_j^{(2)}$ contain the continuous feature variables in $\boldsymbol{y}_j$. Then on the $(k+1)$th iteration of the EM algorithm, the updated estimates are given by

$$\pi_i^{(k+1)} = \sum_{s=1}^{m} \sum_{j=1}^{n} \delta_{js} \tau_{ijs}^{(k)} / n, \tag{5.2}$$

$$p_{is}^{(k+1)} = \sum_{j=1}^{n} \delta_{js} \tau_{ijs}^{(k)} \Big/ \sum_{s=1}^{m} \sum_{j=1}^{n} \delta_{js} \tau_{ijs}^{(k)}, \tag{5.3}$$

$$\boldsymbol{\mu}_{is}^{(k+1)} = \sum_{j=1}^{n} \delta_{js} \tau_{js}^{(k)} \boldsymbol{y}_j^{(2)} \Big/ \sum_{j=1}^{n} \delta_{js} \tau_{ijs}^{(k)}, \tag{5.4}$$

and

$$\boldsymbol{\Sigma}_i^{(k+1)} = \sum_{s=1}^{m} \sum_{j=1}^{n} \delta_{js} \tau_{ijs}^{(k)} (\boldsymbol{y}_j^{(2)} - \boldsymbol{\mu}_{is}^{(k+1)}) (\boldsymbol{y}_j^{(2)} - \boldsymbol{\mu}_{is}^{(k+1)})^T \Big/ \sum_{s=1}^{m} \sum_{j=1}^{n} \delta_{js} \tau_{ijs}^{(k)}, \tag{5.5}$$

where

$$\tau_{ijs}^{(k)} = \frac{\pi_i^{(k)} p_{is}^{(k)} \phi(\boldsymbol{y}_j^{(2)};\, \boldsymbol{\mu}_{is}^{(k)},\, \boldsymbol{\Sigma}_i^{(k)})}{\sum_{h=1}^{g} \pi_h^{(k)} p_{hs}^{(k)} \phi(\boldsymbol{y}_j^{(2)};\, \boldsymbol{\mu}_{hs}^{(k)},\, \boldsymbol{\Sigma}_h^{(k)})} \tag{5.6}$$

for $s = 1, \ldots, m;\ i = 1, \ldots, g$.

5.2.2 Implementation of Location Model

In practice, the number of parameters with the location model approach can be large if the multinomial distribution replacing the categorical variables has many cells and there are several continuous feature variables. We now outline the implementation of the location model approach as adopted by Hunt and Jorgensen (1999) in their MULTMIX program.

Suppose, for the moment, that the first two feature variables are binary variables, taking on without loss of generality the values one or zero. Let $y_{vj} = (\boldsymbol{y}_j)_v$ $(v = 1, \ldots, p)$. Suppose also that Y_{1j} is independent of all the other feature variables, but that Y_{2j} is not independent of the remaining $(p-2)$ continuous variables. Then we partition the feature vector $\boldsymbol{y}_j$ as

$$\boldsymbol{y}_j = (y_{1j},\, y_{2j},\, \boldsymbol{y}_j^{(2)^T})^T, \tag{5.7}$$

where $\boldsymbol{y}_j^{(2)}$ is the subvector containing the $(p-2)$ continuous variables. Then the ith component density of $\boldsymbol{Y}_j$ is modeled as

$$f_i(\boldsymbol{y}_j) = \{\prod_{v=1}^{2} f_B(y_{vj};\, \theta_{vi})\}\phi(\boldsymbol{y}_j^{(2)};\, \boldsymbol{\mu}_{is},\, \boldsymbol{\Sigma}_i), \tag{5.8}$$

where

$$f_B(w) = \theta^w\, (1-\theta)^{1-w}$$

denotes the binomial frequency function for a zero–one variable W taking the values one and zero with probabilities θ and $(1-\theta)$, respectively, and where $\boldsymbol{\mu}_{is}$ $(s = 1, 2)$ denotes the mean of $\boldsymbol{Y}_j^{(2)}$ corresponding to the two distinct values of the binary variable Y_{2j}.

If the preliminary fit suggests that it is unreasonable to take the two binary variables Y_{1j} and Y_{2j} to be independent, then they are replaced by a single multinomial variable $\boldsymbol{Y}_j^{(1)}$ (here with $s = 4$ cells), as in the general case described above. The conditional distribution of the subvector of continuous variables $\boldsymbol{Y}_j^{(2)}$ is then assumed to be normal with covariance matrix $\boldsymbol{\Sigma}_i$ and mean $\boldsymbol{\mu}_{is}$ corresponding to $(\boldsymbol{y}_j^{(1)})_s = 1$ $(s = 1, \ldots, 4)$.

In a general situation involving categorical and continuous variables, the intent in MULTIMIX is to divide the feature vector into as many subvectors as possible that can be taken to be independently distributed. The extreme form would be to take all p feature variables to be independent and to include correlation structure where necessary. The question of whether the correlation structure of an existing model should be expanded to include additional local associations between the variables, categorical or continuous or both, is considered formally in terms of the change in the log likelihood, but informal considerations include the inspection of scatter plots and two-way tables. In contrast to the use of the likelihood ratio test statistic, $-2 \log \lambda$, for the number of components, its asymptotic null distribution with degrees of freedom equal to the difference between the two models should not give misleading results in this role where the models are nested; see Hunt and Jorgensen (1999).

5.3 EXAMPLE 5.1: PROSTATE CANCER DATA

To further illustrate the approach adopted in MULTIMIX, we report in some detail a case study of Hunt and Jorgensen (1999).

5.3.1 Description of Data Set

Hunt and Jorgensen (1999) considered the clustering of patients on the basis of pretrial covariates alone for the prostate cancer clinical trial data of Byar and Green (1980), which is reproduced in Andrews and Herzberg (1985, p. 261–274). This data set was obtained from a randomized clinical trial comparing four treatments for $n = 506$ patients with prostatic cancer grouped on clinical criteria into Stages 3 and 4 of the disease. As reported by Byar and Green (1980), Stage 3 represents local extension of the disease without evidence of distant metastasis, while Stage 4 represents distant metastasis as evidenced by elevated acid phosphatase, X-ray evidence, or both.

Hunt and Jorgensen (1999) assessed the clusters obtained by MULTIMIX in comparison with the clinical stages, and they also considered the trial outcomes for patients in different clusters. The treatments consisted of estrogen therapy at differing rates. Daily pills containing 0.0 (placebo), 0.2, 1.0, and 5.0 mg of diethylstilbestrol were administered in the four treatments. As Byar and Green (1980) noted little difference between the effects of the first two treatments and also between the effects of the last two treatments, Hunt and Jorgensen (1999) called patients in either of the first two treatments "Untreated" and in either of the last two treatments "Treated." Twelve pre-trial covariates (see Table 5.1) were measured on each patient; seven may be taken to be continuous and four to be discrete. The one variable (SG), which could be considered either discrete or continuous, was taken to be continuous. The skewed variables, SZ (the size of the primary tumor) and AP (serum prostatic acid phosphatase), were transformed by using a square root and a logarithm transformation, respectively. Observations that had missing values in any of the twelve pretreatment covariates were omitted from further analysis, leaving 475 out of the original 506 observations available.

Table 5.1 Pretreatment Covariates

Covariate	Abbreviation	Number of Levels (if Categorical)
Age	Age	
Weight index	WtI	
Performance rating	PF	4
Cardiovascular disease history	HX	2
Systolic blood pressure	SBP	
Diastolic blood pressure	DBP	
Electrocardiogram code	EKG	7
Serum hemoglobin	HG	
Size of primary tumor	SZ	
Index of tumor stage and histolic grade	SG	
Serum prostatic acid phosphatase	AP	
Bone metastases	BM	2

Source: From Hunt and Jorgensen (1999).

5.3.2 Fitting Strategy under MULTIMIX

Hunt and Jorgensen (1999) fitted a two-component mixture model, using a fitting strategy that was a form of forward selection of covariances, beginning with Model [ind], in which the $p = 12$ variables were taken to all be independent. Progressively local associations were added to the model by taking coarser and coarser partitions of the set of the $p = 12$ variables. The modifications to the current model were determined by examining correlations, scatter plots, and two-way tables within each of the two clusters formed by allocating each observation according to the current model. Table 5.2 summarizes the results of this fitting process.

Table 5.2 Models and Fits

Model	Variable Groups	No. of Parameters d	Log Likelihood +11386.265
[ind]	–	55	0.000
[2]	{SBP,DBP}	57	117.542
[3, 2]	{BM,WtI,HG},{SBP,DBP}	63	149.419
[5]	{BM,WtI,HG,SBP,DBP}	75	169.163
[9]	Complement of {PF,HX,EKG}	127	237.092

Source: From Hunt and Jorgensen (1999).

When the data had been grouped into two classes following the fitting of Model [ind], correlations between SBP and DBP of about 0.62 were observed within both of the classes. As they appeared to be the strongest associations, they were put together in a bivariate subvector that was taken to have a two-component bivariate normal distribution with an arbitrary covariance matrix (that is, with three parameters for each of the two components in the mixture model). The remaining variables were taken as singleton subvectors. Thus this model labeled here as [2] has two more parameters in it.

The next group of variables chosen was the triple, BM, WtI, HG, giving a location model factor to the mixture densities, because BM is dichotomous while WtI and HG are continuous. The resultant model is denoted here by [3,2], referring to the size of the two subvectors that have more than one variable in them. Six extra parameters are introduced in this change: There are four new mean parameters, because the fitted means of WtI and HG are now specific to each level of BM within each subpopulation, and two new covariance parameters. In Model [5], these two variable groups are combined into one subvector at a cost of introducing 12 new parameters. Finally, in Model [9], all variables, apart from PF, HX, and EKG (which are still taken to be singeltons), are combined into a nine-dimensional subvector. The results are given in Table 5.2, which gives the number of parameters d and the log likelihood for each of the models.

Hunt and Jorgensen (1999) reported that the log likelihoods were obtained from several initial configurations including random groupings of the observations. However, Model [9] proved to be sensitive to the choice of starting configuration and the greatest log likelihood over four runs is shown for this model. Convergence was usually obtained after 60 to 70 iterations although one run for Model [9] reached 200 iterations without converging. They noted that there was little difference between the group allocations determined by [ind], [2], [3,2], and [5], with the allocation of only four patients out of 475 changing between these models. Model [9] allocations were found to be sensitive to the initial classification and did not agree so closely with each other nor with the classifications of the more parsimonious models. A comparison of the clustering under Model [2] with the clinical grouping into Stages 3 and 4 of the disease shows one cluster with 252 Stage 3 and 21 Stage 4 patients and the other cluster with 21 Stage 3 and 181 Stage 4 patients.

Hunt and Jorgensen (1999) compared the posttrial survival status of patients with their pretrial status (Stage 3 or 4) and with their two-cluster allocation implied by the two-component mixture model fitted. This information is reproduced here in Table 5.3 for Model [2]. It can be seen that cluster-one membership and clinical Stage 3 status are associated with a better chance of survival. The patterns of outcomes for the 42 patients whose model and clinical classifications conflict suggest that the mixture model-based classifications are better indicators of prognosis than the clinical criteria used. This is especially noticeable among the "Treated" patients.

Table 5.3 Clusters and Outcomes for Treated and Untreated Patients

Patient Group		Outcome			
		Alive	Prostate Death	Cardio Death	Other Death
		Untreated Patients			
Cluster 1	Stage 3	39	18	37	33
	Stage 4	3	4	3	3
Cluster 2	Stage 3	1	4	2	3
	Stage 4	14	49	18	6
		Treated Patients			
Cluster 1	Stage 3	50	3	52	20
	Stage 4	4	0	1	3
Cluster 2	Stage 3	1	6	3	1
	Stage 4	25	37	22	10

Source: From Hunt and Jorgensen (1999).

5.4 GENERALIZED LINEAR MODEL

5.4.1 Definition

With the generalized linear model (GLM) approach originally proposed by Nelder and Wedderburn (1972), the log density of the (univariate) variable Y_j has the form

$$\log f(y_j;\, \theta_j, \kappa) = m_j \kappa^{-1}\{\theta_j y_j - b(\theta_j)\} + c(y_j;\, \kappa), \tag{5.9}$$

where θ_j is the natural or canonical parameter, κ is the dispersion parameter, and m_j is the prior weight. As only a single random variable is being considered here, the subscript j on y_j for a single observation may seem superfluous, but it is being used to avoid any confusion with y, which is used throughout to denote the observed sample of n observations.

The mean and variance of Y_j are given by

$$E(Y_j) = \mu_j = b'(\theta_j)$$

and

$$\text{var}(Y_j) = \kappa b''(\theta_j),$$

respectively, where the prime denotes differentiation with respect to θ_j. In a GLM, it is assumed that

$$\begin{aligned} \eta_j &= h(\mu_j) \\ &= \boldsymbol{x}_j^T \boldsymbol{\beta}, \end{aligned}$$

where $\boldsymbol{x}_j$ is a vector of covariates or explanatory variables on the jth response y_j, $\boldsymbol{\beta}$ is a vector of unknown parameters, and $h(\cdot)$ is a monotonic function known as the link function. If the dispersion parameter κ is known, then the distribution (5.9) is a member of the (regular) exponential family with natural or canonical parameter θ_j. The distribution may or may not be a member of the two-parameter exponential family if κ is unknown. We are using the notation $h(\cdot)$ for the link function instead of the usual GLM notation of $g(\cdot)$, since g is used throughout this book for the number of components in the mixture model.

The variance of Y_j is the product of two terms, the dispersion parameter κ and the variance function $b''(\theta_j)$, which is usually written in the form

$$V(\mu_j) = \partial\mu_j/\partial\theta_j.$$

So-called natural or canonical links occur when $\eta_j = \theta_j$, which are respectively the log and logit functions for the Poisson and binomial distributions; see Table 2.1 in McCullagh and Nelder (1989, Chapter 2) for the canonical links, dispersion parameters, and variance functions for some commonly used univariate distributions in the exponential family, including the binomial, Poisson, gamma, inverse gamma, and normal.

In the standard form of a GLM, μ_j is modeled as a function of the unknown parameter vector $\boldsymbol{\beta}_j$, assuming κ fixed and with $V(\mu_j)$ containing no unknown parameters. Of distributions of the form (5.9), the Poisson and binomial have $\kappa = 1$ (that is, fixed *a priori* at 1). The negative binomial distribution, whose variance function can be written in the form

$$V(\mu_j) = \mu_j + \mu_j^2/k,$$

is an example of a variance function containing an unknown parameter that is not a dispersion parameter.

5.4.2 ML Estimation for a Single GLM Component

Suppose $y_1, \ldots, y_n$ denote n independent observations, where Y_j has prior weight m_j, canonical parameter θ_j, mean μ_j, and covariate vector $\boldsymbol{x}_j$ $(j = 1, \ldots, n)$. Then the log likelihood for $\boldsymbol{\beta}$ is given by

$$\log L(\boldsymbol{\beta}) = \sum_{j=1}^{n} [m_j\kappa^{-1}\{\theta_j y_j - b(\theta_j)\} + c(y_j;\, \kappa)]. \tag{5.10}$$

On differentiation in (5.10) with respect to $\boldsymbol{\beta}$ using the chain rule (McCullagh and Nelder, 1989, Section 2.5), the likelihood equation for $\boldsymbol{\beta}$ can be expressed as

$$\sum_{j=1}^{n} m_j\, w(\mu_j)(y_j - \mu_j)\eta'(\mu_j)\, \boldsymbol{x}_j = \mathbf{0}, \tag{5.11}$$

where $\eta'(\mu_j) = d\eta_j/d\mu_j$ and $w(\mu_j)$ is the weight function defined by

$$w(\mu_j) = 1/[\{\eta_j'(\mu_j)\}^2 V(\mu_j)].$$

It can be seen that for fixed κ, the likelihood equation for $\boldsymbol{\beta}$ is independent of κ.

The likelihood equation (5.11) can be solved iteratively using Fisher's method of scoring, which for a GLM is equivalent to using iteratively reweighted least squares (IRLS); see Nelder and Wedderburn (1972). On the $(k+1)$th iteration, we form the adjusted response variable $\tilde{y}_j$ as

$$\tilde{y}_j^{(k)} = \eta(\mu_j^{(k)}) + (y_j - \mu_j^{(k)})\eta'(\mu_j^{(k)}). \tag{5.12}$$

These n adjusted responses are then regressed on the covariates $\boldsymbol{x}_1, \ldots, \boldsymbol{x}_n$ using weights $m_1 w(\mu_1^{(k)}), \ldots, m_n w(\mu_n^{(k)})$. This produces an updated estimate $\boldsymbol{\beta}^{(k+1)}$ for $\boldsymbol{\beta}$, and hence updated estimates $\mu_j^{(k+1)}$ for the μ_j, for use in the right-hand side of (5.12) to update the adjusted responses, and so on. This process is repeated until changes in the estimates are sufficiently small.

5.4.3 Quasi-Likelihood Approach

For all GLMs, we have the relation

$$\partial \log f(y_j;\, \theta_j, \kappa)/\partial \mu_j = m_j\,(y_j - \mu_j)/\{\kappa V(\mu_j)\},$$

so that this first derivative depends only on the first two moments of Y. This led Wedderburn (1974) to define a quasi-likelihood approach by the relation

$$\partial q/\partial \mu_j = m_j(y_j - \mu_j)/\{\kappa V(\mu_j)\}.$$

The use of q as a criterion for fitting allows the class of GLMs to be extended to models defined only by the properties of the first two moments. The function q will correspond to a true log density if there is a distribution of the GLM type for which

$$\text{var}(Y_j) = \kappa V(\mu_j). \tag{5.13}$$

For a fixed value of the dispersion parameter κ, the quasi-likelihood approach estimates $\boldsymbol{\beta}$ by the value of $\boldsymbol{\beta}$ that minimizes the sum of weighted squares

$$\sum_{j=1}^{n} m_j(y_j - \mu_j)^2/\{\kappa V(\mu_j)\}. \tag{5.14}$$

A simple moment estimate of κ is obtained by the value of κ that makes the mean deviance equal to one or the expected value of the Pearson statistic equal to its degrees of freedom. With the latter, κ is obtained as a root of the equation

$$\sum_{j=1}^{n} m_j(y_j - \mu_j)^2/\{\kappa V(\mu_j)\} = (n - d), \tag{5.15}$$

where d is the number of parameters in the model.

In the context of allowing for extra-Poisson variation, Breslow (1984) suggested first fitting the ordinary Poisson model with $\kappa = 1$ to obtain an initial estimate of μ_j

for use in the left-hand side of (5.15). The value of κ obtained from (5.15) is then substituted into (5.14) to produce a new estimate of $\boldsymbol{\beta}$ and hence the μ_j, which are substituted into the left-hand side of (5.15) to produce a new estimate of κ, and so on. This process can be continued until convergence. A detailed review of the quasi-likelihood approach may be found in the book of McCullagh and Nelder (1989). It is well known that this approach leads to a consistent and efficient estimate of $\boldsymbol{\beta}$; see also Lawless (1987) and Kim (1994).

5.5 MIXTURES OF GLMs

5.5.1 Specification of Mixture Model

For a mixture of g component distributions of GLMs in proportions $\pi_1, \ldots, \pi_g$, we have that the density of the jth response variable Y_j is given by

$$f(y_j; \boldsymbol{\Psi}) = \sum_{i=1}^{g} \pi_i f(y_j; \theta_{ij}, \kappa_i), \tag{5.16}$$

where for a fixed dispersion parameter κ_i, we obtain

$$\log f(y_j; \theta_{ij}, \kappa_i) = m_j \kappa_i^{-1}\{\theta_{ij} y_j - b_i(\theta_{ij})\} + c_i(y_j; \kappa_i) \tag{5.17}$$

for $i = 1, \ldots, g$. For the ith component GLM, we let μ_{ij} be the mean of Y_j, $h_i(\mu_{ij})$ the link function, and $\eta_i = h_i(\mu_{ij}) = \boldsymbol{\beta}_i^T \boldsymbol{x}_j$ the linear predictor ($i = 1, \ldots, g$).

The fitting of mixtures of GLMs has been considered by Jansen (1993), Wedel and DeSarbo (1995), Aitkin (1996, 1999a), Oskrochi and Davies (1997), and Scallan (1999), among others. These models have been used also in the machine learning literature, where they are referred to as mixtures-of-experts (ME) models; see Jacobs et al. (1991). In Section 5.13 we shall consider an extension of them, called hierarchical mixtures-of-experts (HME). Strictly speaking, the component densities in ME and HME models need not be GLMs, but in practice they generally are.

Typically, in practice, the components of the mixture will be from the same GLM, so that the log density for the ith component can be written as

$$\log f(y_j; \theta_{ij}, \kappa_i) = \kappa_i^{-1}\{\theta_{ij} y_j - b(\theta_{ij})\} + c(y_j; \kappa_i) \tag{5.18}$$

for $i = 1, \ldots, g$.

In some applications, the mixing proportions may be modeled as functions of some vector of covariates associated with the response. This vector of covariates may or may not have some elements in common with the vector of covariates $\boldsymbol{x}$ on which the component means of the mixture depend. Without loss of generality, we shall denote both of these vectors of covariates by $\boldsymbol{x}$ (as irrelevant covariates in the regression forms for the canonical means and mixing proportions can have their coefficients set equal to zero).

A common model for expressing the ith mixing proportion π_i as a function of $\boldsymbol{x}$ is the logistic. Under this model, we have corresponding to the jth observation y_j with covariate vector $\boldsymbol{x}_j$

$$\begin{aligned}\pi_{ij} &= \pi_i(\boldsymbol{x}_j;\,\boldsymbol{\alpha}) \\ &= \exp(\boldsymbol{\omega}_i^T\boldsymbol{x}_j)/\{1+\sum_{h=1}^{g-1}\exp(\boldsymbol{\omega}_h^T\boldsymbol{x}_j)\} \quad (i=1,\,\ldots,\,g),\end{aligned} \tag{5.19}$$

where $\boldsymbol{\omega}_g = \boldsymbol{0}$ and

$$\boldsymbol{\alpha} = (\boldsymbol{\omega}_1^T,\,\ldots,\,\boldsymbol{\omega}_{g-1}^T)^T$$

contains the logistic regression coefficients. The first element of $\boldsymbol{x}_j$ is usually taken to be one, so that the first element of each $\boldsymbol{\omega}_i$ is an intercept. We let $\boldsymbol{\Psi}$ be the vector of unknown parameters, given by

$$\boldsymbol{\Psi} = (\boldsymbol{\alpha}^T,\,\boldsymbol{\beta}^T)^T,$$

where $\boldsymbol{\beta}$ contains the elements of $\boldsymbol{\beta}_1,\,\ldots,\,\boldsymbol{\beta}_g$ known *a priori* to be distinct.

As the mixing proportions are modeled to depend on some or all of the covariates on which the canonical means depend, it means that there may be identifiability problems with some of the parameters in $\boldsymbol{\alpha}$ and $\boldsymbol{\beta}$, in particular with the intercept terms of the $\boldsymbol{\alpha}$ and the elements of $\boldsymbol{\beta}$; see Wang (1994). The question of identifiability is to be examined more closely later in the specific cases of Poisson and binomial components.

5.5.2 ML Estimation via the EM Algorithm

The log likelihood for $\boldsymbol{\Psi}$ that can be formed from these data under the mixture model (5.16) is given by

$$\log L(\boldsymbol{\Psi}) = \sum_{j=1}^{n}\log\sum_{i=1}^{g}\pi_{ij}f(y_j;\,\theta_{ij},\kappa_i), \tag{5.20}$$

where

$$\pi_{ij} = \pi_i(\boldsymbol{x}_j;\,\boldsymbol{\alpha}) \quad (i=1,\,\ldots,\,g;\,j=1,\,\ldots,\,n) \tag{5.21}$$

and all prior weights are unity.

The EM algorithm of Dempster et al. (1977) can be applied to obtain the MLE of $\boldsymbol{\Psi}$ as in the case of a finite mixture of arbitrary distributions described in Chapter 2. The complete-data log likelihood is given by

$$\log L_c(\boldsymbol{\Psi}) = \sum_{i=1}^{g}\sum_{j=1}^{n} z_{ij}\{\log\pi_{ij} + \log f_i(y_j;\,\theta_{ij},\kappa_i)\}, \tag{5.22}$$

where the z_{ij} denote the component-indicator variables as defined in Section 2.8.2.

As the E-step is essentially the same as given in Section 2.8.3 for arbitrary component densities, we move straight to the M-step.

5.5.3 M-Step

The M-step on the $(k+1)$th iteration involves solving the two systems of equations

$$\sum_{i=1}^{g}\sum_{j=1}^{n}\tau_{ij}(y_j;\,\boldsymbol{\Psi}^{(k)})\partial\log\pi_{ij}/\partial\boldsymbol{\alpha}=\mathbf{0} \tag{5.23}$$

and

$$\sum_{i=1}^{g}\sum_{j=1}^{n}\tau_{ij}(y_j;\,\boldsymbol{\Psi}^{(k)})\partial\log f_i(y_j;\,\theta_{ij},\kappa_i)/\partial\boldsymbol{\beta}=\mathbf{0}, \tag{5.24}$$

where

$$\tau_{ij}(y_j;\,\boldsymbol{\Psi}^{(k)})=\frac{\pi_{ij}^{(k)}f_i(y_j;\,\theta_{ij}^{(k)},\kappa_i)}{\sum_{h=1}^{g}\pi_{hj}^{(k)}f_h(y_j;\,\theta_{hj}^{(k)},\kappa_h)}. \tag{5.25}$$

It is assumed that $\boldsymbol{\alpha}$ and $\boldsymbol{\beta}$ have no elements known *a priori* to be in common. This will often be the case in practice. An application where this is not the case concerns the zero-inflated Poisson regression model of Lambert (1992), which is to be discussed later.

Equation (5.23) can be solved using a standard algorithm for logistic regression to produce the updated estimate $\boldsymbol{\alpha}^{(k+1)}$ for the logistic regression coefficients. For $g=2$, $\boldsymbol{\alpha}^{(k+1)}$ can be computed using the GLIM macro for a binomial error structure with the canonical logit transformation as the link.

Concerning the computation of $\boldsymbol{\beta}^{(k+1)}$, it follows from the previous section on the ML fitting of a single GLM that (5.24) can be written as

$$\sum_{i=1}^{g}\sum_{j=1}^{n}\tau_{ij}(y_j;\,\boldsymbol{\Psi}^{(k)})w(\mu_{ij})(y_j-\mu_{ij})\eta_i'(\mu_{ij})\{\partial\eta_i(\mu_{ij})/\partial\boldsymbol{\beta}\}=\mathbf{0}, \tag{5.26}$$

where, for the ith component, μ_{ij} is the mean of Y_j.

If the $\boldsymbol{\beta}_1,\,\ldots,\,\boldsymbol{\beta}_g$ have no elements in common *a priori*, then

$$\begin{aligned}\partial\eta_i(\mu_{ij})/\partial\boldsymbol{\beta}_h &= \boldsymbol{x}_j, && \text{if } h=i.\\ &= 0, && \text{otherwise.}\end{aligned}$$

In this case, (5.26) reduces to solving

$$\sum_{j=1}^{n}\tau_{ij}(y_j;\,\boldsymbol{\Psi}^{(k)})w(\mu_{ij})(y_j-\mu_{ij})\eta_i'(\mu_{ij})\boldsymbol{x}_j=\mathbf{0} \tag{5.27}$$

separately for each $\boldsymbol{\beta}_i$ to produce $\boldsymbol{\beta}_i^{(k+1)}$ $(i=1,\,\ldots,\,g)$.

On contrasting (5.27) with (5.11), it can be seen that it has the same form as for a single GLM fitted to the responses $y_1,\,\ldots,\,y_n$ with prior weights $m_1=\tau_{i1}(y_1;\,\boldsymbol{\Psi}^{(k)}),\,\ldots,\,m_n=\tau_{in}(y_n;\,\boldsymbol{\Psi}^{(k)})$ and fixed dispersion parameter κ_i.

In the general case where $\boldsymbol{\beta}_1,\,\ldots,\,\boldsymbol{\beta}_g$ may have some elements in common, we can still solve (5.27) using the iteratively reweighted least-squares approach for a single GLM. The double summation over i and j in (5.26) can be handled by expanding the response vector to have dimension $g \times n$ by replicating each original observation $(y_j, \boldsymbol{x}_j^T)^T$ g times, with prior weights $\tau_{1j}(y_j; \boldsymbol{\Psi}^{(k)}), \ldots, \tau_{gj}(y_j; \boldsymbol{\Psi}^{(k)})$, fixed dispersion parameters $\kappa_1, \ldots, \kappa_g$, and linear predictors $\boldsymbol{x}_j^T\boldsymbol{\beta}_1, \ldots, \boldsymbol{x}_j^T\boldsymbol{\beta}_g$.

Although there are more efficient methods of solving (5.27), this approach has the advantage that it is easily done in a GLM fitting program, such as GLIM (see, for instance, Aitkin et al. (1989)), the glm( ) function in S-PLUS (see Becker, Chambers, and Wilks (1988)), or the GLM subroutine in SAS (1993). Dietz (1992) has provided a GLIM-macro for the computation of $\boldsymbol{\beta}$. Previously, Hinde (1982) provided the GLIM code for a Poisson model and the modifications needed for the binomial model; see also Anderson (1988) and Aitkin (1996, 1999a, 1999b). Wang et al. (1996) have available FORTRAN codes for algorithms that fit finite mixtures of Poisson regression models.

The response for each entity has been taken to be univariate in the above. The results generalize in a straightforward manner to the case of multivariate responses $\boldsymbol{Y}_j = (Y_{1j}, \ldots, Y_{pj})^T$, if it is assumed that $Y_{1j}, \ldots, Y_{pj}$ are independently distributed when conditioned on their component membership of the mixture model; see Wedel and DeSarbo (1995). The case of component multivariate GLMs where $Y_{1j}, \ldots, Y_{pj}$ are not necessarily independent has been considered by Dietz (1992).

5.5.4 Multicycle ECM Algorithm

We have seen above in the computation of the updated estimate of

$$\boldsymbol{\Psi} = (\boldsymbol{\Psi}_1^T, \boldsymbol{\Psi}_2^T)^T,$$

where $\boldsymbol{\Psi}_1 = \boldsymbol{\alpha}$ and $\boldsymbol{\Psi}_2 = \boldsymbol{\beta}$, that $\boldsymbol{\alpha}^{(k+1)}$ and $\boldsymbol{\beta}^{(k+1)}$ are computed independently of each other on the M-step of the EM algorithm. Therefore, the latter is the same as the expectation-conditional maximization (ECM) algorithm with two CM-steps, where on the first CM-step, $\boldsymbol{\Psi}_1^{(k+1)}$ is calculated with $\boldsymbol{\Psi}_2$ fixed at $\boldsymbol{\Psi}_2^{(k)}$, and where on the second CM-step, $\boldsymbol{\Psi}_2^{(k+1)}$ is calculated with $\boldsymbol{\Psi}_1$ fixed at $\boldsymbol{\Psi}_1^{(k+1)}$.

In order to improve convergence, a multicycle version of the ECM algorithm can be used, where an E-step is performed after the computation of $\boldsymbol{\alpha}^{(k+1)}$ and before the computation of $\boldsymbol{\beta}^{(k+1)}$; see Meng and Rubin (1993) and McLachlan and Krishnan (1997) for further details of the ECM algorithm. The multicycle E-step is effected here by updating $\boldsymbol{\alpha}^{(k)}$ with $\boldsymbol{\alpha}^{(k+1)}$ in $\boldsymbol{\Psi}^{(k)}$ in the right-hand side of the expression (5.25) for $\tau_{ij}(y_j; \boldsymbol{\Psi}^{(k)})$.

5.5.5 Choice of the Number of Components

Up to now, we have considered the fitting of a finite mixture of GLMs for a given value of the number of components g in the mixture model. Typically, in practice where the mixture model is being used to handle overdispersion, the value of g has to be inferred from the data. A guide to the final choice of g can be obtained from monitoring the increase in the log likelihood as g is increased from a single component. Unfortunately, it is difficult to carry out formal tests at any stage of this sequential process for the need of an additional component since, as to be explained in Chapter 6, regularity conditions fail to hold for the likelihood ratio test statistic, $-2\log\lambda$, to have its usual asymptotic null distribution. There is the resampling approach of McLachlan (1987), which was used by Schlattmann and Böhning (1993) to decide on g in their application of Poisson mixtures to disease mapping. Also, Pauler et al. (1996) used this method to decide on the number of Poisson components in the finite mixture modeling of anticipatory saccade counts from schizophrenic patients and controls. In the context of the fitting of mixtures of Poisson regression components to overdispersed count data, Wang et al. (1996) have reported encouraging results for the selection of g based on Akaike's information criterion (AIC) and the Bayesian information criterion (BIC); see also Wang and Puterman (1999). Concerning the significance of the covariates in the mixture of GLMs, Wang et al. (1996) considered the deletion of covariates from the model only after the choice of g had been essentially finalized.

5.6 A GENERAL ML ANALYSIS OF OVERDISPERSION IN A GLM

In an extension to a GLM for overdispersion, a random effect U_j can be introduced additively into a GLM on the same scale as the linear predictor, as proposed by Aitkin (1996). This extension in a two-level variance component GLM has been considered recently by Aitkin (1999a). For an unobservable random effect u_j for the jth response on the same scale as the linear predictor, we have that

$$\eta_j = \boldsymbol{\beta}^T \boldsymbol{x}_j + \sigma u_j,$$

where u_j is realization of a random variable U_j distributed $N(0,\, 1)$ independently of the jth response $Y_j (j = 1,\, \ldots,\, n)$.

The (marginal) log likelihood is thus

$$\log L(\boldsymbol{\Psi}) = \sum_{j=1}^{n} \log \int_{-\infty}^{\infty} f(y_j;\, \boldsymbol{\beta},\, \sigma,\, u)\phi(u)\, du. \tag{5.28}$$

The integral (5.28) does not exist in closed form except for a normally distributed response y_j. Following the development in Anderson and Hinde (1988), Aitkin (1996, 1999a) suggested that it be approximated by Gaussian quadrature, whereby the integral over the normal distribution of U is replaced by a finite sum of g Gaussian quadrature mass-points u_i with masses π_i; the u_i and π_i are given in standard references, for example, Abramowitz and Stegun (1964).

The log likelihood so approximated thus has the form for that of a g-component mixture model,

$$\sum_{j=1}^{n} \log \sum_{i=1}^{g} \pi_i f(y_j;\, \boldsymbol{\beta},\, \sigma,\, u_i),$$

where the masses $\pi_1,\, \ldots,\, \pi_g$ correspond to the (known) mixing proportions, and the corresponding mass points $u_1,\, \ldots,\, u_g$ correspond to the (known) parameter values.

The linear predictor for the jth response in the ith component of the mixture is

$$\eta_{ij} = \boldsymbol{\beta}^T \boldsymbol{x}_j + \sigma u_i \quad (i = 1, \ldots, g).$$

Hence in this formulation, u_i is an intercept term.

The influential paper by Heckman and Singer (1982) showed substantial changes in parameter estimates with quite small changes in the mixing distribution. As noted by Aitkin (1996), a particular disadvantage of the modeling approach described above is the possible sensitivity of conclusions to the choice of a particular distributional form for the random effect U_j; there is a lack of information in the data about this distribution. Another disadvantage is the possible inaccuracy of Gaussian quadrature, where even 20-point integration may not give high accuracy for the logistic/normal model (Crouch and Spiegelman, 1990).

As a consequence, Aitkin (1996, 1999a) suggested treating the masses $\pi_1, \ldots, \pi_g$ as g unknown mixing proportions and treating the mass points $u_1, \ldots, u_g$ as g unknown values of a parameter. This g-component mixture model is then fitted using the EM algorithm, as described in Section 5.5.2. The value of g is increased sequentially until the increase in the likelihood is assessed to be nonsignificant. If $\boldsymbol{\beta}$ were known, then this approach would correspond to finding the nonparametric MLE of the distribution of U (the mixing distribution). The advantage of this approach is that it avoids having to specify the mixing distribution.

In this framework since now u_i is also unknown, we can drop the scale parameter σ and define the linear predictor for the jth response in the ith component of the mixture as

$$\eta_{ij} = \boldsymbol{\beta}^T \boldsymbol{x}_j + u_i.$$

Thus u_i acts as an intercept parameter for the ith component. One of the u_i parameters will be aliased with the intercept term β_0; alternatively, the intercept can be removed from the model.

We shall give an example of this approach with a Poisson mixture model applied to handle overdispersion in some count data in Section 5.8.3. But firstly, we take a brief look at the problem of overdispersion with the use of Poisson regression models.

5.7 POISSON REGRESSION MODEL

5.7.1 Some Standard Modifications for Overdispersed Data

We consider now the Poisson regression model. We shall briefly review some modifications that can be made to it within a single-GLM framework for the modeling of overdispersed count data before proceeding to consider some methodology that can be implemented using a finite mixture of GLMs.

The Poisson regression model is an example of a GLM in which the distribution of the response Y_j with covariate vector $\boldsymbol{x}_j$ is Poisson with density

$$f(y_j; \mu_j) = \{e^{-\mu_j} \mu_j^{y_j} / y_j!\} I_A(y_j), \tag{5.29}$$

which has mean $E(Y_j) = \mu_j$, and the natural link is the log function

$$\begin{aligned} h(\mu_j) &= \log \mu_j \\ &= \boldsymbol{\beta}^T \boldsymbol{x}_j. \end{aligned}$$

In (5.29), $A = \{0, 1, 2, \ldots\}$ is the set of nonnegative integers.

In many situations in practice, the population size or, say, the time of exposure, varies for each subject so that the mean of Y_j is given by $a_j \mu_j$, where a_j denotes the known population size or time of exposure, and μ_j now denotes the mean rate per unit size or time. This can be dealt with in the theory and software for GLMs by either declaring a_j as an offset in the specification of the linear predictor or by redefining the response to be the observed rate y_j / a_j, with a_j^{-1} specified as the prior weight. Hence in the sequel we shall assume without loss of generality that $a_j = 1$ for all subjects.

A consequence of using the Poisson regression model is that the variance equals the mean; that is,

$$\begin{aligned} \text{var}(Y_j) &= E(Y_j) \\ &= \mu_j. \end{aligned}$$

In practice, however, we often have overdispersed data; that is,

$$\text{var}(Y_j) > \mu_j.$$

When Poisson regression model fits the count data poorly, overdispersion is often a cause of the problem.

There are several ways to modify the Poisson regression model. Using the GLM formulation, we can modify it by choosing either an alternative link function or an alternative frequency distribution, or both. Since the log link has properties such as multiplicative effects of covariates on the Poisson mean, researchers have suggested use of alternative link functions. On the other hand, there are a lot of studies of alternative frequency distributions for the Poisson distribution; see, for example, Breslow (1984), Efron (1986, 1992), and Lawless (1987). More recently, Lee and Nelder (1996) have proposed hierarchical GLMs for which $\boldsymbol{\beta}$ is estimated by consideration of the likelihood formed on the basis of the joint distribution of the observed responses and the unobservable random effects. Their approach thus avoids the integration in (5.28) that is necessary with the use of the marginal likelihood.

5.7.2 Gamma-Poisson Mixture Model

A classical approach is to use a continuous Poisson mixture model to adjust for extra-Poisson variation. In this framework in the nonregression case, the Poisson mean μ_j is taken to be a latent variable from a distribution, $H(\mu_j)$, so that the density of Y_j is modeled as

$$f(y_j) = \int_0^\infty \{e^{-\mu} \mu_j^y / y_j!\} I_A(y_j) \, dH(\mu). \tag{5.30}$$

A common choice for $H(\mu)$ in (5.30) is the gamma $(\alpha,\ \beta)$ distribution, which has density function defined by (4.16). This leads to the density of Y_j being modeled as

$$f(y_j;\ \alpha,\ \beta) = \binom{y_j + \alpha - 1}{y_j} \left(\frac{\beta}{\beta + 1}\right)^{\alpha} \left(\frac{1}{\beta+1}\right)^{y_j} I_A(y_j), \qquad (5.31)$$

which is the negative binomial distribution $NB(\alpha,\ \beta/(\beta+1))$. This distribution is the model for the number of tails y_j, in independent flips of a coin with probability of heads equal to $\beta/(\beta+1)$, until one observes α heads. On letting μ now denote the mean of this distribution, we have that

$$\mu = \alpha/\beta,$$

while its variance is

$$\begin{aligned} \text{var}(Y_j) &= (\alpha/\beta)\{(\beta+1)/\beta\} \\ &= \mu_j + k\mu_j^2, \end{aligned} \qquad (5.32)$$

where $k = 1/\alpha$ and where μ is written as μ_j to explicitly denote that it may be a function of the covariate vector $\boldsymbol{x}_j$. Hence this two-parameter model allows the variance to be greater than the mean, with the variance equal to the mean inflated multiplicatively by the factor $(1 + k\mu_j)$. As k tends to zero in (5.32), we obtain the Poisson model. We can rewrite (5.31) as

$$f(y_j;\mu_j,\ k) = \binom{y_j + k^{-1} - 1}{y_j} \left(\frac{k^{-1}}{\mu_j + k^{-1}}\right)^{k^{-1}} \left(\frac{\mu_j}{\mu_j + k^{-1}}\right)^{y_j} I_A(y_j). \qquad (5.33)$$

This is the standard negative binomial model for extra-Poisson variation, and it can be seen that it arises by assuming that α is fixed as μ_j varies. If, however, we assume that μ_j varies with α and that β remains constant, we obtain a negative binomial distribution with

$$\text{var}(Y_j) = \mu_j(1 + k), \qquad (5.34)$$

where $k = 1/\beta$. This distribution does not have the form of a standard GLM; see Nelder and Lee (1992).

For the negative binomial distribution (5.33), the MLE of $\boldsymbol{\Psi} = (\boldsymbol{\beta}^T,\ k)^T$ can be obtained as described in Lawless (1987) for the log linear model

$$\log \mu_j = \boldsymbol{\beta}^T \boldsymbol{x}_j.$$

It was noted there that the results for other regression specifications are qualitatively similar.

Within the GLM framework, $\boldsymbol{\Psi}$ can be estimated as described in McCullagh and Nelder (1989, Section 11.2). For fixed k, the negative binomial distribution (5.33) has the form of a GLM with canonical link

$$\eta(\mu_j) = \log\{\mu_j/(\mu_j + k^{-1})\},$$

and variance function

$$V(\mu_j) = \mu_j + k\mu_j^2.$$

Ordinarily, k is unknown. Using the method of moments, it can be computed as a solution of

$$\sum_{j=1}^{n} \frac{(y_j - \hat{\mu}_j)^2}{\hat{\mu}_j(1 + k\hat{\mu}_j)} = n - d, \tag{5.35}$$

where $\hat{\mu}_j$ is the current estimate of μ_j. Hence $\boldsymbol{\beta}$ can be estimated by a combined quasi-likelihood and method of moments approach.

Alternatively within the GLM framework, $\boldsymbol{\beta}$ can be estimated by using the Poisson error function with the log link function and defining the prior weights m_j as

$$m_j = (1 + k\hat{\mu}_j)^{-1}.$$

The value of k is obtained iteratively from (5.35). The initial fit is made with unit prior weights $m_j = 1$; see Breslow (1984, 1987).

5.7.3 Multiplicative Random Effects Model

Another way of viewing the gamma-Poisson mixture model (5.30) is to write the Poisson parameter as

$$\mu_j = u_j \mu_{0j},$$

where μ_{0j} is an unknown parameter and u_j is a value of the random effect U_j taken to have some distribution H, which without loss of generality, can be assumed to have mean one. This is the multiplicative random-effects model; see Brillinger (1986) and Manton, Woodbury, and Stallard (1981).

If we take the random effect U_j to have the gamma (α, β) distribution with $\alpha = 1/\beta = k^{-1}$, we obtain the negative binomial distribution as given by (5.31). The mean–variance relationship (5.32) will hold for any mixing distribution H for U that has mean 1 and variance k. Other choices of the distribution of U include the inverse Gaussian (Folks and Chhikara, 1978) as adopted by Dean, Lawless, and Willmot (1989), and the log normal (Hinde, 1982).

5.7.4 Additive Random Effects Model

As seen in Section 5.6 for an arbitrary GLM, one way to handle overdispersion is to introduce a random effect U_j additively on the same scale as the linear predictor. If for the log link function, the distribution of $\exp(U_j)$ is taken to be gamma, then it corresponds to the multiplicative random effects model given in the previous section for overdispersed Poisson data; if it has a log normal distribution, then it corresponds to the log normal model considered in Hinde (1982).

This additive random effects model leads to a finite mixture model as demonstrated in Section 5.6 for an arbitrary GLM. We shall give an example on its use in Section 5.8.3.

5.8 FINITE MIXTURE OF POISSON REGRESSION MODELS

5.8.1 Mean and Variance

We consider now the finite mixture model (5.16) of arbitrary component GLMs in the case where the component GLMs are Poisson regression models with means specified by a log linear model. That is,

$$f(y_j;\,\theta_{ij}) = \{e^{-\mu_{ij}}\mu_{ij}^{y_j}/y_j!\}I_A(y_j), \tag{5.36}$$

where $\theta_{ij} = \log \mu_{ij} = \eta_{ij}$, and

$$\begin{aligned}\eta_{ij} &= \boldsymbol{\beta}_i^T \boldsymbol{x}_j \\ &= (\beta_{0i},\, \boldsymbol{\beta}_{1i}^T)^T(1,\, x_{2j},\, \ldots,\, x_{pj})^T \qquad (i = 1,\, \ldots,\, g).\end{aligned} \tag{5.37}$$

1) If we set $g = 1$, then we obtain the Poisson regression model, which reduces to the Poisson model if $\boldsymbol{\beta}_{11}$ is specified as the null vector.
2) If we specify $\boldsymbol{\beta}_{1i}$ as the null vector, we obtain the g-component Poisson mixture model.
3) If we set $g = 2$ and $\mu_{1j} = 0$ for all j, then we obtain a Poisson regression model with an extra mass at $Y_j = 0$.

For the g-component mixture (5.36) of Poisson regression models, the mean and variance of Y_j are equal to

$$E(Y_j) = \sum_{i=1}^{g} \pi_i \mu_{ij}$$

and

$$\begin{aligned}\text{var}(Y_j) &= E\{\text{var}(Y_j \mid \boldsymbol{Z}_j\} + \text{var}\{E(Y_j \mid \boldsymbol{Z}_j\} \\ &= E(Y_j) + v_{ij},\end{aligned}$$

respectively, where

$$v_{ij} = \sum_{i=1}^{g} \pi_i \mu_{ij}^2 - \left(\sum_{i=1}^{g} \pi_i \mu_{ij}\right)^2.$$

Here

$$\mu_{ij} = \exp(\boldsymbol{\beta}_i^T \boldsymbol{x}_j)$$

is the mean of the jth response conditional on its membership of the ith component of the mixture, and $\boldsymbol{Z}_j$ is the component-indicator vector of zeros and ones, where $Z_{ij} = (\boldsymbol{Z}_j)_i$ is one or zero, according to whether $\boldsymbol{y}_j$ is viewed as having come from the ith component or not $(i = 1,\, \ldots,\, g;\; j = 1,\, \ldots,\, n)$.

Obviously, $v_{ij} = 0$ if and only if

$$\mu_{1j} = \mu_{2j} = \cdots = \mu_{gj}.$$

Hence the mixture model is able to cope better than the one-component (homogeneous) model with excess variation among $Y_1,\, \ldots,\, Y_n$.

5.8.2 Identifiability

In order to be able to estimate $\boldsymbol{\Psi}$, we require the mixture to be identifiable; that is, two sets of parameters which do not agree after permutation cannot yield the same mixture distribution. Without covariates, Teicher (1960) proved that the class of finite mixtures of Poisson distributions is identifiable. As noted by Wang et al. (1996), a sufficient condition for the class of Poisson regression mixtures to be identifiable is that the matrix $(\boldsymbol{x}_1, \ldots, \boldsymbol{x}_n)$ be of full rank. Wang et al. (1996) also considered the residual analysis and goodness-of-fit statistics for this class of mixture regression models.

Applications of mixtures of Poisson regression models to biological data sets are given in Wang et al. (1996), who used this methodology to analyze epileptic seizure frequency and Ames salmonella assay data; see Wedel et al. (1993) for an application in marketing.

5.8.3 Example 5.2: Fabric Faults Data Set

We report the analysis by Aitkin (1996), who fitted a Poisson mixture regression model to the data set on fabric faults, as discussed previously by Hinde (1982) and Nelder (1985). The response variable is the number of faults y_j in a bolt of fabric of length ℓ_j. The data from Hinde (1982) are reproduced in Table 5.4.

Table 5.4 Fabric Data

Length of Roll	Number of Faults	Length of Roll	Number of Faults
551	6	543	8
651	4	842	9
832	17	905	23
375	9	542	9
715	14	522	6
868	8	122	1
271	5	657	9
630	7	170	4
491	7	738	9
372	7	371	14
645	6	735	17
441	8	749	10
895	28	495	7
458	4	716	3
642	10	952	9
492	4	417	2

Source: From Aitkin (1996).

Initially, Aitkin (1996) fitted the Poisson model $\log \mu_j = \beta_0 + \beta_1 \log \ell_j$, which gave a deviance of 64.54 with 30 degrees of freedom, and parameter estimates (standard errors) of $\hat{\beta}_0 = 4.173$ (1.135), $\hat{\beta}_1 = 0.997$ (0.176). Thus proportionality

($\beta_1 = 1$) is obviously well supported, but the residual deviance shows substantial overdispersion. The deviance for the negative binomial model was 54.09. Using the approach discussed in Section 5.6 for handling overdispersion in a GLM, Aitkin (1996) subsequently fitted the g-component Poisson mixture model (5.36) for $g = 2$ to 6, where

$$\eta_{ij} = \beta_0 + \beta_1 \ell_j + u_i \quad (i = 1, \ldots, g),$$

where the component GLMs (here Poisson components) differ in their intercepts with the introduction of the u_i. Since either the regression intercept β_0 or one of the intercepts u_i is redundant, Aitkin (1996) effectively fitted the model without β_0 and then took

$$\hat{\beta}_0 = -\sum_{i=1}^{g} \hat{u}_i / g$$

so that the estimated intercepts $\hat{u}_i$ are centered to allow direct comparison with the simple Poisson and Poisson/normal models.

The results of these fits for $g = 2$ and 3 are given in Table 5.5. The standard errors were computed using the method described in Section 2.16.1. It can be seen from Table 5.5 that clearly only $g = 2$ components are needed in the Poisson mixture model.

Table 5.5 Results of Fitting Mixtures of Poisson Regression Models

g	β_0	β_1 (SE)	u_1 (π_1)	u_2 (π_2)	u_3 (π_3)	Deviance
2	−2.979	0.800	0.609	−0.156		49.364
		(0.201)	(0.203)	(0.797)		
3	−2.972	0.799	0.611	−0.154	−0.165	49.364
		(0.201)	(0.202)	(0.711)	(0.087)	

Source: Adapted from Aitkin (1996).

The posterior probabilities of membership of the first component of the mixture are shown with the standardized residuals from the ordinary Poisson fit in Table 5.6. The posterior probability increases almost monotonically with the residual, with only five observations having a posterior probability of more than 0.5 of being in component 1; these observations have standardized residuals 1.255, 2.530, 3.541, 1.773 and 3.946. Thus component one contains the observations with large positive residuals; these are "downweighted" in their effect on the fitted model by the intercept term $\hat{u}_1 = 0.609$ compared with $\hat{u}_2 = -0.156$ for the second component with small positive, or negative, residuals.

Table 5.6 Posterior Probabilities of First Component Membership from the Poisson Mixture Model, and Pearson Residuals from the Poisson Model

Probability	Residual	Probability	Residual
0.0102	−0.806	0.0496	−0.071
0.0007	−1.860	0.0044	−1.039
0.6887	1.255	0.9903	2.530
0.4476	1.397	0.1019	0.283
0.4349	0.976	0.0142	−0.672
0.0016	−1.408	0.0506	−0.626
0.1237	0.443	0.0305	−0.292
0.0091	−0.815	0.2065	0.886
0.0423	−0.155	0.0129	−0.641
0.1540	0.579	0.9750	3.541
0.0036	−1.198	0.8604	1.773
0.1456	0.516	0.0244	−0.388
0.9998	3.946	0.0404	−0.176
0.0065	−1.111	0.0002	−2.375
0.0739	0.098	0.0014	−1.414
0.0044	−1.260	0.0023	−1.715

Source: From Aitkin (1996).

5.8.4 Components and Mixing Proportions Without Covariates

Böhning, Schlattmann, and Lindsay (1992, 1998) have provided an excellent account of the use of Poisson mixture models where the mixing proportions and the components do not depend on any covariates. They also gave several examples of applications of the Poisson mixture model in medical problems; see also Lindsay (1995). Albert (1991), Le, Leroux, and Puterman (1992), and Leroux and Puterman (1992) have presented examples in the context of modeling epileptic seizure counts and fetal movements by Poisson mixtures where the assumption of independence of the data has been relaxed, as in hidden Markov chain models to be discussed in Chapter 13.

In the absence of covariates, we can write the Poisson mixture model as

$$\begin{aligned} f(y_j;\, \boldsymbol{\Psi}) &= \sum_{i=1}^{g} \pi_i f(y_j;\, \mu_i) \\ &= \int f(y_j;\, \theta)\, dH(\theta), \end{aligned}$$

where

$$f(y_j; \theta) = \{e^{-\theta} \theta^{y_j} / y_j!\} I_A(y_j)$$

and $H(\theta)$ is the measure that puts mass π_i at the point $\theta = \mu_i$ $(i = 1, \ldots, g)$.

In the work up to now, we have effectively considered the estimation of the mixing distribution H in the fixed support case. However, we can treat g itself as unknown, which is the flexible support size case. In this latter case, we can approach the problem by considering the so-called nonparametric MLE (NPMLE) of H, $\hat{H}$, which is the probability measure that maximizes

$$l(H) = \sum_{j=1}^{n} \log \int f(y_j;\, \theta)\, dH(\theta),$$

where $H(\theta)$ is now allowed to be any mixing distribution. As discussed in Section 1.12, Lindsay (1983) showed that $\hat{H}$ is a discrete measure with at most a finite number of support points; see also Lindsay (1995) and Lindsay and Roeder (1992a), who have considered residual diagnostics for mixture models.

5.8.5 Algorithms for NPMLE of a Mixing Distribution

Böhning et al. (1992, 1998) have provided the computer package C.A.MAN (Computer-Assisted Mixture Analysis) for computing $\hat{H}$. It includes an algorithmic menu with choices of the EM algorithm, the vertex exchange algorithm, a combination of both, as well as the vertex direction method. The package C.A.MAN has the option to work with fixed support size; that is, when the number of components is known *a priori*. In the latter case, the EM algorithm is used. The C.A.MAN program can handle also component distributions that include the normal and exponential. In the former case, the variance has to be specified in the flexible support case; see Böhning et al. (1992, 1998) for further discussion of this. More recently, Böhning (1995) has reviewed reliable algorithms for the ML fitting of mixture models; see also Böhning (1999).

5.8.6 Disease Mapping

Poisson mixtures have played a very useful role in disease mapping. The analysis of the geographic variation of disease and its representation on a map is an important topic in epidemiological research. Identification of high-risk groups provides valuable hints for possible experience and targets for subsequent analytical studies; see Schlattmann and Böhning (1993), Schlattmann, Dietz, and Böhning (1996), and Böhning (1999). A measure often used is the standardized mortality rate (SMR). For a given area, SMR_j is defined as

$$\begin{aligned} \text{SMR}_j &= y_j / e_j \\ &= y_j \Big/ \sum_h A_{hj}\omega_h, \end{aligned} \tag{5.38}$$

where for the jth regional area, y_j is the number of observed cases, e_j is the expected number based on an external reference, A_{hj} is the person years in the hth age stratum, and ω_h is the age-specific mortality rate, which is assumed to be known.

A common approach to map construction in the literature is based on the assumption that y_j is the realization of the random variable Y_j which has a Poisson distribution with parameter $\mu_j = \lambda e_j$. Here λ denotes the relative risk of disease due to living within the study area.

The assumption (5.38) implies that all geographical areas have the same relative risk λ. This homogeneous model of a single Poisson distribution is often too simple with overdispersion frequently occurring. One approach for more flexibility has been to be adopt a random effects model, where λ is gamma or log normal; see Clayton and Kaldor (1987), Mollie and Richardson (1991), and Breslow and Clayton (1993).

Schlattmann and Böhning (1993) modeled the distribution of Y_j by the Poisson mixture distribution

$$f(y_j) = \sum_{i=1}^{g} \pi_i f(y_j;\ \lambda_i e_j),$$

where the relative risk λ_i is specific to the ith component of the mixture ($i = 1, \ldots, g$). More recently, Schlattmann et al. (1996) proposed the Poisson mixture regression model

$$f(y_j) = \sum_{i=1}^{g} \pi_i f(y_j;\ \mu_{ij})$$

for the distribution of Y_j, where

$$\mu_{ij} = \lambda_i e_j \exp(\boldsymbol{\beta}_i^T \boldsymbol{x}_j)$$

and $\boldsymbol{x}_j$ is a vector of covariates associated with the jth region, and the parameters λ_i and $\boldsymbol{\beta}_i$ are specific to the ith component of the mixture ($i = 1, \ldots, g$).

5.9 COUNT DATA WITH EXCESS ZEROS

Two-component mixture models are frequently used to model data that appear to have an excess of zeros. For example, when a reliable manufacturing process is in control, the number of defects on an item should be Poisson distributed. If the Poisson mean is λ, the expected number of items without defects is $ne^{-\lambda}$. In practice, however, there are situations where the observed number of items without defects is much lower than expected. One possible explanation is that slight, unobserved changes in the environment cause the process to swerve randomly back and forth between a perfect state in which defects are impossible or at least extremely rare and an imperfect state in which defects are possible but not inevitable; see Lambert (1992). The transient perfect state increases the number of zeros in the data.

In a medical context, a possible explanation for the excess of zeros might be due to the fact that the patient is cured after the treatment and so no realization of the symptom being monitored will occur. This phenomenon can be handled by a two-component mixture where one of the components is taken to be a degenerate distribution, having mass 1 at $y_j = 0$. The other component is a Poisson (or binomial) regression model, depending on the situation. A mixture model of this type with a degenerate component is sometimes referred to as a nonstandard mixture model. A detailed account of the use of nonstandard mixture models in auditing may be found in the report by the Panel on Nonstandard Mixtures of Distributions (1989).

5.9.1 History of Problem

The fascinating history of attempts to analyze count data with excess zeros by a Poisson distribution has been given recently by Meng (1997). He noted the link of the paper by McKendrick (1926), which was the earliest cited reference in Dempster et al. (1977), to the use of the EM algorithm through an improvement of McKendrick's method suggested in Irwin (1963). The data set was on an epidemic of cholera in an Indian village. The iterative solution proposed by Irwin (1963) was shown by Meng (1997) to converge to the MLE obtained by fitting a zero-truncated Poisson model via the EM algorithm.

As discussed in Meng (1997), an alternative model, perhaps more direct for many modern statisticians, is to fit the nonstandard mixture model,

$$f(y_j;\, \boldsymbol{\Psi}) = \pi_1 I_{[0]}(y_j) + \pi_2 f_2(y_j;\, \mu), \tag{5.39}$$

where the first component is the degenerate distribution with mass one at $y_j = 0$ and the second component is the Poisson with mean μ,

$$f_2(y_j;\, \mu) = \{e^{-\mu} \mu^{y_j} / y_j!\} I_A(y_j).$$

As demonstrated by Meng (1997), the iterative process for the MLE of μ as obtained under this nonstandard mixture model via the EM algorithm converges to the MLE as obtained (via the EM algorithm) under the zero-truncated Poisson model. However, the number of iterations of the EM algorithm is essentially tripled, a key reason being that in this second application of the EM algorithm, there is more data augmentation than in the first application.

5.9.2 Zero-Inflated Poisson Regression

The model (5.39) with μ depending on some covariates forms the basis of the zero-inflated Poisson (ZIP) regression technique proposed by Lambert (1992) for the handling of zero-inflated count data with covariates; see also Yip (1988, 1991) and Fong and Yip (1993, 1995).

5.10 LOGISTIC REGRESSION MODEL

We shall briefly review some modifications that can be made to logistic regression for the modeling of overdispersed count data before proceeding to consider (finite) mixtures of this model.

For the (single-component) binomial model, the frequency function for the jth response Y_j is given by

$$f(y_j;\, \theta_j) = \binom{N_j}{y_j} \theta_j^{y_j} (1 - \theta_j)^{N_j - y_j} I_{A_j}(y_j), \tag{5.40}$$

where $A_j = \{0,\, 1,\, \ldots,\, N_j\}$. That is, the response y_j denotes the number of successes in a series of N_j independent Bernoulli trials on which the probability of success on each Bernoulli trial is θ_j. In our notation here, we are suppressing the dependence of $f(y_j;\, \theta_j)$ on N_j.

In the case of logistic regression, θ_j is postulated to depend on the vector $\boldsymbol{x}_j$ of covariates through the logit function,

$$\log\{\theta_j/(1 - \theta_j)\} = \boldsymbol{\beta}^T \boldsymbol{x}_j \quad (j = 1,\, \ldots,\, n), \tag{5.41}$$

or equivalently,

$$\theta_j = \exp(\boldsymbol{\beta}^T \boldsymbol{x}_j)/\{1 + \exp(\boldsymbol{\beta}^T \boldsymbol{x}_j)\}.$$

It is given within the GLM framework by taking the response variable to be y_j/N_j, specifying the error function to be the binomial, and using the canonical logit link.

Logistic regression is a common method for analyzing the effect of a vector of covariates on the number of successes in a series of N_j independent Bernoulli trials. Overdispersion relative to the binomial distribution may occur if the N_j trials in a set are positively correlated, an important covariate is omitted, or $\boldsymbol{x}_j$ is measured with error.

Other link functions include the probit, which gives similar results to the logit, and the complementary log–log function, which is limited to situations where it is appropriate to deal with the probability parameter θ in an asymmetric manner; see McCullagh and Nelder (1989) for a comparison of these link functions.

A classical approach in the case of no covariates is to use a continuous binomial mixture model,

$$\int_0^1 \binom{N_j}{y_j} \theta^{y_j} (1 - \theta)^{N_j - y_j} I_{A_j}(y_j) dH(\theta), \tag{5.42}$$

where $H(\theta)$ is taken to be the beta $(\alpha_1,\, \alpha_2)$ distribution, which has density

$$\{u^{\alpha_1 - 1}(1 - u)^{\alpha_2 - 1}/B(\alpha_1, \alpha_2)\} I_{(0,1)}(u),$$

and

$$B(\alpha_1, \alpha_2) = \Gamma(\alpha_1)\Gamma(\alpha_2)/\Gamma(\alpha_1 + \alpha_2).$$

This leads to the beta–binomial distribution

$$f(y_j;\, \alpha_1, \alpha_2) = \binom{N_j}{y_j} \{B(\alpha_1 + y_j,\, \alpha_2 + N_j - y_j)/B(\alpha_1, \alpha_2)\} I_{A_j}(y_j); \tag{5.43}$$

see Williams (1975).

If we now let $\theta = \alpha_1/\alpha_2$, we have that

$$E(Y_j) = N_j \theta$$

and

$$\text{var}(Y_j) = N_j\theta(1-\theta)\{1+(N_j-1)\kappa\},$$

where $\kappa = (\alpha_1 + \alpha_2 + 1)^{-1}$.

A beta–binomial regression model can be defined by postulating parametric forms for θ and κ. Applications of this type of regression model appear to have been limited mainly to the special cases of one- and two-way ANOVA designs; see McLachlan (1997).

As in the case of the Poisson distribution, a quasi-likelihood approach can be used to deal with overdispersion through the use of the binomial regression model (Williams, 1982). With this approach, only the first two moments of the distribution of Y_j have to be specified. Two such specifications have

$$E(Y_j) = \theta_j \tag{5.44}$$

with either

$$\text{var}(Y_j) = \kappa_j\theta_j(1-\theta_j) \tag{5.45}$$

or

$$\text{var}(Y_j) = N_j\theta_j(1-\theta_j)\{1+(N_j-1)\kappa_j\}, \tag{5.46}$$

where

$$\log\{\theta_j/(1-\theta_j)\} = \boldsymbol{\beta}^T\boldsymbol{x}_j \quad (j = 1, \ldots, n).$$

As Anderson (1988) noted, it is interesting that the assumptions (5.45) and (5.46) for the first two moments of Y_j are satisfied by the beta–binomial distribution (5.43) if $\alpha_1 = (1-\kappa_j)\kappa_j^{-1}\theta_j$ and $\alpha_2 = (1-\kappa_j)\kappa_j^{-1}(1-\theta_j)$.

5.11 FINITE MIXTURES OF LOGISTIC REGRESSIONS

5.11.1 Mean and Variance

We consider now the mixture of GLMs model (5.16) in the case where the component GLMs belong to the binomial family. That is, the ith component frequency function $f(y_j;\, \theta_{ij})$ for the jth response Y_j is given by

$$f(y_j;\, \theta_{ij}) = \binom{N_j}{y_j}\theta_{ij}^{y_j}(1-\theta_{ij})^{N_j-y_j}I_{A_j}(y_j), \tag{5.47}$$

where $A_j = \{0, 1, \ldots, N_j\}$. Under the logistic regression model, θ_{ij} is postulated to depend on the covariates so that

$$\log\{\theta_{ij}/(1-\theta_{ij})\} = \boldsymbol{\beta}_i^T\boldsymbol{x}_j \quad (i = 1, \ldots, g;\ j = 1, \ldots, n). \tag{5.48}$$

We consider now the mean and variance of the logistic regression mixture model

$$\sum_{i=1}^{g} \pi_{ij} f(y_j;\, \theta_{ij}). \tag{5.49}$$

The mean of (5.49) is

$$E(Y_j) = \sum_{i=1}^{g} \pi_{ij}\theta_{ij}$$

and its variance is

$$\begin{aligned}\text{var}(Y_j) &= E\{\text{var}(Y_j \mid \boldsymbol{Z}_j)\} + \text{var}\{E(Y_j \mid \boldsymbol{Z}_j)\} \\ &= N_j(\sum_{i=1}^{g} \pi_{ij}\theta_{ij})(1 - \sum_{i=1}^{g} \pi_{ij}\theta_{ij}) \\ &+ \quad \{(N_j - 1)/N_j\}\text{var}\{E(Y_j \mid \boldsymbol{Z}_j)\},\end{aligned}$$

where

$$\text{var}\{E(Y_j \mid \boldsymbol{Z}_j)\} = N_j^2\{\sum_{i=1}^{g} \pi_{ij}\theta_{ij}^2 - (\sum_{i=1}^{g} \pi_{ij}\theta_{ij})^2\}.$$

For $N_j > 1$, $\text{var}\{E(Y_j \mid \boldsymbol{Z}_j)\} = 0$ holds if and only if $E(Y_j \mid \boldsymbol{Z}_j)$ is constant. Hence for each j $(j = 1, \ldots, n)$,

$$\text{var}(Y_j) = N_j \left(\sum_{i=1}^{g} \theta_{ij}\right)\left(1 - \sum_{i=1}^{g} \theta_{ij}\right)$$

if and only if

$$\theta_{1j} = \theta_{2j} = \cdots = \theta_{gj}$$

for $1 \leq j \leq n$. This implies that the proposed mixture model is able to cope better than the one-component model with extra-binomial variation among $Y_1, \ldots, Y_n$ due to heterogeneity in the population.

5.11.2 Mixing at the Binary Level

For binary data $(N_j = 1)$, we can rewrite (5.49) as

$$\left(\sum_{i=1}^{g} \pi_{ij}\theta_{ij}\right)^{y_j} \left(1 - \sum_{i=1}^{g} \pi_{ij}\theta_{ij}\right)^{1-y_j},$$

so that the g-component mixture of Bernoulli distributions is itself a Bernoulli distribution with probability parameter

$$\theta_{.j} = \sum_{i=1}^{g} \pi_{ij}\theta_{ij}. \tag{5.50}$$

This model is given within the GLM framework by still specifying the binomial as the error function, but now specifying the link function according to

$$\log\{\theta_{.j}/(1 - \theta_{.j})\} = \boldsymbol{\beta}^T \boldsymbol{x}_j \quad (j = 1, \ldots, n).$$

In the parlance of GLMs, mixing at the binary level changes the link but not the frequency function or dispersion, whereas mixing at the binomial level ($N_j > 1$) changes both the link and the frequency function and introduces overdispersion. Caution has to be exercised in using (5.49) with $N_j = 1$, as the model may not be identifiable without imposing some unrealistic restrictions on the covariates; see Follmann and Lambert (1989, 1991) and Wang (1994). Brooks et al. (1998) have considered a mixture of a beta–binomial and a binomial in handling outlying litters in the analysis of data sets recording fetal control mortality in mouse litters.

5.11.3 Identifiability

Teicher (1960, 1963), Blischke (1978), and Margolin, Kim, and Risko (1989) have given necessary and sufficient conditions for the identifiability of the finite binomial mixture

$$f(y_j; \boldsymbol{\Psi}) = \sum_{i=1}^{g} \pi_i \binom{N}{y_j} \theta_i^{y_j} (1 - \theta_i)^{N - y_j} I_{A_N}(y_j), \tag{5.51}$$

where $A_N = \{0, 1, \ldots, N\}$. Their results may be summarized as follows. The g-component binomial mixture model (5.51) with $0 < \theta_i < 1\,(i = 1, \ldots, g)$ is identifiable if and only if

$$g \leq \tfrac{1}{2}(N + 1).$$

Wang (1994) has considered the identifiability of the collection of logistic regression models

$$f(y_j; \boldsymbol{\Psi}) = \sum_{i=1}^{g} \pi_{ij} \binom{N_j}{y_j} \theta_{ij}^{y_j} (1 - \theta_{ij})^{N_j - y_j} I_{A_j}(y_j) \quad (j = 1, \ldots, n), \tag{5.52}$$

where the π_{ij} and the θ_{ij} are specified as functions of the covariates by (5.21) and (5.48), respectively. In the case where the number of Bernoulli trials N_j are all equal ($N_j = N$ for all j), sufficient conditions for the identifiability of (5.52) are that $g \leq \frac{1}{2}(N + 1)$ and that the matrix $(\boldsymbol{x}_1, \ldots, \boldsymbol{x}_n)$ is of full rank. In the case of unequal N_j, the restriction on g in these sufficient conditions is specified in terms of the minimum number of trials for all proper subsets of the observations.

Previously, Follmann and Lambert (1989, 1991) considered sufficient conditions for the identifiability of (5.52) in the special case where the mixing proportions are not functions of any covariates and $\boldsymbol{\beta}_1, \ldots, \boldsymbol{\beta}_g$ have no common elements apart from the first; that is, the logistic components differ only in their intercepts. They showed that for a binary response the number of components g in the mixture must be bounded by a function of the number of covariate vectors that agree except for one element; and for a binomial response, g must satisfy the same bound or be bounded by a function of the largest number of trials per response.

Examples on the fitting of mixtures of logistic regressions to biological data may be found in Follmann and Lambert (1989, 1991) and Wang (1994), while Farewell and Sprott (1988) gave an example on the fitting of a mixture of binomial distributions. Overdispersion in the case of the multinomial distribution has been considered by

Mosiman (1962), Paul, Liang, and Self (1989), Kim and Margolin (1992), Morel and Nagaraj (1993), and Banjeree and Paul (1999), among others.

5.11.4 Example 5.3: Beta-Blockers Data Set

We now present an example on the use of mixtures of logistic regressions from Aitkin (1999b). It concerns a problem in meta-analysis, which is widely documented in an increasing number of papers; see, for example, Olkin (1995) for a recent review.

The example is the 22-center clinical trial of beta-blockers for reducing mortality after myocardial infarction, described by Yusuf et al. (1985) and analyzed in detail in Gelman et al. (1995). The data are given in Table 5.7, adapted from Gelman et al. (1995, p. 149) and are represented by a two-level model, with centers at the upper level and patients at the lower level. There is only one explanatory variable, the treatment assignment at the lower level.

Table 5.7 Results of 22 Clinical Trials of Beta-Blockers

Center	Control		Treated	
j	Deaths	Total	Deaths	Total
1	3	39	3	38
2	14	116	7	114
3	11	93	5	69
4	127	1520	102	1533
5	27	365	28	355
6	6	52	4	59
7	152	939	98	945
8	48	471	60	632
9	37	282	25	278
10	188	1921	138	1916
11	52	583	64	873
12	47	266	45	263
13	16	293	9	291
14	45	883	57	858
15	31	147	25	154
16	38	213	33	207
17	12	122	28	251
18	6	154	8	151
19	3	134	6	174
20	40	218	32	209
21	43	364	27	391
22	39	674	22	680

Source: From Aitkin (1999b).

Aitkin (1999b) fitted a logistic regression model with treatment as a fixed effect, ignoring the center classification. This gave a deviance of 305.76 with 42 degrees of freedom, indicating large variations in intercept (response under the standard treatment) among the centers. Fitting the fixed treatment model with a random intercept term and successively more components g gives deviances of 145.23 ($g = 2$), 101.29 ($g = 3$), and 101.29 ($g = 4$). For the three-component mixture model, the fixed treatment effect estimate is 0.258 with standard error 0.050. The estimates of the random effects nearly form a symmetric three-point distribution located at the points 1.61, 2.25, and 2.83, with respective masses 0.249, 0.512, and 0.239. The standard deviation of the mixing distribution is 0.43, representing substantial variation on the logit scale. This distribution is quite close to a discretized normal; the mass points are located similarly to those in the three-point Gaussian quadrature analysis (which are at 0 and $\pm\sqrt{3}\sigma$ relative to the intercept), but the probability masses are somewhat different, being 2/3, 1/6, and 1/6 for the Gaussian model. The deviance for the three-point Gaussian quadrature analysis is 114.35, and the MLE of the standard deviation for the Gaussian random effect model is 0.40.

Aitkin (1999b) also fitted the model with the treatment effect in the random part of the model; that is, fitting a full random slope and intercept model. This gave a deviance of 99.12, which is a reduction of only 2.17 for the two additional parameters relative to the fixed treatment model. The three-component solution has component (intercept, slope) at $(-1.58, -0.32)$, $(-2.25, -0.26)$, and $(-2.92, -0.08)$ with respective mixing proportions 0.249, 0.511, and 0.240. There is very little evidence of any variation in the treatment effect over centers.

Similar results were given by Gelman et al. (1995), who gave a posterior median for the treatment effect of -0.25 with an approximate posterior standard deviation of about 0.07, though the posterior median of the standard deviation for their Gaussian random effect model was substantially smaller, 0.13, than the standard deviation 0.43 for the estimated random effect distribution. This may be a consequence of the additional information in the prior, or of the use of a normal approximation for the likelihood contributions of the individual centers.

5.12 LATENT CLASS MODELS

Latent class models constitute another class of models that can be easily fitted when formulated as a mixture of GLMs. Suppose that the feature vector $\boldsymbol{y}_j = (y_{1j}, \ldots, y_{pj})^T$ consists of p binary variables. Then a common approach to the mixture analysis of such data is to assume that these variables are conditionally independent given their membership of a component in the mixture model. That is, the density (frequency function) of an observation $\boldsymbol{Y}_j$ can be expressed as

$$f(\boldsymbol{y}_j) = \sum_{i=1}^{g} \pi_i \, f(\boldsymbol{y}_j;\, \boldsymbol{\theta}_i), \tag{5.53}$$

where the ith component density is given by

$$f(\boldsymbol{y}_j;\, \boldsymbol{\theta}_i) = \prod_{v=1}^{p} \theta_{vi}^{y_{vj}} (1 - \theta_{vi})^{1-y_{vj}}$$

and where $\boldsymbol{\theta}_i = (\theta_{1i}, \ldots, \theta_{pi})^T$ and θ_{vi} is the conditional probability that $Y_{vj} = 1$ $(v = 1, \ldots, p)$ given its membership of the ith component of the mixture.

This is the basic model of latent class analysis with its fundamental assumption of local independence, which goes back to Lazarsfeld (1950) and Lazarsfeld and Henry (1968). Since then, it has attracted considerable interest and has been widely used in the social sciences and in biostatistics; see Formann and Kohlmann (1996) for a recent review. For an interesting example of latent class analysis, the reader is referred to McLachlan and Basford (1988), who reported the use of this model by Aitkin et al. (1981) in their analysis of an extensive body of educational research data on teaching. Celeux and Govaert (1991) have considered clustering criteria for discrete data and latent class models. With latent class models, the component-indicator variables are not actually missing in the sense that the components in the mixture model correspond to physically existing groups. The components do represent underlying "classes" without precise physical definition, but which often turn out to have meaningful physical interpretation.

The basic latent class model can be easily fitted as a mixture of GLMs using the general formulation via the EM algorithm, as described in Section 5.5; see also Everitt (1984). Early attempts using conventional iterative schemes date from the 1950s (McHugh, 1956; Lazarsfeld and Henry, 1968). The rigorous solution to the problem of parameter estimation was found by Goodman (1974a, 1974b) about 20 years later. In his seminal papers, he proposed the use of the iterative proportional fitting algorithm. This algorithm was implemented by Clogg (1977) in his computer program MLLSA; see Clogg (1995) and Formann and Kohlmann (1996). It subsequently became apparent that the proportional fitting algorithm is a special variant of the EM algorithm.

Over the years, there have been many variants and extensions of the basic latent class model; see, for example, McCutcheon (1987), Lindsay, Clogg, and Grego (1991), Uebersax and Grove (1993), Wedel and DeSarbo (1993), Formann (1994), and Muthén and Shedden (1999), along with the aforementioned references in this section.

5.13 HIERARCHICAL MIXTURES-OF-EXPERTS MODEL

5.13.1 Mixtures-of-Experts Model

Suppose that we observe n independent observations $\boldsymbol{y}_1, \ldots, \boldsymbol{y}_n$ with associated covariates, $\boldsymbol{x}_1, \ldots, \boldsymbol{x}_n$, respectively. Then under the ME model as proposed by Jacobs et al. (1991), the density of $\boldsymbol{Y}_j$ given covariate $\boldsymbol{x}_j$ is modeled by

$$f(\boldsymbol{y}_j \mid \boldsymbol{x}_j, \boldsymbol{\Psi}) = \sum_{i=1}^{g} \pi_i(\boldsymbol{x}_j; \alpha) f_i(\boldsymbol{y}_j \mid \boldsymbol{x}_j, \boldsymbol{\theta}_i), \tag{5.54}$$

where $\boldsymbol{\Psi} = (\alpha^T, \boldsymbol{\xi}^T)^T$ and where $\boldsymbol{\xi}$ contains the elements of $\boldsymbol{\theta}_1, \ldots, \boldsymbol{\theta}_g$ known *a priori* to be distinct.

Figure 5.1 presents a graphical representation of a ME model ($g = 2$) where, for economy of notation, $f(\boldsymbol{y}_j \mid \boldsymbol{x}_j, \boldsymbol{\Psi})$ is written as f_j, $\pi_i(\boldsymbol{x}_j;\, \boldsymbol{\alpha})$ as π_{ij}, and $f_i(\boldsymbol{y}_j \mid \boldsymbol{x}_j, \boldsymbol{\theta}_i)$ as f_{ij}. In the machine learning literature, the architecture underlying ME is described in terms of two types of networks. There are the g modules, referred to as expert networks. These networks approximate the distribution of the response vector $\boldsymbol{y}_j$ within each region of the covariate space. The expert network maps its input, the covariate vector $\boldsymbol{x}_j$, to an output, the density $f_i(\boldsymbol{y}_j;\, \boldsymbol{\theta}_i)$. It is assumed that different experts are appropriate in different regions of the covariate space. Consequently, the model requires a module, referred to as a gating network, that identifies for any covariate vector $\boldsymbol{x}_j$, the expert or blend of experts whose output is most likely to approximate the corresponding density $f(\boldsymbol{y}_j;\, \boldsymbol{\Psi})$ of the response vector $\boldsymbol{y}_j$. The gating network outputs are a set of scalar coefficients $\pi_i(\boldsymbol{x}_j;\, \boldsymbol{\alpha})$ that weight the contributions of the various experts.

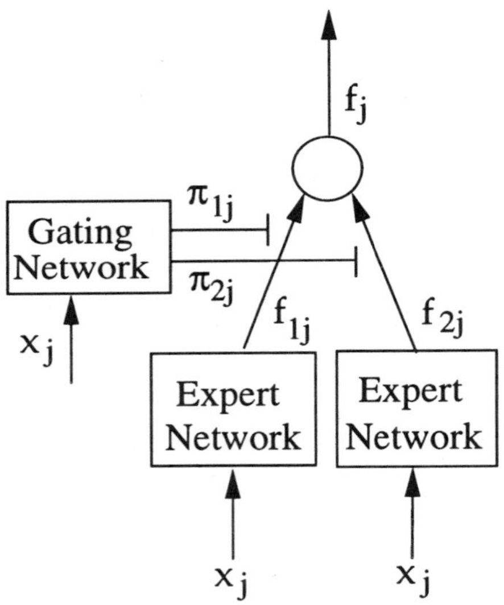

Fig. 5.1 A mixture-of-experts model.

From the perspective of statistical mixture modeling, the gating network models the input-dependent mixing proportions $\pi_i(\boldsymbol{x}_j;\, \boldsymbol{\alpha})$, which are the probabilities in a multinomial distribution consisting of one draw on g categories. The expert networks model the input-dependent component densities $f_i(\boldsymbol{y}_j \mid \boldsymbol{x}_j, \boldsymbol{\theta}_i)$. It can be seen that (5.54) will have the form of a mixture of GLMs if the component densities $f_i(\boldsymbol{y}_j \mid \boldsymbol{x}_j, \boldsymbol{\theta}_i)$ belong to the exponential family. In practice, the idea is to choose the number of components g to be sufficiently large so that the component densities need be relatively simple functions. In which case, they will typically belong to the exponential family.

5.13.2 Hierarchical Mixtures-of-Experts

With the ME model, there is only one level of experts; that is, there is only one gating network. For the modeling of complicated random phenomena, we can improve the flexibility of the ME model by increasing the level of experts. Jordan and Jacobs (1994) have empirically found that models with a nested structure often outperform single-level models with an equivalent number of free parameters. For a two-level network of experts, we have the hierarchical mixtures-of-experts (HME) model proposed in Jordan and Jacobs (1992). It has the form

$$f(\boldsymbol{y}_j \mid \boldsymbol{x}_j, \boldsymbol{\Psi}) = \sum_{i=1}^{g} \pi_i(\boldsymbol{x}_j;\, \boldsymbol{\alpha}) \sum_{h=1}^{g_i} \pi_{hi}(\boldsymbol{x}_j;\, \boldsymbol{\alpha}_i) f_{hi}(\boldsymbol{y}_j \mid \boldsymbol{x}_j,\, \boldsymbol{\theta}_{hi}), \tag{5.55}$$

where $\pi_i(\boldsymbol{x}_j;\, \boldsymbol{\alpha})$ is the probability that the observation $\boldsymbol{Y}_j$ (with covariate $\boldsymbol{x}_j$) belongs to the ith component at the first level, and $\pi_{hi}(\boldsymbol{x}_j;\, \boldsymbol{\alpha}_i)$ is the conditional probability that $\boldsymbol{Y}_j$ (with covariate $\boldsymbol{x}_j$) belongs to the hth component of the first-level ith component (that is, the (i, h)th component) given that it belongs to the ith component at the first level.

The mixing proportions π_i and the π_{hi} are usually modeled as functions of the covariate $\boldsymbol{x}_j$ by the logistic function. Thus,

$$\pi_i(\boldsymbol{x}_j;\, \boldsymbol{\alpha}) = \exp(\boldsymbol{\omega}_i^T \boldsymbol{x}_j) / \sum_{s=1}^{g} \exp(\boldsymbol{\omega}_s^T \boldsymbol{x}_j) \quad (i = 1, \ldots, g), \tag{5.56}$$

where $\boldsymbol{\omega}_g = \boldsymbol{0}$ and where it is implicitly assumed that the first element of $\boldsymbol{x}_j$ is one. Here $\boldsymbol{\alpha}$ contains the elements in the $\boldsymbol{\omega}_i$ $(i = 1, \ldots, g-1)$. Similarly,

$$\pi_{hi}(\boldsymbol{x}_j;\, \boldsymbol{\alpha}_i) = \exp(\boldsymbol{\omega}_{hi}^T \boldsymbol{x}_j) / \sum_{s=1}^{g_i} \exp(\boldsymbol{\omega}_{si}^T \boldsymbol{x}_j) \quad (h = 1, \ldots, g_i),$$

where $\boldsymbol{\omega}_{g_i, i} = \boldsymbol{0}$ and $\boldsymbol{\alpha}_i$ contains the elements in the $\boldsymbol{\omega}_{hi}$ $(h = 1, \ldots, g_i - 1)$.

The vector of unknown parameters is given by

$$\boldsymbol{\Psi} = (\boldsymbol{\alpha}^T, \boldsymbol{\alpha}_1^T, \ldots, \boldsymbol{\alpha}_g^T, \boldsymbol{\xi}^T)^T,$$

where $\boldsymbol{\xi}$ contains the elements of the $\boldsymbol{\theta}_{hi}$ known *a priori* to be distinct. The extension to trees of arbitrary depth and width is straightforward.

The HME architecture can be described also in terms of a tree, where each terminal node is an expert network, and each nonterminal node is a root of a subtree that itself corresponds to an HME architecture. At every nonterminal node in the tree, there is a gating network that is responsible for the topmost split of the HME architecture rooted at that node. Thus these gating networks implement the recursive splitting of the covariate space. All of the expert and gating networks in the architecture have the same input $\boldsymbol{x}_j$. Figure 5.2 presents a graphical representation of a two-level $(i = 2)$ HME model with $g_i = 2$.

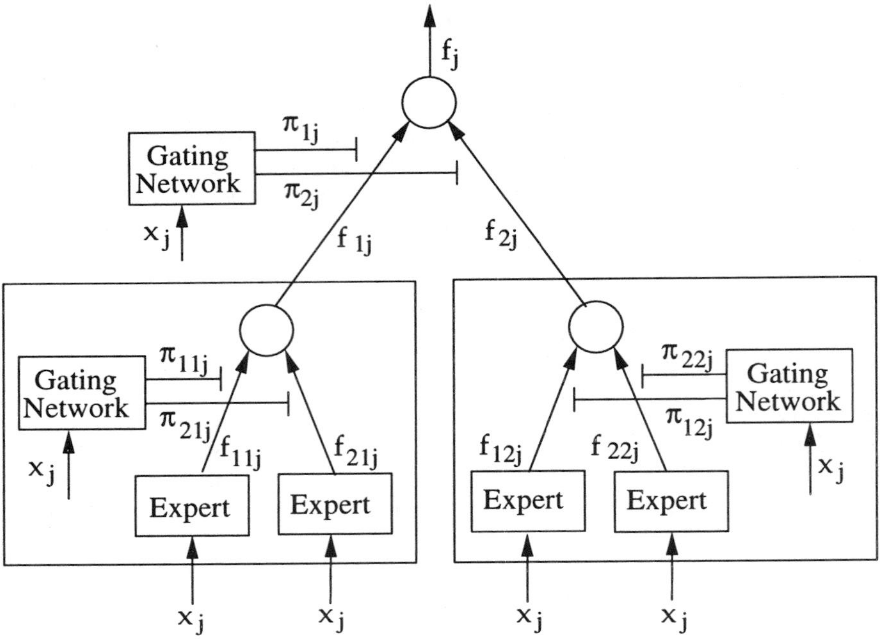

Fig. 5.2 A two-layer hierarchical mixtures-of-experts model.

The probabilistic interpretation of the HME model can be given in terms of a nested sequence of decisions, each based on the covariate $\boldsymbol{x}_j$. In the two-level tree above, a component label i is selected with probability $\pi_i(\boldsymbol{x}_j;\, \boldsymbol{\alpha})$ from the multinomial distribution

$$\text{Mult}_g(1,\, \boldsymbol{\pi}_j), \tag{5.57}$$

consisting of one draw on g categories with probabilities

$$\boldsymbol{\pi}_j = (\pi_1(\boldsymbol{x}_j;\, \boldsymbol{\alpha}),\, \ldots,\, \pi_g(\boldsymbol{x}_j;\, \boldsymbol{\alpha}))^T. \tag{5.58}$$

Next a subcomponent label h is chosen with probability $\pi_{hi}(\boldsymbol{x}_j;\, \boldsymbol{\alpha}_i)$ from the multinomial distribution

$$\text{Mult}_{g_i}(1,\, \boldsymbol{\pi}_{ij}), \tag{5.59}$$

consisting of one draw on g_i categories with probabilities

$$\boldsymbol{\pi}_{ij} = (\pi_{1i}(\boldsymbol{x}_j;\, \boldsymbol{\alpha}_i),\, \ldots,\, \pi_{g_i,i}(\boldsymbol{x}_j;\, \boldsymbol{\alpha}_i))^T. \tag{5.60}$$

The $(i,\, h)$th process generates a vector $\boldsymbol{y}_j$ from the $(i,\, h)$th density $f_{hi}(\boldsymbol{y}_j \mid \boldsymbol{x}_j,\, \boldsymbol{\theta}_i)$.

It can be seen that HME models provide a "divide and conquer" strategy: Complex tasks are decomposed into simpler subtasks which are, in turn, decomposed into even simpler sub-subtasks. Because HME models have a tree-structured approach to piecewise approximation, they have similarities to classification and regression tree models that have previously been proposed in the statistics literature, such as CART in Breiman et al. (1984) and MARS (Friedman, 1991). As explained in Jacobs, Tanner, and Peng (1996), they differ in the way they divide the covariate space into regions. With the HME architecture, a probabilistic (soft) subdivision is employed. In contrast, CART and MARS use hard boundaries with each data point lying in exactly one of the regions. In addition, MARS is restricted to forming region boundaries that are perpendicular to one of the covariate space axes. Thus MARS is coordinate-dependent. As HME is a parametric model (CART and MARS are both nonparametric techniques), probability-based methods including posterior simulation can be used in the fitting of the model and in assessing the uncertainty in its predictions (Jacobs et al., 1991). Comparisons between CART, MARS, and an HME on a robot dynamics task were performed by Jordan and Jacobs (1994), and the HME (using the EM algorithm to estimate the parameters) yielded a modest improvement.

5.13.3 Application of EM Algorithm to HME Model

To apply the EM algorithm to this model, we introduce the component-indicator variables z_{hij}, where z_{hij} is one or zero according to whether $\boldsymbol{y}_j$ belongs or does not belong to the (i, h)th component of the HME model. We let $\boldsymbol{z}$ be the vector containing all these component indicators. The probability that Z_{hij} is one, given the covariate $\boldsymbol{x}_j$, is

$$\begin{aligned}\pi_{hij} &= \text{pr}_{\boldsymbol{\Psi}}\{Z_{hij} = 1 \mid \boldsymbol{x}_j\} \\ &= \pi_i(\boldsymbol{x}_j;\, \boldsymbol{\alpha})\, \pi_{hi}(\boldsymbol{x}_j;\, \boldsymbol{\alpha}_i) \quad (h = 1,\, \ldots,\, g_i;\, i = 1,\, \ldots,\, g). \qquad (5.61)\end{aligned}$$

The conditional expectation of Z_{hij}, given the response $\boldsymbol{y}_j$ with covariate $\boldsymbol{x}_j$, is given by

$$\begin{aligned}\tau_{hij} &= \text{pr}_{\boldsymbol{\Psi}}\{Z_{hij} = 1 \mid \boldsymbol{y}_j,\, \boldsymbol{x}_j\} \\ &= \frac{\pi_i(\boldsymbol{x}_j;\, \boldsymbol{\alpha})\, \pi_{hi}(\boldsymbol{x}_j;\, \boldsymbol{\alpha}_i)\, f_{hi}(\boldsymbol{y}_j \mid \boldsymbol{x}_j,\, \boldsymbol{\theta}_{hi})}{f(\boldsymbol{y}_j \mid \boldsymbol{x}_j,\, \boldsymbol{\Psi})} \quad (h = 1,\, \ldots,\, g_i;\, i = 1,\, \ldots,\, g).\end{aligned}$$

(5.62)

As with the fitting of a mixture of GLMs, if $\boldsymbol{z}$ containing all the component-indicators were known, then the computation of the MLE would separate out into regression problems for each expert network and a multiway classification problem for the multinomials, which can be solved independently of each other. To see this, the complete-data log likelihood is given by

$$\begin{aligned}\log L_c(\boldsymbol{\Psi}) &= \sum_{h=1}^{g_i} \sum_{i=1}^{g} \sum_{j=1}^{n} z_{hij} \{\log \pi_i(\boldsymbol{x}_j;\, \boldsymbol{\alpha}) + \log \pi_{hi}(\boldsymbol{x}_j;\, \boldsymbol{\alpha}_i) \\ &\qquad + \log f_{hi}(\boldsymbol{y}_j \mid \boldsymbol{x}_j,\, \boldsymbol{\theta}_{hi})\}. \qquad (5.63)\end{aligned}$$

The EM algorithm on the $(k + 1)$th iteration reduces to the following:

E-step. Replace z_{hij} in (5.63) by its current conditional expectation $\tau_{hij}^{(k)}$ obtained by using $\boldsymbol{\Psi}^{(k)}$ for $\boldsymbol{\Psi}$ in (5.62).

M-step. The M-step consists of three separate maximization problems. The updated estimate of $\boldsymbol{\alpha}^{(k+1)}$ is obtained by solving

$$\sum_{i=1}^{g}\sum_{h=1}^{g_i}\sum_{j=1}^{n}\tau_{hij}^{(k)}\partial\log\pi_{hi}(\boldsymbol{x}_j;\ \boldsymbol{\alpha})/\partial\boldsymbol{\alpha}=\mathbf{0}. \tag{5.64}$$

The updated estimate of $\boldsymbol{\alpha}_i^{(k+1)}$ is obtained by solving

$$\sum_{h=1}^{g_i}\sum_{j=1}^{n}\tau_{hij}^{(k)}\partial\log\pi_{hi}(\boldsymbol{x}_j,\ \boldsymbol{\alpha}_i)/\partial\boldsymbol{\alpha}_i=\mathbf{0}\quad(i=1,\ \ldots,\ g). \tag{5.65}$$

Assuming that the $\boldsymbol{\theta}_{hi}$ have no elements known *a priori* to be in common, the updated estimate of $\boldsymbol{\theta}_{hi}^{(k+1)}$ is obtained by solving

$$\sum_{j=1}^{n}\tau_{hij}^{(k)}\partial\log f_{hi}(\boldsymbol{y}_j\mid\boldsymbol{x}_j,\ \boldsymbol{\theta}_{hi})/\partial\boldsymbol{\theta}_{hi}=\mathbf{0} \tag{5.66}$$

for each (i, h) $(i=1,\ \ldots,\ g;\ h=1,\ \ldots,\ g_i)$.

All these equations require iterative methods; (5.64) and (5.65) can be solved by IRLS for a GLM. So too can (5.66) if $f_{hi}(\boldsymbol{y}_j \mid \boldsymbol{x}_j,\ \boldsymbol{\theta}_{hi})$ belongs to the exponential family.

Jacobs et al. (1991) and Jordan and Jacobs (1992) showed empirically for the ME and HME models, respectively, that the EM algorithm yields significantly faster convergence than gradient ascent. Jordan and Xu (1995) subsequently provided a theoretical analysis. Jacobs (1997) has investigated the bias and variance of HME architectures, while Jiang and Tanner (1999a, 1999b) have considered the approximation rate of HME for GLMs. Jiang and Tanner (1999c) have obtained conditions for the identifiability of the ME model, which they showed held for Poisson, gamma, normal, and binomial experts. Chen, Xu, and Chi (1999) have investigated the use of the IRLS algorithm on the M-step in the EM-fitting of ME models where the response variable $\boldsymbol{y}_j$ is multivariate, corresponding to multiclass classification problems.

5.13.4 Example 5.4: Speech Recognition Problem

We illustrate the use of the HME model in its application to a speech recognition task as reported in Peng, Jacobs, and Tanner (1996). The values of the covariate vector $\boldsymbol{x}_j$ are the first and second formant values extracted from ten classes of spoken vowels. The response variable $\boldsymbol{y}_j$ is the class indicator of the spoken vowel with y_{vj} being one or zero, according to whether it belongs or does not belong to the vth class $(v=1,\ \ldots,\ 10)$. The covariates $\boldsymbol{x}_j$ are plotted in Figure 5.3, and the value of the response vector $\boldsymbol{y}_j$ is represented by the digit 0–9. The data were segmented from utterances of 10 words, each of which began with an "h", contained a vowel in the middle, and ended with a "d." In all, there were $n = 75$ speakers (32 male adults, 28 female adults, and 15 children). The training set consisted of 149 items randomly selected from the available data, leaving 1,345 items for a test set.

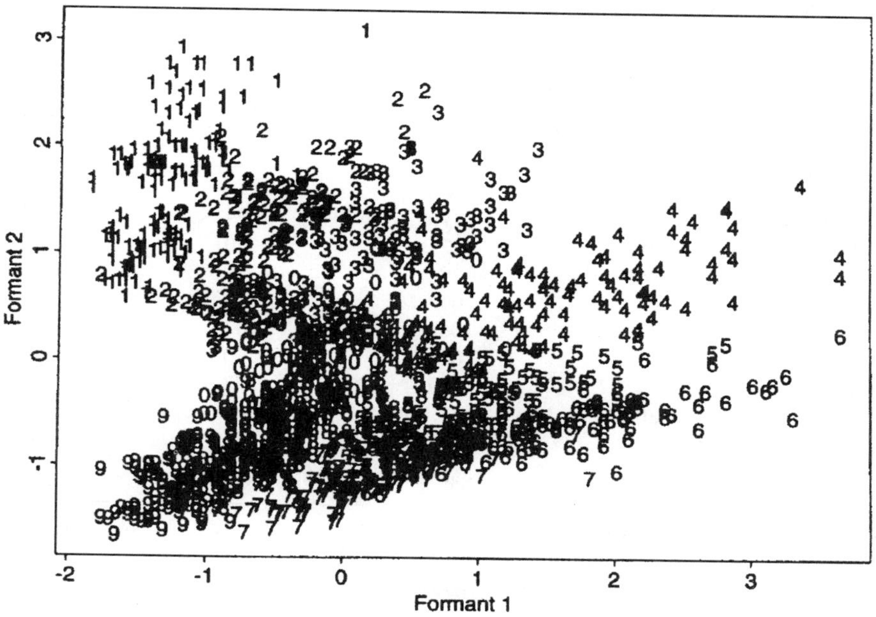

Fig. 5.3 The horizontal and vertical axes give the first and second formants, respectively. Each data point is labeled with a digit (0–9) that indicates the vowel class to which the data point belongs. From Peng et al. (1996).

Peng et al. (1996) fitted a two-level HME with 10 experts in the second level arranged in two modules of five each. Thus in the definition (5.55) of the HME model, $g = 2$ and $g_i = 5$. The (i, h)th component density was taken to be the multinomial consisting of one draw on $p = 10$ categories,

$$\mathrm{Mult}_p(1, \boldsymbol{\theta}_{hi}), \tag{5.67}$$

where

$$\boldsymbol{\theta}_{hi} = (\theta_{1hi}, \ldots, \theta_{phi})^T$$

and θ_{vhi} is the conditional probability that $Y_{vj} = 1$, given $\boldsymbol{x}_j$ and its membership of the (i, h)th component in the HME model $(v = 1, \ldots, p)$. Thus the probability that the spoken vowel for the jth speaker belongs to the class denoted by $\boldsymbol{y}_j$ is modeled as

$$f(\boldsymbol{y}_j \mid \boldsymbol{x}_j, \boldsymbol{\Psi}) = \sum_{i=1}^{2} \sum_{h=1}^{5} \prod_{v=1}^{10} \theta_{vhi}^{y_{vj}} (1 - \theta_{vhi})^{1-y_{vj}}. \tag{5.68}$$

Peng et al. (1996) adopted a Bayesian approach to the fitting of this model, using MCMC methods that included Gibbs sampling. The EM algorithm was used to provide an initial value of $\boldsymbol{\Psi}$ in these simulations. They used normal priors for all the unknown parameters; see Peng et al. (1996) for details on the implementation of their MCMC methods. The performance of the classification rule so obtained for their HME model is given in Table 5.8.

Table 5.8 Classification Performances of Four Systems on the Data in the Prediction Set

Total Category	Gibbs # Correct	EM # Correct	CART I # Correct	CART II # Correct
127	123.3	123.3	114	124
132	96.3	97.9	94	112
139	95.9	92.0	30	60
137	84.1	79.2	90	109
133	104.6	100.9	73	117
136	109.1	106.3	102	109
128	88.5	84.4	90	96
132	74.5	78.0	72	71
140	80.7	77.1	49	38
141	58.0	50.2	87	60
1345	915.0	889.3	801	896

Source: From Peng et al. (1996).

5.13.5 Pruning HME Tree Structures

Jacobs, Peng, and Tanner (1997) have considered ways of assessing whether a given HME architecture is over- or under-specified for a given data set. They suggest that a guide to the value of an expert (i, h) is given by

$$\sum_{j=1}^{n} z_{hij}/n, \tag{5.69}$$

and the value of the ith branch by

$$\sum_{h=1}^{g_i}\sum_{j=1}^{n} z_{hij}/n, \tag{5.70}$$

where in the Bayesian framework adopted by Jacobs et al. (1997), the unknown z_{hij} in (5.69) and (5.70) can be estimated by the average of its generated values on a specified number of simulations. In a frequentist framework, it can be replaced by its estimated conditional expectation, which is the limiting value of $\tau_{hij}^{(k)}$ as $k \rightarrow \infty$. Jacobs et al. (1997) suggest some guidelines as to the value of (5.69) for an expert to be deleted. They applied their pruning methods to the speech recognition problem discussed above, starting with a 5–5 HME architecture (that is, $g_i = 5,\ i = 1, 2$). As their indicators suggested that four experts would be useful for this problems, they tried a 2–2 architecture and then finally a 1–1 architecture.

6

Assessing the Number of Components in Mixture Models

6.1 INTRODUCTION

6.1.1 Some Practical Issues

Testing for the number of components g in a mixture is an important but very difficult problem which has not been completely resolved. We have seen in the previous chapters that finite mixture distributions are employed in the modeling of data with two main purposes in mind. One is to provide an appealing semiparametric framework in which to model unknown distributional shapes, as an alternative to, say, the kernel density method; see Escobar and West (1995), Robert (1996), and Roeder and Wasserman (1997). The other is to use the mixture model to provide a model-based clustering. In both situations, there is the question of how many components to include in the mixture.

In the former situation of density estimation, the commonly used criteria of AIC and BIC would appear to be adequate for choosing the number of components g for a suitable density estimate. In particular, Leroux (1992a) established under mild conditions that certain penalized log likelihood criteria, including AIC and BIC, do not underestimate the true number of components, asymptotically. Roeder and Wasserman (1997) have shown that when a normal mixture model is used to estimate a density "nonparametrically," the density estimate that uses BIC to select the number of components in the mixture is consistent. Other satisfactory conclusions for the use of AIC or BIC in this situation are discussed in Biernacki, Celeux, and Govaert (1998), Ćwik and Koronacki (1997), and Solka et al. (1998).

In this chapter we shall concentrate on the choice of the number of components in the latter situation where the mixture model is used in a clustering context. In cluster or latent class analyses, the choice of the number of components arises with the question of how many clusters or classes there are.

As discussed in Roeder (1994), if no prior information is available concerning the component densities, then arguably this problem is more appropriately considered in terms of an assessment of the number of modes. Inferential procedures for assessing the number of modes of a distribution are described by Titterington et al. (1985, Section 5.6), including the univariate technique of Silverman (1981, 1986) which uses the kernel method to estimate the density function nonparametrically and which permits the assessment of modality under certain circumstances. Other papers on tests for the number of modes include Hartigan and Hartigan (1985), Wong (1985), Hartigan (1988), Hartigan and Mohanty (1992), and Fisher, Mammen, and Marron (1994). However, a drawback of this approach is that the components of the mixture have to be fairly wide apart in order to be detected. As seen in Section 1.5.1, a mixture distribution can be unimodal if the components are not sufficiently far apart.

The specification of a parametric family for the component densities may have a major impact on the clustering so obtained. In Section 6.2 we shall present an analysis of some stamp data, where the number of types of paper used in the production of these stamps depends on whether the component variances should be comparable in size.

Another practical issue arises with the parametric specification of the component densities when the number of components in a mixture model are being taken to reflect the number of distinct groups in a population. As noted in Section 1.5.2, normal mixture densities can play a useful role in modeling the distribution of data that have asymmetrical distributions. Indeed, any continuous distribution can be approximated arbitrarily well by a finite mixture of normal densities with common variance (or covariance matrix in the multivariate case). Thus if a normal mixture model is being used to detect the presence of grouping in some data, then there may not be a one-to-one correspondence between the mixture components and the groups if the data have a skewed distribution within some of the groups. This is because more than one normal component may be needed to model a skewed group-conditional distribution (Section 1.6). We shall illustrate this problem in Section 6.1.3, using the case study of Gutierrez et al. (1995), in which there is an adjustment for skewness in the fitting of the normal mixture model.

6.1.2 Order of a Mixture Model

Cutler and Windham (1994) explain some of the difficulties in discussing the choice of functionals, which are measures of goodness-of-fit of a model for a given number of components g. They note that one difficulty has been discussed by Donoho (1988), who points out that, "near any distribution of interest, there are empirically indistinguishable distributions (indistinguishable at a given sample size) where the functional takes on arbitrarily large values." A mixture density with g components might be empirically indistinguishable from one with either fewer than g components or more than g components. It is therefore sensible in practice to approach the question of the number of components in a mixture model in terms of an assessment of the smallest number of components in the mixture compatible with the data. To this end, the true order g_o of the g-component mixture model

$$f(\boldsymbol{y};\,\boldsymbol{\Psi}) = \sum_{i=1}^{g} \pi_i\, f_i(\boldsymbol{y};\,\boldsymbol{\theta}_i) \tag{6.1}$$

is defined to be the smallest value of g such that all the components $f_i(\boldsymbol{y};\,\boldsymbol{\theta}_i)$ are different and all the associated mixing proportions π_i are nonzero.

6.1.3 Example 6.1: Adjusting for Effect of Skewness on the LRT

The effect of skewness on hypothesis tests for the number of components in normal mixture models has been discussed in Section 1.6. In the work of Maclean et al. (1976), Schork and Schork (1988), and Gutierrez et al. (1995) on this problem, the Box–Cox (1964) transformation is employed initially in an attempt to obtain normal components. Hence to model some univariate data $y_1, \ldots, y_n$ by a two-component mixture distribution, the density of Y_j is taken to be

$$f(y_j;\,\boldsymbol{\Psi}) = \{\pi_1\phi(y_j^{(\zeta)};\,\mu_1,\,\sigma_1^2) + \pi_2\phi(y_j^{(\zeta)};\,\mu_2,\,\sigma_2^2)\}\, y_j^{\zeta-1}, \tag{6.2}$$

where

$$\begin{aligned} y^{(\zeta)} &= (y_j^\zeta - 1)/\zeta \quad \text{if}\ \zeta \neq 0, \\ &= \log y_j \qquad\quad\ \text{if}\ \zeta = 0, \end{aligned} \tag{6.3}$$

and where the last term on the right-hand side of (6.2) corresponds to the Jacobian of the transformation from $y_j^{(\zeta)}$ to y_j.

Gutierrez et al. (1995) adopted this mixture model of transformed normal components in an attempt to identify the number of underlying physical phenomena behind tomato root initiation. The observation y_j corresponds to the inverse proportion of the jth lateral root which expresses GUS ($j = 1, \ldots, 40$). This measurement is a possible indicator of the number of initial cells in the lateral root. In Figure 6.1 we report the kernel density estimate and the normal Q–Q plot of the y_j. It shows that the data are clearly skewed with a mode at $y_j = 2$ initial cells. However, Gutierrez et al. (1995) noted that if the reciprocals of the data are used, then the population appears to be normally distributed; see Figure 6.2. They confirmed this by the application of the likelihood ratio test (LRT) for homogeneity versus a two-component normal model applied to the data, firstly in their original form and then to the reciprocals of the data. The P-values for these two tests were equal to 0.012 and 0.955, respectively. The former P-value was obtained through their simulation of the null distribution of the likelihood ratio test statistic (LRTS), $-2\log\lambda$, which yielded a distribution bounded between χ_4^2 and χ_6^2 while resembling the latter in the upper tail. As the primary interest of Gutierrez et al. (1995) was in determining whether the observed data were

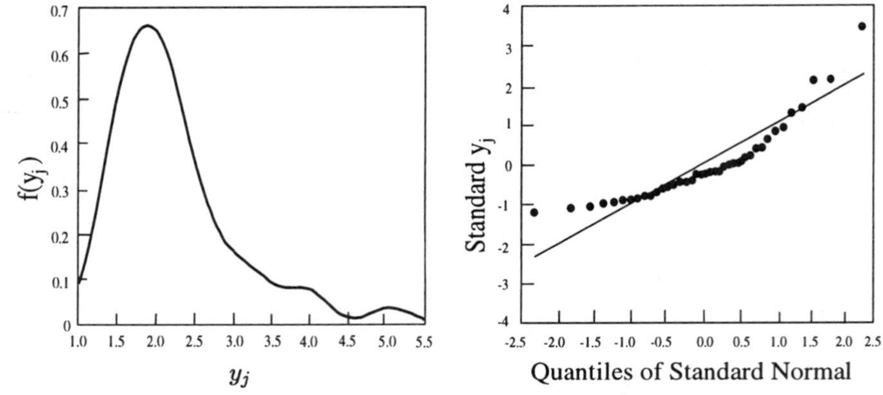

Fig. 6.1 Kernel density estimate and normal Q–Q plot of the y_j. From Gutierrez et al. (1995).

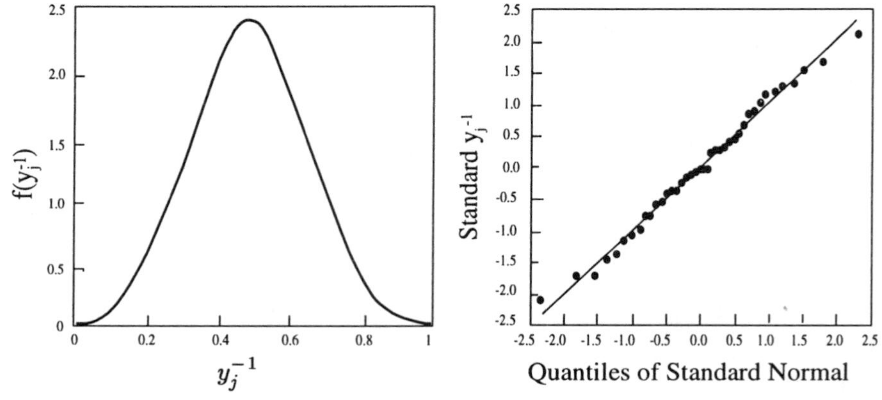

Fig. 6.2 Kernel density estimate and normal Q–Q plot of the y_j^{-1}. From Gutierrez et al. (1995).

the result of one or possibly two underlying physical phenomena, they reanalyzed the data, adjusting for skewness through the use of the mixture model (6.2). For fixed ζ this model can be fitted by ML estimation as for a two-component normal mixture, and so it was fitted unconditionally through a grid search over ζ. The parameter estimates so obtained were $\hat{\zeta} = -0.8$, $\hat{\mu}_1 = 0.52$, $\hat{\sigma}_1^2 = 0.02$, $\hat{\mu}_2 = 0.82$, $\hat{\sigma}_2^2 = 0.004$, and $\hat{\pi}_1 = 0.91$. That $\hat{\zeta}$ is closer to -1 than 1 reveals that y_j^{-1} is more conducive to the normal mixture model (6.2). The P-value for the LRT of $g = 1$ versus $g = 2$ was found by simulation to be 0.957. The null distribution of the LRTS was virtually unaffected by the transformation, which was anticipated from usual likelihood theory given that a free parameter was added under both the null and alternative hypotheses. Thus when the Box–Cox adjustment is made for skewness, the LRT is nonsignificant, and so there is little evidence of any major secondary underlying phenomena. Subsequent simulations by Gutierrez et al. (1995) indicated that the power of the LRT here is less than 10% and that a sample size of $n \geq 350$ would be needed to attain a power of 50%. These simulations also revealed that there is little drop in power in using the Box–Cox transformation.

6.2 EXAMPLE 6.2: 1872 HIDALGO STAMP ISSUE OF MEXICO

Izenman and Sommer (1988) considered the modeling of the distribution of stamp thickness for the printing of a given stamp issue from different types of paper. Their main concern was the application of the nonparametric approach to identify components by the resulting placement of modes in the density estimate. The specific example of a philatelic mixture, the 1872 Hidalgo issue of Mexico, was used as a particularly graphic demonstration of the combination of a statistical investigation and extensive historical data to reach conclusions regarding the mixture components.

For comparison with a parametric approach, Izenman and Sommer (1988) fitted by maximum likelihood a mixture of normal components with unequal variances to the data. They used the LRTS, as modified by the rule of thumb of Wolfe (1971), to test for the smallest number g of components compatible with the data. It yielded $g = 3$ components, which is in apparent conflict with their nonparametric solution of seven modes. In an attempt to resolve this conflict, Basford, McLachlan, and York (1997b) considered further the fitting of a normal mixture model to the data. They demonstrated that by restricting the component variances to be the same, the LRT for the smallest number of components in the mixture model ultimately leads to results consistent with the nonparametric approach.

The data as analyzed by Izenman and Sommer (1988) are displayed in histogram form in Figure 6.3. They are the thicknesses in mm of 485 unwatermarked used white wove stamps. Izenman and Sommer (1988) tested for multimodality, using nonparametric kernel density estimation techniques (Silverman, 1981) to determine the most probable number of modes in the underlying density. They concluded that the stamp thickness data were consistent with an underlying density having seven modes. Using the extensive historical information, plus an analysis of some related data (the 1868 issue), they concluded that seven modes was a sensible description of these data; that is, it is plausible that paper of seven different thicknesses was used in the production of this stamp issue.

For comparison with the nonparametric mode fitting technique, Izenman and Sommer (1988) fitted a mixture of seven normal distributions with unrestricted variances; see Figure 6.3. This produced a consistent result in the sense that the parametrically fitted mixture density had modes at almost the same locations as the seven modes previously determined.

Actually, Basford et al. (1997b) located another $g = 7$ solution corresponding to a larger local maximum of the likelihood function. It differs from the latter by devoting a component to model the outlying observations in the lower tail of the data. It thus

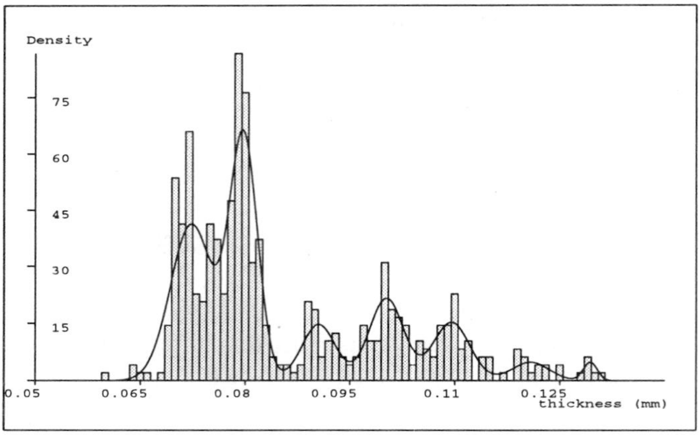

Fig. 6.3 Plot of $g = 7$-component normal mixture fit with unrestricted variances given by Izenman and Sommer (1988). From Basford et al. (1997b).

uses only one rather than two components to model the modes in the histogram at 0.10 and 0.11mm.

By applying Wolfe's (1971) test to sequentially test g against $g + 1$, starting at $g = 1$, Izenman and Sommer (1988) determined that only $g = 3$ components in the heteroscedastic normal mixture model were needed. This fit (Figure 6.4) is given by

$$0.196N(0.0712,\ 0.00000176) + 0.367N(0.0786,\ 0.00000564)$$
$$+ 0.437N(0.0989,\ 0.00019666).$$

The values of the log likelihood for $g = 1$ *to* 9 are listed in Table 6.1 for both

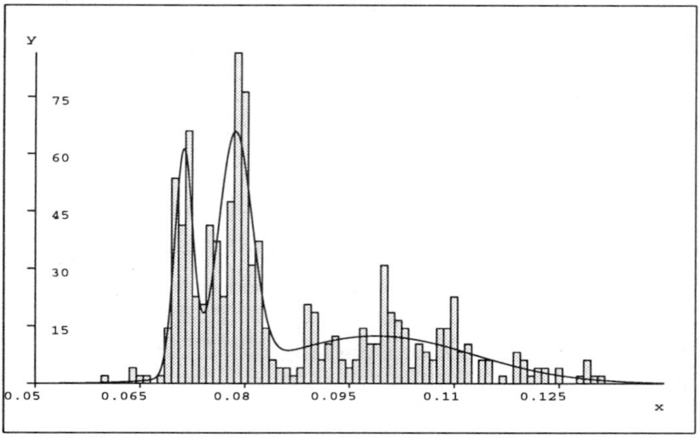

Fig. 6.4 Plot of $g = 3$-component normal mixture fit with unrestricted variances. From Basford et al. (1997b).

unrestricted and equal variances for the normal components.

Table 6.1 Value of the Log Likelihood for $g = 1$ to 9 Normal Components

Number of Components	Unrestricted Variances	Equal Variances
1	1350.3	1350.3
2	1484.8	1442.6
3	1518.8	1475.7
4	1521.9	1487.4
5	1527.1	1489.5
6	1535.2	1512.9
7	1538.7	1525.3
8	1544.1	1535.1
9	1552.2	1536.5

Source: From Basford et al. (1997b).

Besides the assessment of g by testing, as above, for its smallest value compatible with the data, other practical considerations are important if a sensible description or summary of the data is to be reached. For instance, if a mixture model is used to describe a univariate multimodal density, it may be unacceptable from a practical point of view to have the tail of one component (say, the one centered on the highest mean) accounting for data at the lower end (below that of the smallest mean) by allowing its variance to be much greater than the other component variances. This is the case here where the variance of the third component in the $g = 3$ heteroscedastic normal mixture solution of Izenman and Sommer (1988) is appreciably greater than the variances for the other two components. As a consequence of the long left-hand tail of the third component, the four thinnest stamps (ranging from 0.60 mm to 0.66 mm) are assigned to this component (which has a mean of 0.0989 mm) and not the nearest one (which has a mean of 0.0712 mm); see Table 6.2. As argued by Basford et al. (1997b), it seems reasonable to assume that the variability of stamp thickness would be consistent about any particular mean in the production process across the range of measurements observed. Thus they concluded that the $g = 3$ heteroscedastic normal mixture model is impractical for physical considerations, and so they considered the fitting of a normal mixture model with the components restricted to having a common variance.

It can be seen from Table 6.1 that there is only a slight increase in the log likelihood in preceding from $g = 4$ to $g = 5$. This is because the common variance of the components in the mixture model is not sufficiently small in the case of $g = 5$ to provide any further modes in the fit. Thus, if on the basis of $-2 \log \lambda$ we were to test g versus $g + 1$ sequentially, starting with $g = 1$, then testing would terminate at $g = 4$, no matter what sensible criterion were used to decide on the value of g.

If we continue on past $g = 5$ and monitor the increase in the log likelihood, then it is clear from Table 6.1 that there is no need to proceed beyond $g = 8$. On whether there is a need to proceed from $g = 7$ to $g = 8$, the value of $-2 \log \lambda$ is 19.6,

Table 6.2 Fitted Posterior Probabilities of Component Membership

y_j	$\hat{\tau}_1(y_j)$	$\hat{\tau}_2(y_j)$	$\hat{\tau}_3(y_j)$
.060			1.00
.064			1.00
.065			1.00
.066	.03		.97
.068	.73		.27
.069	.92		.08
.070	.96		.04
.071	.96	.01	.03
.072	.94	.02	.04
.073	.80	.12	.08
.074	.36	.50	.14
.075	.05	.83	.12
.076		.91	.09
.077		.93	.07
.078		.94	.06
.079		.93	.07
.080		.91	.09
.081		.87	.13
.082		.79	.21
.083		.63	.37
.084		.40	.60
.085		.18	.82
.086		.06	.94
.087		.01	.99
.088			1.00
.089 − .112			1.00
.114 − .115			1.00
.117			1.00
.119 − .123			1.00
.125			1.00
.128 − .131			1.00

Source: From Basford et al. (1997b).

which would appear to be significant. For example, if Wolfe's approximation were adopted, which would take $-2\log\lambda$ to be χ^2_2, then the P-value is practically zero. The $g = 7$ and $g = 8$ homoscedastic normal mixture fits for the density are displayed in Figures 6.5 and 6.6, respectively. It can be seen that the additional component in the $g = 8$-component normal mixture model exists to accommodate the outlying observations in the lower tail of the data. Elsewhere, it practically gives the same fit as the $g = 7$ homoscedastic solution.

From Basford et al. (1997b), the chi-squared goodness-of-fit test statistic for the $g = 8$ homoscedastic normal mixture fit is 57.4. While this is highly significant, it

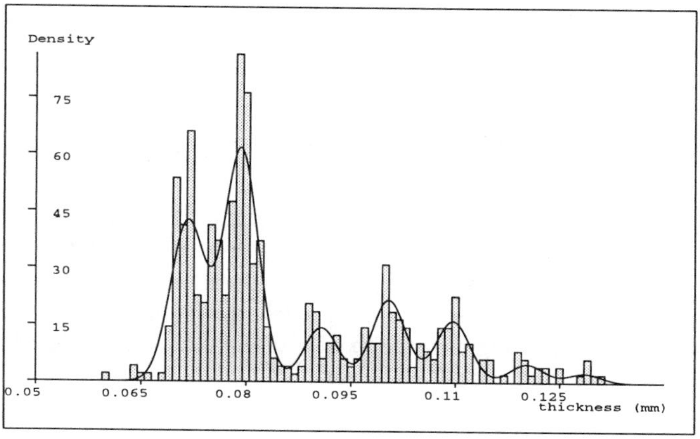

Fig. 6.5 Plot of $g = 7$-component normal mixture fit with equal variances. From Basford et al. (1997b).

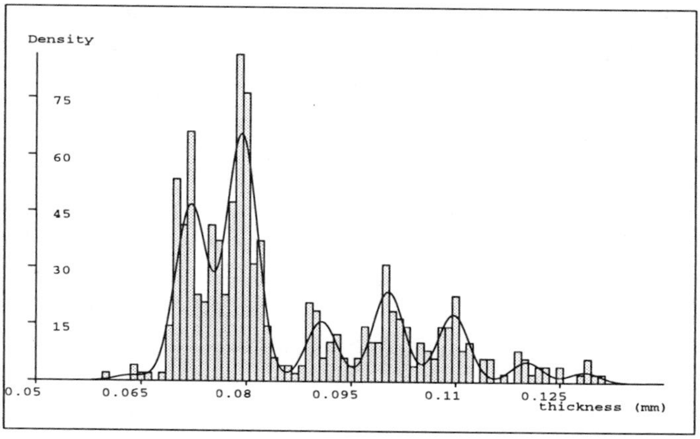

Fig. 6.6 Plot of $g = 8$-component normal mixture fit with equal variances. From Basford et al. (1997b).

has fallen from a value of 158.14 for the $g = 4$ fit. This high value would appear to be due to the rounding of the stamp measurements to the third decimal place, there being only 72 distinct values for the 485 measurements (effectively, bins of width 0.001 mm). When Basford et al. (1997b) made allowance for this rounding by the judicious grouping of intervals in forming the chi-squared goodness-of-fit statistic, the latter was reduced to 20.1 with a P-value of 27%.

If the assumption of equal variances is relaxed in fitting the $g = 8$-component mixture model, then we obtain essentially the same fit as under homoscedasticity.

The only difference, really, is that a smaller variance is obtained for the component with the largest fitted mean.

6.3 APPROACHES FOR ASSESSING MIXTURE ORDER

In most of the work on inference on the number of components g in a mixture model, Bayesian or otherwise, the approach has been to separate the problem of testing for g from the fitting of the mixture model, and hence estimation, for fixed g. However, as discussed in Section 4.7 on the Bayesian approach, the more direct line of modeling the unknown g case by mixing over the fixed g case has been considered by Phillips and Smith (1996) and Richardson and Green (1997).

6.3.1 Main Approaches

The estimation of the order of a mixture model has been considered mainly by consideration of the likelihood, using two main ways. One way is based on a penalized form of the likelihood. As the likelihood increases with the addition of a component to a mixture model, the likelihood (usually, the log likelihood) is penalized by the subtraction of a term that "penalizes" the model for the number of parameters in it. This leads to a penalized log likelihood, yielding what are called information criteria for the choice of g.

The other main way for deciding on the order of a mixture model is to carry out a hypothesis test, using the likelihood ratio as the test statistic. In this chapter we shall consider these two ways in some detail, commencing with the LRT in Section 6.4.

We shall see that penalized likelihood criteria, like AIC and BIC, are less demanding than the LRT, which requires bootstrapping in order to obtain an assessment of the P-value. However, they produce no number that quantifies the confidence in the result, such as a P-value.

6.3.2 Nonparametric Methods

As mentioned in Section 6.1.1, the order of a mixture model can be investigated nonparametrically by tests on the number of modes. Other nonparametric methods that have been used for this problem include the work of Henna (1985) and a number of graphical displays such as the normal scores plot (Cassie, 1954; Harding, 1948), the modified percentile plot of Fowlkes (1979), and the more recent approach of Roeder (1994). In the latter paper, it is demonstrated that a mixture of two normals divided by a normal density having the same mean and variance as the mixture density is always bimodal. This analytic result and other related results form the basis of a diagnostic test for the number of components in a mixture of normals. Previously, Lindsay and Roeder (1992a) had proposed the use of residual diagnostics for determining the number of components.

6.3.3 Method of Moments

Heckman, Robb, and Walker (1990), Furman and Lindsay (1994a), and Dacunha-Castelle and Gassiat (1997b) have used the method of moments to test for the number of components in a mixture model. Also, Vlassis and Likas (1999) and Vlassis, Papakonstantinou, and Tsanakas (1999) have considered a kurtosis-based approach to this problem.

6.4 LIKELIHOOD RATIO TEST STATISTIC

6.4.1 Introduction

An obvious way of approaching the problem of testing for the smallest value of the number of components in a mixture model is to use the LRTS, $-2 \log \lambda$. Suppose we wish to test the null hypothesis,

$$H_0 : g = g_0 \tag{6.4}$$

versus

$$H_1 : g = g_1 \tag{6.5}$$

for some $g_1 > g_0$. Usually, $g_1 = g_0 + 1$ in practice as it is common to keep adding components until the increase in the log likelihood starts to fall away as g exceeds some threshold. The value of this threshold is often taken to be the g_0 in H_0. Of course it can happen that the log likelihood may fall away for some intermediate values of g only to increase sharply at some larger value of g, as in Example 6.2.

We let $\hat{\boldsymbol{\Psi}}_i$ denote the MLE of $\boldsymbol{\Psi}$ calculated under H_i $(i = 0, 1)$. Then the evidence against H_0 will be strong if λ is sufficiently small, or equivalently, if $-2 \log \lambda$ is sufficiently large, where

$$-2 \log \lambda = 2\{\log L(\hat{\boldsymbol{\Psi}}_1) - \log L(\hat{\boldsymbol{\Psi}}_0)\} \tag{6.6}$$

is twice the increase in the log likelihood or the decrease in deviance. Unfortunately with mixture models, regularity conditions do not hold for $-2 \log \lambda$ to have its usual asymptotic null distribution of chi-squared with degrees of freedom equal to the difference between the number of parameters under the null and alternative hypotheses.

To briefly explain why this is so, suppose that the component densities are completely specified. Then the parameter vector $\boldsymbol{\Psi}$ consists of just the mixing proportions. Thus, as $g_0 < g_1$ in (6.4) and (6.5), the null hypothesis is specified by the true value of $\boldsymbol{\Psi}$ being on the boundary of the parameter space (with one or more of the mixing proportions specified as zero). Further, if the component densities belong to, say, the same parametric family $f(\boldsymbol{y}_j; \boldsymbol{\theta})$ with $\boldsymbol{\theta}$ unspecified, then H_0 will hold also if $\boldsymbol{\theta}_h = \boldsymbol{\theta}_i$ for some $h \neq i$. That is, H_0 corresponds to a nonidentifiable subset of the parameter space. Thus with the true value of the parameter vector under H_0 lying on the boundary of the parameter space and also in a nonidentifiable subset if the component densities depend on unknown parameters, the classic regularity conditions (Cramér, 1946) about the asymptotic properties of the MLE are not valid under the null hypothesis H_0. In particular, the asymptotic distribution of the MLE in the nonidentifiable case under H_0 is unknown. The lack of identifiability leads to a degeneracy in the information matrix when considering the asymptotic null distribution of the (normalized) log likelihood formed under the alternative distribution H_1. As a consequence, when using classical Taylor series expansions for the LRTS, the remainder terms may not be bounded uniformly.

6.4.2 Example 6.3: Breakdown in Regularity Conditions

We now present an example to illustrate the breakdown in the standard regularity conditions. Consider a mixture of $g = 2$ univariate normal densities in proportions π_1 and $\pi_2 = 1 - \pi_1$ with means $\mu_1 = 0$ and μ_2 and common unit variances, where π_1 and μ_2 are unspecified. Suppose we wish to test

$$H_0 : f(y_j;\, \boldsymbol{\Psi}) = \phi(y_j;\, 0, 1) \tag{6.7}$$

versus

$$H_1 : f(y_j;\, \boldsymbol{\Psi}) = \pi_1\phi(y_j;\, 0, 1) + \pi_2\phi(y_j;\, \mu_2,\, 1). \tag{6.8}$$

The parameter space is given by

$$\boldsymbol{\Omega} = \{\boldsymbol{\Psi} = (\pi_1,\, \mu_2)^T : [0, 1] \times (-\infty, \infty)\}.$$

The subspace $\boldsymbol{\Omega}_0$ of $\boldsymbol{\Omega}$ that specifies H_0 is the entire μ_2-axis when $\pi_1 = 0$ and the line segment [0,1] on the π_1-axis when $\mu_2 = 0$; that is,

$$\boldsymbol{\Omega}_0 = \{\boldsymbol{\Psi} = (\pi_1,\, \mu_2)^T : ([1] \times (-\infty, \infty)) \cup ([0, 1] \times [0])\}. \tag{6.9}$$

When the null hypothesis H_0 holds, the parameter vector $\boldsymbol{\Psi}$ is on the boundary of the parameter space $\boldsymbol{\Omega}$ and is in the nonidentifiable subspace $\boldsymbol{\Omega}_0$.

We let $\hat{\boldsymbol{\Psi}}_i$ be the MLE of $\boldsymbol{\Psi} = (\pi_1,\, \mu_2)^T$ formed under H_i $(i = 0, 1)$. Then it follows from the result of Feng and McCulloch (1996) discussed in Section 2.5 that $\hat{\boldsymbol{\Psi}}_1$ converges in probability to $\boldsymbol{\Omega}_0$. That is, for fixed y_j, the density $f(y_j;\, \hat{\boldsymbol{\Psi}}_1)$ converges to the common value of the density of Y_j for $\boldsymbol{\Psi} \in \boldsymbol{\Omega}_0$, which is the standard normal as specified under H_0. In this sense, one might anticipate that $-2\log\lambda$ would be reasonably behaved as $n \to \infty$. But Hartigan (1985a, 1985b) showed that $-2\log\lambda$ is asymptotically unbounded above in probability at a very slow rate ($\frac{1}{2}\log(\log n)$) when H_0 is true. The null distribution of $-2\log\lambda$ does exist for a finite sample size n. In practice, we can approximate it by the bootstrap, as to be discussed in Section 6.6. But firstly, we consider some theoretical results that have been obtained for the null distribution of $-2\log\lambda$ in special cases.

6.5 DISTRIBUTIONAL RESULTS FOR THE LRTS

A lot of theoretical conjectures and simulations have been published on the null distribution of the LRTS, $-2 \log \lambda$, for inference on the number of components in a finite mixture model. We consider here some of the theoretical results that have been derived in special cases.

6.5.1 Some Theoretical Results

The article by Ghosh and Sen (1985) provides a comprehensive account of the breakdown in regularity conditions for the classical asymptotic theory to hold for the LRTS, $-2 \log \lambda$. For a mixture of two known but general univariate densities in unknown proportions, Titterington (1981) and Titterington et al. (1985) considered the LRT of $H_0 : g = 1 \ (\pi_1 = 1)$ versus $H_1 : g = 2 \ (\pi_1 < 1)$. They showed asymptotically under H_0 that $-2 \log \lambda$ is zero with probability 0.5 and, with the same probability, is distributed as chi-squared with one degree of freedom. Another way of expressing this is that the asymptotic null distribution of $-2 \log \lambda$ is the same as the distribution of

$$\{\max(0, W)\}^2, \tag{6.10}$$

where W is a standard normal random variable. A further way of expressing this is to say that

$$-2 \log \lambda \sim \tfrac{1}{2}\chi_0^2 + \tfrac{1}{2}\chi_1^2$$

under H_0, where χ_0^2 denotes the degenerate distribution that puts mass 1 at zero. In his monograph, Lindsay (1995, Section 4.2) referred to this distribution as a chi-bar squared; that is, a mixture of chi-squared distributions.

Hartigan (1985a, 1985b) obtained the same result for the asymptotic null distribution of $-2 \log \lambda$ in the case of the two-component normal mixture,

$$f(y_j; \boldsymbol{\Psi}) = \pi_1 \, \phi(y_j; \mu_1, \sigma_1^2) + \pi_2 \, \phi(y_j; \mu_2, \sigma_2^2) \tag{6.11}$$

with unspecified π_1 but known common variance and known means μ_1 and μ_2 where, as in the previous example, the null hypothesis $H_0 : g = 1$ was specified by $\pi_1 = 1$. This example was considered also by Ghosh and Sen (1985) in the course of their development of asymptotic theory for the distribution of the LRTS for mixture models. They were able to derive the limiting null distribution of $-2 \log \lambda$ for unknown but identifiable μ_1 and μ_2, where μ_2 lies in a compact set. They showed in the limit, that $-2 \log \lambda$ is distributed as a certain functional,

$$\left[\max\left\{0, \sup_{\mu_2} W(\mu_2)\right\}\right]^2, \tag{6.12}$$

where $W(\cdot)$ is a Gaussian process with zero mean and covariance kernel depending on the true value of μ_1 under H_0, and the variance of $W(\mu_2)$ is unity for all μ_2. Ghosh and Sen (1985) established a similar result for component densities from a general parametric family under certain conditions. For the case where the vector of parameters $\boldsymbol{\Psi}$ was not assumed to be identifiable, they imposed a separation condition on the values of $\boldsymbol{\Psi}$ under H_0 and H_1. Hartigan (1985a, 1985b) showed that if μ_2 is unknown with no restrictions on it, then $-2\log\lambda$ is asymptotically infinite. Also, Bickel and Chernoff (1993) investigated the null behavior of the LRTS for this model.

Berdai and Garel (1996) established the result corresponding to (6.12) when all the parameters in (6.11) were unknown, including the common variance σ^2, but with the condition that $0 < a_1 \leq \Delta \leq a_2$ for fixed a_1 and a_2, where

$$\Delta = \mid \mu_1 - \mu_2 \mid / \sigma$$

is the Mahalanobis distance. By making use of a bound suggested in Davies (1977), they calculated the upper percentiles of this asymptotic distribution of the LRTS and compared them with the corresponding percentiles of the χ_2^2 distribution, which is Wolfe's approximation for this problem. The χ_2^2 distribution was confirmed in the simulations of McLachlan (1987) and Thode, Finch, and Mendell (1988) to provide a reasonable approximation in this case. Berdai and Garel (1996) found that there was reasonable agreement between the asymptotic percentiles and the χ_2^2 percentiles, provided that a_2 was not too large. As the range of Δ is allowed to increase by increasing a_2, the percentiles increase beyond the χ_2^2 percentiles, albeit very slowly. For example, the 95th percentile of the χ_2^2 distribution is 5.99, and the asymptotic value is 5.89 when Δ is restricted to $[0.1, 6]$, increasing to 7.717, when Δ is restricted to $[0.1, 15]$. This behavior of the percentiles increasing with the range of Δ is consistent with the above result of Hartigan (1985a, 1985b) that the LRTS goes to infinity in probability (but at a very slow rate) in the case of an unbounded parameter space.

Garel (1998) claimed that the separation condition of Ghosh and Sen (1985) can be removed. The removal of this condition has been investigated also by Dacunha-Castelle and Gassiat (1997a) and Lemdani and Pons (1999). All these authors agree that it is possible to remove the separation condition. But what are the relevant hypotheses to assume is still an open question. Following on from the work of Leroux (1992a), Keribin (1998) recently considered the form of a penalty term so that the penalized form of the log likelihood does not overestimate asymptotically the order of a mixture model.

In an attempt to overcome the shortcomings of the LRT for the number of components in a mixture model in a frequentist framework, Bayesian approaches have been suggested. For example, Aitkin and Rubin (1985) adopted an approach which places a prior distribution on the vector of mixing proportions $\boldsymbol{\pi}$. An advantage of this proposal is that any null hypothesis about the number of components is specified in the interior of the parameter space. However, Quinn, McLachlan, and Hjort (1987) showed that the asymptotic null distribution of $-2\log\lambda$ will not necessarily be chi-squared, as regularity conditions still do not hold. In particular, for the test of $H_0 : g = 1$ versus $H_1 : g = g_1 \ (> 1)$ for component densities belonging to the same parametric family, they showed that $1/n$ times the observed information matrix has negative eigenvalues with nonzero probability under H_0, as $n \rightarrow \infty$.

In some other work, Feng and McCulloch (1992) showed that the LRTS has its classical asymptotic properties if the parameter space is extended. Böhning et al.

(1994) have examined the general case of exponential family component densities and showed that the asymptotic distribution of the relevant test statistic is not chi-squared. They also gave an interesting geometrical interpretation of the failure of regularity conditions which lead to the rejection of the asymptotic result. More recently, Wu and Xu (1999) have studied local sequential testing procedures in a normal mixture model. In a variation of the normal mixture model, Nettleton (1999) has derived the asymptotic null distribution of versions of the LRTS for hypotheses that place linear inequality constraints on the component means.

A number of papers have been written on the LRTS and modifications to it in special cases where some of the parameters are known, including those by Durairajan and Kale (1979, 1982), Goffinet, Loisel, and Laurent (1992), Mangin, Goffinet, and Elsen (1993), Chen (1994), Mangin and Goffinet (1995), Chen and Cheng (1997), and Polymenis and Titterington (1999). For example, Goffinet et al. (1992) established for the model (6.11) that the asymptotic distribution of $-2\log\lambda$ is χ_1^2 for fixed $\pi_1 \neq 0.5$ and $\frac{1}{2}\chi_0^2 + \frac{1}{2}\chi_1^2$ for $\pi_1 = 0.5$.

Concerning mixtures of discrete component distributions, Shoukri and Lathrop (1993), Chernoff and Lander (1995), Lemdani and Pons (1995, 1997), and Liang and Rathouz (1999) have considered testing for homogeneity in the binomial mixture model with applications to genetic linkage analysis. Simar (1976), Symons, Grimson, and Yuan (1983), Böhning et al. (1994), Leroux and Puterman (1992), and Karlis and Xekalaki (1999) have considered the problem for mixtures of Poisson distributions. Chen (1998) proposed adding a particular type of penalty function to the LRTS. He showed that the resulting penalized LRTS has a simple limiting distribution for mixture models with multinomial components.

6.5.2 Some Simulation Results

There have been several simulation results reported in the literature on the null distribution of the LRTS and its power. These results have been reviewed in Titterington et al. (1985), McLachlan (1987), McLachlan and Basford (1988), Thode et al. (1988), Mendell, Thode, and Finch (1991), Mendell, Finch, and Thode (1993), and Chuang and Mendell (1997).

One of the initial simulation studies on the LRTS for mixtures was by Wolfe (1971). He considered the LRT for assessing the null hypothesis that the data arise from a normal mixture of g_0 components versus the alternative that they arise from g_1 components $(g_0 < g_1)$. He put forward a recommendation on the basis of a small scale simulation study performed for $g_0 = 1$ and $g_1 = 2$ in the case of multivariate normal component densities with a common covariance matrix.

It follows from his proposal that the null distribution of $-2\log\lambda$ would be approximated as

$$-2C\log\lambda \sim \chi_{d_1}^2, \tag{6.13}$$

where the degrees of freedom, d_1, is taken to be twice the difference in the number of parameters in the two hypotheses, not including the mixing proportions. His suggested value of C is

$$(n - 1 - p - \tfrac{1}{2}g_1)/n,$$

which is similar to the correction factor

$$\{n - 1 - \tfrac{1}{2}(p + g_1)\}/n \tag{6.14}$$

derived by Bartlett (1938), for improving the asymptotic chi-squared distribution of $-2\log\lambda$ for the problem of testing the equality of g_1 means in a multivariate analysis of variance; see McLachlan and Basford (1988, Section 1.10).

Some simulations performed by Everitt (1981) for testing $g = 1$ versus $g = 2$ normal components with equal covariance matrices suggested that the ratio n/p needs to be at least five if Wolfe's (1971) approximation is to be applicable for the determination of P-values. Even then the simulated power of the test was low when the Mahalanobis distance Δ was not greater than two.

6.5.3 Mixtures of Two Unrestricted Normal Components

We consider some results for the null distribution of the LRTS in testing the null hypothesis $H_0 : g = 1$ versus the alternative hypothesis $H_1 : g = 2$ in the g-component univariate normal mixture model

$$f(y_j; \boldsymbol{\Psi}) = \pi_1\,\phi(y_j;\, \mu_1,\, \sigma_1^2) + \pi_2\,\phi(y_j;\, \mu_2,\, \sigma_2^2) \tag{6.15}$$

with no restrictions on the component variances.

Some initial simulations for this model were performed by McLachlan (1987). He concluded that the χ_6^2 distribution provided a better fit to the null distribution of $-2\log\lambda$ than the χ_4^2 distribution using the approximation of Wolfe (1971). Feng and McCulloch (1994) performed some simulations to demonstrate the dependence of the null distribution of the LRTS on the constraints placed on the minimum value C of the fitted component variances $\hat{\sigma}_1^2$ and $\hat{\sigma}_2^2$. In their simulations, they found the null distribution of $-2\log\lambda$ to lie between a χ_4^2 and χ_5^2 distribution for $C = 10^{-6}$, to lie between a χ_5^2 and χ_6^2 for $C = 10^{-10}$, and to be closer to a χ_6^2 for $C = 10^{-20}$. Another source of difference between some of the reported results in the literature is the extent of the search for local maxima when finding the MLE under the alternative hypothesis. To illustrate this point, we carried out some simulations of $-2\log\lambda$ for the present model with $C = 10^{-10}$ for data generated from a single normal distribution, having the same mean of zero and variance 4, as in Feng and McCulloch (1994). In fitting the two-component mixture model under the alternative hypothesis, we endeavored to find the largest local maximum, irrespective of whether it was a spurious local maximizer or not. The consequent simulated null distribution was closer to being a χ_{16}^2.

Brooks and Morgan (1995) compared the results of their hybrid algorithm, which combines simulated annealing with traditional techniques for this same problem. The simulated plot so produced for the null distribution of $-2\log\lambda$ was to the left of the χ^2_{10} up to the 90th percentiles. At the time, they consequently concluded that their hybrid algorithm was more effective than the EM algorithm in fitting the two-component heteroscedastic normal mixture model specified under H_1. But our current simulations demonstrate that the EM algorithm can actually find larger local maxima if need be, although they are of limited practical interest given that they may correspond to spurious solutions.

6.5.4 Mixtures of Two Exponentials

Seidel et al. (2000a) studied the two-component exponential model,

$$f(y_j;\, \boldsymbol{\Psi}) = \pi_1 f(y_j;\, \theta_1) + \pi_2 f(y_j;\, \theta_2), \tag{6.16}$$

where

$$f(y_j;\, \theta) = \exp(-\theta y_j),$$

and $\boldsymbol{\Psi} = (\pi_1, \theta_1, \theta_2)^T$. They demonstrated how different starting strategies and stopping rules with the implementation of the EM algorithm can lead to completely different results for the LRT of homogeneity against the two-component model; see also Seidel et al. (2000b).

The MLE of $\boldsymbol{\Psi}$ under the null hypothesis $g_0 = 1$ is given in closed form by

$$\hat{\Psi}_0 = \hat{\theta}_0 = \{\sum_{j=1}^{n} y_j/n)\}^{-1}.$$

Under the alternative hypothesis H_1, the MLE $\hat{\boldsymbol{\Psi}}_1$ of $\boldsymbol{\Psi}$ has to be computed iteratively. Seidel et al. (2000a) used the EM algorithm with two different starting strategies. With the first, START-A, the initial value $\theta_1^{(0)}$ of θ_1 was taken to be $y_{(1)} + 0.5$, $\theta_2^{(0)} = y_{(n)} - 0.5$, and $\pi_1^{(0)} = 0.5$, where $y_{(1)}, \ldots, y_{(n)}$ denote the order statistics for the observed sample. With the second starting strategy, START-B, $\theta_1^{(0)} = \bar{y} - 0.5\theta_0$ and $\theta_2^{(0)} = \bar{y} + 0.5\theta_0$, where θ_0 denotes the true value of θ under the null hypothesis.

Seidel et al. (2000a) used two stopping criteria. With the first, STOP-A, the EM algorithm in fitting the MLE under H_1 was terminated when $\log L(\boldsymbol{\Psi}^{(k+1)}) - \log L(\boldsymbol{\Psi}^{(k)})$ was less than $n \times acc$ for some specified value of acc. In Table 6.3, acc was set at 10^{-5}. Their second stopping rule, STOP-B, was based on the directional derivative,

$$D_{\boldsymbol{\Psi}}(\theta) = \sum_{j=1}^{n} \frac{f(y_j; \theta) - f(y_j; \boldsymbol{\Psi})}{f(y_j;\, \boldsymbol{\Psi})}.$$

The EM algorithm in fitting $\hat{\boldsymbol{\Psi}}$ under H_1 was terminated when

$$\max\{D_{\boldsymbol{\Psi}^{(k)}}(\theta_1^{(k)}),\, D_{\boldsymbol{\Psi}^{(k)}}(\theta_2^{(k)})\} < n \cdot acc.$$

In Table 6.3 we report the simulated values obtained by Seidel et al. (2000a) for the upper percentiles of the null distribution of $-2\log\lambda$ for the various combinations of

Table 6.3 Quantiles of the Null Distribution of the LRTS for Homogeneity under Different Starting and Termination Strategies for Exponential Components

Percentiles	Published in Böhning et al. (1994)	START-A		START-B	
		STOP-A	STOP-B	STOP-A	STOP-B
		$n = 100$			
90	1.69	3.36	3.36	2.22	2.44
95	3.26	4.73	4.76	3.64	3.80
97.5	4.67	6.13	6.19	5.04	5.23
99	6.33	8.15	8.01	6.94	7.09
		$n = 1000$			
90	1.49	3.49	3.52	1.50	2.60
95	2.59	4.95	4.99	2.55	4.09
97.5	3.76	6.42	6.42	3.71	5.52
99	5.48	8.30	8.40	5.34	7.41
		$n = 10,000$			
90	0.50	3.23	3.41	1.09	1.57
95	1.86	4.65	4.84	2.22	2.73
97.5	3.19	6.06	6.19	3.39	4.01
99	4.94	7.93	8.16	5.10	5.73

Source: Adapted from Seidel et al. (2000a).

starting and stopping strategies versus the corresponding values published in Böhning et al. (1994) for three different levels of the sample size n (the null distribution of the LRTS does not depend on θ). It can be seen that the simulated percentiles differ markedly, depending on how the EM algorithm was started and terminated. In an attempt to explain these differences, Seidel et al. (2000a) noticed that the value of $\log L(\hat{\boldsymbol{\Psi}})$ was very similar for the various starting/stopping strategies, but that it was only slightly larger than the log likelihood under the null. Hence when the latter is subtracted from $\log L(\hat{\boldsymbol{\Psi}})$ to form $\log \lambda$, large differences can result; that is, small relative errors in $\log L(\hat{\boldsymbol{\Psi}})$ can produce large relative errors in $\log \lambda$.

6.6 BOOTSTRAPPING THE LRTS

6.6.1 Implementation

McLachlan (1987) proposed a resampling approach to the assessment of the P-value of the LRTS in testing

$$H_0 : g = g_0 \quad \text{v} \quad H_1 : g = g_1 \tag{6.17}$$

for a specified value of g_0. Previously, Aitkin et al. (1981) had adopted a resampling approach in the context of a latent class analysis. Bootstrap samples are generated from the mixture model fitted under the null hypothesis of g_0 components. That is, the bootstrap samples are generated from the mixture model with the vector $\boldsymbol{\Psi}$ of unknown parameters replaced by its MLE $\hat{\boldsymbol{\Psi}}_{g_0}$ computed by consideration of the log likelihood formed from the original data under H_0. The value of $-2\log\lambda$ is computed for each bootstrap sample after fitting mixture models for $g = g_0$ and g_1 in turn to it. The process is repeated independently a number of times B, and the replicated values of $-2\log\lambda$ formed from the successive bootstrap samples provide an assessment of the bootstrap, and hence of the true, null distribution of $-2\log\lambda$. It enables an approximation to be made to the achieved level of significance P corresponding to the value of $-2\log\lambda$ evaluated from the original sample. The jth-order statistic of the B bootstrap replications can be used to estimate the quantile of order $j/(B+1)$. A preferable alternative would be to use the jth-order statistic as an estimate of the quantile of order $(3j-1)/(3B+1)$; see Hoaglin (1985).

If a very accurate estimate of the P-value were required, then B may have to be very large (Efron and Tibshirani, 1993). Usually, however, there is no interest in estimating a P-value with high precision. Even with a limited replication number B, the amount of computation involved is still considerable, in particular for values of g_0 and g_1 not close to one. However, as noted by Smyth (2000), the process can be easily and efficiently implemented on parallel computing hardware, for example, by using B parallel processors (Smyth, 2000).

As discussed earlier in the chapter, the distribution of the LRTS obtained via simulation will depend to varying degrees on the adopted parametric model, the starting strategy, the stopping rule, and the handling of the occurrence of spurious local maximizers in the fitting process. In using the bootstrap to assess the null distribution of the LRTS, it is important that the same procedures be used in fitting the mixture model to the bootstrap data as to the original sample. In the case of normal components, the bootstrap-based LRT can be carried out using the EMMIX program of McLachlan et al. (1999); see also McLachlan and Peel (1996).

In the narrower sense where the decision to be made concerns solely the rejection or retention of the null hypothesis at a specified significance level α, Aitkin et al. (1981) noted how, analogous to the Monte Carlo test procedure of Barnard (1963) and Hope (1968), the bootstrap replications can be used to provide a test of approximate size α. The test that rejects H_0 if $-2\log\lambda$ for the original data is greater than the jth smallest of its K bootstrap replications has size

$$\alpha = 1 - j/(B+1) \tag{6.18}$$

approximately. For if any difference between the bootstrap and true null distributions of $-2\log\lambda$ is ignored, then the original and subsequent bootstrap values of $-2\log\lambda$ can be treated as the realizations of a random sample of size $B+1$, and the probability that a specified member is greater than j of the others is $1 - j/(B+1)$. For some hypotheses the null distribution of λ will not depend on any unknown parameters, and so then there will be no difference between the bootstrap and true null distribution of

$-2 \log \lambda$. An example is the case of normal populations with all parameters unknown where $g_0 = 1$ under H_0. The normality assumption is not crucial in this example.

Note that the result (6.18) applies to the unconditional size of the test and not to its size conditional on the B bootstrap values of $-2 \log \lambda$. For a specified significance level α, the values of j and B can be appropriately chosen according to (6.18). For example, for $\alpha = 0.05$, the smallest value of B needed is 19 with $j = 19$. As cautioned above on the estimation of the P-value for the LRT, B needs to be very large to ensure an accurate assessment. In the present context the size of B manifests itself in the power of the test; see Hope (1968). Although the test may have essentially the prescribed size for small B, its power may be well below its limiting value as $B \to \infty$. For the 0.05 level test of a single normal population versus a mixture of two normal homoscedastic populations, McLachlan (1987) performed some simulations to demonstrate the improvement in the power as B was increased from 19 through 39 to 99.

In general, the use of the estimate $\hat{\boldsymbol{\Psi}}_{g_0}$, in place of the unknown value of $\boldsymbol{\Psi}$ under the null hypothesis, will affect the accuracy of the P-values assessed on the basis of the bootstrap replications of $-2 \log \lambda$. In Section 6.7 we report some simulation results from McLachlan and Peel (1997a) to demonstrate this effect.

6.6.2 Application to Three Real Data Sets

McLachlan and Peel (1997b) fitted a mixture of normal components with unrestricted variances to the so-called acidity, enzyme, and galaxy data sets, which were analyzed in Richardson and Green (1997). The first data set concerns an acidity index measured in a sample of 155 lakes in north-central Wisconsin, and it has been analyzed previously (on the log scale) by Crawford et al. (1992) and Crawford (1994). The second data set concerns the distribution of enzymatic activity in the blood, for an enzyme involved in the metabolism of carcinogenic substances, among a group of 245 unrelated individuals. This data set was analyzed by Bechtel et al. (1993), who identified a mixture of two skewed distributions using the maximum likelihood technique implemented in the program of Maclean et al. (1976). The third data set (galaxy data set) was introduced in Section 3.10.6. Histograms of the three data sets are shown in Figure 6.7.

In Table 6.4 we report the assessed P-values for these three real data sets. It can been seen that in a number of cases in Table 6.4 the assessed P-value lies between 5% and 10%. Performing further bootstrap replications in these cases did not clarify the situation. Also, it should be noted that these assessed P-values should not be interpreted too finely, as it is the bootstrap rather than the actual P-value that is being assessed in these cases since the value of g under the null hypothesis is greater than one.

On the basis of these P-values interpreted rigidly at the 5% level of significance, g would be chosen to be equal to 2, 3, and 6, for the acidity, enzyme, and galaxy data sets, respectively. If the significance level were increased to 10%, it would result in g being chosen equal to 3, 4, and 6 for the acidity, enzyme, and galaxy data sets, respectively. This is in general agreement with the results obtained with the

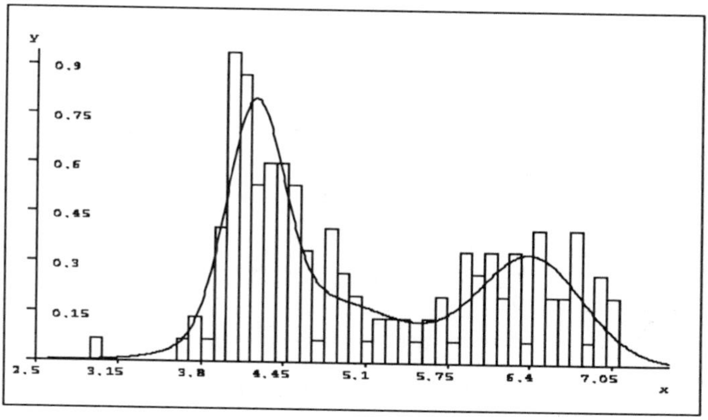

acidity data set

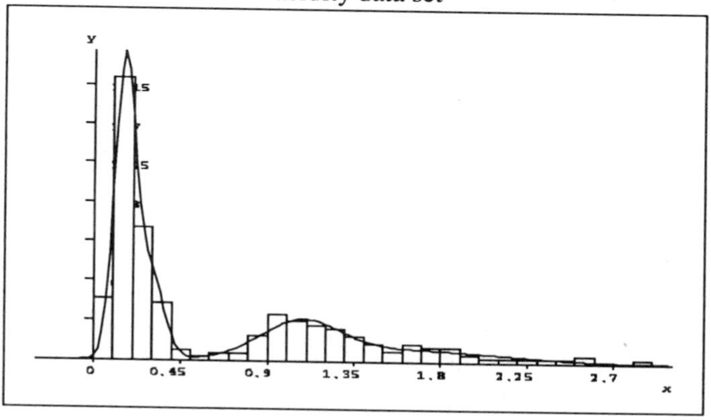

enzyme data set

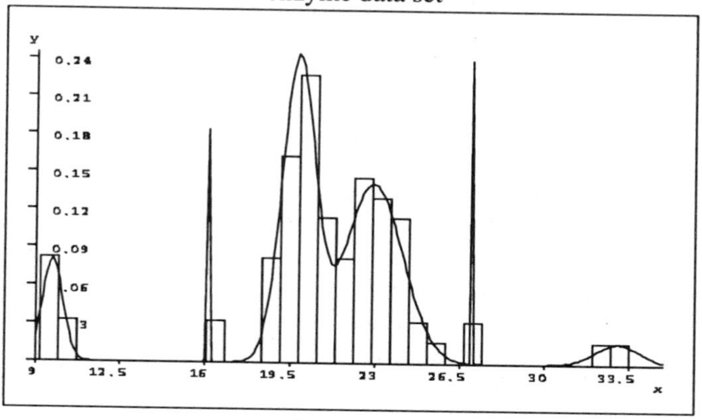

galaxy data set

Fig. 6.7 Three real data sets.

Table 6.4 *P*-Values Using Bootstrap LRT

	P-Value for g (versus $g+1$)					
Data Set	1	2	3	4	5	6
Acidity	0.01	0.08	0.44	—	—	—
Enzyme	0.01	0.02	0.06	0.39	—	—
Galaxy	0.01	0.01	0.01	0.04	0.02	0.22

Source: From McLachlan and Peel (1997b).

fully Bayesian framework for g in Richardson and Green (1997), except in part for the acidity data set. For this set, their Bayesian approach provides essentially equal support for $g = 2\text{-}6$, whereas according to the P-values assessed by resampling, one would not go beyond $g = 3$ components. A Bayesian analysis of the galaxy data set has been undertaken also by Carlin and Chib (1995) and Chib (1995). In the latter, normal mixture models with $g = 2$ or 3 components were preferred.

6.6.3 Applications in Astronomy

In the previous section, we considered a data set on the velocities of galaxies, which has been used by various statisticians for testing their clustering techniques. The structure of such sets is of much interest in astronomy. As explained by Kriessler and Beers (1997), it was once assumed that most clusters were relaxed systems that could be adequately modeled by a simple set of parameters, such as a single-core radius and the velocity dispersion of neighboring galaxies. However, numerous recent studies have concluded that many, perhaps even most, clusterings are far from being in dynamical equilibrium. Evidence cited includes the existence of: (a) "clumpy" distributions of galaxies seen in the projection on the sky, (b) apparent structure in the distribution of radial velocities for cluster membership, and (c) multiple centers of X-ray-derived temperature profiles, suggestive of ongoing collisions.

The desire to identify substructure in clusters of galaxies has led to the bootstrap form of the LRT for the number of components in a mixture model being applied in studies in astronomy. Recent papers in the astronomical literature in which this method has been used for investigating substructure in galaxy clusters include those by Ashman and Bird (1993, 1994), Bird (1994a, 1994b, 1995), Beers and Sommer-Larsen (1995), Bird, Davis, and Beers (1995), Davis et al. (1995), Zepf, Ashman, and Geisler (1995), and Bridges et al. (1997),

For example, Kriessler and Beers (1997) concluded from their use of this statistical test that 57% of the Dressler (1980) morphological-sample clusters have statistically significant substructure. Figure 6.8, which is taken from Kriessler and Beers (1997), gives the contour plots of bivariate normal mixtures fitted to the positions of some of the galaxy clusters.

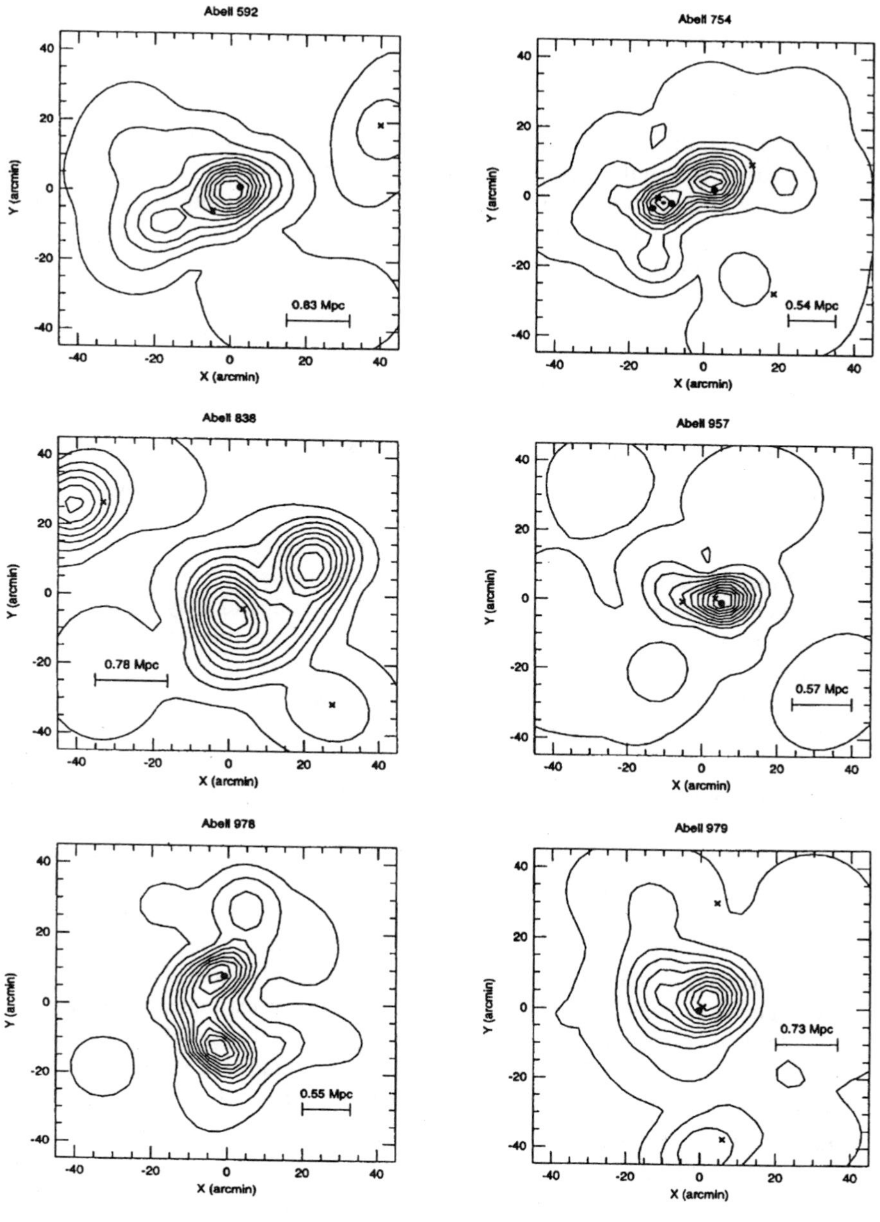

Fig. 6.8 Adaptive-kernel density contour maps of galaxy positions in Dressler's morphological sample. The filled circles indicate the positions of galaxies identified by Dressler (1980) as D or cD. The crosses mark the average positions identified as significant in the normal mixture fit. Adapted from Kriessler and Beers (1997).

6.7 EFFECT OF ESTIMATES ON P-VALUES OF BOOTSTRAPPED LRTS

6.7.1 Some Simulation Results

As noted above, the bootstrap replications of the LRTS are actual replications when the value of g_0 is 1 in the case of normal components. McLachlan and Peel (1997a) performed some simulations to investigate the effect of assessing the P-value of the LRT on the basis of the bootstrap rather than actual replicates of $-2\log\lambda$. Attention was confined to mixtures of normal components with equal covariance matrices. For simplicity, the focus was on the case of $g_0 = 2$ versus $g_1 = 3$. In this special case of two normal components with means $\boldsymbol{\mu}_1$ and $\boldsymbol{\mu}_2$ and common covariance matrix $\boldsymbol{\Sigma}$, it can be assumed without loss of generality that

$$\boldsymbol{\Sigma}_1 = \boldsymbol{\Sigma}_2 = \boldsymbol{I}_p,$$

$$\boldsymbol{\mu}_1 = \mathbf{0}_{1,p} \text{ and } \boldsymbol{\mu}_2 = [\Delta, \mathbf{0}_{1,p-1}]^T,$$

where $\boldsymbol{I}_p$ is the $p \times p$ identity matrix, $\mathbf{0}_{q,p}$ is the $q \times p$ zero matrix, and

$$\Delta = \{(\boldsymbol{\mu}_1 - \boldsymbol{\mu}_2)^T \boldsymbol{\Sigma}^{-1}(\boldsymbol{\mu}_1 - \boldsymbol{\mu}_2)\}^{1/2}$$

is the Mahalanobis distance between the two components.

In this case, the null distribution of $-2\log\lambda$ will depend only on n, π_1, and Δ. In their simulations, McLachlan and Peel (1997a) took the mixing proportions to be equal ($\pi_1 = \pi_2 = 0.5$) and took the sample size n to be 100. There were two levels of Δ ($\Delta = 2$ and 3), corresponding to moderately and widely separated components, respectively. The dimension of the observation vector ranged from $p = 1$ to $p = 8$. The percentiles of the null distribution of $-2\log\lambda$ were based on 999 simulation trials. Concerning the effectiveness of the bootstrapping of $-2\log\lambda$, 50 simulation trials were performed for each combination of the parameters Δ and p with $\pi_1 = 0.5$ and $n = 100$. On each trial, $B = 99$ bootstrap replications were used to assess the percentiles of the null distribution of $-2\log\lambda$.

Corresponding to the (simulated) 90th and 95th percentiles of the null distribution of $-2\log\lambda$, we list in Table 6.5 the averages of the bootstrap-based estimates of these percentiles extracted from the results of McLachlan and Peel (1997a). The percentage areas to the left of these average percentile estimates under the actual (simulated) null distribution of $-2\log\lambda$ (that is, the actual order of these percentiles) are given in parentheses alongside them; see McLachlan and Peel (1997a) for the root mean squared errors of these bootstrap-based estimates of the percentiles. In Table 6.5, $\overline{\hat{\Delta}}$ is the mean of $\hat{\Delta}$ over the 50 simulation trials. We give in Table 6.6 the empirical results obtained by McLachlan and Peel (1997a) in each series of 50 simulation trials for the Type I error of the resampling approach when applied at a nominal level of $\alpha = 10\%, 5\%,$ and 1%, respectively. The results in Tables 6.5 and 6.6 provide some guide as to the effect of using bootstrap replications in place of actual replicates of the test statistic $-2\log\lambda$ under the null hypothesis. As to be expected, the effect is more marked for $\Delta = 2$ than for $\Delta = 3$, and it increases as the dimension p of the observation vector increases for a given level of Δ.

Table 6.5 Summary of Simulation Results for the Null Distribution of the LRTS

			Simulated Percentile			Bootstrap Estimate of Percentile (Actual Percentage)		
p	Δ	$\overline{\hat{\Delta}}$	90%	95%	99%	90%	95%	99%
1	2	3.08	5.0	6.2	9.6	5.1 (91.0%)	6.7 (96.0%)	10.7 (99.4%)
	3	3.12	4.9	6.5	9.8	4.8 (89.6%)	6.3 (94.5%)	10.6 (99.6%)
4	2	2.89	20.9	23.0	28.1	17.8 (79.1%)	20.3 (89.1%)	26.8 (98.6%)
	3	3.34	17.9	20.2	26.1	16.6 (87.1%)	18.8 (92.4%)	24.6 (98.7%)
8	2	3.38	40.2	44.3	48.9	32.7 (67.2%)	36.3 (81.2%)	45.1 (95.8%)
	3	3.67	36.1	39.0	47.1	30.6 (72.1%)	33.7 (82.7%)	41.4 (96.7%)

Source: Adapted from McLachlan and Peel (1997a).

Table 6.6 Percentage of the 50 Simulation Trials for which a Type I Error Occurred

		Percentage of Observed Type I Errors		
p	Δ	10%	5%	1%
1	2	14%	8%	4%
	3	14%	0%	0%
2	2	18%	10%	0%
	3	16%	6%	4%
3	2	14%	4%	0%
	3	8%	6%	0%
4	2	18%	6%	2%
	3	4%	4%	0%
5	2	22%	16%	0%
	3	10%	4%	0%
6	2	28%	16%	6%
	3	24%	8%	2%
7	2	20%	10%	2%
	3	24%	12%	0%
8	2	42%	18%	4%
	3	32%	14%	4%

Source: From McLachlan and Peel (1997a).

As Table 6.5 indicates, there does seem to be a tendency for the resampling approach using bootstrap replications to underestimate the upper percentiles of the null distribution of $-2\log\lambda$, and hence overestimate the P-value of tests based on this statistic. The associated bias is in the direction that we would anticipate. For it is well known that the plug-in estimate of Δ that effectively is being used here in the generation of the bootstrap samples tends to overestimate the true value of Δ. Hence the bootstrap data are from a normal mixture with components more widely separated than in the original mixture, and so the null hypothesis of $g_0 = 2$ would be favored by this wider separation. This leads to the bootstrap replicates of $-2\log\lambda$ tending to be smaller than what the actual replicates would be. This tendency is confirmed in the simulation results in Table 6.5.

6.7.2 Double Bootstrapping

In fitting a mixture density $f(\boldsymbol{y};\, \boldsymbol{\Psi}_g)$, some guide to the sensitivity of the assessed P-value to the use of $\hat{\boldsymbol{\Psi}}_{g_0}$ for $\boldsymbol{\Psi}_{g_0}$ in $f(\boldsymbol{y}\,;\, \boldsymbol{\Psi}_{g_0})$ in the generation of the bootstrap replications can be obtained by performing another layer of bootstrapping as follows. Replications of $-2\log\lambda$ can be formed from bootstrap samples drawn from $f(\boldsymbol{y}\,;\, \hat{\boldsymbol{\Psi}}^*_{g_0})$, where $\hat{\boldsymbol{\Psi}}^*_{g_0}$ denotes the MLE calculated from a bootstrap sample drawn from $f(\boldsymbol{y}\,;\, \hat{\boldsymbol{\Psi}}_{g_0})$. These latter replications of $-2\log\lambda$ and the implied P-value can be contrasted with the original bootstrap replications of $-2\log\lambda$ and the implied P-value.

To illustrate the use of the double bootstrap applied to this problem, we report the results of McLachlan and Peel (1997a), who considered two of the aforementioned combinations of the parameters Δ and p ($\Delta = 2$ with $p = 4$ and 8). We plot in Figures 6.9 and 6.10, corresponding to these two combinations, the inverse of the simulated null distribution of $-2\log\lambda$ versus the two bootstrap analogues obtained by resampling from the normal mixture model with parameter vector equal to $\hat{\boldsymbol{\Psi}}_{g_0}$ and $\hat{\boldsymbol{\Psi}}^*_{g_0}$, respectively.

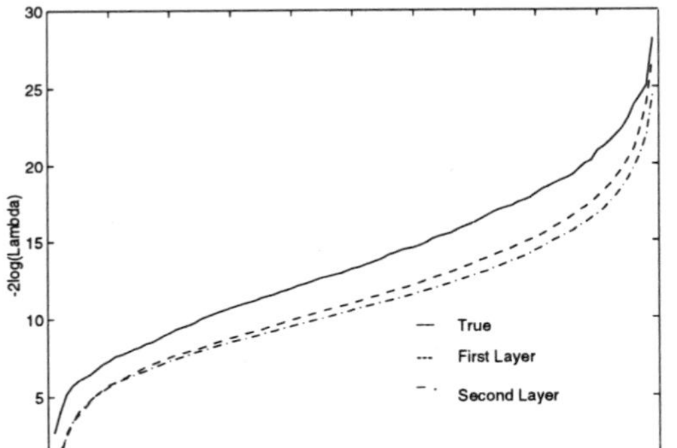

Fig. 6.9 Plot of the inverse of the bootstrap-based estimate of the null distribution of $-2\log\lambda$ for $p = 4$ and $\Delta = 2$. From McLachlan and Peel (1997a).

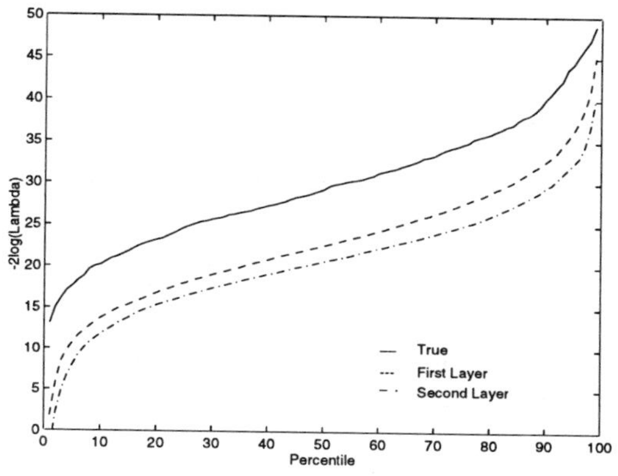

Fig. 6.10 Plot of the inverse of the bootstrap-based estimate of the null distribution of $-2\log\lambda$ for $p = 8$ and $\Delta = 2$. From McLachlan and Peel (1997a).

The second layer of bootstrapping can be used to adjust the estimates of the percentiles based on the first layer of bootstrap replicates. These adjusted estimates of the percentiles are reported in Table 6.7, along with the corresponding unadjusted values from Table 6.5, where $\overline{\hat{\Delta}^*}$ denotes the average value of $\hat{\Delta}^*$ for each series of 50 simulation trials and $\hat{\Delta}^*$ is the MLE of Δ obtained from a bootstrap sample drawn from $f(\boldsymbol{y}\,;\,\hat{\boldsymbol{\Psi}}_{g_0})$. It can be seen that by performing a second layer of bootstrapping a more accurate assessment of the percentiles of the null distribution of $-2\log\lambda$ can be obtained. However, the double bootstrap only partially removes the bias in the bootstrap estimates obtained from the first layer. This can be explained by an examination of Figures 6.9 and 6.10, where it can be seen that the difference between the true and first layered-bootstrap assessment of the percentiles is more marked than between the first and second layered bootstrap assessments.

Table 6.7 Summary of Simulation Results for the Adjusted Percentile

p	Δ	$\overline{\hat{\Delta}}$	$\overline{\hat{\Delta}^*}$	Unadjusted Percentile (Actual Percentage)		Adjusted Percentile (Actual Percentage)	
				90%	95%	90%	95%
4	2	2.89	3.19	17.8 (79.1%)	20.3 (89.1%)	19.0 (84.1%)	21.6 (92.2%)
8	2	3.38	3.89	32.7 (67.2%)	36.3 (81.2%)	35.8 (79.7%)	40.2 (90.1%)

Source: From McLachlan and Peel (1997a).

In this particular case of $g_0 = 2$, where the null distribution of $-2\log\lambda$ depends on the component parameters through just the Mahalanobis distance Δ, it would be straightforward to produce a suitable adjustment by working with the canonical form of the problem in terms of Δ and using a less biased estimate of Δ (McLachlan, 1992, Chapter 10) in the production of the first layer of bootstrap replications.

6.8 INFORMATION CRITERIA IN MODEL SELECTION

6.8.1 Bias Correction of the Log Likelihood

Model selection can be approached in terms of the Kullback–Leibler (1951) information of the true model with respect to the fitted model. If $f(\boldsymbol{w})$ denotes the true density, then the Kullback–Leibler information of $f(\boldsymbol{w})$ with respect to an estimate $f(\boldsymbol{w};\, \hat{\boldsymbol{\Psi}})$ is

$$I\{f(\boldsymbol{w});\, f(\boldsymbol{w};\, \hat{\boldsymbol{\Psi}})\} = \int f(\boldsymbol{w}) \log f(\boldsymbol{w})\, d\boldsymbol{w} - \int f(\boldsymbol{w}) \log f(\boldsymbol{w};\, \hat{\boldsymbol{\Psi}})\, d\boldsymbol{w}, \quad (6.19)$$

which is a measure of the divergence of $f(\boldsymbol{w})$ relative to $f(\boldsymbol{w};\, \hat{\boldsymbol{\Psi}})$. The aim is to make the Kullback–Leibler information (6.19) small. As the first term on the right-hand side of (6.19) does not depend on the model, only the second term is relevant. It can be expressed as

$$\begin{aligned} \eta(\boldsymbol{y};\, F) &= \int f(\boldsymbol{w}) \log f(\boldsymbol{w};\, \hat{\boldsymbol{\Psi}})\, d\boldsymbol{w} \\ &= \int \log f(\boldsymbol{w};\, \hat{\boldsymbol{\Psi}})\, dF(\boldsymbol{w}), \end{aligned} \quad (6.20)$$

where F denotes the true distribution and $\boldsymbol{y} = (\boldsymbol{y}_1^T,\, \ldots,\, \boldsymbol{y}_n^T)^T$ contains the observed data.

A simple estimator of $\eta(\boldsymbol{y};\, F)$ is given by

$$\begin{aligned} \eta(\boldsymbol{y};\, \hat{F}_n) &= \tfrac{1}{n} \sum_{j=1}^{n} \log f(\boldsymbol{y}_j;\, \hat{\boldsymbol{\Psi}}) \\ &= \tfrac{1}{n} \log L(\hat{\boldsymbol{\Psi}}), \end{aligned} \quad (6.21)$$

obtained by replacing F in (6.20) by the empirical distribution function $\hat{F}_n$, which places mass $1/n$ at each observation $\boldsymbol{y}_j$ $(j = 1,\, \ldots,\, n)$. Usually this provides an overestimate of the expected log density

$$\int \log f(\boldsymbol{w})\, dF(\boldsymbol{w}), \quad (6.22)$$

since the empirical distribution function $\hat{F}_n$ is generally closer to the fitted distribution function $F_{\hat{\boldsymbol{\Psi}}}$ than the true one F. The bias of $\eta(\boldsymbol{y};\, \hat{F}_n)$ as an estimator of (6.22) is the functional

$$\begin{aligned} b(F) &= E_F\{\eta(\boldsymbol{Y};\, \hat{F}_n) - \eta(\boldsymbol{Y};\, F)\} \\ &= E_F\{\tfrac{1}{n} \sum_{j=1}^{n} \log f(\boldsymbol{Y}_j;\, \hat{\boldsymbol{\Psi}}) - \int \log f(\boldsymbol{w};\, \hat{\boldsymbol{\Psi}})\, dF(\boldsymbol{w})\}, \end{aligned} \quad (6.23)$$

where E_F denotes expectation using F as the common distribution function of the (independent) $\boldsymbol{Y}_1, \ldots, \boldsymbol{Y}_n$.

An information criterion for model selection can be based on the the bias-corrected log likelihood given by

$$\log L(\hat{\boldsymbol{\Psi}}) - b(F), \tag{6.24}$$

using an appropriate estimate of the bias term $b(F)$. The intent is to select the model (that is, the number of components in the present context) to maximize (6.24), and thus to minimize the Kullback–Leibler information (6.19).

In the literature, the information criteria so formed are generally expressed in terms of twice the negative value of this difference, so that they are of the form

$$-2 \log L(\boldsymbol{\Psi}) + 2C, \tag{6.25}$$

where the first term on the right-hand side of (6.25) measures the lack of fit and the second term C is the penalty term that measures the complexity of the model. The intent therefore is to choose a model to minimize the criterion (6.25).

6.8.2 Akaike's Information Criterion

Akaike (1973, 1974) showed that $b(F)$ is asymptotically equal to d, where d is equal to the total number of parameters in the model. Thus from (6.24), Akaike's Information Criterion (AIC) selects the model that minimizes

$$-2 \log L(\hat{\boldsymbol{\Psi}}) + 2d; \tag{6.26}$$

see Bozdogan and Sclove (1984) and Sclove (1987) on the use of AIC in the present context of selecting the number of components in a mixture.

Konishi and Kitagawa (1996) derived the corresponding asymptotic bias where the true density $f(\boldsymbol{y}_j)$ does not belong to the postulated parametric family and where the parameter vector is not necessarily estimated by maximum likelihood. However, the validity of these asymptotic expansions for $b(F)$ depend on the same regularity conditions needed for the usual asymptotic theory for the null distribution of the LRTS to hold (Titterington et al., 1985). As discussed in Section 6.4, these conditions break down for tests on the number of components in a mixture model.

However, in spite of this, the AIC criterion is still often used to assess the order of a mixture model. Many authors (for example, Koehler and Murphee (1988)) observed that AIC is order inconsistent and tends to overfit models. In the mixture context, it means that AIC tends to overestimate the correct number of components (Soromenho, 1993; Celeux and Soromenho, 1996).

6.8.3 Bootstrap-Based Information Criterion

Ishiguro, Sakamoto, and Kitagawa (1997) proposed that the bias term in (6.24) be estimated using Efron's (1979) bootstrap; see also Pan (1999). Their Efron (bootstrapped) information criterion, which they called EIC, chooses the number of components g

on the basis of

$$-2\log L(\hat{\boldsymbol{\Psi}}) + 2b(\hat{F}_n), \tag{6.27}$$

where the (nonparametric) bootstrap bias $b(\hat{F}_n)$ is approximated by Monte Carlo methods on the basis of B bootstrap samples. From (6.23),

$$b(\hat{F}_n) = E_{\hat{F}_n}\{\tfrac{1}{n}\sum_{j=1}^{n}\log f(\boldsymbol{Y}_j^*;\, \hat{\boldsymbol{\Psi}}^*) - \tfrac{1}{n}\sum_{j=1}^{n}\log f(\boldsymbol{Y}_j;\, \hat{\boldsymbol{\Psi}}^*)\}, \tag{6.28}$$

where $\boldsymbol{\Psi}^*$ denotes the MLE formed from the bootstrap sample

$$\boldsymbol{Y}_1^*, \ldots, \boldsymbol{Y}_n^* \overset{\text{i.i.d.}}{\sim} \hat{F}_n.$$

We can approximate this bootstrap bias on the basis of B independent bootstrap samples

$$\boldsymbol{Y}_{1b}^*, \ldots, \boldsymbol{Y}_{nb}^* \overset{\text{i.i.d.}}{\sim} \hat{F}_n \quad (b = 1, \ldots, B),$$

where we let $\hat{\boldsymbol{\Psi}}_b^*$ denote the MLE formed from the bth bootstrap sample $(b = 1, \ldots, B)$. This gives

$$b(\hat{F}_n) \approx \tfrac{1}{B}\sum_{b=1}^{B}\{\tfrac{1}{n}\sum_{j=1}^{n}\log f(\boldsymbol{y}_{jb}^*;\, \hat{\boldsymbol{\Psi}}_b^*) - \tfrac{1}{n}\sum_{j=1}^{n}\log f(\boldsymbol{y}_j;\, \hat{\boldsymbol{\Psi}}_b^*)\}. \tag{6.29}$$

Konishi and Kitagawa (1996) showed that the number of bootstrap samples can be greatly reduced by using a variance-reduction technique in the bootstrap simulation. They achieved this by noting that the expectation on the right-hand side of (6.23) can be decomposed into three parts as follows:

$$b(F) = b_1(F) + b_2(F) + b_3(F), \tag{6.30}$$

where

$$b_1(F) = E_F\{\tfrac{1}{n}\sum_{j=1}^{n}\log f(\boldsymbol{Y}_j;\, \hat{\boldsymbol{\Psi}}) - \tfrac{1}{n}\sum_{j=1}^{n}\log f(\boldsymbol{Y}_j;\, \boldsymbol{\Psi})\} \tag{6.31}$$

$$b_2(F) = E_F\{\tfrac{1}{n}\sum_{j=1}^{n}\log f(\boldsymbol{Y}_j;\, \boldsymbol{\Psi}) - \int \log f(\boldsymbol{w};\, \boldsymbol{\Psi})\, dF(\boldsymbol{w})\} \tag{6.32}$$

$$b_3(F) = E_F\{\int \log f(\boldsymbol{w};\, \boldsymbol{\Psi})\, dF(\boldsymbol{w}) - \int \log f(\boldsymbol{w};\, \hat{\boldsymbol{\Psi}})\, dF(\boldsymbol{w})\}. \tag{6.33}$$

As $b_2(F)$ is zero, only the first and third terms on the right-hand side of (6.30) have to be estimated to give

$$b(\hat{F}_n) = b_1(\hat{F}_n) + b_3(\hat{F}_n), \tag{6.34}$$

where

$$b_1(\hat{F}_n) \approx \tfrac{1}{B}\sum_{b=1}^{B}\{\tfrac{1}{n}\sum_{j=1}^{n}\log f(\boldsymbol{y}_{jb}^*;\, \hat{\boldsymbol{\Psi}}^*) - \tfrac{1}{n}\sum_{j=1}^{n}\log f(\boldsymbol{y}_{jb}^*;\, \hat{\boldsymbol{\Psi}}_b)\} \tag{6.35}$$

and

$$b_3(\hat{F}_n) \approx \frac{1}{B}\sum_{b=1}^{B}\{\frac{1}{n}\sum_{j=1}^{n}\log f(\boldsymbol{y}_j;\, \hat{\boldsymbol{\Psi}}) - \frac{1}{n}\sum_{j=1}^{n}\log f(\boldsymbol{y}_j;\, \hat{\boldsymbol{\Psi}}_b^*)\}. \tag{6.36}$$

With the ordinary version of the bootstrap used here, we are effectively estimating all three terms in (6.30). Now, although the bootstrap estimate of the second term $b_2(\hat{F}_n)$ will be small, Konishi and Kitagawa (1996) observed from their experiments that it had a much larger variance than that of the bootstrap estimates of $b_1(F)$ and $b_2(F)$; hence the saving in bootstrap replications when using the modified bootstrap approximation that uses only estimates of $b_1(F)$ and $b_3(F)$.

6.8.4 Cross-Validation-Based Information Criterion

The bias correction of the log likelihood can be undertaken using cross-validation as in Smyth (2000). This cross-validation-based information criterion (CVIC) chooses g on the basis of the cross-validated log likelihood,

$$\sum_{j=1}^{n}\log f(\boldsymbol{y}_j;\, \hat{\boldsymbol{\Psi}}_{(j)}), \tag{6.37}$$

where $\hat{\boldsymbol{\Psi}}_{(j)}$ denotes the MLE of $\boldsymbol{\Psi}$ formed from the observed sample $\boldsymbol{y}_1, \ldots, \boldsymbol{y}_n$, after first deleting the jth observation $\boldsymbol{y}_j$ $(j = 1, \ldots, n)$. The use of cross-validation in this role can be viewed as an alternative method to evaluating the fitted model on a test sample of the same size as the original one (the training sample) on which $\boldsymbol{\Psi}$ is based.

This "leave-one-out" cross-validated form would be very time-consuming, given that only one observation is deleted at a time. Hence consideration might be given to v-fold cross-validation where $v > 1$ observations are deleted at a time. The data set is divided up into v disjoint subsets each of size n/v. Another way known as Monte Carlo cross-validation generates B independent partitions of the data set into a test sample of size γn and a training sample of size $(1-\gamma)n$ for the estimation of $\boldsymbol{\Psi}$ for some fixed value of γ. The main difference between this method and the conventional v-fold cross-validation method is that each data point may be used more than once in the test set. Smyth (2000) suggests that the choice of $\gamma = 0.5$ appears to be reasonably robust across a variety of problems, while values of B between 20 and 50 appear to be adequate for most applications. Smyth (2000) reports some simulation results in which CVIC implemented via Monte Carlo methods was comparable with the Autoclass algorithm of Cheeseman and Stutz (1996), but the 10-fold cross-validation version was often unreliable.

Smyth (2000) compared CVIC with BIC and the bootstrap LRTS in their application to the diabetes data analyzed by Reaven and Miller (1979) and subsequently by McLachlan (1992, Section 6.8), Banfield and Raftery (1993), and Fraley and Raftery (1998), among others. They also applied these criteria to decide on the number of clusters in the clustering of Northern Hemisphere geopotential height data; see Smyth,

Ide, and Ghil (1999) for details. The three criteria led to the same decision on the number of components in the normal mixture model.

6.8.5 Minimum Information Ratio Criterion

As discussed in Section 2.14. the rate of convergence of the EM algorithm is determined by the largest eigenvalue of the rate matrix,

$$\boldsymbol{I}_d - \mathcal{I}_c^{-1}(\hat{\boldsymbol{\Psi}};\, \boldsymbol{y})\boldsymbol{I}(\hat{\boldsymbol{\Psi}};\, \boldsymbol{y}), \tag{6.38}$$

or, equivalently, by the smallest eigenvalue of the information rate matrix,

$$\mathcal{I}_c^{-1}(\hat{\boldsymbol{\Psi}};\, \boldsymbol{y})\boldsymbol{I}(\hat{\boldsymbol{\Psi}};\, \boldsymbol{y}). \tag{6.39}$$

Windham and Cutler (1992) proposed basing the choice of the number of components on (6.39). Their motivation was that, heuristically, a large value of this smallest eigenvalue is suggestive of a good fit, whereas a small value is not. We let $\hat{\boldsymbol{\Psi}}_g$ be the MLE of $\boldsymbol{\Psi}$ for a g-component mixture model, and let e_g be the smallest eigenvalue of (6.39).

With the minimum information ratio (MIR) criterion of Windham and Cutler (1992), which is based on the magnitude of the smallest eigenvalue of the information rate matrix, we choose g to maximize e_g over g. The value of e_g can be computed making use of the result that it is equal to one minus the rate of convergence of the EM algorithm, which can be calculated numerically using (2.46).

Polymenis and Titterington (1998) proposed a modification of the MIR criterion, which was motivated by the remark of Windham and Cutler (1992) that as soon as a mixture model with too many components is fitted, the observed information matrix $\boldsymbol{I}(\hat{\boldsymbol{\Psi}};\, \boldsymbol{y})$ will be close to singular with the result that the corresponding e_g will be close to zero. The idea of Polymenis and Titterington (1998) therefore is to detect the smallest value of g (g_o) for which e_g is "close to zero," and select g to be $g_o - 1$. In order to quantify at what point an observed value of e_g is close to zero, Polymenis and Titterington (1998) propose a Monte Carlo approach as follows:

1. Commencing with $g = 2$, determine the fits $\hat{\boldsymbol{\Psi}}_g$ and $\hat{\boldsymbol{\Psi}}_{g+1}$ and evaluate $r_g = e_{g+1}/e_g$.

2. Generate 99 bootstrap samples of size n from the two-component mixture model with parameter $\hat{\boldsymbol{\Psi}}_g$ and calculate a value of r_g^* for each of them.

3. If the observed value of r_g is greater than at least 94 of the bootstrap values of r_g^*, increase g by one and repeat steps 1 and 2; otherwise, assume that the number of components is g and stop.

   In a simulation experiment performed by Polymenis and Titterington (1998), their modified method outperformed the basic method of Windham and Cutler (1992), and it was equivalent to the bootstrap LRT when the components were well separated. When the sample size was increased from $n = 11$ to $n = 300$, the modified method became equivalent to the bootstrap LRT for those combinations of the parameters where the components were not well separated.

6.8.6 Informational Complexity Criterion

Bozdogan (1990, 1993) proposed the informational complexity (ICOMP) criterion in an attempt to improve on the performance of AIC. This criterion has the form

$$ICOMP(g) = -2\log L(\hat{\boldsymbol{\Psi}}) + C_1 - C_2, \tag{6.40}$$

where

$$C_1 = d\log\{d^{-1}\text{tr}\mathcal{I}^{-1}(\hat{\boldsymbol{\Psi}})\} \tag{6.41}$$

and

$$C_2 = \log \mid \mathcal{I}^{-1}(\hat{\boldsymbol{\Psi}}) \mid, \tag{6.42}$$

where d is the number of parameters in the model. A concern with this criterion is that it is not invariant under a reparameterization of the model.

Note that in the definitions of C_1 and C_2 we could have replaced $\mathcal{I}(\boldsymbol{\Psi})$, which is the expected information in the sample, with the expected information in a single observation, since this change would simply result in $d\log n$ being subtracted from both terms. To apply this criterion in practice, the expected information matrix $\mathcal{I}(\boldsymbol{\Psi})$ must be estimated. Given the difficulties that would be involved in doing this for a mixture model, Bozdogan (1993) approximates this matrix by using instead the (estimated) expected information matrix for a classified sample. For a g-component normal mixture model, this leads to C_1 and C_2 being estimated by

$$C_1 = d\log[d^{-1}\sum_{i=1}^{g}\{\hat{\pi}_i^{-1}\text{tr}\hat{\boldsymbol{\Sigma}}_i + \tfrac{1}{2}\text{tr}(\hat{\boldsymbol{\Sigma}}_i\hat{\boldsymbol{\Sigma}}_i) + \tfrac{1}{2}(\text{tr}\hat{\boldsymbol{\Sigma}}_i)^2 + \sum_{v=1}^{p}(\hat{\boldsymbol{\Sigma}}_i)_{vv}^2\}]$$

and

$$C_2 = (p+2)\sum_{i=1}^{g}\log \mid \hat{\boldsymbol{\Sigma}}_i \mid - p\sum_{i=1}^{g}\log(n\hat{\pi}_i) + gp\log(2n),$$

where $d = gp + \frac{1}{2}gp(p+1)$.

In their simulations with normal mixture models, Celeux and Soromenho (1996) found that when the component-covariance matrices were very different, ICOMP had a tendency, more marked than AIC, to overestimate the order of the mixture model.

6.9 BAYESIAN-BASED INFORMATION CRITERIA

6.9.1 Bayesian Approach

We now consider some criteria that have been derived within a Bayesian framework for model selection, but can be applied also in a non-Bayesian framework, and hence to the choice of the number of components in mixture models considered from either a Bayesian or frequentist perspective. We shall also describe some criteria that apply only within a Bayesian framework.

The main Bayesian-based information criteria use an approximation to the integrated likelihood, as in the original proposal by Schwarz (1978) leading to his

Bayesian information criterion (BIC). Available general theoretical justifications of this approximation rely on the same regularity conditions that break down for inference on the number of components in a frequentist framework. However, we shall still sketch the derivation of this approximation, assuming that the prerequisite regularity conditions do hold, as use is made of this approximation and results used in its derivation in forming information criteria for model selection regardless of the breakdown in regularity conditions.

6.9.2 Laplace's Method of Approximation

The Bayes factor for one model against another model is the posterior odds for that model against the other when neither model is favored over the other *a priori*. It is equal to the ratio of the marginal or integrated likelihood for each model. The reader is referred to Kass and Raftery (1995) for an excellent account of Bayes factors.

We let $p(\boldsymbol{\Psi})$ denote the prior density for $\boldsymbol{\Psi}$. The integrated likelihood $p(\boldsymbol{y})$ is defined to be

$$\begin{aligned} p(\boldsymbol{y}) &= \int p(\boldsymbol{\Psi}, \boldsymbol{y})\, d\boldsymbol{\Psi}, \\ &= \int \exp\{\log p(\boldsymbol{\Psi}, \boldsymbol{y})\}\, d\boldsymbol{\Psi}, \end{aligned} \tag{6.43}$$

where

$$p(\boldsymbol{\Psi}, \boldsymbol{y}) = p(\boldsymbol{\Psi})\, L(\boldsymbol{\Psi}).$$

The results here are conditional on the adopted model for the component densities and the number of components g, but we have suppressed this dependence in our notation here.

We let $\tilde{\boldsymbol{\Psi}}$ denote the posterior mode, satisfying

$$\partial \log p(\tilde{\boldsymbol{\Psi}}, \boldsymbol{y})/\partial \boldsymbol{\Psi} = \boldsymbol{0}, \tag{6.44}$$

where $\partial \log p(\tilde{\boldsymbol{\Psi}}, \boldsymbol{y})/\partial \boldsymbol{\Psi}$ denotes the gradient of $\log p(\boldsymbol{\Psi}, \boldsymbol{y})$ evaluated at $\boldsymbol{\Psi} = \tilde{\boldsymbol{\Psi}}$. The negative Hessian matrix of $\log p(\boldsymbol{\Psi}, \boldsymbol{y})$ evaluated at $\boldsymbol{\Psi} = \tilde{\boldsymbol{\Psi}}$ is denoted by $\boldsymbol{H}(\tilde{\boldsymbol{\Psi}})$.

To approximate the integral (6.43), the integrand is expanded in a second-order Taylor series about the point $\boldsymbol{\Psi} = \tilde{\boldsymbol{\Psi}}$ to give

$$\log p(\boldsymbol{\Psi}, \boldsymbol{y}) \approx \log p(\tilde{\boldsymbol{\Psi}}, \boldsymbol{y}) - \tfrac{1}{2}(\boldsymbol{\Psi} - \tilde{\boldsymbol{\Psi}})^T \boldsymbol{H}(\tilde{\boldsymbol{\Psi}})(\boldsymbol{\Psi} - \tilde{\boldsymbol{\Psi}}), \tag{6.45}$$

on noting from (6.44) that the first-order term vanishes. On substituting this expansion for the integrand into (6.43) and realizing that, apart from a normalizing constant, it is equal to a normal density with mean $\tilde{\boldsymbol{\Psi}}$ and covariance matrix equal to the inverse of $\boldsymbol{H}(\tilde{\boldsymbol{\Psi}})$, we obtain

$$\begin{aligned} p(\boldsymbol{y}) &= \exp\{\log p(\tilde{\boldsymbol{\Psi}}, \boldsymbol{y})\} \int \exp\{-\tfrac{1}{2}(\boldsymbol{\Psi} - \tilde{\boldsymbol{\Psi}})^T \boldsymbol{H}(\tilde{\boldsymbol{\Psi}})(\boldsymbol{\Psi} - \tilde{\boldsymbol{\Psi}})\}\, d\boldsymbol{\Psi} \\ &= p(\tilde{\boldsymbol{\Psi}}, \boldsymbol{y})(2\pi)^{\frac{1}{2}d} \mid \boldsymbol{H}(\tilde{\boldsymbol{\Psi}}) \mid^{-1/2}. \end{aligned} \tag{6.46}$$

Thus from (6.46), the integrated log likelihood is approximated as

$$\log p(\boldsymbol{y}) \approx \log L(\tilde{\boldsymbol{\Psi}}) + \log p(\tilde{\boldsymbol{\Psi}}) - \tfrac{1}{2}\log \mid \boldsymbol{H}(\tilde{\boldsymbol{\Psi}}) \mid + \tfrac{1}{2} d \log(2\pi). \qquad (6.47)$$

This approximation is known as Laplace's method or the saddle-point approximation.

Laplace's method may be applied in alternative forms by omitting part of the integrand from the exponent when performing the expansion; see Kass and Raftery (1995). An important variant on (6.47) is

$$\log p(\boldsymbol{y}) = \log L(\hat{\boldsymbol{\Psi}}) + \log p(\hat{\boldsymbol{\Psi}}) - \tfrac{1}{2}\log \mid \boldsymbol{I}(\hat{\boldsymbol{\Psi}};\, \boldsymbol{y}) \mid + \tfrac{1}{2} d \log(2\pi), \qquad (6.48)$$

where the posterior mode is replaced by the MLE $\hat{\boldsymbol{\Psi}}$ and $\boldsymbol{H}(\tilde{\boldsymbol{\Psi}})$ is replaced by the observed information matrix $\boldsymbol{I}(\hat{\boldsymbol{\Psi}};\, \boldsymbol{y})$. This approximation thus assumes that the prior is very diffuse so that its effect can be effectively ignored. As cautioned by Ripley (1996, Section 2.6), the assumption that the prior can be neglected is a strong one.

6.9.3 Bayesian Information Criterion

The Bayesian information criterion (BIC) of Schwarz (1978) is obtained by ignoring terms of $O(1)$ in (6.48) and noting that

$$\mid \boldsymbol{I}(\hat{\boldsymbol{\Psi}};\, \boldsymbol{y}) \mid = O(d \log n) \qquad (6.49)$$

to give

$$-2 \log L(\hat{\boldsymbol{\Psi}}) + d \log n \qquad (6.50)$$

as the penalized likelihood to be maximized in model selection, including the present situation for the number of components g in a mixture model.

The approximation (6.48) requires the parameters to be identifiable. Hence both this approximation and the expansion (6.47) depend on regularity conditions that do not hold for finite mixture models. However, as Fraley and Raftery (1998) note, there is considerable support for use of BIC in this context. As mentioned previously in this chapter, Leroux (1992a) has shown that BIC does not underestimate the true number of components, asymptotically. And Roeder and Wasserman (1997) have shown that when a normal mixture model is used to estimate a density "nonparametrically," the density estimate that uses BIC to select the number of components in the mixture is consistent. They also reported a simulation study in which BIC performed very well. Also, Campbell et al. (1997) and Dasgupta and Raftery (1998) have reported encouraging results for BIC applied to mixture models.

As $\log n > 2$ for $n > 8$, it can be seen that the penalty term of BIC penalizes complex models more heavily than AIC, whose penalty term of $2d$ does not depend on the sample size. As a consequence, it does reduce the tendency of the AIC criterion to fit too many components. On the other hand, it has been found to fit too few components when the model for the component densities is valid and the sample size is not very large; see, for example, Celeux and Soromenho (1996). If the model for the component densities is not valid, then Biernacki et al. (1998) have found that it tends to fit too many components.

Note that BIC can be used not only to choose the number of components in the mixture model, but also to decide on the adopted model, say, for the component-covariance matrices in the parameterization considered in Section 3.12; see, for example, Biernacki and Govaert (1999).

The criterion BIC has been derived also by Rissanen (1986, 1989) from another perspective based on coding theory.

We have presented BIC in a non-Bayesian framework in the above. Recently, Roeder and Wasserman (1997) used it in a Bayesian framework to construct an estimate of $\mathrm{pr}\{g \mid \boldsymbol{y}\}$ of the form

$$\hat{m}_g / \sum_{h=1}^{g_u} \hat{m}_h, \tag{6.51}$$

where

$$\hat{m}_h = n^{-\frac{1}{2}d_h} L(\tilde{\boldsymbol{\Psi}}_h), \tag{6.52}$$

and $\tilde{\boldsymbol{\Psi}}_h$ is the posterior mode for the h-component mixture model and d_h denotes the number of unknown parameters in it $(h = 1, \ldots, g_u)$. Here g_u is some fixed upper bound for g. Roeder and Wasserman (1997) suggested that $g_u = 10$ should suffice in practice. They considered the extension to a nonparametric framework of the mixture model, where g_u is allowed to grow with n, creating what is called a sieve; see also Genovese and Wasserman (1999). As stressed in Roeder and Wasserman (1997), it is important not to let g grow too quickly, or the density estimate may be inconsistent. They established that if $g_u = o(n/logn)$, then the posterior estimate is consistent. The same proof of this result also shows consistency of the posterior for fixed g as long as the true density is a mixture of g normals for $1 \leq g \leq g_u$.

6.9.4 Laplace–Metropolis Criterion

The Bayes factor B_{10} for a model H_1 against another model H_0 given the data $\boldsymbol{y}$ is the ratio of the posterior to prior odds, namely

$$B_{10} = p(\boldsymbol{y} \mid H_1)/p(\boldsymbol{y} \mid H_0), \tag{6.53}$$

which is the ratio of the integrated likelihoods where, for a given model H_i,

$$p(\boldsymbol{y} \mid H_i) = \int p(\boldsymbol{\Psi}_i \mid H_i) L(\boldsymbol{\Psi}_i \mid H_i)\, d\boldsymbol{\Psi}_i \quad (i = 0, 1), \tag{6.54}$$

where $\boldsymbol{\Psi}_i$ is the vector of parameters under the model H_i $(i = 0, 1)$. As advocated by Raftery (1996), the Bayes factor is a summary of the evidence for H_1 against H_0. In Table 6.8 we give the rounded scale for interpreting B_{10} as provided by Raftery (1996), based on that of Jeffreys (1961).

Raftery (1996) proposed choosing both the model and the number of components g in the mixture by using posterior simulation to calculate the required quantities in the Laplace approximation (6.47) to the integrated likelihood. The idea is to use the simulated posterior values for $\boldsymbol{\Psi}$ to avoid having to calculate the posterior mode $\tilde{\boldsymbol{\Psi}}$

Table 6.8 Calibration of the Bayes Factor B_{10}

B_{10}	$2 \log B_{10}$	Evidence for H_1
< 1	< 0	Negative (supports H_0)
1 to 3	0 to 2	Barely worth mentioning
2 to 12	2 to 5	Positive
12 to 150	5 to 10	Strong
> 150	> 10	Very strong

Source: From Raftery (1996).

and, in particular, the derivatives for the Hessian matrix $\boldsymbol{H}(\tilde{\boldsymbol{\Psi}})$ needed to form the Laplace approximation (6.47). They can be approximated by a robust version of location and scale from the simulated posterior replicates of $\boldsymbol{\Psi}$; see Raftery (1996).

Recently, Aitkin et al. (1996) considered a test for the number of components, using the posterior Bayes factors, as a variation of the "prior" Bayes factors defined above; that is, one uses the posterior density of $\boldsymbol{\Psi}$ in place of its prior density in (6.54).

6.9.5 Laplace–Empirical Criterion

Rather than use the approximation (6.49) which gives the final form of BIC, we might wish to choose the number of components g on the basis of (6.48) directly. It is thus the analogue of the Laplace–Metropolis criterion in a non-Bayesian framework. The use of (6.48) requires the calculation of the observed information matrix $\boldsymbol{I}(\hat{\boldsymbol{\Psi}}; \boldsymbol{y})$. As discussed in Section 2.15.3, it can be conveniently approximated by the empirical information matrix $\boldsymbol{I}_e(\hat{\boldsymbol{\Psi}})$. We shall call the criterion that uses this approximation in (6.48), the Laplace–Empirical (matrix-based) criterion (LEC). This criterion therefore chooses g to minimize

$$-2 \log L(\hat{\boldsymbol{\Psi}}) - 2 \log p(\hat{\boldsymbol{\Psi}}) + \log \mid \boldsymbol{I}_e(\hat{\boldsymbol{\Psi}}) \mid - d \log(2\pi). \qquad (6.55)$$

Roberts et al. (1998) considered this criterion in the form where the observed information matrix $\boldsymbol{I}(\hat{\boldsymbol{\Psi}}; \boldsymbol{y})$ was calculated directly. They initially calculated the second-order partial derivatives of the log likelihood in the simplified case of diagonal component-covariance matrices. They then performed a transformation to convert their results to the general case of unrestricted component-covariance matrices. However, we have been unable to confirm the validity of this conversion.

Concerning the specification of a prior for use in this criterion, we can first transform the data so that the sample covariance matrix is the identity matrix. Roberts et al. (1998) then take the ith component-mean elements $(\boldsymbol{\mu}_i)_1, \ldots, (\boldsymbol{\mu}_i)_p$ to be independent and uniformly distributed on the interval $[-\omega, \omega]$ $(i = 1, \ldots, g)$. They take the diagonal elements of the component-covariance matrices to be independent and uniformly distributed on the interval $[0, \beta]$. Note that with β fixed, the posterior is not robust; see Robert et al. (2000), who use a hyperprior on the component variances to make the posterior robust.

Roberts et al. (1998) did not specify a prior for the off-diagonal elements, but they could be taken to be independently distributed on the interval $[-\beta, \beta]$. They adopted the Dirichlet distribution $\mathcal{D}(\alpha_1, \ldots, \alpha_g)$, defined by (4.5) with $\alpha_i = 1$ ($i = 1, \ldots, g$), as the prior for the mixing proportions π_i. For this specification, the log of the prior density $p(\boldsymbol{\Psi})$ for $\boldsymbol{\Psi}$ is given by

$$\begin{aligned}\log p(\boldsymbol{\Psi}) &= \log(g-1)! - gp\log(2\omega) - gp\log\beta - \tfrac{1}{2}gp(p-1)\log(2\beta) \\ &= \log(g-1)! - gp\log(\omega) - \tfrac{1}{2}gp(p+1)\log(2\beta). \qquad (6.56)\end{aligned}$$

To make this criterion fully consistent with a frequentist setting, we would need to ignore this log prior term in (6.55). But if we do this, it is no longer invariant under a reparameterization. However, its presence has no effect on estimation for a fixed value of g.

6.9.6 Reversible Jump Method

To conclude this section on Bayesian-based methods for choosing the number of components g, we mention the fully Bayesian approaches of Noble (1994), Phillips and Smith (1996), and Richardson and Green (1997) in which the number of components g is treated as an unknown parameter in the formulation of the Bayesian model. As seen in Section 4.7, Richardson and Green (1997) use reversible jump MCMC methods to handle this case where the dimension of the parameter space is of varying dimension. The results of Phillips and Smith (1996) and Richardson and Green (1997) were presented for univariate component densities. A marked increase in computational effort would be needed for multivariate components.

6.9.7 MML Principle

It can be seen from (6.47) that criteria based on this approximation are very similar to the MML approach defined in Section 4.11, whereby g is chosen to minimize the minimum message length. Apart from the lattice constant term, the minimum message length has the same form as (6.47).

6.10 CLASSIFICATION-BASED INFORMATION CRITERIA

We consider now some criteria that have been developed by consideration either from a frequentist or Bayesian perspective of the so-called classification likelihood $L_c(\boldsymbol{\Psi})$, which is the complete-data likelihood (2.26) within the EM framework for the fitting of a mixture model.

6.10.1 Classification Likelihood Criterion

Biernacki and Govaert (1997) made use of the relationship linking the likelihood $L(\boldsymbol{\Psi})$ for the mixture model and the complete-data likelihood $L_c(\boldsymbol{\Psi})$ to propose a criterion for selecting the number of clusters arising from the fitting of a normal mixture model. Although $L_c(\boldsymbol{\Psi})$ defined by (2.26) is referred to as the complete-data likelihood within the EM framework, it is sometimes called the classification likelihood in a classification context, as discussed in Section 2.21.

As noted by Hathaway (1986a), among others, we can express the mixture log likelihood, $\log L(\boldsymbol{\Psi})$, as

$$\log L(\boldsymbol{\Psi}) = \log L_c(\boldsymbol{\Psi}) - \log k(\boldsymbol{\Psi}), \tag{6.57}$$

where

$$\log k(\boldsymbol{\Psi}) = \sum_{i=1}^{g} \sum_{j=1}^{n} z_{ij} \log \tau_{ij}$$

and where $\tau_{ij} = \tau_i(\boldsymbol{y}_j;\ \boldsymbol{\Psi})$ is the posterior probability of ith component membership defined by (1.19). That is, $k(\boldsymbol{\Psi})$ is the conditional density of the vector of component-indicator variables

$$\boldsymbol{z} = (\boldsymbol{z}_1^T, \ldots, \boldsymbol{z}_n^T)^T,$$

given the observed data $\boldsymbol{y} = (\boldsymbol{y}_1^T, \ldots, \boldsymbol{y}_n^T)^T$.

The conditional mean of $\log k(\boldsymbol{\Psi})$ given the observed data $\boldsymbol{y}$ is equal to $-EN(\boldsymbol{\tau})$, where

$$EN(\boldsymbol{\tau}) = -\sum_{i=1}^{g} \sum_{j=1}^{n} \tau_{ij} \log \tau_{ij}$$

is the entropy of the fuzzy classification matrix $\boldsymbol{C} = ((\tau_{ij}))$ and where

$$\boldsymbol{\tau} = (\boldsymbol{\tau}_1^T, \ldots, \boldsymbol{\tau}_n^T)^T, \tag{6.58}$$

and

$$\boldsymbol{\tau}_j = (\tau_1(\boldsymbol{y}_j;\ \boldsymbol{\Psi}), \ldots, \tau_g(\boldsymbol{y}_j;\ \boldsymbol{\Psi}))^T \tag{6.59}$$

is the vector of posterior probabilities of component membership of $\boldsymbol{y}_j$ $(j = 1, \ldots, n)$.

We now write the complete-data likelihood as $L_c(\boldsymbol{\Psi};\ \boldsymbol{z})$ to explicitly denote that it is formed on the basis of $\boldsymbol{z}$ containing the component indicators, in addition to $\boldsymbol{y}$. Then it follows from (6.57) that if we put $\boldsymbol{z} = \hat{\boldsymbol{\tau}}$ in $L_c(\boldsymbol{\Psi};\ \boldsymbol{z})$, we have

$$\log L_c(\hat{\boldsymbol{\Psi}};\ \hat{\boldsymbol{\tau}}) = \log L(\hat{\boldsymbol{\Psi}}) - EN(\hat{\boldsymbol{\tau}}), \tag{6.60}$$

where $\hat{\boldsymbol{\tau}}$ is the MLE of $\boldsymbol{\tau}$ formed by replacing τ_{ij} by

$$\hat{\tau}_{ij} = \tau_i(\boldsymbol{y}_j;\ \hat{\boldsymbol{\Psi}}) \quad (i = 1, \ldots, g;\ j = 1, \ldots, n) \tag{6.61}$$

in (6.59). From (6.60), we can form the classification likelihood information criterion (CLC), where g is chosen to minimize

$$-2 \log L(\hat{\boldsymbol{\Psi}}) + 2EN(\hat{\boldsymbol{\tau}}), \tag{6.62}$$

where the estimated entropy $EN(\hat{\boldsymbol{\tau}})$ is used as the term that penalizes the model for its complexity.

If the components of the mixture are well separated, then $EN(\hat{\boldsymbol{\tau}})$ will be close to its minimum value of zero. But if the mixture components are poorly separated, then $EN(\hat{\boldsymbol{\tau}})$ will have a large value. Hence how severely this criterion penalizes the log likelihood depends on how well separated the fitted components are. According to Biernacki, Celeux, and Govaert (1999), this criterion works well when the mixing proportions are restricted to being equal. But it tends to overestimate the correct number of clusters when no restriction is placed on the mixing proportions.

Banfield and Raftery (1993) suggested an approximate Bayesian solution to the choice of the number of clusters using the classification ML approach. Their approximation, which is a crude approximation to twice the log Bayes factor for g clusters, leads to the approximate weight of evidence (AWE) criterion having the form

$$AWE(g) = -2 \log L_c + 2d(3/2 + \log n).$$

When the mixture components are well separated, we have seen above that $L_c(\hat{\boldsymbol{\Psi}}) \approx L(\hat{\boldsymbol{\Psi}})$, and thus it can then be expected to be similar to BIC. When the clusters are not well separated, it has been noted in Section 2.21 that the classification likelihood approach to model fitting leads to severely biased estimates of the parameters.

6.10.2 Normalized Entropy Criterion

Celeux and Soromenho (1996) proposed using the estimated entropy $EN(\hat{\boldsymbol{\tau}})$ (after normalization) as a criterion in its own right for choosing the number of clusters. This criterion is known as the normalized entropy criterion (NEC). The estimated entropy $EN(\hat{\boldsymbol{\tau}})$ cannot be used directly to assess the number of components in a mixture model, since $\log L(\hat{\boldsymbol{\Psi}})$ is an increasing function of g. The normalized form is given by

$$NEC(g) = \frac{EN(\hat{\boldsymbol{\tau}})}{\log L(\hat{\boldsymbol{\Psi}}) - \log L(\hat{\boldsymbol{\Psi}}^*)}, \tag{6.63}$$

where $\hat{\boldsymbol{\Psi}}^*$ denotes the MLE of $\boldsymbol{\Psi}$ in the case of a single ($g = 1$) component. The entropy for $g = 1$ is zero. As it stands, this criterion is unable to decide between $g = 1$ and a value of g greater than one. Celeux and Soromenho (1996) proposed a rule of thumb, but their procedure was restricted to normal mixtures and had performed disappointingly (Biernacki et al., 1999). In the latter paper, a general procedure was proposed to deal with this problem. Effectively, they define $NEC(g)$ to be one for $g = 1$. The modified criterion simply then consists of choosing g to minimize $NEC(g)$. The justification for setting $NEC(1) = 1$ is as follows. When comparing two values of g, g_0 and g_1 ($g_0 < g_1$), it is reasonable from a parsimony perspective to prefer g_0 if $L_c(\hat{\boldsymbol{\Psi}}_1; \hat{\boldsymbol{\tau}}_1) < L_c(\hat{\boldsymbol{\Psi}}_0; \hat{\boldsymbol{\tau}}_0)$, where $\hat{\boldsymbol{\Psi}}_i$ and $\hat{\boldsymbol{\tau}}_i$ denote the MLEs for $g = g_i$ ($i = 0, 1$). This is because one would expect that the model with the larger g would provide a better fit as measured by $L_c(\hat{\boldsymbol{\Psi}}_i; \hat{\boldsymbol{\tau}}_i)$. As a consequence, to choose $g > 1$ rather than $g = 1$, it is natural to require that $L_c(\hat{\boldsymbol{\Psi}}_i; \hat{\boldsymbol{\tau}}_i)$ for $g > 1$ to be greater than its value for $g = 1$. If the latter holds, then it follows from (6.60) that $0 \leq NEC(g) \leq 1$. Hence the only values of $g > 1$ of interest must satisfy this condition. If there is no value of g such that $NEC(g) < 1$, then there is no reason to choose $g > 1$. According to Biernacki et al. (1999), this improved version of the NEC criterion corrects for the tendency of the original version to prefer $g > 1$ clusters when the true number is one.

A similar type criterion is the partition coefficient (PC) of Bezdek (1981), where

$$PC(g) = \sum_{i=1}^{g} \sum_{j=1}^{n} \hat{\tau}_{ij}^2.$$

Numerical experiments reported by Windham and Cutler (1992) clearly show that the PC criterion tends to underestimate the order of the mixture model.

6.10.3 Integrated Classification Likelihood Criterion

We now sketch the development of the integrated classification likelihood (ICL) criterion as proposed by Biernacki et al. (1998). They introduced this criterion in an attempt to overcome the shortcomings of BIC and CLC.

The integrated classification likelihood is given by

$$p(\boldsymbol{y}, \boldsymbol{z}) = \int L_c(\boldsymbol{\Psi}) p(\boldsymbol{\Psi}) \, d\boldsymbol{\Psi}. \tag{6.64}$$

Suppose that the prior density $p(\boldsymbol{\Psi})$ for $\boldsymbol{\Psi} = (\boldsymbol{\pi}^T, \boldsymbol{\xi}^T)^T$ factors as

$$p(\boldsymbol{\Psi}) = p(\boldsymbol{\pi}) \, p(\boldsymbol{\xi}), \tag{6.65}$$

where $\boldsymbol{\xi} = (\boldsymbol{\theta}_1^T, \ldots, \boldsymbol{\theta}_g^T)^T$. Then

$$p(\boldsymbol{y}, \boldsymbol{z}) = p(\boldsymbol{z}) \, p(\boldsymbol{y} \mid \boldsymbol{z}), \tag{6.66}$$

where

$$p(\boldsymbol{y} \mid \boldsymbol{z}) = \int p(\boldsymbol{y} \mid \boldsymbol{z}, \boldsymbol{\xi}) \, p(\boldsymbol{\xi}) \, d\boldsymbol{\xi},$$

and

$$p(\boldsymbol{z}) = \int p(\boldsymbol{\pi}) \, p(\boldsymbol{z} \mid \boldsymbol{\pi}) \, d\boldsymbol{\pi}.$$

Biernacki et al. (1998) adopted the Dirichlet distribution $\mathcal{D}(\alpha_1, \ldots, \alpha_g)$ defined by (4.5) for $\boldsymbol{\pi}$ with $\alpha_i = \alpha \; (i = 1, \ldots, g)$. For this prior, the log integrated likelihood for $\boldsymbol{\pi}$ is given by

$$\begin{aligned} \log p(\boldsymbol{z}) &= \log \int_{\boldsymbol{A}} \pi_1^{\alpha + n_1} \cdots \pi_g^{\alpha + n_g} \{\Gamma(g\alpha)/\Gamma(\alpha)^g\} I_{\boldsymbol{A}}(\boldsymbol{\pi}) \, d\boldsymbol{\pi} \\ &= K(n_1, \ldots, n_g), \end{aligned} \tag{6.67}$$

where

$$K(n_1, \ldots, n_g) = \sum_{i=1}^{g} \log \Gamma(n_i + \alpha) - \log \Gamma(n + g\alpha) - g \log \Gamma(\alpha) + \log \Gamma(g\alpha) \tag{6.68}$$

and

$$n_i = \sum_{j=1}^{n} z_{ij} \quad (i = 1, \ldots, g).$$

On substituting (6.67) into (6.66), we obtain

$$\log p(\boldsymbol{y}, \boldsymbol{z}) = \int \log p(\boldsymbol{y} \mid \boldsymbol{z}, \boldsymbol{\xi})\, d\boldsymbol{\xi} + K(n_1, \ldots, n_g). \tag{6.69}$$

Biernacki et al. (1998) then applied the BIC approximation to the first term on the right-hand side of (6.69) to give

$$\log p(\boldsymbol{y}, \boldsymbol{z}) \approx \max_{\boldsymbol{\xi}} \log p(\boldsymbol{y} \mid \boldsymbol{z}, \boldsymbol{\xi}) - \tfrac{1}{2} d_1 \log n + K(n_1, \ldots, n_g), \tag{6.70}$$

where d_1 is the number of unknown parameters in $\boldsymbol{\xi}$.

If we maximize $p(\boldsymbol{y} \mid \boldsymbol{z}, \boldsymbol{\xi})$ simultaneously over $\boldsymbol{z}$ and $\boldsymbol{\xi}$, then we will not obtain the MLE $\hat{\boldsymbol{\xi}}$ of $\boldsymbol{\xi}$ as obtained by consideration of the mixture likelihood $L(\boldsymbol{\Psi})$. We proceed here by putting $\boldsymbol{z} = \hat{\boldsymbol{\tau}}$, as defined by (6.61), in $p(\boldsymbol{y} \mid \boldsymbol{z}, \boldsymbol{\xi})$. That is, we replace the component-indicator vector of $\boldsymbol{y}_j$ by its estimated conditional expectation $\hat{\boldsymbol{\tau}}_j$, which is the vector containing the MLEs of the posterior probabilities of component membership of $\boldsymbol{y}_j$ $(j = 1, \ldots, n)$.

It follows then from Section 2.8.4 that the maximum over $p(\boldsymbol{y} \mid \boldsymbol{z}, \boldsymbol{\xi})$ with $\boldsymbol{z} = \hat{\boldsymbol{\tau}}$ occurs at $\boldsymbol{\xi} = \hat{\boldsymbol{\xi}}$. Since

$$p(\boldsymbol{y} \mid \boldsymbol{z}, \boldsymbol{\xi}) = \log L_c(\boldsymbol{\Psi};\, \boldsymbol{z}) - \sum_{i=1}^{g} \sum_{j=1}^{n} z_{ij} \log \pi_i,$$

the maximum value of $p(\boldsymbol{y} \mid \boldsymbol{z}, \boldsymbol{\xi})$ with $\boldsymbol{z} = \hat{\boldsymbol{\tau}}$ is equal to

$$\log L_c(\hat{\boldsymbol{\Psi}};\, \hat{\boldsymbol{\tau}}) - n \sum_{i=1}^{g} \hat{\pi}_i \log \hat{\pi}_i, \tag{6.71}$$

on noting that

$$\sum_{i=1}^{g} \sum_{j=1}^{n} \hat{\tau}_{ij} \log \hat{\pi}_i = n \sum_{i=1}^{g} \hat{\pi}_i \log \hat{\pi}_i.$$

From (6.60), we can express (6.71) in terms of the mixture log likelihood as

$$\log L(\hat{\boldsymbol{\Psi}}) - EN(\hat{\boldsymbol{\tau}}) - n \sum_{i=1}^{g} \hat{\pi}_i \log \hat{\pi}_i. \tag{6.72}$$

Thus from (6.70) and (6.72), twice the negative of the log integrated classification likelihood with $\boldsymbol{z} = \hat{\boldsymbol{\tau}}$ can be approximated as

$$-2 \log L(\hat{\boldsymbol{\Psi}}) + 2EN(\hat{\boldsymbol{\tau}}) + 2n \sum_{i=1}^{g} \hat{\pi}_i \log \hat{\pi}_i + d_1 \log n - 2K(n\hat{\pi}_1, \ldots, n\hat{\pi}_g). \tag{6.73}$$

The ICL criterion chooses the number of components g to minimize (6.73).

Biernacki et al. (1998) derived an approximation to (6.73) when the $n\hat{\pi}_i$ are sufficiently large to approximate the Gamma function by Stirling's formula

$$\Gamma(u) \approx u^{u+\frac{1}{2}} \exp(-u)(2\pi)^{1/2}.$$

On using this approximation in $K(n\hat{\pi}_1, \ldots, n\hat{\pi}_g)$, we obtain the following on setting $\alpha = 1$ and neglecting terms of order $O(1)$:

$$K(n\hat{\pi}_1, \ldots, n\hat{\pi}_g) \approx n \sum_{i=1}^{g} \hat{\pi}_i \log \hat{\pi}_i - \tfrac{1}{2}(g-1) \log n. \tag{6.74}$$

On substituting (6.74) into (6.73), this expression can be written as

$$-2 \log L(\hat{\boldsymbol{\Psi}}) + 2EN(\hat{\boldsymbol{\tau}}) + d_1 \log n + (g-1) \log n,$$

which reduces to

$$-2 \log L(\hat{\boldsymbol{\Psi}}) + 2EN(\hat{\boldsymbol{\tau}}) + d \log n, \tag{6.75}$$

where $d = d_1 + (g-1)$ denotes the number of unknown parameters in $\boldsymbol{\Psi}$.

We shall refer to this version of the ICL criterion that chooses g by minimizing (6.75) as the ICL–BIC criterion. Although it is only appropriate for large cluster sizes, Biernacki et al. (1998) have found that its performance differs little from the more accurate version using (6.73).

As pointed out by Biernacki et al. (1998), the ICL criterion has a close link with the CS criterion proposed by Cheeseman and Stutz (1996); see also Cheeseman and Heckerman (1997). This criterion is based on (6.73) without the entropy term $EN(\hat{\boldsymbol{\tau}})$. On using the approximation (6.74), it becomes equivalent to ordinary BIC.

6.11 AN EMPIRICAL COMPARISON OF SOME CRITERIA

We report here some simulation results of McLachlan and Ng (2000b) to compare the performances of some of the more recently suggested criteria with classical procedures such as BIC for the selection of the number of components in a mixture model. Seven selection criteria were compared, namely AIC, BIC, EIC [Efron (bootstrap-based) information criterion], CLC (classification likelihood criterion), ICL (integrated classification likelihood) criterion, its BIC-type approximation, ICL–BIC, and LEC (Laplace–Empirical criterion). The latter was implemented with $\omega = 2$ and $\beta = 1$ in (6.56).

The EMMIX algorithm of McLachlan et al. (1999) was used to fit normal mixture models with a varying number of components to three simulated data sets. Ten random starts were performed for each set to initialize the EM algorithm. The solution corresponding to the largest local maximum of the log likelihood located was taken as the MLE after elimination of local maximizers considered to be spurious on the basis of the relative sizes of the fitted generalized component variances.

6.11.1 Simulated Set 1

Set 1 consisted of $n = 625$ four-dimensional observations obtained by generating samples separately from each of five normal distributions. The component-sample sizes, means, and covariance matrices, which were those adopted in Bozdogan (1993) and Celeux and Soromenho (1996), are displayed below:

$$\boldsymbol{\mu}_1 = (10, 12, 10, 12)^T \qquad \boldsymbol{\Sigma}_1 = \boldsymbol{I}_p \qquad n_1 = 75$$

$$\boldsymbol{\mu}_2 = (8.5, 10.5, 8.5, 10.5)^T \qquad \boldsymbol{\Sigma}_2 = \boldsymbol{I}_p \qquad n_2 = 100$$

$$\boldsymbol{\mu}_3 = (12, 14, 12, 14)^T \qquad \boldsymbol{\Sigma}_3 = \boldsymbol{I}_p \qquad n_3 = 125$$

$$\boldsymbol{\mu}_4 = (13, 15, 7, 9)^T \qquad \boldsymbol{\Sigma}_4 = 4\boldsymbol{I}_p \qquad n_4 = 150$$

$$\boldsymbol{\mu}_5 = (7, 9, 13, 15)^T \qquad \boldsymbol{\Sigma}_5 = 9\boldsymbol{I}_p \qquad n_5 = 175$$

A g-component normal mixture model with diagonal component-covariance matrices was fitted. The results are given in Table 6.9, where the superscript $*$ denotes the true value of g and also, for a given criterion, the value of g selected.

Table 6.9 Determination of the Number of Components: Set 1

g	AIC	EIC	BIC	CLC	ICL–BIC	ICL	LEC
2	12375	12377	12451	12430	12540	12437	12430
3	12022	12026	12138	12220	12388	12232	12080
4	11379	11379	11535	11457	11682	11475	11482
5*	11283*	11285*	11479*	11365*	11648*	11387*	11422*
6	11287	11309	11522	11396	11737	11423	11426
7	11296	11320	11571	11542	11941	11573	11449

Source: Adapted from McLachlan and Ng (2000b).

6.11.2 Simulated Set 2

Set 2 consisted of $n = 300$ trivariate observations obtained by generating samples separately from each of three normal distributions with component-sample sizes, means, and covariance matrices as displayed below:

$$\boldsymbol{\mu}_1 = (0, 0, 0)^T \qquad \boldsymbol{\Sigma}_1 = \begin{pmatrix} 9 & 0 & 0 \\ 0 & 4 & 0 \\ 0 & 0 & 1 \end{pmatrix} \qquad n_1 = 200$$

$$\boldsymbol{\mu}_2 = (-6, 3, 6)^T \quad \boldsymbol{\Sigma}_2 = \begin{pmatrix} 4 & -3.2 & -0.2 \\ -3.2 & 4 & 0 \\ -0.2 & 0 & 1 \end{pmatrix} \quad n_2 = 50$$

$$\boldsymbol{\mu}_3 = (6, 6, 4)^T \quad \boldsymbol{\Sigma}_3 = \begin{pmatrix} 4 & 3.2 & 2.8 \\ 3.2 & 4 & 2.4 \\ 2.8 & 2.4 & 2 \end{pmatrix} \quad n_3 = 50$$

The results are given in Table 6.10 for the fitting of a g-component normal mixture model with unrestricted component-covariance matrices.

Table 6.10 Determination of the Number of Components: Set 2

g	AIC	EIC	BIC	CLC	ICL–BIC	ICL	LEC
2	4239	4245	4309	4203	4311	4312	4300
3*	3978	3982	4085*	3933	4098*	4098*	4085*
4	3968	3982*	4112	3905*	4128	4127	4126
5	3967*	3982	4149	4019	4298	4292	–

* True value of g or value of g given by criterion.
Source: Adapted from McLachlan and Ng (2000b).

6.11.3 Simulated Set 3

The third simulated set consisted of a sample of $n = 200$ bivariate observations obtained by generating a sample of $n_1 = 100$ observations from a bivariate normal distribution with mean $(3.3, 0)^T$ and identity covariance matrix, along with a further sample of $n_2 = 100$ observations from the density

$$f(\boldsymbol{w}) = 0.25 I_{(-1,1)}(w_1) I_{(-1,1)}(w_2).$$

The results are given in Table 6.11 for fitting a normal mixture model with unrestricted component-covariance matrices.

Table 6.11 Determination of the Number of Components: Set 3

g	AIC	EIC	BIC	CLC	ICL–BIC	ICL	LEC
1	1315	1315	1308	1305	1332	1720	1302
2*	1207	1205	1244	1204	1262*	1263*	1239*
3	1178	1179*	1234*	1175*	1265	1265	1240
4	1175	1184	1251	1184	1306	1305	1249
5	1174*	1185	1269	1208	1362	1357	1273

* True value of g or value of g given by criterion.
Source: Adapted from McLachlan and Ng (2000b).

6.11.4 Conclusions from Simulations

It can be seen that the integrated classification likelihood (ICL), its large cluster size approximation, ICL–BIC, and the Laplace–Empirical criterion (LEC) are the only criteria to correctly select the true number of components in all three simulated data sets. The criterion of ICL–BIC is the easiest to apply of these criteria, while LEC is the hardest in the sense that it requires the evaluation of the determinant of the observed information matrix. The tendency of the AIC criterion to select too many components (clusters) is evident in the results for Sets 2 and 3. The bootstrap version of this criterion, EIC, does do a better job for these two sets, but it still selects too many components.

7

Multivariate t Mixtures

7.1 INTRODUCTION

For many applied problems, the tails of the normal distribution are often shorter than required. Also, the estimates of the component means and covariance matrices can be affected by observations that are atypical of the components in the normal mixture model being fitted. The problem of providing protection against outliers in multivariate data is a very difficult problem and increases in difficulty with the dimension of the data (Rocke and Woodruff, 1997; Kosinski, 1999).

In this chapter we consider the fitting of mixtures of (multivariate) t distributions, as proposed in McLachlan and Peel (1998a) and Peel and McLachlan (2000). The t distribution provides a longer-tailed alternative to the normal distribution. Hence it provides a more robust approach to the fitting of normal mixture models, as observations that are atypical of a component are given reduced weight in the calculation of its parameters. Also, the use of t components gives less extreme estimates of the posterior probabilities of component membership of the mixture model.

With this t mixture model-based approach, the normal distribution for each component in the mixture is embedded in a wider class of elliptically symmetric distributions with an additional parameter called the degrees of freedom ν. As ν tends to infinity, the t distribution approaches the normal distribution. Hence this parameter ν may be viewed as a robustness tuning parameter. It can be fixed in advance or it can be inferred from the data for each component, thereby providing an *adaptive* robust procedure, as explained in Lange, Little, and Taylor (1989), who considered the use of a single component t distribution in linear and nonlinear regression problems; see also Rubin (1983) and Sutradhar and Ali (1986). An early reference for using the t distribution to model data in the case of a single component distribution is Jeffreys

(1932). Of course, as noted by Lange et al. (1989), the use of the t distribution is not a panacea for all forms of robustness. Data with shorter-than-normal tails, asymmetric distributions, varying degrees of long-tailedness among the feature variables, or with extreme outliers will not be able to be modeled adequately by a mixture of t distributions.

7.2 PREVIOUS WORK

Robust estimation in the context of mixture models has been considered in the past by Campbell (1984), McLachlan and Basford (1988, Chapter 3), and De Veaux and Kreiger (1990), among others, using M-estimates of the means and covariance matrices of the normal components of the mixture model. This line of approach is to be discussed in Section 7.6.

Recently, Markatou (1998) has provided a formal approach to robust mixture estimation by applying weighted likelihood methodology in the context of mixture models. With this methodology, an estimate of the vector of unknown parameters is obtained as a solution of the equation

$$\sum_{j=1}^{n} w(\boldsymbol{y}_j)\partial \log f(\boldsymbol{y}_j;\, \boldsymbol{\Psi})/\partial \boldsymbol{\Psi} = \mathbf{0}, \tag{7.1}$$

where $f(\boldsymbol{y}_j;\, \boldsymbol{\Psi})$ denotes the specified parametric form for the density of $\boldsymbol{Y}_j$. The weight function $w(\boldsymbol{y}_j)$ is defined in terms of the Pearson residuals; see Markatou, Basu, and Lindsay (1998) and the previous work of Green (1984). The weighted likelihood methodology provides robust and first-order efficient estimators in general, and Markatou (1998) has established these results in the context of univariate mixture models.

Also, Tibshirani and Knight (1999) have proposed the technique of bootstrap "bumping," which can be used for resistant fitting. They demonstrated its use in fitting a two-component univariate normal mixture model in the presence of an outlier. The target criterion to be minimized was taken to be the median of the negative of the log mixture density at each of the observations $\boldsymbol{y}_j$, while the working criterion was taken to be the negative of the log likelihood. Corresponding to this choice for the working criterion, the usual ML solution $\hat{\boldsymbol{\Psi}}_b^*$ was obtained via the EM algorithm for each of a number B of bootstrap samples $\boldsymbol{y}_b^*$ $(b = 1, \ldots, B)$ obtained by sampling with replacement from the original sample $\boldsymbol{y}$. The fit was then taken to be that estimate among the $\hat{\boldsymbol{\Psi}}_b^*$ and the MLE $\hat{\boldsymbol{\Psi}}$ for the original sample that minimized the target criterion.

7.3 ROBUST CLUSTERING

One useful application of normal mixture models has been in the important field of cluster analysis. Besides having a sound mathematical basis, this approach is not confined to the production of spherical clusters, such as with k-means-type algorithms that use Euclidean distance rather than the Mahalanobis distance metric, which allows for within-cluster correlations between the variables in the feature vector $\boldsymbol{Y}_j$. Moreover, unlike clustering methods defined solely in terms of the Mahalanobis distance, the normal mixture-based clustering takes into account the normalizing term $| \boldsymbol{\Sigma}_i |^{-1/2}$ in the estimate of the multivariate normal density adopted for the component distribution of $\boldsymbol{Y}$ corresponding to the ith cluster. This term can make an important contribution in the case of disparate group-covariance matrices (McLachlan, 1992; Chapter 2).

Although even a crude estimate of the within-cluster covariance matrix $\boldsymbol{\Sigma}_i$ often suffices for clustering purposes (Gnanadesikan, Harvey, and Kettenring, 1993), it can be severely affected by outliers. Hence it is highly desirable for methods of cluster analysis to provide robust clustering procedures. The problem of making clustering algorithms more robust has received much attention recently as, for example, in Jolion, Meer, and Bataouche (1991), Smith, Bailey, and Munford (1993), Frigui and Krishnapuram (1996), Kharin (1996), Rousseeuw, Kaufman, and Trauwaert (1996), Zhuang et al. (1996), and Davé and Krishnapuram (1997).

7.4 MULTIVARIATE t DISTRIBUTION

One way to broaden the normal parametric family for potential outliers or data with longer-than-normal tails is to adopt the two-component normal mixture density

$$(1-\epsilon)\phi(\boldsymbol{y}_j;\ \boldsymbol{\mu},\ \boldsymbol{\Sigma}) + \epsilon\phi(\boldsymbol{y}_j;\ \boldsymbol{\mu},\ k\boldsymbol{\Sigma}), \tag{7.2}$$

where k is large and ϵ is small, representing the small proportion of observations that have a relatively large variance. Huber (1964) subsequently considered more general forms of contamination of the normal distribution in the development of his robust M-estimators of a location parameter, as to be discussed further in Section 7.6.

The normal scale mixture model (7.2) can be written as

$$\int \phi(\boldsymbol{y}_j;\ \boldsymbol{\mu},\ \boldsymbol{\Sigma}/u)\, dH(u), \tag{7.3}$$

where H is the probability distribution that places mass $(1-\epsilon)$ at the point $u = 1$ and mass ϵ at the point $u = 1/k$. Suppose we now replace H by the distribution of a chi-squared random variable on its degrees of freedom ν; that is, by the random variable U distributed as

$$U \sim \text{ gamma}\left(\tfrac{1}{2}\nu,\ \tfrac{1}{2}\nu\right), \tag{7.4}$$

where the gamma $(\alpha,\ \beta)$ density function is defined by (4.16). We then obtain the t distribution with location parameter $\boldsymbol{\mu}$, positive definite inner product matrix $\boldsymbol{\Sigma}$, and ν degrees of freedom,

$$f(\boldsymbol{y}_j;\ \boldsymbol{\mu},\ \boldsymbol{\Sigma},\ \nu) = \frac{\Gamma(\frac{\nu+p}{2})\,|\boldsymbol{\Sigma}|^{-1/2}}{(\pi\nu)^{\frac{1}{2}p}\Gamma(\frac{\nu}{2})\{1+\delta(\boldsymbol{y}_j,\ \boldsymbol{\mu};\ \boldsymbol{\Sigma})/\nu\}^{\frac{1}{2}(\nu+p)}}, \tag{7.5}$$

where

$$\delta(\boldsymbol{y}_j,\, \boldsymbol{\mu};\, \boldsymbol{\Sigma}) = (\boldsymbol{y}_j - \boldsymbol{\mu})^T \boldsymbol{\Sigma}^{-1}(\boldsymbol{y}_j - \boldsymbol{\mu}) \tag{7.6}$$

denotes the Mahalanobis squared distance between $\boldsymbol{y}_j$ and $\boldsymbol{\mu}$ (with $\boldsymbol{\Sigma}$ as the covariance matrix). If $\nu > 1$, $\boldsymbol{\mu}$ is the mean of $\boldsymbol{Y}_j$, and if $\nu > 2$, $\nu(\nu-2)^{-1}\boldsymbol{\Sigma}$ is its covariance matrix. As ν tends to infinity, U converges to one with probability one, and so $\boldsymbol{Y}_j$ becomes marginally multivariate normal with mean $\boldsymbol{\mu}$ and covariance matrix $\boldsymbol{\Sigma}$. The family of t distributions thus provides a heavy-tailed alternative to the normal family with mean $\boldsymbol{\mu}$ and covariance matrix that is equal to a scalar multiple of $\boldsymbol{\Sigma}$ (if $\nu > 2$).

7.5 ML ESTIMATION OF MIXTURE OF t DISTRIBUTIONS

7.5.1 Application of EM Algorithm

A brief history of the development of ML estimation of a single-component t distribution is given in Liu and Rubin (1995). An account of more recent work is given in Liu (1997). Liu and Rubin (1994, 1995) have shown that the MLEs can be found much more efficiently by using an extension of the EM algorithm called the expectation–conditional maximization either (ECME) algorithm. Meng and van Dyk (1997) demonstrated that the more promising versions of the ECME algorithm for the t distribution can be obtained using alternative data augmentation schemes. They called this algorithm the alternating expectation–conditional maximization (AECM) algorithm. Following Meng and van Dyk (1997), Liu (1997) considered a class of data augmentation schemes even more general than the class of Meng and van Dyk (1997). This led to new versions of the ECME algorithm for ML estimation of the t distribution with possible missing values, corresponding to applications of the parameter-expanded EM (PX-EM) algorithm (Liu et al., 1998).

We consider now ML estimation for a g-component mixture of t distributions, given by

$$f(\boldsymbol{y}_j;\, \boldsymbol{\Psi}) = \sum_{i=1}^{g} \pi_i f(\boldsymbol{y}_j;\, \boldsymbol{\mu}_i,\, \boldsymbol{\Sigma}_i,\, \nu_i), \tag{7.7}$$

where

$$\boldsymbol{\Psi} = (\pi_1,\, \ldots,\, \pi_{g-1},\, \boldsymbol{\xi}^T,\, \boldsymbol{\nu}^T)^T,$$

$\boldsymbol{\nu} = (\nu_1,\, \ldots,\, \nu_g)^T$, and $\boldsymbol{\xi}$ contains the elements of the $\boldsymbol{\mu}_i$ and the distinct elements of $\boldsymbol{\Sigma}_i$ $(i = 1,\, \ldots,\, g)$. The application of the EM algorithm for ML estimation in the case of a single component t distribution has been described in McLachlan and Krishnan (1997, Sections 2.6 and 5.8). The results there can be extended to cover the present case of a g-component mixture of multivariate t distributions.

In the EM framework, the complete-data vector is given by

$$\boldsymbol{y}_c = (\boldsymbol{y}^T,\, \boldsymbol{z}_1^T,\, \ldots,\, \boldsymbol{z}_n^T,\, u_1,\, \ldots,\, u_n)^T, \tag{7.8}$$

where $\boldsymbol{y} = (\boldsymbol{y}_1^T,\, \ldots,\, \boldsymbol{y}_n^T)^T$ denotes the observed-data vector, $\boldsymbol{z}_1,\, \ldots,\, \boldsymbol{z}_n$ are the component-label vectors defining the component of origin of $\boldsymbol{y}_1,\, \ldots,\, \boldsymbol{y}_n$, respectively, and $z_{ij} = (\boldsymbol{z}_j)_i$ is one or zero, according as to whether $\boldsymbol{y}_j$ belongs or does not belong to the ith component. In the light of the above characterization of the t distribution, it is convenient to view the observed data augmented by the $\boldsymbol{z}_j$ as still being incomplete and introduce into the complete-data vector the additional missing data, $u_1, \ldots, u_n$, which are defined so that given $z_{ij} = 1$, we obtain

$$\boldsymbol{Y}_j \mid u_j, z_{ij} = 1 \sim N(\boldsymbol{\mu}_i, \boldsymbol{\Sigma}_i/u_j), \tag{7.9}$$

independently for $j = 1, \ldots, n$, and

$$U_j \mid z_{ij} = 1 \sim \text{gamma}\,(\tfrac{1}{2}\nu_i, \tfrac{1}{2}\nu_i). \tag{7.10}$$

Given $\boldsymbol{z}_1, \ldots, \boldsymbol{z}_n$, the $U_1, \ldots, U_n$ are independently distributed according to (7.10).

The complete-data likelihood $L_c(\boldsymbol{\Psi})$ can be factored into the product of the marginal densities of the $\boldsymbol{Z}_j$, the conditional densities of the U_j given the $\boldsymbol{z}_j$, and the conditional densities of the $\boldsymbol{Y}_j$ given the u_j and the $\boldsymbol{z}_j$. Accordingly, the complete-data log likelihood can be written as

$$\log L_c(\boldsymbol{\Psi}) = \log L_{1c}(\boldsymbol{\pi}) + \log L_{2c}(\boldsymbol{\nu}) + \log L_{3c}(\boldsymbol{\xi}), \tag{7.11}$$

where

$$\log L_{1c}(\boldsymbol{\pi}) = \sum_{i=1}^{g}\sum_{j=1}^{n} z_{ij} \log \pi_i, \tag{7.12}$$

$$\begin{aligned}\log L_{2c}(\boldsymbol{\nu}) &= \sum_{i=1}^{g}\sum_{j=1}^{n} z_{ij}\{-\log\Gamma(\tfrac{1}{2}\nu_i) + \tfrac{1}{2}\nu_i\log(\tfrac{1}{2}\nu_i) \\ &\qquad + \tfrac{1}{2}\nu_i(\log u_j - u_j) - \log u_j\},\end{aligned} \tag{7.13}$$

and

$$\begin{aligned}\log L_{3c}(\boldsymbol{\xi}) &= \sum_{i=1}^{g}\sum_{j=1}^{n} z_{ij}\{-\tfrac{1}{2}p\log(2\pi) - \tfrac{1}{2}\log|\boldsymbol{\Sigma}_i| \\ &\qquad - \tfrac{1}{2}u_j(\boldsymbol{y}_j - \boldsymbol{\mu}_i)^T\boldsymbol{\Sigma}_i^{-1}(\boldsymbol{y}_j - \boldsymbol{\mu}_i)\}.\end{aligned} \tag{7.14}$$

In (7.11), $\boldsymbol{\pi} = (\pi_1, \ldots, \pi_g)^T$ and $\boldsymbol{\xi} = (\boldsymbol{\theta}_1^T, \ldots, \boldsymbol{\theta}_g^T)^T$, and $\boldsymbol{\theta}_i$ contains the elements of $\boldsymbol{\mu}_i$ and the distinct elements of $\boldsymbol{\Sigma}_i$ $(i = 1, \ldots, g)$.

7.5.2 E-Step

Now the E-step on the $(k+1)$th iteration of the EM algorithm requires the calculation of $Q(\boldsymbol{\Psi}; \boldsymbol{\Psi}^{(k)})$, the current conditional expectation of the complete-data log likelihood function, $\log L_c(\boldsymbol{\Psi})$. This E-step can be effected by first taking the expectation of $\log L_c(\boldsymbol{\Psi})$ conditional on $\boldsymbol{z}_1, \ldots, \boldsymbol{z}_n$, as well as $\boldsymbol{y}$, and then finally over the $\boldsymbol{z}_j$ given $\boldsymbol{y}$. It can be seen from (7.12) to (7.14) that in order to do this, we need to calculate

$$E_{\boldsymbol{\Psi}^{(k)}}(Z_{ij} \mid \boldsymbol{y}_j),$$

$$E_{\boldsymbol{\Psi}^{(k)}}(U_j \mid \boldsymbol{y}_j, \boldsymbol{z}_j),$$

and

$$E_{\Psi^{(k)}}(\log U_j \mid \boldsymbol{y}_j, \boldsymbol{z}_j) \tag{7.15}$$

for $i = 1, \ldots, g;\ j = 1, \ldots, n$.

It follows that

$$E_{\Psi^{(k)}}(Z_{ij} \mid \boldsymbol{y}_j) = \tau_{ij}^{(k)}, \tag{7.16}$$

where

$$\tau_{ij}^{(k)} = \frac{\pi_i^{(k)} f(\boldsymbol{y}_j;\ \boldsymbol{\mu}_i^{(k)},\ \boldsymbol{\Sigma}_i^{(k)},\ \nu_i^{(k)})}{f(\boldsymbol{y}_j;\ \boldsymbol{\Psi}^{(k)})} \tag{7.17}$$

is the posterior probability that $\boldsymbol{y}_j$ belongs to the ith component of the mixture, using the current fit $\boldsymbol{\Psi}^{(k)}$ for $\boldsymbol{\Psi}$ $(i = 1, \ldots, g;\ j = 1, \ldots, n)$.

Since the gamma distribution is the conjugate prior distribution for U_j, it is not difficult to show that the conditional distribution of U_j given $\boldsymbol{Y}_j = \boldsymbol{y}_j$ and $Z_{ij} = 1$ is

$$U_j \mid \boldsymbol{y}_j,\ z_{ij} = 1 \sim \text{ gamma}\,(m_{1i},\ m_{2i}), \tag{7.18}$$

where

$$m_{1i} = \tfrac{1}{2}(\nu_i + p)$$

and

$$m_{2i} = \tfrac{1}{2}\{\nu_i + \delta(\boldsymbol{y}_j,\ \boldsymbol{\mu}_i;\ \boldsymbol{\Sigma}_i)\}. \tag{7.19}$$

From (7.18), we have that

$$E(U_j \mid \boldsymbol{y}_j,\ z_{ij} = 1) = \frac{\nu_i + p}{\nu_i + \delta(\boldsymbol{y}_j,\ \boldsymbol{\mu}_i;\ \boldsymbol{\Sigma}_i)}. \tag{7.20}$$

Thus from (7.20), we obtain

$$E_{\Psi^{(k)}}(U_j \mid \boldsymbol{y}_j, z_{ij} = 1) = u_{ij}^{(k)}, \tag{7.21}$$

where

$$u_{ij}^{(k)} = \frac{\nu_i^{(k)} + p}{\nu_i^{(k)} + \delta(\boldsymbol{y}_j,\ \boldsymbol{\mu}_i^{(k)};\ \boldsymbol{\Sigma}_i^{(k)})}. \tag{7.22}$$

To calculate the conditional expectation (7.15), we need the result that if a random variable R has a gamma $(\alpha,\ \beta)$ distribution, then

$$E(\log R) = \psi(\alpha) - \log \beta, \tag{7.23}$$

where

$$\psi(s) = \{\partial\Gamma(s)/\partial s\}/\Gamma(s)$$

is the Digamma function. Applying the result (7.23) to the conditional density of U_j given $\boldsymbol{y}_j$ and $z_{ij} = 1$, as specified by (7.18), it follows that

$$\begin{aligned} E_{\Psi^{(k)}}(\log U_j \mid \boldsymbol{y}_j,\ z_{ij} = 1) &= \psi\left(\frac{\nu_i^{(k)} + p}{2}\right) - \log[\tfrac{1}{2}\{\nu_i^{(k)} + \delta(\boldsymbol{y}_j,\ \boldsymbol{\mu}_i^{(k)};\ \boldsymbol{\Sigma}_i^{(k)})\}] \\ &= \log u_{ij}^{(k)} + \left\{\psi\left(\frac{\nu_i^{(k)} + p}{2}\right) - \log\left(\frac{\nu_i^{(k)} + p}{2}\right)\right\} \end{aligned} \tag{7.24}$$

for $j = 1, \ldots, n$. The last term on the right-hand side of (7.24),

$$\psi\left(\frac{\nu_i^{(k)} + p}{2}\right) - \log\left(\frac{\nu_i^{(k)} + p}{2}\right),$$

can be interpreted as the correction for just imputing the conditional mean value $u_{ij}^{(k)}$ for u_j in $\log u_j$.

On using the results (7.16), (7.21), and (7.24) to calculate the conditional expectation of the complete-data log likelihood from (7.11), we have that $Q(\boldsymbol{\Psi}; \boldsymbol{\Psi}^{(k)})$ is given by

$$Q(\boldsymbol{\Psi}; \boldsymbol{\Psi}^{(k)}) = Q_1(\boldsymbol{\pi}; \boldsymbol{\Psi}^{(k)}) + Q_2(\boldsymbol{\nu}; \boldsymbol{\Psi}^{(k)}) + Q_3(\boldsymbol{\xi}; \boldsymbol{\Psi}^{(k)}), \tag{7.25}$$

where

$$Q_1(\boldsymbol{\pi}; \boldsymbol{\Psi}^{(k)}) = \sum_{i=1}^{g}\sum_{j=1}^{n} \hat{\tau}_{ij}^{(k)} \log \pi_i, \tag{7.26}$$

$$Q_2(\boldsymbol{\nu}; \boldsymbol{\Psi}^{(k)}) = \sum_{i=1}^{g}\sum_{j=1}^{n} \hat{\tau}_{ij}^{(k)} Q_{2j}(\nu_i; \boldsymbol{\Psi}^{(k)}), \tag{7.27}$$

and

$$Q_3(\boldsymbol{\xi}; \boldsymbol{\Psi}^{(k)}) = \sum_{i=1}^{g}\sum_{j=1}^{n} \hat{\tau}_{ij}^{(k)} Q_{3j}(\boldsymbol{\theta}_i; \boldsymbol{\Psi}^{(k)}), \tag{7.28}$$

and where, on ignoring terms not involving the ν_i,

$$\begin{aligned} Q_{2j}(\nu_i; \boldsymbol{\Psi}^{(k)}) = & -\log\Gamma(\tfrac{1}{2}\nu_i) + \tfrac{1}{2}\nu_i \log(\tfrac{1}{2}\nu_i) \\ & + \tfrac{1}{2}\nu_i\left\{\sum_{j=1}^{n}(\log u_{ij}^{(k)} - u_{ij}^{(k)}) + \psi\left(\frac{\nu_i^{(k)} + p}{2}\right) - \log\left(\frac{\nu_i^{(k)} + p}{2}\right)\right\} \end{aligned} \tag{7.29}$$

and

$$\begin{aligned} Q_{3j}(\boldsymbol{\theta}_i; \boldsymbol{\Psi}^{(k)}) = & \{-\tfrac{1}{2}p\log(2\pi) - \tfrac{1}{2}\log|\boldsymbol{\Sigma}_i| + \tfrac{1}{2}p\log u_{ij}^{(k)} \\ & -\tfrac{1}{2}u_{ij}(\boldsymbol{y}_j - \boldsymbol{\mu}_i)^T \boldsymbol{\Sigma}_i^{-1}(\boldsymbol{y}_j - \boldsymbol{\mu}_i)\}. \end{aligned} \tag{7.30}$$

7.5.3 M-Step

On the M-step at the $(k+1)$th iteration of the EM algorithm, it follows from (7.25) that $\boldsymbol{\pi}^{(k+1)}$, $\boldsymbol{\xi}^{(k+1)}$, and $\boldsymbol{\nu}^{(k+1)}$ can be computed independently of each other, by separate consideration of (7.26), (7.27), and (7.28), respectively. The solutions for $\pi_i^{(k+1)}$ and $\boldsymbol{\theta}_i^{(k+1)}$ exist in closed form. Only the updates $\nu_i^{(k+1)}$ for the degrees of freedom ν_i need to be computed iteratively.

The mixing proportions are updated by consideration of the first term $Q_1(\boldsymbol{\pi};\, \boldsymbol{\Psi}^{(k)})$ on the right-hand side of (7.25). This leads to $\pi_i^{(k+1)}$ being given by the average of the posterior probabilities of component membership of the mixture. That is,

$$\pi_i^{(k+1)} = \sum_{j=1}^{n} \tau_{ij}^{(k)}/n \quad (i = 1,\, \ldots,\, g). \tag{7.31}$$

To update the estimates of $\boldsymbol{\mu}_i$ and $\boldsymbol{\Sigma}_i$ $(i = 1,\, \ldots,\, g)$, we need to consider

$$Q_{3j}(\boldsymbol{\theta}_i;\, \boldsymbol{\Psi}^{(k)}).$$

This is easily undertaken on noting that it corresponds to the log likelihood function formed from n independent observations $\boldsymbol{y}_1,\, \ldots,\, \boldsymbol{y}_n$ with common mean $\boldsymbol{\mu}_i$ and covariance matrices $\boldsymbol{\Sigma}_i/u_1^{(k)},\, \ldots,\, \boldsymbol{\Sigma}_i/u_n^{(k)}$, respectively. It is thus equivalent to computing the weighted sample mean and sample covariance matrix of $\boldsymbol{y}_1,\, \ldots,\, \boldsymbol{y}_n$ with weights $u_1^{(k)},\, \ldots,\, u_n^{(k)}$. Hence

$$\boldsymbol{\mu}_i^{(k+1)} = \sum_{j=1}^{n} \tau_{ij}^{(k)} u_{ij}^{(k)} \boldsymbol{y}_j \Big/ \sum_{j=1}^{n} \tau_{ij}^{(k)} u_{ij}^{(k)} \tag{7.32}$$

and

$$\boldsymbol{\Sigma}_i^{(k+1)} = \frac{\sum_{j=1}^{n} \tau_{ij}^{(k)} u_{ij}^{(k)} (\boldsymbol{y}_j - \boldsymbol{\mu}_i^{(k+1)})(\boldsymbol{y}_j - \boldsymbol{\mu}_i^{(k+1)})^T}{\sum_{j=1}^{n} \tau_{ij}^{(k)}}. \tag{7.33}$$

It can be seen that the EM process effectively chooses $\boldsymbol{\mu}_i^{(k+1)}$ and $\boldsymbol{\Sigma}_i^{(k+1)}$ by IRLS. The E-step updates the weights $u_{ij}^{(k)}$, while the M-step chooses $\boldsymbol{\mu}_i^{(k+1)}$ and $\boldsymbol{\Sigma}_i^{(k+1)}$ by weighted least-squares estimation. From the form of the equation (7.32) derived for the MLE of $\boldsymbol{\mu}_i$, we have that, as $\nu_i^{(k)}$ decreases, the degree of downweighting of an outlier increases. For finite $\nu_i^{(k)}$ as $\|\, \boldsymbol{y}_j \,\| \rightarrow \infty$, the effect on the ith component location parameter estimate goes to zero, whereas the effect on the ith component scale estimate remains bounded but does not vanish.

Following the proposal of Kent, Tyler, and Vardi (1994) in the case of a single-component t distribution, we can replace the divisor $\sum_{j=1}^{n} \tau_{ij}^{(k)}$ in (7.33) by

$$\sum_{j=1}^{n} \tau_{ij}^{(k)} u_{ij}^{(k)}.$$

This modified algorithm, however, converges faster than the conventional EM algorithm, as reported by Kent et al. (1994) and Meng and van Dyk (1997) in the case of a single-component t distribution $(g = 1)$. In the latter situation, Meng and van Dyk (1997) showed that this modified EM algorithm is optimal among EM algorithms generated from a class of data augmentation schemes. More recently, in the case of $g = 1$, Liu (1997) and Liu et al. (1998) have derived this modified EM algorithm using the PX-EM algorithm.

It can be seen that if the degrees of freedom ν_i is fixed in advance for each component, then the M-step exists in closed form. In this case where ν_i is fixed beforehand, the estimation of the component parameters is a form of M-estimation; see Lange et al. (1989, p. 882). However, an attractive feature of the use of the t distribution to model the component distributions is that the degrees of robustness as controlled by ν_i can be inferred from the data by computing its MLE. In this case, we have to compute also on the M-step the updated estimate $\nu_i^{(k+1)}$ of ν_i. On calculating the left-hand side of the equation

$$\sum_{j=1}^{n} \partial Q_{2j}(\nu_i;\, \boldsymbol{\Psi}^{(k)})/\partial \nu_i = 0,$$

it follows that $\nu_i^{(k+1)}$ is a solution of the equation

$$\left\{ -\psi(\tfrac{1}{2}\nu_i) + \log(\tfrac{1}{2}\nu_i) + 1 + \frac{1}{n_i^{(k)}} \sum_{j=1}^{n} \tau_{ij}^{(k)} (\log u_{ij}^{(k)} - u_j^{(k)}) \right.$$
$$\left. + \psi\left(\frac{\nu_i^{(k)} + p}{2}\right) - \log\left(\frac{\nu_i^{(k)} + p}{2}\right) \right\} = 0, \qquad (7.34)$$

where $n_i^{(k)} = \sum_{j=1}^{n} \tau_{ij}^{(k)}$ $(i = 1, \ldots, g)$.

7.5.4 Application of ECM Algorithm

For ML estimation of a single t component, Liu and Rubin (1995) noted that the convergence of the EM algorithm is slow for unknown ν and the one-dimensional search for the computation of $\nu^{(k+1)}$ is time-consuming. Consequently, they considered extensions of the EM algorithm in the form of the ECM and ECME algorithms; see McLachlan and Krishnan (1997, Section 5.8) and Liu (1997).

We consider now the ECM algorithm for the present problem of more than one t component, where $\boldsymbol{\Psi}$ is partitioned as $(\boldsymbol{\Psi}_1^T, \boldsymbol{\Psi}_2^T)^T$, with $\boldsymbol{\Psi}_1 = (\pi_1, \ldots, \pi_{g-1}, \boldsymbol{\xi}^T)^T$ and with $\boldsymbol{\Psi}_2$ equal to $\boldsymbol{\nu} = (\nu_1, \ldots, \nu_g)^T$. On the $(k+1)$th iteration of the ECM algorithm, the E-step is the same as given above for the EM algorithm, but the M-step of the latter is replaced by two CM-steps, as follows:

CM-Step 1. Calculate $\boldsymbol{\Psi}_1^{(k+1)}$ by maximizing $Q(\boldsymbol{\Psi};\, \boldsymbol{\Psi}^{(k)})$ with $\boldsymbol{\Psi}_2$ fixed at $\boldsymbol{\Psi}_2^{(k)}$; that is, $\boldsymbol{\nu}$ fixed at $\boldsymbol{\nu}^{(k)}$.

CM-Step 2. Calculate $\boldsymbol{\Psi}_2^{(k+1)}$ by maximizing $Q(\boldsymbol{\Psi};\, \boldsymbol{\Psi}^{(k)})$ with $\boldsymbol{\Psi}_1$ fixed at $\boldsymbol{\Psi}_1^{(k+1)}$.

But as seen above, $\boldsymbol{\Psi}_1^{(k+1)}$ and $\boldsymbol{\Psi}_2^{(k+1)}$ are calculated independently of each other on the M-step, and so these two CM-steps of the ECM algorithm are equivalent to the M-step of the EM algorithm. Hence there is no difference between this ECM and the EM algorithms here. However, Liu and Rubin (1995) used the ECM algorithm to give two modifications that are different from the EM algorithm. These two modifications are a multicycle version of the ECM algorithm and an ECME extension. The multicycle version of the ECM algorithm has an additional E-step between the two CM-steps. That is, after the first CM-step, the E-step is taken with

$$\boldsymbol{\Psi} = (\boldsymbol{\Psi}_1^{(k+1)^T}, \boldsymbol{\Psi}_2^{(k)^T})^T,$$

instead of with $\boldsymbol{\Psi} = (\boldsymbol{\Psi}_1^{(k)^T}, \boldsymbol{\Psi}_2^{(k)^T})^T$ as on the commencement of the $(k+1)$th iteration of the ECM algorithm. The EMMIX algorithm of McLachlan et al. (1999) has an option for the fitting of mixtures of multivariate t components either with or without the specification of the component degrees of freedom.

For a single-component t distribution, Liu and Rubin (1994, 1995), Kowalski et al. (1997), Liu (1997), Meng and van Dyk (1997), and Liu et al. (1998) have considered further extensions of the ECM algorithm corresponding to various versions of the ECME algorithm. However, the implementation of the ECME algorithm for mixtures of t distributions is not as straightforward, and so it is not applied here.

7.6 PREVIOUS WORK ON M-ESTIMATION OF MIXTURE COMPONENTS

A common way in which robust fitting of normal mixture models has been undertaken in the past has been to use M-estimates to update the component estimates on the M-step of the EM algorithm, as in Campbell (1984) and McLachlan and Basford (1988). With M-estimation, the updated component means $\boldsymbol{\mu}_i^{(k+1)}$ are given by (7.32), but where now the weights $u_{ij}^{(k)}$ are defined as

$$u_{ij}^{(k)} = \psi(d_{ij}^{(k)})/d_{ij}^{(k)} \tag{7.35}$$

and where

$$d_{ij}^{(k)} = \{(\boldsymbol{y}_j - \boldsymbol{\mu}_i^{(k)})^T \boldsymbol{\Sigma}_i^{(k)^{-1}} (\boldsymbol{y}_j - \boldsymbol{\mu}_i^{(k)})\}^{1/2}$$

and $\psi(s) = -\psi(-s)$ is Huber's (1964) ψ-function defined as

$$\begin{aligned} \psi(s) &= s, & \mid s \mid \leq a, \\ &= \text{sign}(s)\, a, & \mid s \mid > a, \end{aligned} \tag{7.36}$$

for an appropriate choice of the tuning constant a. The ith component-covariance matrix $\boldsymbol{\Sigma}_i^{(k+1)}$ can be updated as (7.33), where $u_{ij}^{(k)}$ is replaced by $\{\psi(d_{ij}^{(k)})/d_{ij}^{(k)}\}^2$. An alternative to Huber's ψ-function is a redescending ψ-function, for example, Hampel's (1973) piecewise linear function. However, there can be problems in forming the posterior probabilities of component membership, as there is the question as to which parametric family to use for the component densities (McLachlan and Basford, 1988; Section 2.8). One possibility is to use the form of the density corresponding to the ψ-function adopted. However, in the case of any redescending ψ-function with finite rejection points, there is no corresponding density. In Campbell (1984), the normal density was used, while in the related univariate work in De Veaux and Kreiger (1990), the t density with three degrees of freedom was used, with the location and scale component parameters estimated by the (weighted) median and mean absolute deviations, respectively.

It can be therefore seen that the use of mixtures of t distributions provides a sound statistical basis for formalizing and implementing the somewhat *ad hoc* approaches that have been proposed in the past. It also provides a framework for assessing the degree of robustness to be incorporated into the fitting of the mixture model through the specification or estimation of the degrees of freedom ν_i in the t component densities.

As noted in the introduction, the use of t components in place of the normal components will generally give less extreme estimates of the posterior probabilities of component membership of the mixture model. The use of the t distribution in place of the normal distribution leading to less extreme posterior probabilities of group membership was noted in a discriminant analysis context, where the group-conditional densities correspond to the component densities of the mixture model (Aitchison and Dunsmore, 1975, Chapter 2). If a Bayesian approach is adopted and the conventional improper or vague prior specified for the mean and the inverse of the covariance matrix in the normal distribution for each group-conditional density, it leads to the so-called predictive density estimate, which has the form of the t distribution; see McLachlan (1992, Section 3.5).

7.7 EXAMPLE 7.1: SIMULATED NOISY DATA SET

One way in which the presence of atypical observations or background noise in the data has been handled when fitting mixtures of normal components has been to include an additional component having a uniform distribution. The support of the latter component is generally specified by the upper and lower extremities of each dimension defining the rectangular region that contains all the data points. Typically, the mixing proportion for this uniform component is left unspecified to be estimated from the data, as in Example 3.11 on the minefield data set as analyzed by Dasgupta and Raftery (1998). Another example may be found in Schroeter et al. (1998), who fitted a mixture of three normal components and a uniform distribution to segment magnetic resonance images of the human brain into three regions (gray matter, white matter, and cerebrospinal fluid) in the presence of background noise arising from instrument irregularities and tissue abnormalities.

Here we consider a sample consisting initially of 100 simulated points from a three-component bivariate normal mixture model, to which 50 noise points were added from a uniform distribution over the range -10 to 10 on each variate. The parameters of the mixture model were

$$\boldsymbol{\mu}_1 = \begin{pmatrix} 0 & 3 \end{pmatrix}^T, \boldsymbol{\mu}_2 = \begin{pmatrix} 3 & 0 \end{pmatrix}^T, \boldsymbol{\mu}_3 = \begin{pmatrix} -3 & 0 \end{pmatrix}^T,$$

$$\boldsymbol{\Sigma}_1 = \begin{pmatrix} 2 & 0.5 \\ 0.5 & .5 \end{pmatrix}, \ \boldsymbol{\Sigma}_2 = \begin{pmatrix} 1 & 0 \\ 0 & .1 \end{pmatrix}, \ \boldsymbol{\Sigma}_3 = \begin{pmatrix} 2 & -0.5 \\ -0.5 & .5 \end{pmatrix},$$

with mixing proportions $\pi_1 = \pi_2 = \pi_3 = \frac{1}{3}$. The true grouping is shown in Figure 7.1. We now consider the clustering obtained by fitting a mixture of three t components with unequal scale matrices but equal degrees of freedom ($\nu_1 = \nu_2 = \nu_3 = \nu$). The values of the weights $u_{ij}^{(k)}$ at convergence, $\hat{u}_{ij}$, were examined. The noise points (points 101–150) generally produced much lower $\hat{u}_{ij}$ values. In this application, an observation $\boldsymbol{y}_j$ is treated as an outlier (background noise) if $\sum_{i=1}^{g} \hat{z}_{ij}\hat{u}_{ij}$ is sufficiently small; or, equivalently,

$$\sum_{i=1}^{g} \hat{z}_{ij}\delta(\boldsymbol{y}_j,\ \hat{\boldsymbol{\mu}}_i;\ \hat{\boldsymbol{\Sigma}}_i) \tag{7.37}$$

is sufficiently large, where

$$\hat{z}_{ij} = \arg\max_h \hat{\tau}_{hj} \quad (i = 1,\ \ldots,\ g;\ j = 1,\ \ldots,\ n),$$

and $\hat{\tau}_{ij}$ denotes the estimated posterior probability that $\boldsymbol{y}_j$ belongs to the ith component of the mixture.

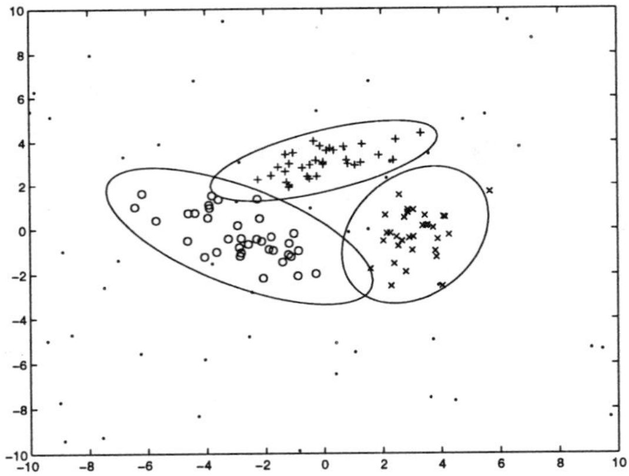

Fig. 7.1 Plot of the true grouping of the simulated noisy data set. From Peel and McLachlan (2000).

To decide on how large the statistic (7.37) must be in order for $\boldsymbol{y}_j$ to be classified as noise, we compared it to the 95th percentile of the chi-squared distribution with p degrees of freedom, where the latter is used to approximate the distribution of $\delta(\boldsymbol{Y}_j,\ \hat{\boldsymbol{\mu}}_i;\ \hat{\boldsymbol{\Sigma}}_i)$.

The clustering so obtained is displayed in Figure 7.2. It compares well with the true grouping in Figure 7.1 and the clustering in Figure 7.3 obtained by fitting a mixture of three normal components and an additional uniform component. In this particular example the model of three normal components with an additional uniform component to model the noise works well since it is the same model used to generate the data in the first instance. However, this model, unlike the t mixture model, cannot be expected to work as well in situations when the noise is not uniform or is unable to be modeled adequately by the uniform distribution.

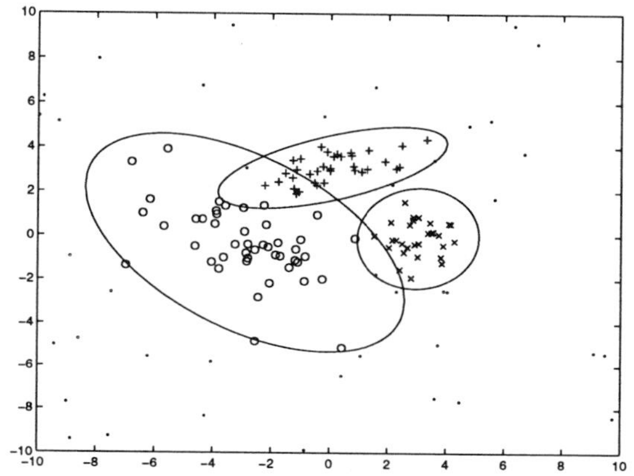

Fig. 7.2 Plot of the result of fitting a mixture of t distributions (with a classification of noise at a significance level of 5%) to the simulated noisy data. From Peel and McLachlan (2000).

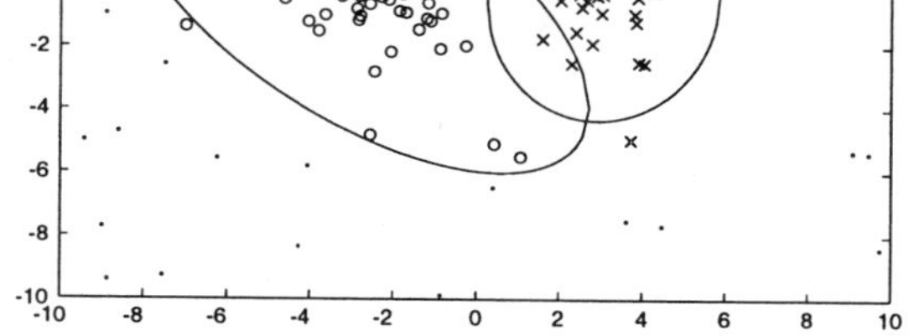

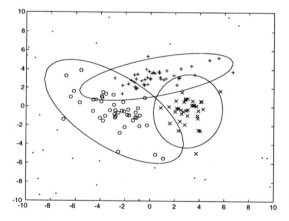

Fig. 7.3 Plot of the result of fitting a three-component normal mixture plus a uniform component model to the simulated noisy data. From Peel and McLachlan (2000).

It is of interest to note that if the number of groups is treated as unknown and a normal mixture is fitted, then the number of groups selected via AIC, BIC, and the bootstrap LRT is four for each of these criteria. The result of fitting a mixture of four normal components is displayed in Figure 7.4. Obviously, the additional fourth component is attempting to model the background noise. However, it can be seen from Figure 7.4 that this normal mixture-based clustering is still affected by the noise.

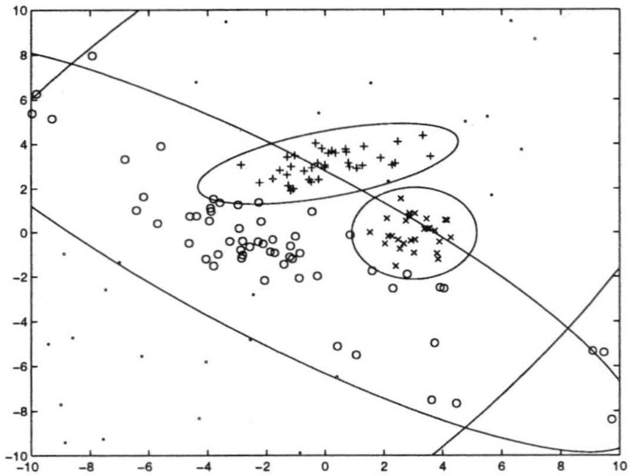

Fig. 7.4 Plot of the result of fitting a four-component normal mixture model to the simulated noisy data. From Peel and McLachlan (2000).

7.8 EXAMPLE 7.2: CRAB DATA SET

To further illustrate the t mixture model-based approach to clustering, we return to the crab data set of Campbell and Mahon (1974) on the genus *Leptograpsus*, as introduced in Section 3.7.1; see Figure 3.2. As explained there, it consists of $n = 100$ four-dimensional observations on $n_1 = 50$ males and $n_2 = 50$ females, which we shall refer to as groups G_1 and G_2, respectively. It was noted for these data that Hawkins' (1981) simultaneous test for multivariate normality and equal covariance matrices (homoscedasticity) suggests that it is reasonable to assume that the group-conditional distributions are normal with a common covariance matrix. Consistent with this, fitting a mixture of two t components (with equal scale matrices and equal degrees of freedom) gives only a slightly improved outright clustering over that obtained using a mixture of two normal homoscedastic components. The t mixture model-based clustering results in one cluster containing 32 observations from G_1 and another containing all 50 observations from G_2, along with the remaining 18 observations from G_1; the normal mixture model leads to one additional member of G_1 being assigned to the cluster corresponding to G_2. It was seen in Section 3.7.1 that although the groups are homoscedastic, a much improved clustering was obtained with a normal mixture model without restrictions on the component matrices. This same pattern of behavior is observed here with the t mixture model, with the latter giving 7 fewer misallocations when the scale matrices are unrestricted.

In this example, where the normal model for the components appears to be a reasonable assumption, the estimated degrees of freedom for the t components should be large, which they are. The estimate of their common value ν in the case of equal scale matrices and equal degrees of freedom ($\nu_1 = \nu_2 = \nu$) is $\hat{\nu} = 22.5$; the estimates of ν_1 and ν_2 in the case of unequal scale matrices and unequal degrees of freedom are $\hat{\nu}_1 = 23.0$ and $\hat{\nu}_2 = 120.3$.

The likelihood function can be fairly flat near the MLEs of the degrees of freedom of the t components. To illustrate this, we have plotted in Figure 7.5 the profile likelihood function in the case of equal scale matrices and equal degrees of freedom ($\nu_1 = \nu_2 = \nu$), while in Figure 7.6 we have plotted the profile likelihood function in the case of unequal scale matrices and unequal degrees of freedom.

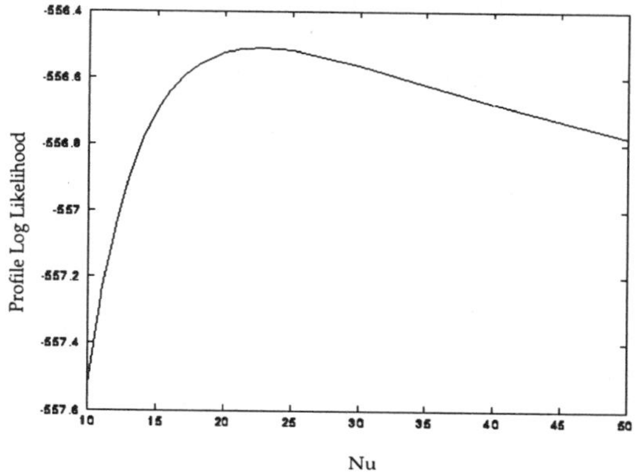

Fig. 7.5 Plot of the profile log likelihood for various values of ν for the crab data set with equal scale matrices and equal degrees of freedom ($\nu_1 = \nu_2 = \nu$). From Peel and McLachlan (2000).

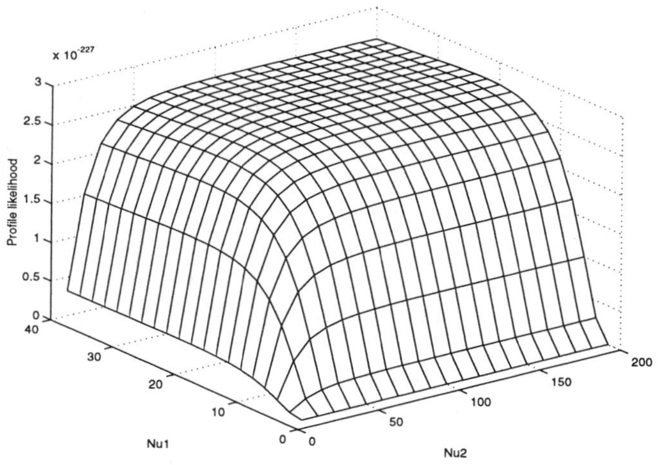

Fig. 7.6 Plot of the profile likelihood for ν_1 and ν_2 for the crab data set with unequal scale matrices and unequal degrees of freedom ν_1 and ν_2. From Peel and McLachlan (2000).

Finally, we now compare the t and normal mixture-based clusterings after some outliers are introduced into the original data set. This was done by adding various values to the second variate of the 25th point. In Table 7.1 we report the overall misallocation rate of the normal and t mixture-based clusterings for each perturbed version of the original data set. It can be seen that the t mixture-based clustering is robust to these perturbations, unlike the normal mixture-based clustering. In the cases where the constant equals -15, -10 and 20, it should be noted that fitting a normal mixture model results in an outright classification of the outlier into one cluster. The remaining points are allocated to the second cluster, giving an error rate of 49. In practice the user should identify this situation when interpreting the results and hence remove the outlier, giving an error rate of 20%. However, when the fitting is part of an automatic procedure this would not be possible.

Table 7.1 Summary of Comparison of Error Rates When Fitting Normal and t Distributions with the Modification of a Single Point

Constant	Normal Error Rate	t Component Error Rate	$\hat{\nu}$	$\hat{u}_{1,25}$	$\hat{u}_{2,25}$
-15	49	19	5.76	0.0154	0.0118
-10	49	19	6.65	0.0395	0.0265
-5	21	20	13.11	0.1721	0.3640
0	19	18	23.05	0.8298	1.1394
5	21	20	13.11	0.1721	0.3640
10	50	20	7.04	0.0734	0.0512
15	47	20	5.95	0.0138	0.0183
20	49	20	5.45	0.0092	0.0074

7.9 EXAMPLE 7.3: OLD FAITHFUL GEYSER DATA SET

As another demonstration of the use of the t mixture approach to the handling of outliers, we now consider the Old Faithful geyser data set consisting of 272 univariate measurements of eruption lengths of the Old Faithful geyser. The data set is originally from Silverman (1986). However, in this example we consider a bivariate version analyzed in García-Escudero and Gordaliza (1999), where the data have been lagged.

Figure 7.7 shows the clustering obtained by fitting a mixture of three t components with unequal scale matrices, but equal degrees of freedom ($\nu_1 = \nu_2 = \nu_3 = \nu$). The values of the weights $u_{ij}^{(k)}$ at convergence, $\hat{u}_{ij}$, were examined in a manner similar to that in Section 7.7 to classify points as outliers. The outliers determined at the 1% level are indicated by dots.

Figure 7.8 gives the corresponding results when the restriction of equal component-scale matrices is applied. These last results are in general agreement with those obtained by García-Escudero and Gordaliza (1999), who provided a robust clustering of these data using a trimmed k-means procedure. The latter assumes also that the equal component-covariance matrices are spherical. However, this would appear to be a reasonable assumption for this data set. The solutions differ on the conclusion of whether two of the observations (labeled A and B in Figure 7.8) are atypical; the latter are considered atypical on the basis of the t mixture model.

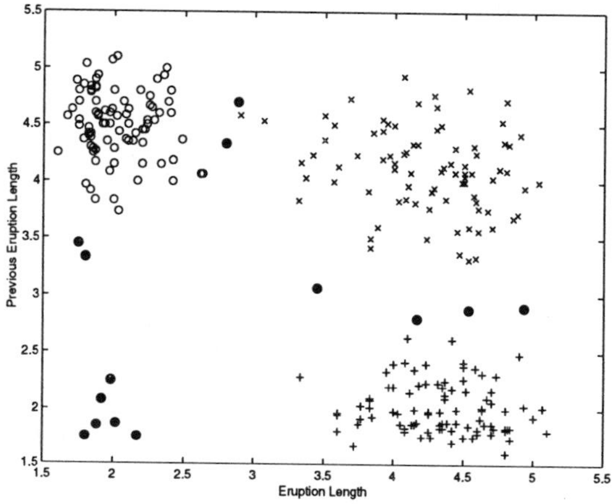

Fig. 7.7 Result of fitting a three-component t mixture model with equal ν_i and unequal scale matrices to the Geyser data set; dots denote those observations assessed as atypical at the 1% level.

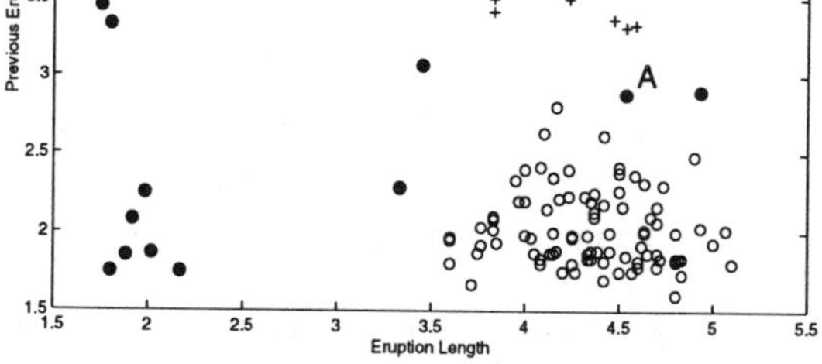

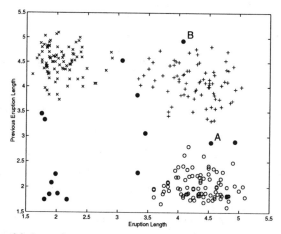

Fig. 7.8 Result of fitting a three component t mixture model with equal ν_i and equal scale matrices to the Geyser data set; dots denote those observations assessed as atypical at the 1% level.

8

Mixtures of Factor Analyzers

8.1 INTRODUCTION

Factor analysis is commonly used for explaining data, in particular, correlations between variables in multivariate observations. It can be used also for dimensionality reduction, although the method of principal component analysis is more widely used in this role. However, the effectiveness of these two statistical techniques is limited by their global linearity. The applicability of these two methods can be widened by combining local models of them in the form of a finite mixture, along the lines proposed by Ghahramani and Hinton (1997) and Tipping and Bishop (1997).

In this chapter we shall focus on mixtures of factor analyzers from the perspective of both a method for model-based density estimation for high-dimensional data (and hence for the clustering of such data) and a method for local dimensionality reduction. We shall also discuss the close link of mixtures of factor analyzers with mixtures of probabilistic principal component analyzers. The mixtures of factor analyzers model enables a normal mixture model to be fitted to high-dimensional data. The number of free parameters is controlled through the dimension q of the latent factor space. It allows an interpolation in model complexities from isotropic to full covariance structures without any restrictions.

Before we proceed to introduce mixtures of factor analyzers, we briefly look at the use of principal components for revealing group structure and dimension reduction.

8.2 PRINCIPAL COMPONENT ANALYSIS

In exploring high-dimensional data sets for group structure, it is typical to rely on "second-order" multivariate techniques, in particular, principal component analysis (PCA); see Huber (1985), Friedman (1987), and Jones and Sibson (1987) and the subsequent discussions for an excellent account of available exploratory multivariate techniques. Here we briefly discuss a PCA on the sample covariance matrix

$$V = \sum_{j=1}^{n} (\boldsymbol{y}_j - \overline{\boldsymbol{y}})(\boldsymbol{y}_j - \overline{\boldsymbol{y}})^T / n. \tag{8.1}$$

We let $\boldsymbol{a}_1, \ldots, \boldsymbol{a}_p$ be the (unit) eigenvectors, corresponding to the eigenvalues $\lambda_1 \geq \lambda_2 \geq \cdots \geq \lambda_p$ of $\boldsymbol{V}$. In the case where the variates are measured on disparate scales, we may wish to replace $\boldsymbol{V}$ by the sample correlation matrix.

If there are only a few groups and they are well-separated, and the between-group variation dominates the within-group variation, then projections of the feature data $\boldsymbol{y}_j$ onto the first few principal axes should portray the group structure. However, a PCA of $\boldsymbol{V}$ may not always be useful. This point was stressed by Chang (1983), who showed in the case of two groups that the principal component of the feature vector that provides the best separation between the two groups in terms of Mahalanobis distance is not necessarily the first component $\boldsymbol{a}_1^T \boldsymbol{y}_j$; see McLachlan (1992, Section 6.6).

To illustrate this, we generated a sample of size $n = 100$ five-dimensional observations from two groups G_1 and G_2 with means

$$\boldsymbol{\mu}_1 = (5, 0, 0, 0, 0)^T \quad \text{and} \quad \boldsymbol{\mu}_2 = \boldsymbol{0}$$

and covariance matrices

$$\boldsymbol{\Sigma}_1 = \text{diag}(1, 10, 10, 10, 10)$$

and

$$\boldsymbol{\Sigma}_2 = \begin{pmatrix} 1 & & & & \\ 3 & 10 & & & \\ 0 & 0 & 10 & & \\ 0 & 0 & 0 & 10 & \\ 0 & 0 & 0 & 0 & 10 \end{pmatrix}.$$

We then obtained the feature data, $\boldsymbol{y}_1, \ldots, \boldsymbol{y}_n$, by premultiplying each of these generated 100 observations by an orthogonal matrix,

$$\boldsymbol{H} = \begin{pmatrix} \cos(\pi/4) & -\sin(\pi/4) & 0 & 0 & 0 \\ \sin(\pi/4) & \cos(\pi/4) & 0 & 0 & 0 \\ 0 & 0 & 1 & 0 & 0 \\ 0 & 0 & 0 & \cos(\pi/4) & -\sin(\pi/4) \\ 0 & 0 & 0 & \sin(\pi/4) & \cos(\pi/4) \end{pmatrix}.$$

These two groups are well separated. However, the distribution of the feature vector $\boldsymbol{Y}_j$ has been designed so that the differences between its group means are limited to one direction in the feature space with the within-group variation in the other orthogonal directions being relatively very large. As a consequence, a PCA fails to locate the optimal viewing direction for group structure. It can be seen from Figure 8.1 that the plot of the first two principal components (based on the correlation matrix) provides no evidence of the presence of two groups in the data. Indeed, if we were to proceed to fit a mixture of $g = 2$ normal components to just these first two principal components, we would obtain a clustering that would misallocate 44% of the data points. In contrast, if we were to fit a mixture of $g = 2$ factor analyzers with $q = 2$ factors, as to be considered shortly in Section 8.11, we would obtain a clustering that misallocates only one point.

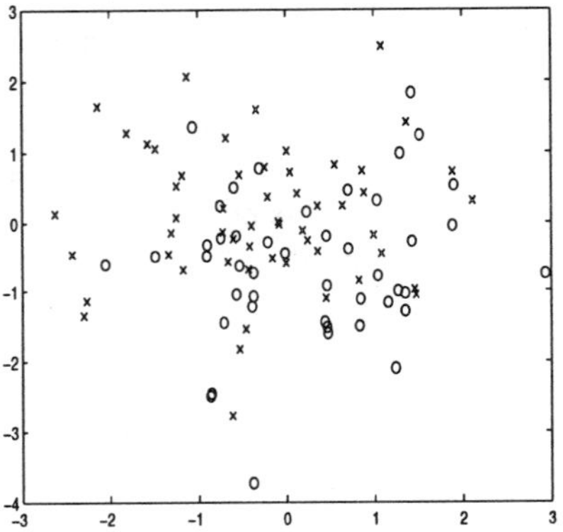

Fig. 8.1 Plot of the first two principal components with true group of origin shown (o and x denote membership of groups G_1 and G_2, respectively).

8.3 SINGLE-FACTOR ANALYSIS MODEL

We let $\boldsymbol{Y}_1, \ldots, \boldsymbol{Y}_n$ denote a random sample of size n on a p-dimensional random vector. In a typical factor analysis model, each observation $\boldsymbol{Y}_j$ is modeled as

$$\boldsymbol{Y}_j = \boldsymbol{\mu} + \boldsymbol{B}\boldsymbol{U}_j + \boldsymbol{e}_j \quad (j = 1, \ldots, n), \tag{8.2}$$

where $\boldsymbol{U}_j$ is a q-dimensional ($q < p$) vector of latent or unobservable variables called factors and $\boldsymbol{B}$ is a $p \times q$ matrix of factor loadings (parameters). It is assumed that

$$(\boldsymbol{Y}_1^T, \boldsymbol{U}_1^T)^T, \ldots, (\boldsymbol{Y}_n^T, \boldsymbol{U}_n^T)^T$$

are i.i.d. The $\boldsymbol{U}_j$ are assumed to be i.i.d. as $N(\boldsymbol{0}, \boldsymbol{I}_q)$, independently of the errors $\boldsymbol{e}_j$, which are assumed to be i.i.d. as $N(\boldsymbol{0}, \boldsymbol{D})$, where $\boldsymbol{D}$ is a diagonal matrix,

$$\boldsymbol{D} = \text{diag}\,(\sigma_1^2, \ldots, \sigma_p^2),$$

and where I_q denotes the $q \times q$ identity matrix. The σ_i^2 are called the uniquenesses. Thus, conditional on the u_j, the Y_j are independently distributed as $N(\mu + B u_j, D)$. Unconditionally, the Y_j are i.i.d. according to a normal distribution with mean μ and covariance matrix

$$\Sigma = BB^T + D. \tag{8.3}$$

Under the model (8.2), the variables in Y_j are conditionally independent given u_j. Thus the factors in u_j are intended to explain the correlations between the variables in Y_j, while the error terms e_j represent the unexplained noise unique to a particular y_j $(j = 1, \ldots, n)$. Note that in the case of $q > 1$, there is an infinity of choices for B, since this model is still satisfied if we replace u_j by Hu_j and B by BH^T, where H is any orthogonal matrix of order q. As $\frac{1}{2}q(q-1)$ constraints are needed for B to be uniquely defined, the number of free parameters is

$$pq + p - \tfrac{1}{2}q(q-1); \tag{8.4}$$

see Lawley and Maxwell (1971, Chapter 1). If q is chosen sufficiently smaller than p so that the difference

$$\begin{aligned} C &= \tfrac{1}{2}p(p+1) - pq - p + \tfrac{1}{2}q(q-1) \\ &= \tfrac{1}{2}\{(p-q)^2 - (p+q)\} \end{aligned} \tag{8.5}$$

is positive, then the representation (8.3) imposes some constraints on the covariance matrix Σ and thus reduces the number of free parameters to be estimated.

8.4 EM ALGORITHM FOR A SINGLE-FACTOR ANALYZER

The factor analysis model (8.2) can be fitted by maximum likelihood, although the solution has to be computed iteratively as no closed-form expressions exist for the MLEs of B and D. They can be computed iteratively via the EM algorithm as considered in Dempster et al. (1977); see Rubin and Thayer (1982) and McLachlan and Krishnan (1997, Chapter 5) for further details. The MLE of the mean μ is obviously the sample mean $\overline{y}$ of the n observed values $y_1, \ldots, y_n$ corresponding to the random sample $Y_1, \ldots, Y_n$. Hence in the sequel, μ can be set equal to $\overline{y}$ without loss of generality. Accordingly, we let the parameter vector Ψ of unknown parameters consist of the elements of B and the diagonal elements of D. The (incomplete-data) log likelihood for Ψ that can be formed from the observed data $y = (y_1^T, \ldots, y_n^T)^T$ is, apart from an additive constant,

$$\log L(\Psi) = -\tfrac{1}{2}n\{\log \mid BB^T + D \mid + \sum_{j=1}^{m}(y_j - \overline{y})^T(BB^T + D)^{-1}(y_j - \overline{y})\}. \tag{8.6}$$

In order to apply the EM algorithm and its variants to this problem, we follow Dempster et al. (1977) and formulate

$$y_c = (y^T, u_1^T, \ldots, u_n^T)^T$$

as the complete-data vector, where where $\boldsymbol{u}_j$ corresponds to $\boldsymbol{U}_j$. The complete-data log likelihood is, but for an additive constant,

$$\log L_c(\boldsymbol{\Psi}) = -\tfrac{1}{2} n \log \mid \boldsymbol{D} \mid - \tfrac{1}{2} \sum_{j=1}^{n} \{(\boldsymbol{y}_j - \overline{\boldsymbol{y}} - \boldsymbol{B}\boldsymbol{u}_j)^T \boldsymbol{D}^{-1} (\boldsymbol{y}_j - \overline{\boldsymbol{y}} - \boldsymbol{B}\boldsymbol{u}_j) + \boldsymbol{u}_j^T \boldsymbol{u}_j\}.$$

The complete-data density belongs to the exponential family, and the complete-data sufficient statistics are $\boldsymbol{C}_{yy}$, $\boldsymbol{C}_{yu}$, and $\boldsymbol{C}_{uu}$, where

$$\boldsymbol{C}_{yy} = \sum_{j=1}^{n} (\boldsymbol{y}_j - \overline{\boldsymbol{y}})(\boldsymbol{y}_j - \overline{\boldsymbol{y}})^T; \; \boldsymbol{C}_{yu} = \sum_{j=1}^{n} (\boldsymbol{y}_j - \overline{\boldsymbol{y}})\boldsymbol{u}_j^T; \; \boldsymbol{C}_{uu} = \sum_{j=1}^{n} \boldsymbol{u}_j \boldsymbol{u}_j^T.$$

To calculate the conditional expectations of these sufficient statistics given the observed data, we need to use the result that the random vector $(\boldsymbol{Y}_j^T, \boldsymbol{U}_j^T)^T$ has a multivariate normal distribution with mean

$$\begin{pmatrix} \boldsymbol{\mu} \\ \boldsymbol{0} \end{pmatrix} \tag{8.7}$$

and covariance matrix

$$\begin{pmatrix} \boldsymbol{B}\boldsymbol{B}^T + \boldsymbol{D} & \boldsymbol{B} \\ \boldsymbol{B}^T & \boldsymbol{I}_q \end{pmatrix}. \tag{8.8}$$

It thus follows that the conditional distribution of $\boldsymbol{U}_j$ given $\boldsymbol{y}_j$ is given by

$$\boldsymbol{U}_j \mid \boldsymbol{y}_j \sim N(\boldsymbol{\gamma}^T(\boldsymbol{y}_j - \boldsymbol{\mu}), \; \boldsymbol{I}_q - \boldsymbol{\gamma}^T \boldsymbol{B}) \tag{8.9}$$

for $j = 1, \ldots, n$, where

$$\boldsymbol{\gamma} = (\boldsymbol{B}\boldsymbol{B}^T + \boldsymbol{D})^{-1} \boldsymbol{B}. \tag{8.10}$$

The EM algorithm is implemented as follows on the $(k+1)$th iteration.

E-Step. Given the current fit $\boldsymbol{\Psi}^{(k)}$ for $\boldsymbol{\Psi}$, calculate as follows the conditional expectation of these sufficient statistics given the observed data $\boldsymbol{y}$:

$$E_{\boldsymbol{\Psi}^{(k)}}(\boldsymbol{C}_{yy} \mid \boldsymbol{y}) = \boldsymbol{C}_{yy},$$

$$E_{\boldsymbol{\Psi}^{(k)}}(\boldsymbol{C}_{yu} \mid \boldsymbol{y}) = \boldsymbol{C}_{yy} \boldsymbol{\gamma}^{(k)},$$

and

$$E_{\boldsymbol{\Psi}^{(k)}}(\boldsymbol{C}_{uu} \mid \boldsymbol{y}) = \boldsymbol{\gamma}^{(k)^T} \boldsymbol{C}_{yy} \boldsymbol{\gamma}^{(k)} + n\boldsymbol{\omega}^{(k)},$$

where

$$\boldsymbol{\gamma}^{(k)} = \{\boldsymbol{B}^{(k)} \boldsymbol{B}^{(k)^T} + \boldsymbol{D}^{(k)}\}^{-1} \boldsymbol{B}^{(k)}$$

and

$$\boldsymbol{\omega}^{(k)} = \boldsymbol{I}_q - \boldsymbol{\gamma}^{(k)^T} \boldsymbol{B}^{(k)}.$$

M-Step. Calculate

$$B^{(k+1)} = C_{yy}\gamma^{(k)}(\gamma^{(k)^T} C_{yy}\gamma^{(k)} + n\omega^{(k)})^{-1}$$

and

$$\begin{aligned} D^{(k+1)} & \\ &= n^{-1}\,\text{diag}\{C_{yy} - C_{yy}\gamma^{(k)}(\gamma^{(k)^T} C_{yy}\gamma^{(k)} + n\omega^{(k)})^{-1}\gamma^{(k)^T} C_{yy}\} \\ &= n^{-1}\,\text{diag}\{C_{yy} - C_{yy}\gamma^{(k)} B^{(k+1)^T}\}. \end{aligned}$$

The inversion of the current value of the $p \times p$ matrix $(BB^T + D)$ on each iteration can be undertaken using the result that

$$(BB^T + D)^{-1} = D^{-1} - D^{-1}B(I_q + B^T D^{-1}B)^{-1}B^T D^{-1}, \qquad (8.11)$$

where the right-hand side of (8.11) involves only the inverses of $q \times q$ matrices, since D is a diagonal matrix. The determinant of $(BB^T + D)$ can then be calculated as

$$\mid BB^T + D \mid = \mid D \mid / \mid I_q - B^T(BB^T + D)^{-1}B \mid .$$

Liu and Rubin (1994, 1998) have considered the application of the ECME algorithm to this problem. The M-step is replaced by two CM-steps. On the first CM-step, $B^{(k+1)}$ is calculated as on the M-step above, while on the second CM-step the diagonal matrix $D^{(k+1)}$ is obtained by using an algorithm such as Newton-Raphson to maximize the actual log likelihood with B fixed at $B^{(k+1)}$.

Dunmur and Titterington (1998b) have considered latent profile models, which include the standard factor analysis model above but with each latent vector U_j being categorical rather than normally distributed. They noted that this categorical-factor analysis model has links with factorial models as proposed in Hinton and Zeal (1994) and Ghahramani (1995).

8.5 DATA VISUALIZATION IN LATENT SPACE

An original data point y_j can be represented in q-dimensions by the posterior distribution of its associated q-dimensional latent factor U_j. A convenient summary of this distribution is its mean. Hence we can portray the y_j in q-dimensional space by plotting the estimated conditional expectation of each U_j given y_j, that is, the (estimated) posterior mean of the factor U_j $(j = 1, \ldots, n)$. We have that

$$\begin{aligned} \hat{u}_j &= E_{\hat{\Psi}}\{U_j \mid y_j\} \\ &= \hat{\gamma}^T(y_j - \overline{y}), \end{aligned} \qquad (8.12)$$

where $E_{\hat{\Psi}}$ denotes expectation using the estimate $\hat{\Psi}$ in place of Ψ.

8.6 MIXTURES OF FACTOR ANALYZERS

As the single-factor analysis model (8.2) provides only a global linear model for the representation of the data in a lower-dimensional subspace, the scope of its application is limited. A global nonlinear approach can be obtained by postulating a finite mixture of linear submodels for the distribution of the full observation vector $\boldsymbol{Y}_j$ given the (unobservable) factors $\boldsymbol{u}_j$. That is, we can provide a local dimensionality reduction method by assuming that the distribution of the observation $\boldsymbol{Y}_j$ can be modeled as

$$\boldsymbol{Y}_j = \boldsymbol{\mu}_i + \boldsymbol{B}_i \boldsymbol{U}_{ij} + \boldsymbol{e}_{ij} \qquad \text{with prob. } \pi_i \quad (i = 1, \ldots, g) \tag{8.13}$$

for $j = 1, \ldots, n$, where the factors $\boldsymbol{U}_{i1}, \ldots, \boldsymbol{U}_{in}$ are distributed independently $N(\boldsymbol{0}, \boldsymbol{I}_q)$, independently of the $\boldsymbol{e}_{ij}$, which are distributed independently $N(\boldsymbol{0}, \boldsymbol{D}_i)$, where $\boldsymbol{D}_i$ is a diagonal matrix $(i = 1, \ldots, g)$. The so-called mixing proportions π_i are nonnegative and sum to one.

Thus, unconditionally, the density of each observation $\boldsymbol{Y}_j$ is a mixture of g normal densities in proportions $\pi_1, \ldots, \pi_g$; that is,

$$f(\boldsymbol{y}_j; \boldsymbol{\Psi}) = \sum_{i=1}^{g} \pi_i \, \phi(\boldsymbol{y}_j; \boldsymbol{\mu}_i, \boldsymbol{\Sigma}_i), \tag{8.14}$$

where

$$\boldsymbol{\Sigma}_i = \boldsymbol{B}_i \boldsymbol{B}_i^T + \boldsymbol{D}_i \quad (i = 1, \ldots, g). \tag{8.15}$$

The parameter vector $\boldsymbol{\Psi}$ now consists of the elements of the $\boldsymbol{\mu}_i$, the $\boldsymbol{B}_i$, and the $\boldsymbol{D}_i$, along with the mixing proportions π_i $(i = 1, \ldots, g-1)$, on putting $\pi_g = 1 - \sum_{i=1}^{g-1} \pi_i$.

In Ghahramani and Hinton (1997), $\boldsymbol{D}_i$ is taken to be the same for each component i; see also Yung (1997). More recently, Ghahramani and Beal (2000) have considered a Bayesian approach, using a deterministic variational approximation to full Bayesian integration over model parameters. In work on related models, Jedidi, Jagpai, and DeSarbo (1997) have considered the fitting of a general finite mixture structural equation. Shi and Lee (2000) have considered a single-component model with mixed continuous and polytomous latent variables. Attias (1999b, 2000) has recently proposed a technique that he calls independent factor analysis, which further extends ordinary factor analysis.

The mixtures of factor analyzers model (8.14) is also useful in the modeling of high-dimensional data by mixtures of normal components, particularly in situations where there are limited data. With the fitting of a mixture of normal components with unrestricted covariance matrices $\boldsymbol{\Sigma}_i$, there are $\frac{1}{2}p(p+1)$ parameters for each $\boldsymbol{\Sigma}_i$ $(i = 1, \ldots, g)$. This means that as the number of components g in the mixture model grows, the total number of parameters can quickly become very large relative to the sample size n, leading to overfitting. The mixture of factor analyzers model provides a way of controlling the number of parameters through the reduced model (8.15) for the component-covariance matrices. It thus provides a model intermediate between the independent and unrestricted models. The adequacy of the fit of a mixture of factor analyzers with q factors can be tested using the likelihood ratio statistic.

8.7 AECM ALGORITHM FOR FITTING MIXTURES OF FACTOR ANALYZERS

8.7.1 AECM Framework

The log likelihood for $\boldsymbol{\Psi}$ that can be formed from the observed data $\boldsymbol{y}$ under model (8.14) is

$$\log L(\boldsymbol{\Psi}) = \sum_{j=1}^{n} \log\{\sum_{i=1}^{g} \pi_i \phi(\boldsymbol{y}_j;\, \boldsymbol{\mu}_i,\, \boldsymbol{\Sigma}_i)\}. \tag{8.16}$$

We can use the alternating expectation–conditional maximization (AECM) algorithm as developed by Meng and van Dyk (1997) to fit the mixture of factor analyzers model (8.14) by maximum likelihood. The expectation–conditional maximization (ECM) algorithm proposed by Meng and Rubin (1993) replaces the M-step of the EM algorithm by a number of computationally simpler conditional maximization (CM) steps. The AECM algorithm is an extension of the ECM algorithm, where the specification of the complete data is allowed to be different on each CM-step.

To apply the AECM algorithm to the fitting of the mixture model (8.14), we partition the vector of unknown parameters $\boldsymbol{\Psi}$ as $(\boldsymbol{\Psi}_1^T,\, \boldsymbol{\Psi}_2^T)^T$, where $\boldsymbol{\Psi}_1$ contains the mixing proportions π_i $(i = 1, \ldots, g-1)$ and the elements of the component means $\boldsymbol{\mu}_i$ $(i = 1, \ldots, g)$. The subvector $\boldsymbol{\Psi}_2$ contains the elements of the $\boldsymbol{B}_i$ and the $\boldsymbol{D}_i$ $(i = 1, \ldots, g)$. Concerning the specification of the incomplete data, we have noted in Chapter 2 that for the fitting of mixture models in general, it is useful to conceptualize each observation $\boldsymbol{y}_j$ as having arisen from one of the components of the mixture and then to declare the component-indicator vector $\boldsymbol{z}_j$ so associated with $\boldsymbol{y}_j$ as missing data. In this framework, $z_{ij} = (\boldsymbol{z}_j)_i$ is one or zero, according to whether $\boldsymbol{y}_j$ arose or did not arise from the ith component $(i = 1, \ldots, g;\, j = 1, \ldots, n)$. The conditional expectation of Z_{ij} given $\boldsymbol{y}_j$ is the posterior probability that the jth observation comes from the ith component, given by

$$\begin{aligned} \tau_i(\boldsymbol{y}_j;\, \boldsymbol{\Psi}) &= \operatorname{pr}\{Z_{ij} = 1 \mid \boldsymbol{y}_j\} \\ &= \pi_i\, \phi(\boldsymbol{y}_j;\, \boldsymbol{\mu}_i,\, \boldsymbol{\Sigma}_i) / \sum_{h=1}^{g} \pi_h \phi(\boldsymbol{y}_j;\, \boldsymbol{\mu}_h,\, \boldsymbol{\Sigma}_h), \end{aligned} \tag{8.17}$$

where $\boldsymbol{\Sigma}_i$ has the form (8.15) $(i = 1, \ldots, g;\, j = 1, \ldots, n)$.

We let $\boldsymbol{\Psi}^{(k)} = (\boldsymbol{\Psi}_1^{(k)^T},\, \boldsymbol{\Psi}_2^{(k)^T})^T$ be the value of $\boldsymbol{\Psi}$ after the kth iteration of the AECM algorithm. For this application of the AECM algorithm, one iteration consists of two cycles, and there is one E-step and one CM-step for each cycle. The two CM-steps correspond to the partition of $\boldsymbol{\Psi}$ into the two subvectors $\boldsymbol{\Psi}_1$ and $\boldsymbol{\Psi}_2$.

8.7.2 First Cycle

For the first cycle of the AECM algorithm, we specify the missing data to be just the component-indicator vectors, $\boldsymbol{z}_1, \ldots, \boldsymbol{z}_n$. The complete-data log likelihood is then given by

$$\log L_c(\boldsymbol{\Psi}) = \sum_{i=1}^{g}\sum_{j=1}^{n} z_{ij} \log\{\pi_i \phi(\boldsymbol{y}_j;\ \boldsymbol{\mu}_i,\ \boldsymbol{\Sigma}_i)\}. \tag{8.18}$$

Hence the E-step on the first cycle on the $(k+1)$th iteration requires the calculation of $Q_1(\boldsymbol{\Psi};\ \boldsymbol{\Psi}^{(k)})$, where

$$Q_1(\boldsymbol{\Psi};\ \boldsymbol{\Psi}^{(k)}) = E_{\boldsymbol{\Psi}^{(k)}}\{\log L_c(\boldsymbol{\Psi}) \mid \boldsymbol{y}\}$$

is the conditional expectation of the complete-data log likelihood (8.18) given $\boldsymbol{y}$, using $\boldsymbol{\Psi}^{(k)}$ for $\boldsymbol{\Psi}$. This E-step is achieved simply by replacing each z_{ij} in (8.18) by its current conditional expectation given the observed data (effectively $\boldsymbol{y}_j$); that is, we replace z_{ij} by $\tau_i(\boldsymbol{y}_j;\ \boldsymbol{\Psi}^{(k)})$. The first CM-step is implemented by maximizing $Q_1(\boldsymbol{\Psi};\ \boldsymbol{\Psi}^{(k)})$ over $\boldsymbol{\Psi}$ with $\boldsymbol{\Psi}_2$ held fixed at $\boldsymbol{\Psi}_2^{(k)}$. The updated estimate $\boldsymbol{\Psi}_1^{(k+1)}$ of $\boldsymbol{\Psi}_1$ so obtained contains the new estimates of the π_i and $\boldsymbol{\mu}_i$ given by

$$\pi_i^{(k+1)} = \sum_{j=1}^{n} \tau_i(\boldsymbol{y}_j;\ \boldsymbol{\Psi}^{(k)})/n \tag{8.19}$$

and

$$\boldsymbol{\mu}_i^{(k+1)} = \sum_{j=1}^{n} \tau_i(\boldsymbol{y}_j;\ \boldsymbol{\Psi}^{(k)})\, \boldsymbol{y}_j / \sum_{j=1}^{n} \tau_i(\boldsymbol{y}_j;\ \boldsymbol{\Psi}^{(k)}) \tag{8.20}$$

for $i = 1,\ \ldots,\ g$. We now set $\boldsymbol{\Psi}^{(k+1/2)}$ equal to $(\boldsymbol{\Psi}_1^{(k+1)^T},\ \boldsymbol{\Psi}_2^{(k)^T})^T$.

8.7.3 Second Cycle

For the second cycle for the updating of $\boldsymbol{\Psi}_2$, which contains the elements of the $\boldsymbol{B}_i$ and the $\boldsymbol{D}_i$, we specify the missing data to be the factors $\boldsymbol{u}_1,\ \ldots\ \boldsymbol{u}_n$, as well as the component-indicator vectors, $\boldsymbol{z}_1,\ \ldots,\ \boldsymbol{z}_n$. The complete-data log likelihood is then given by

$$\log L_c(\boldsymbol{\Psi}) = \sum_{i=1}^{g}\sum_{j=1}^{n} z_{ij}\{\pi_i\, \phi(\boldsymbol{y}_j;\ \boldsymbol{\mu}_i + \boldsymbol{B}_i \boldsymbol{u}_{ij},\ \boldsymbol{\Sigma}_i)\}. \tag{8.21}$$

The E-step on the second cycle on the $(k+1)$th iteration therefore requires the calculation of $Q_2(\boldsymbol{\Psi};\ \boldsymbol{\Psi}^{(k+1/2)})$, which denotes the conditional expectation of (8.18) given the observed data $\boldsymbol{y}$, using $\boldsymbol{\Psi}^{(k+1/2)}$ for $\boldsymbol{\Psi}$. In addition to updating the posterior probabilities of component membership to $\tau_i(\boldsymbol{y}_j;\ \boldsymbol{\Psi}^{(k+1/2)})$, it requires calculating the conditional expectation of

$$E_{\boldsymbol{\Psi}^{(k+1/2)}}\{Z_{ij}(\boldsymbol{U}_{ij} - \boldsymbol{\mu}_i) \mid \boldsymbol{y}_j\} \tag{8.22}$$

and

$$E_{\boldsymbol{\Psi}^{(k+1/2)}}\{Z_{ij}(\boldsymbol{U}_{ij} - \boldsymbol{\mu}_i)(\boldsymbol{U}_{ij} - \boldsymbol{\mu}_i)^T \mid \boldsymbol{y}_j\}, \tag{8.23}$$

which are given by

$$\tau_i(\boldsymbol{y}_j;\, \boldsymbol{\Psi}^{(k+1/2)})\, \boldsymbol{\gamma}_i^{(k)^T}(\boldsymbol{y}_j - \boldsymbol{\mu}_i) \tag{8.24}$$

and

$$\tau_i(\boldsymbol{y}_j;\, \boldsymbol{\Psi}^{(k+1/2)})\{\boldsymbol{\gamma}_i^{(k)^T}(\boldsymbol{y}_j - \boldsymbol{\mu}_i)(\boldsymbol{y}_j - \boldsymbol{\mu}_i)^T\boldsymbol{\gamma}_i^{(k)} + \boldsymbol{\omega}_i^{(k)}\}, \tag{8.25}$$

respectively, where

$$\boldsymbol{\gamma}_i^{(k)} = (\boldsymbol{B}_i^{(k)}\boldsymbol{B}_i^{(k)^T} + \boldsymbol{D}_i^{(k)})^{-1}\boldsymbol{B}_i^{(k)} \tag{8.26}$$

and

$$\boldsymbol{\omega}_i^{(k)} = \boldsymbol{I}_q - \boldsymbol{\gamma}_i^{(k)^T}\boldsymbol{B}_i^{(k)} \tag{8.27}$$

for $i = 1, \ldots, g$.

The CM-step on this second cycle is implemented by the maximization of $Q_2(\boldsymbol{\Psi};\, \boldsymbol{\Psi}^{(k+1/2)})$ over $\boldsymbol{\Psi}$ with $\boldsymbol{\Psi}_1$ set equal to $\boldsymbol{\Psi}_1^{(k+1)}$. This yields the updated estimates $\boldsymbol{B}_i^{(k+1)}$ and $\boldsymbol{D}_i^{(k+1)}$ for $\boldsymbol{B}_i$ and $\boldsymbol{D}_i$, given by

$$\boldsymbol{B}_i^{(k+1)} = \boldsymbol{V}_i^{(k+1/2)}\boldsymbol{\gamma}_i^{(k)}(\boldsymbol{\gamma}_i^{(k)^T}\boldsymbol{V}_i^{(k+1/2)}\boldsymbol{\gamma}_i^{(k)} + \boldsymbol{\omega}_i^{(k+1/2)})^{-1} \tag{8.28}$$

and

$$\boldsymbol{D}_i^{(k+1)} = \operatorname{diag}\{\boldsymbol{V}_i^{(k+1/2)} - \boldsymbol{V}_i^{(k+1/2)}\boldsymbol{\gamma}_i^{(k)}\boldsymbol{B}_i^{(k+1)^T}\}, \tag{8.29}$$

where

$$\boldsymbol{V}_i^{(k+1/2)} = \frac{\sum_{j=1}^n \tau_i(\boldsymbol{y}_j;\, \boldsymbol{\Psi}^{(k+1/2)})\,(\boldsymbol{y}_j - \boldsymbol{\mu}_i^{(k+1)})(\boldsymbol{y}_j - \boldsymbol{\mu}_i^{(k+1)})^T}{\sum_{j=1}^n \tau_i(\boldsymbol{y}_j;\, \boldsymbol{\Psi}^{(k+1/2)})}. \tag{8.30}$$

By construction of this AECM algorithm,

$$Q_1(\boldsymbol{\Psi}^{(k+1/2)};\, \boldsymbol{\Psi}^{(k)}) \geq Q_1(\boldsymbol{\Psi}^{(k)};\, \boldsymbol{\Psi}^{(k)}),$$

and

$$Q_2(\boldsymbol{\Psi}^{(k+1)};\, \boldsymbol{\Psi}^{(k+1/2)}) \geq Q_2(\boldsymbol{\Psi}^{(k+1/2)};\, \boldsymbol{\Psi}^{(k+1/2)}),$$

which ensures that

$$L(\boldsymbol{\Psi}^{(k+1/2)}) \geq L(\boldsymbol{\Psi}^{(k)})$$

and

$$L(\boldsymbol{\Psi}^{(k+1)}) \geq L(\boldsymbol{\Psi}^{(k+1/2)}),$$

respectively; see Meng and Rubin (1993) and Meng and van Dyk (1997). Thus the (incomplete-data) likelihood $L(\boldsymbol{\Psi})$ is not decreased after each cycle, and hence after each iteration, of the AECM algorithm.

8.7.4 Representation of Original Data

With this g-component mixture model of factor analyzers, the dimension of the data is being reduced locally by fitting g normal component distributions with covariance matrices of the form (8.15), where the contribution of a point to the process is weighted by its posterior probabilities of component membership.

To consider the representation of an original data point $\boldsymbol{y}_j$ in q-dimensions, we have that the estimated conditional expectation of the unobservable factor $\boldsymbol{U}_{ij}$ given $\boldsymbol{y}_j$ and also its membership of the ith component is from (8.12) equal to

$$\begin{aligned} \hat{\boldsymbol{u}}_{ij} &= E_{\hat{\boldsymbol{\Psi}}}\{\boldsymbol{U}_{ij} \mid \boldsymbol{y}_j,\, z_{ij} = 1\} \\ &= \hat{\gamma}_i^T(\boldsymbol{y}_j - \hat{\boldsymbol{\mu}}_i). \end{aligned} \tag{8.31}$$

If we average these (estimated) conditional expected component scores $\hat{\boldsymbol{u}}_{ij}$ over the estimated posterior distribution of the component membership of $\boldsymbol{y}_j$, we obtain

$$\hat{\boldsymbol{u}}_j = \sum_{i=1}^{g} \tau_i(\boldsymbol{y}_j;\, \hat{\boldsymbol{\Psi}})\, \hat{\gamma}_i^T(\boldsymbol{y}_j - \hat{\boldsymbol{\mu}}_i) \tag{8.32}$$

as the estimated factor score corresponding to $\boldsymbol{y}_j$ $(j = 1, \ldots, n)$.

We can thus display $\boldsymbol{y}_j$ in two-dimensional space by mapping it either onto the corresponding posterior mean $\hat{\boldsymbol{u}}_j$ for $q = 2$ factors or onto $\hat{\boldsymbol{u}}_{ij}$ for that value of i for which $\hat{z}_{ij} = 1$. These plots of the estimated posterior factor means may be useful for exploring group structure in a data set, as demonstrated in the series of papers by Tipping and Bishop (1997, 1999a), Bishop (1998), and Bishop and Tipping (1998) on the use of the related model of mixtures of probabilistic PCAs. However, it is noted that these plots are not always useful in exhibiting group structure in data, as will be demonstrated in Section 8.11.

8.8 LINK OF FACTOR ANALYSIS WITH PROBABILISTIC PCA

Principal component analysis (PCA) is a popular technique for reducing the dimension of data, in particular for visualizing the data in two or three dimensions. However, its effectiveness is limited by its global linearity. One nonlinear extension of this technique can be obtained by attempting to combine separate principal component analyses in the form of a finite mixture model. But conventional PCA does not correspond to an underlying density function for the data, and so there has been no useful way to combine the separate analyses.

Bishop and Tipping (1998) and Tipping and Bishop (1997, 1999a) formulated a mixture of probabilistic principal component analyzers by adopting a model that is closely related to the factor analysis model (8.13). To discuss now this link between factor analysis and PCA, we firstly consider the case of a single-factor analyzer and a single-principal component analyzer.

In factor analysis, the subspace defined by the columns of $\boldsymbol{B}$ will generally not correspond to the principal subspace. However, as noted by Tipping and Bishop

(1999a, 1999b), it had been observed in the literature that the factor loadings in $\boldsymbol{B}$ in a single-factor analyzer (8.2) were quite similar to the principal component axes in situations where the estimates of the diagonal matrix $\boldsymbol{D}$ were approximately equal; see, for example, Rao (1955), Anderson (1963), and Basilevsky (1994).

In their model for a probabilistic PCA, Tipping and Bishop (1997, 1999a) proposed taking the diagonal covariance matrix $\boldsymbol{D}$ for the error terms to have the isotropic structure,

$$\boldsymbol{D} = \sigma^2 \boldsymbol{I}_p. \tag{8.33}$$

In this isotropic case, they showed that the MLEs $\hat{\boldsymbol{B}}$ and $\hat{\sigma}^2$ are related to a PCA, since $\hat{\boldsymbol{B}}$ is the matrix whose columns are scaled eigenvectors of the sample covariance matrix $\boldsymbol{V}$ defined by (8.1), and $\hat{\sigma}^2$ is the average of the variance in the discarded dimensions; see also Tipping and Bishop (1999b). These derivations were undertaken previously by Lawley (1953) and Anderson and Rubin (1956), but it was not noted explicitly as in Tipping and Bishop (1997) that the principal component eigenvectors correspond to the global maximum of the likelihood. More specifically, the likelihood function (8.16) with $\boldsymbol{D}$ specified by (8.33) is maximized by

$$\hat{\boldsymbol{B}} = \boldsymbol{A}(\boldsymbol{\Lambda} - \hat{\sigma}^2 \boldsymbol{I}_q)^{1/2} \boldsymbol{R}, \tag{8.34}$$

where

$$\begin{aligned}
\boldsymbol{A} &= (\boldsymbol{a}_1, \ldots, \boldsymbol{a}_q), \\
\boldsymbol{\Lambda} &= \operatorname{diag}(\lambda_1, \ldots, \lambda_q), \\
\hat{\sigma}^2 &= \sum_{h=q+1}^{p} \lambda_h/(p-q),
\end{aligned}$$

and $\boldsymbol{a}_1, \ldots, \boldsymbol{a}_q$ are the (unit) eigenvectors, corresponding to the eigenvalues $\lambda_1 \geq \lambda_2 \geq \cdots \geq \lambda_q$ of $\boldsymbol{V}$, and $\boldsymbol{R}$ is an arbitrary $q \times q$ orthogonal matrix.

From (8.12) and (8.34), we have for the probabilistic PCA model that the estimated conditional expectation of $\boldsymbol{U}_j$ corresponding to $\boldsymbol{y}_j$, $\hat{\boldsymbol{u}}_j$, is given by

$$\begin{aligned}
\hat{\boldsymbol{u}}_j &= \hat{\boldsymbol{B}}^T(\hat{\boldsymbol{B}}\hat{\boldsymbol{B}}^T + \hat{\sigma}^2 \boldsymbol{I}_p)^{-1}(\boldsymbol{y}_j - \overline{\boldsymbol{y}}) &\quad (8.35)\\
&= (\hat{\boldsymbol{B}}^T \hat{\boldsymbol{B}} + \hat{\sigma}^2 \boldsymbol{I}_q)^{-1} \hat{\boldsymbol{B}}^T (\boldsymbol{y}_j - \overline{\boldsymbol{y}}), &\quad (8.36)
\end{aligned}$$

where (8.36) follows from (8.35) on noting that

$$\hat{\boldsymbol{B}}^T(\hat{\boldsymbol{B}}\hat{\boldsymbol{B}}^T + \hat{\sigma}^2 \boldsymbol{I}_p)^{-1} = (\hat{\boldsymbol{B}}^T \hat{\boldsymbol{B}} + \hat{\sigma}^2 \boldsymbol{I}_q)^{-1} \hat{\boldsymbol{B}}^T.$$

Although the columns of $\hat{\boldsymbol{B}}$ in a probabilistic PCA will span the principal subspace, they need not be an orthogonal projection in the factor space, since

$$\hat{\boldsymbol{B}}^T \hat{\boldsymbol{B}} = \boldsymbol{R}^T(\boldsymbol{\Lambda} - \hat{\sigma}^2 \boldsymbol{I}_q)\boldsymbol{R},$$

which is not diagonal for $\boldsymbol{R} \neq \boldsymbol{I}_q$ (Tipping and Bishop, 1999a). As $\hat{\sigma}^2 \rightarrow 0$, we see from (8.36) that the probabilistic PCA becomes equivalent in the limit to standard PCA. On further distinctions between factor analysis and probabilistic PCA, neither of the factors found by, say, a two-factor model is necessarily the same as that found by a single-factor model. However, with probabilistic PCA, we see that the principal axes can be found incrementally. On the effect of transformations of the original data, $\boldsymbol{H}\boldsymbol{y}_j$, the form of the factor analysis solution is preserved under diagonal $\boldsymbol{H}$; that is, under scaling of each variable in the observation vector $\boldsymbol{y}_j$. On the other hand, $\boldsymbol{H}$ needs to be orthogonal (corresponding to a rotation of the axes) for the probabilistic PCA solution to be preserved. This is where factor analysis and standard PCA fundamentally differ. With the factor analysis model (8.13), the latent variables are intended to explain the correlations between the variables in the observation vector $\boldsymbol{Y}_j$, while the error term $\boldsymbol{e}_j$ represents variability unique to a particular $\boldsymbol{Y}_j$. With standard PCA, covariance and variance terms are treated identically.

8.9 MIXTURES OF PROBABILISTIC PCAs

The mixtures of PCAs, as proposed by Tipping and Bishop (1997), has the form (8.14) and (8.15) with each $\boldsymbol{D}_i$ now having the isotropic structure

$$\boldsymbol{D}_i = \sigma_i^2 \boldsymbol{I}_p \quad (i = 1, \ldots, g). \tag{8.37}$$

From the previous section, it follows that under this isotropic restriction assumption (8.37) the CM-step for the updating of $\boldsymbol{B}_i$ and $\boldsymbol{D}_i$ is not necessary since, given the component membership of the mixture of PCAs, $\boldsymbol{B}_i^{(k+1)}$ and $\sigma_i^{(k+1)^2}$ are given explicitly by an eigenvalue decomposition of the $p \times p$ component-weighted covariance matrix, $\boldsymbol{V}_i^{(k+1/2)}$. That is, application of the EM algorithm will suffice with just the component-indicator vectors as the declared missing data. However, we may still wish to use the AECM algorithm as described in Section 8.7, since it necessitates only the inversion of $q \times q$ matrices and is of computational complexity $O(q^3)$. Under the additional isotropic restriction (8.37), the common diagonal element σ_i^2 of $\boldsymbol{D}_i$ is estimated by the trace of the right-hand side of (8.30) divided by p.

Hinton, Dayan, and Revow (1997) have given an excellent discussion of the use of mixtures of factor and probabilistic analyzers in the context of the clustering of handwritten digits. More recently, Ghahramani and Hinton (1998) and Bishop, Svensén, and Williams (1999) have considered a nonlinear form of factor analysis by replacing the factor vector $\boldsymbol{U}_{ij}$ in (8.13) by a nonlinear function so as to give a nonlinear mapping from the observed data space to the latent space. They call their latent model the generative topographic mapping; see also Bishop (1998).

8.10 INITIALIZATION OF AECM ALGORITHM

We now make use of the link of factor analysis with the probabilistic PCA model (8.37) to specify an initial value $\boldsymbol{\Psi}^{(0)}$ for $\boldsymbol{\Psi}$ in the ML fitting of the mixture of factor analyzers via the AECM algorithm, as described in Section 8.7. We proceed on the basis that under the mixture of factor analyzers model (8.14), if $\boldsymbol{y}_j$ is transformed as $\boldsymbol{D}_i^{-1/2}\boldsymbol{y}_j$, then its ith component-covariance matrix has the form

$$\boldsymbol{D}_i^{-1/2}\boldsymbol{\Sigma}_i\boldsymbol{D}_i^{-1/2} = (\boldsymbol{D}_i^{-1/2}\boldsymbol{B}_i)(\boldsymbol{D}_i^{-1/2}\boldsymbol{B}_i)^T + \boldsymbol{I}_p, \tag{8.38}$$

which corresponds to the probabilistic PCA model (8.37) with $\sigma_i^2 = 1$ $(i = 1, \ldots, g)$. Hence by using some initial value $\boldsymbol{D}_i^{(0)}$ and $\boldsymbol{\Sigma}_i^{(0)}$ for $\boldsymbol{D}_i$ and $\boldsymbol{\Sigma}_i$, we can obtain an initial value $\boldsymbol{B}_i^{(0)}$ for $\boldsymbol{B}_i$ by applying (8.34) to give

$$\boldsymbol{B}_i^{(0)} = \boldsymbol{D}_i^{(0)^{1/2}}\boldsymbol{A}_i(\boldsymbol{\Lambda}_i - \boldsymbol{I}_q)^{1/2} \quad (i = 1, \ldots, g), \tag{8.39}$$

where the q columns of the matrix $\boldsymbol{A}_i$ are the eigenvectors corresponding to the eigenvalues $\lambda_{i1} \geq \lambda_{i2} \geq \cdots \geq \lambda_{iq}$ of

$$\boldsymbol{D}_i^{(0)^{-1/2}}\boldsymbol{\Sigma}_i^{(0)}\boldsymbol{D}_i^{(0)^{-1/2}}, \tag{8.40}$$

and $\boldsymbol{\Lambda}_i = \text{diag}(\lambda_{i1}, \ldots, \lambda_{iq})$. As we are using estimates for the unknown $\boldsymbol{D}_i$ and $\boldsymbol{\Sigma}_i$ in (8.39), we can get a negative value for the term

$$(\boldsymbol{\Lambda}_i - \boldsymbol{I}_q) \tag{8.41}$$

in (8.39); that is, $\lambda_{ih} < 1$ for $h \geq q^*$, for some $q^* \leq q$. To avoid this situation since the square root of (8.41) is required, we can either restrict q to be less than q^* or replace the term (8.41) by $(\boldsymbol{\Lambda}_i - \tilde{\sigma}_i^2\boldsymbol{I}_q)$ to give

$$\boldsymbol{B}_i^{(0)} = \boldsymbol{D}_i^{(0)^{1/2}}\boldsymbol{A}_i(\boldsymbol{\Lambda}_i - \tilde{\sigma}_i^2\boldsymbol{I}_q)^{1/2} \quad (i = 1, \ldots, g), \tag{8.42}$$

where $\tilde{\sigma}_i^2$ is defined by

$$\tilde{\sigma}_i^2 = \sum_{h=q+1}^{p} \lambda_{ih}/(p - q).$$

To specify $\boldsymbol{\Sigma}_i^{(0)}$ for use in (8.40), we can randomly assign the data into g groups and take $\boldsymbol{\Sigma}_i^{(0)}$ to be the sample covariance matrix of the ith group $(i = 1, \ldots, g)$. Concerning the choice of $\boldsymbol{D}_i^{(0)}$, we can take $\boldsymbol{D}_i^{(0)}$ to be the diagonal matrix formed from the diagonal elements of $\boldsymbol{\Sigma}_i^{(0)}$ $(i = 1, \ldots, g)$. In this case, the matrix (8.40) has the form of a correlation matrix. This process may be repeated a number of times to provide a range of starting values.

Another way of proceeding, if p is not too large relative to n for there to be problems with the inversion of the estimated component-covariance matrices, is to fit a mixture of normal components with, say, equal component-covariance matrices, using a number of random starts. The estimate of $\boldsymbol{\Sigma}_i$ so obtained can then be used to specify $\boldsymbol{\Sigma}_i^{(0)}$ for use in (8.40). If p is very large relative to n, we can restrict the component-covariance matrices to being diagonal. That is, we take $\boldsymbol{\Sigma}_i^{(0)} = \boldsymbol{V}_i^{(0)}$, where

$$\boldsymbol{V}_i^{(0)} = \sum_{j=1}^{n} \tau_i(\boldsymbol{y}_j;\, \tilde{\boldsymbol{\Psi}})\,(\boldsymbol{y}_j - \tilde{\boldsymbol{\mu}}_i)(\boldsymbol{y}_j - \tilde{\boldsymbol{\mu}}_i)^T \Big/ \sum_{j=1}^{n} \tau_i(\boldsymbol{y}_j;\, \tilde{\boldsymbol{\Psi}}) \quad (i = 1, \ldots, g), \quad (8.43)$$

and where $\tilde{\boldsymbol{\Psi}}$ denotes the estimate of $\boldsymbol{\Psi}$ using $\tilde{\pi}_i$, $\tilde{\boldsymbol{\mu}}_i$, and $\tilde{\boldsymbol{\Sigma}}_i$ for the estimates of π_i, $\boldsymbol{\mu}_i$, and $\boldsymbol{\Sigma}_i$, respectively $(i = 1, \ldots, g)$, obtained by fitting a mixture of g normal components with diagonal component-covariance matrices $\boldsymbol{\Sigma}_i$ from a number of random starts. We take $\tilde{\pi}_i$ and $\tilde{\boldsymbol{\mu}}_i$ as the initial values for π_i and $\boldsymbol{\mu}_i$, respectively; that is,

$$\pi_i^{(0)} = \tilde{\pi}_i \quad \text{and} \quad \boldsymbol{\mu}_i^{(0)} = \tilde{\boldsymbol{\mu}}_i. \quad (8.44)$$

Concerning the choice of $\boldsymbol{D}_i^{(0)}$, we can take $\boldsymbol{D}_i^{(0)}$ to be the diagonal matrix formed from the diagonal elements of $\boldsymbol{V}_i^{(0)}$ $(i = 1, \ldots, g)$.

An alternative way of specifying an initial value $\boldsymbol{B}_i^{(0)}$ for $\boldsymbol{B}_i$ is to randomly specify its elements as, for example, in Ghahramani and Hinton (1997). In their program, they specify $\boldsymbol{B}_i^{(0)}$ as

$$\boldsymbol{B}_i^{(0)} = \boldsymbol{W}_i \,(|\,\boldsymbol{V}|^{1/p}\,/q)^{1/2} \quad (i = 1, \ldots, g),$$

where the elements of the matrix $\boldsymbol{W}_i$ are generated randomly from the standard normal distribution and $\boldsymbol{V}$ is the sample covariance matrix of the data set. They form the (common) initial value for $\boldsymbol{D}_i$ from the diagonal elements of $\boldsymbol{V}$, while $\boldsymbol{\mu}_i^{(0)}$ is selected randomly from a normal distribution with mean $\overline{\boldsymbol{y}}$ and covariance matrix $\boldsymbol{V}\,(i = 1, \ldots, g)$.

8.11 EXAMPLE 8.1: SIMULATED DATA

We consider the simulated data used in Section 8.2 as an example of a data set for which a plot of the first two principal components does not reveal the underlying group structure known to be present. This data set also can be used to show that a plot of the estimated posterior means of $q = 2$ factors does not always give an indication of any group structure. In Figure 8.2 we give the plot of the estimated posterior factor mean $\boldsymbol{u}_j$ for a single-component (g=1) factor analyzer with $q = 2$ factors. It can be seen that this plot gives no indication of the two groups from which the data were generated.

Although a single-component factor analyzer model (with $q = 2$ factors) fails as does a PCA in revealing the group structure of this artificial data set, it can still be used to cluster it into two clusters corresponding to the two groups of origin with only one error of misallocation. On the other hand, if we were using a PCA to reduce the dimension of this problem to $q = 2$ dimensions, we would obtain a very poor clustering of these data into two clusters with an overall misallocation rate of 44%. The advantage of the factor analysis-based approach is that we are able to fit a

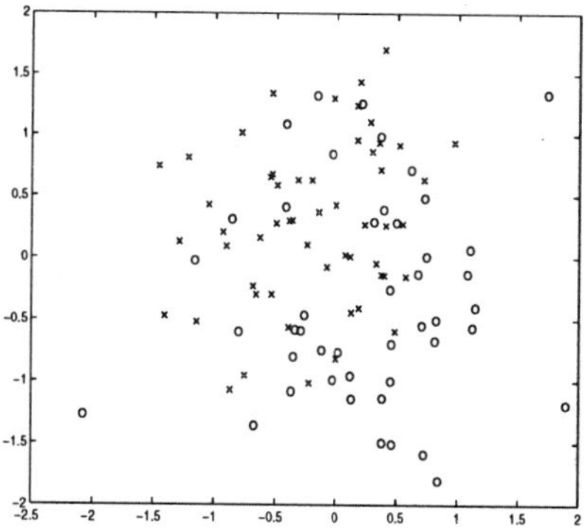

Fig. 8.2 Plot of the estimated posterior means of $q = 2$ factors for a single-factor analyzer model (o and x denote membership of groups G_1 and G_2, respectively).

mixture of factor analyzers, whereas we are unable to consider a mixture of standard PCAs. We can only fit a mixture model to the principal components obtained globally through the standard (single) PCA.

The increase in the log likelihood as a consequence of fitting $g = 2$ components instead of a single component ($g = 1$) is 82.96 (corresponding to a value of 165.92 for $-2 \log \lambda$), which is strongly indicative of the presence of at least two groups in the data. In Figure 8.3 we have plotted the estimated posterior factor means $\hat{\boldsymbol{u}}_{1j}$ and $\hat{\boldsymbol{u}}_{2j}$ for those observations assigned to the first and second components, respectively, of the mixture of factor analyzers model. It can be seen that there is no evidence of group structure, which is what we would expect given that the data have been generated from two normal distributions.

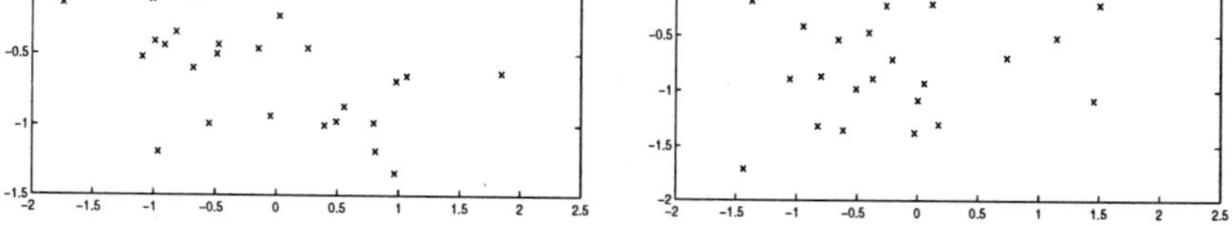

(a) $i = 1$ (b) $i = 2$

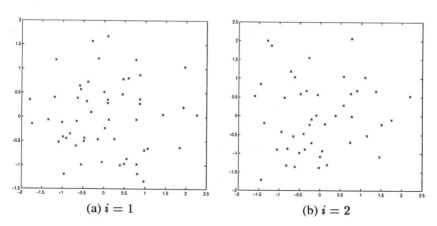

Fig. 8.3 Plot of the estimated posterior mean $\hat{\boldsymbol{u}}_{ij}$ for those observations assigned outright to the ith component of the mixture.

8.12 EXAMPLE 8.2: WINE DATA

To illustrate the use of the mixture of factor analyzers model, we report the analysis of McLachlan and Peel (2000) of the so-called wine data set, as available at the machine learning repository of the University of California, Irvine (Merz, Murphy, and Aha, 1997). These data give the results of a chemical analysis of wines grown in the same region in Italy, but derived from three different cultivars. The analysis determined the quantities of $p = 13$ constituents found in each of $n = 178$ wines. Mixtures of factor analyzers were fitted to this data set via the AECM algorithm with the parameters specified initially according to (8.42) to (8.44).

In Figure 8.4 we have plotted the values of $\hat{\boldsymbol{u}}_{ij}$, the estimated posterior means of the $q = 2$ factors following a single-component (g=1) factor analysis of the wine data. These posterior means have been plotted with their true group labels corresponding to the three different cultivars displayed. But it can be seen that if we were to ignore the group labels, the plot still exhibits the group structure in the data to a good degree, with three clusters clearly evident. To further examine this group structure, we fitted a mixture of $g = 3$ factor analyzers with $q = 2$ factors. In Figure 8.5(a) we have plotted the estimated posterior means $\hat{\boldsymbol{u}}_{1j}$ of the factor scores for those observations $\boldsymbol{y}_j$ assigned outright to the first component on the basis of the estimated posterior probabilities of component membership. Similarly, the estimated posterior means, $\hat{\boldsymbol{u}}_{2j}$ and $\hat{\boldsymbol{u}}_{3j}$, for those observations assigned outright to the second and third components are plotted in Figures 8.5(b) and 8.5(c), respectively. The plots of these estimated posterior means in Figures 8.5(a) to 8.5(c) do not exhibit any group structure, which is consistent with the fact that this wine data set contains only three different types of wine.

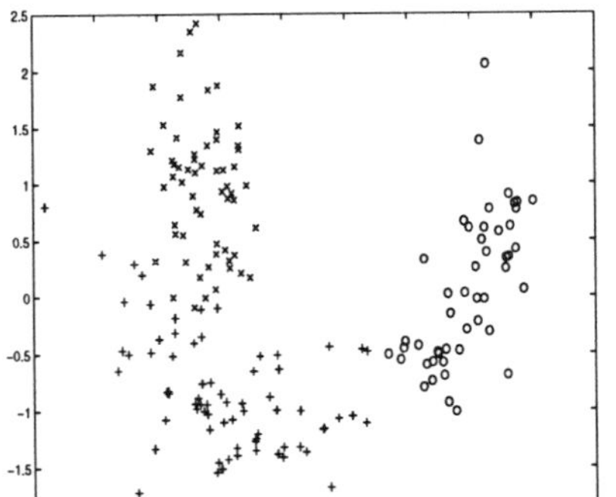

Fig. 8.4 Plot of the estimated posterior means of the $q = 2$ factors following a single-component (g=1) factor analysis of the wine data set (o, +, and x denote true component membership).

Hence the mixtures of factor analyzers model has been useful here in exploring the group structure of the data and illustrating the structure found. We now consider its usefulness as a model for density estimation for providing a model-based approach

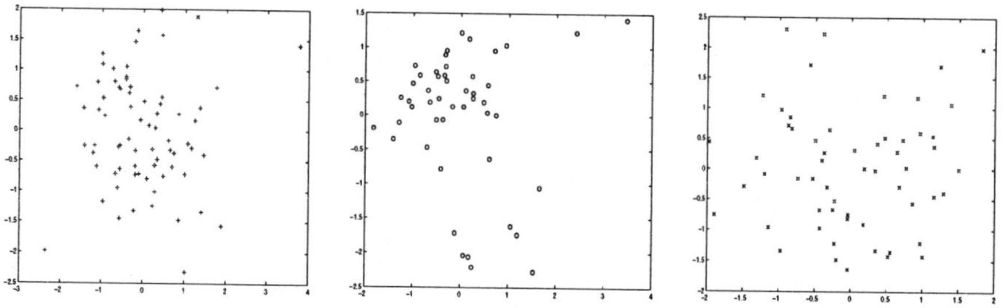

(a) $i = 1$ (b) $i = 2$ (c) $i = 3$

Fig. 8.5 Plot of the estimated posterior means of the factor scores for those observations assigned outright to the ith component of the mixture fitted to the wine data.

to clustering. To cluster this data set, we can model the density of each $p = 13$-dimensional observation $\boldsymbol{y}_j$ by a $g = 3$-component normal mixture model. A soft or probabilistic clustering of the data can be obtained in terms of the estimated posterior probabilities of component membership, $\tau_i(\boldsymbol{y}_j;\, \hat{\boldsymbol{\Psi}})$; a hard or outright clustering is obtained by assigning each observation to the component to which it has the highest estimated posterior probability of belonging. As $p = 13$ in this application, the covariance matrix $\boldsymbol{\Sigma}_i$ in the ith component normal distribution has 91 parameters for each $i\ (i = 1, 2, 3)$, which means that the total number of parameters is very large relative to the sample size of $n = 178$. Hence we consider fitting a mixture of factor analyzers for various q as a way of reducing the number of parameters to be fitted.

In Table 8.1, we report the overall error rate of the normal mixture model-based approach using restricted forms of the $\boldsymbol{\Sigma}_i$ through the adoption of factor analyzer components for various levels of the number of factors q from $q = 1$ to $q = 8$, and for the full model (no restrictions on the $\boldsymbol{\Sigma}_i$). In the latter case, we have also given the results obtained when the AECM algorithm is started from the true classification of these data. We have also reported the value of minus twice the LRTS λ (that is, twice the increase in the log likelihood), as we proceed from fitting a mixture of q factor analyzers to one with $q + 1$ component factors.

Table 8.1 Overall Error Rates for Values of q

q	$\log L$	Err	%Err	$-2\log\lambda$
1	−3102.254	2	1.12	—
2	−2995.334	1	0.56	213.840
3	−2913.122	1	0.56	164.424
4	−2871.655	3	1.69	82.934
5	−2831.860	4	2.25	79.590
6	−2811.290	4	2.25	41.140
7	−2799.204	4	2.25	24.172
8	−2788.542	4	2.25	21.324
Full	−2813.153	7	3.93	—
True	−2781.244	1	0.56	—

It can be seen from Table 8.1 that the error rate of the clustering based on the mixture model with unrestricted component-covariance matrices is increased when the known group membership of the data is not used to start the EM algorithm. This lack of knowledge affects the clustering for all but $q = 2$ and 3 factors, as we can achieve an error rate of one for the other values of q when this knowledge is used to start the AECM algorithm.

The log likelihood reported in Table 8.1 in the full model case is smaller than the log likelihood for $6 \leq q \leq 8$, indicating that there is a larger local maximum to be found in the unrestricted case. Indeed, when we start the EM algorithm in the latter case using the known classification of the data, we do find a local maximum that is greater than the values reported in Table 8.1 for $6 \leq q \leq 8$. This knowledge of the true classification is not used in starting the AECM algorithm to fit the mixture of factor analyzers for $q \leq 8$.

In practice in a typical cluster analysis, the true classification of the data would not be available. It can be seen from Table 8.1 that without this knowledge to start the AECM algorithm, the error rate of the outright clustering tends to increase as the number of factors q increases. This apparent error rate is smallest for $q = 2$ and 3. Concerning the use of the LRTS to decide on the number of factors q, the test of $q = q_0 = 6$ versus $q = q_0 + 1 = 7$ is not significant at a conventional level of significance ($P = 0.28$), on taking $-2 \log \lambda$ to be chi-squared with $d = g(p - q_0) = 21$ degrees of freedom under the null hypothesis that $q = q_0 = 6$.

In Table 8.2 we list the eigenvalues λ_{ih} $(h = 1, \ldots, 13)$ of (8.39). These eigenvalues were used to form the initial value $\boldsymbol{B}_i^{(0)}$ from (8.39) for a specified q. It can be seen that for each i, the λ_{ih} start to fall away after $h = 5$ or 6, which is consistent with the above result based on the LRT that $q = 6$ factors would appear to be adequate. It suggests that the relative sizes of the λ_{ih} $(h = 1, \ldots, p)$ for each i $(i = 1, \ldots, g)$ can be used as guide to an initial choice for the number of factors q. A more informed choice if desired can be made then on the basis of the log likelihood ratio statistic applied for nearby values of q.

Table 8.2 Eigenvalues of the Initial Estimates of the Component-Correlation Matrices

h	λ_{1h}	λ_{2h}	λ_{3h}
1	13.604	7.149	17.060
2	7.411	4.985	8.176
3	4.184	4.185	7.497
4	3.215	2.126	4.547
5	2.392	1.849	2.936
6	2.098	1.417	2.195
7	1.687	0.952	2.012
8	1.178	0.670	0.866
9	0.997	0.564	0.678
10	0.831	0.483	0.541
11	0.622	0.322	0.424
12	0.482	0.196	0.279
13	0.367	0.137	0.198

9

Fitting Mixture Models to Binned Data

9.1 INTRODUCTION

In this chapter we consider the fitting of finite mixture models to binned and truncated data by maximum likelihood via the EM algorithm. Binning can occur systematically when a measurement instrument has finite resolution; for example, a digital camera with finite precision for pixel intensity. Binning may also occur intentionally when real-valued variables are quantized to simplify data collection. Truncation can also occur in data collection, whether due to fundamental limitations of the range of the measurement process or intentionally for other reasons.

McLachlan and Jones (1988) considered the fitting of mixtures of univariate (log) normal components to binned and truncated data collected in the form of red blood cell (RBC) volume counts on cows exposed to the tick-borne parasite *Anaplasma marginale* in a laboratory trial. Also, Jones and McLachlan (1989) considered the fitting of mixtures of two log skew Laplace components in the modeling of the mass-size particle frequencies of soil samples in an attempt to determine their depositional environments; see also McLachlan (1992, Section 7.9) and McLachlan (1994). Further, the results of Jones and McLachlan (1991) on PTC sensitivity data as discussed in Section 3.6.2 were obtained from binned data.

We present here a general solution to the problem of fitting a mixture density,

$$f(\boldsymbol{y}_j;\, \boldsymbol{\Psi}) = \sum_{i=1}^{g} \pi_i f_i(\boldsymbol{y}_j;\, \boldsymbol{\theta}_i), \tag{9.1}$$

to binned and truncated data by maximum likelihood. The solution is then specialized to the case of bivariate normal mixtures, as considered by Cadez et al. (2000). Their work was motivated by an application in medical diagnosis (Cadez et al., 1999). For diagnostic evaluation of anemia and monitoring the response to therapy, blood samples from patients are routinely analyzed to determine the volume of RBCs and the amount of hemoglobin, the oxygen-transporting protein of the red cell. Typically, each sample contains about 40,000 different RBCs. The volume and hemoglobin concentration of RBCs are measured by a cytometric blood cell counter. In addition, it produces a bivariate histogram on a 100×100 grid in volume (V) and hemoglobin concentration (HC) space. Figure 9.1 shows a two-dimensional histogram of the RBC counts in V-HC space for a healthy individual (control) and for an iron-deficient patient. Each bin contains a count of the number of RBCs whose volume and hemoglobin concentration fall into that bin. It is known that the data can be truncated; that is, the range of machine measurement is less than the actual possible range of volume and hemoglobin concentration values. The results of Cadez et al. (1999) on the fitting of a bivariate normal mixture to these data are to be given in Section 9.6. We shall also illustrate how the fitting of this normal mixture model leads to a classification procedure useful in the diagnosis of iron-deficient anemia.

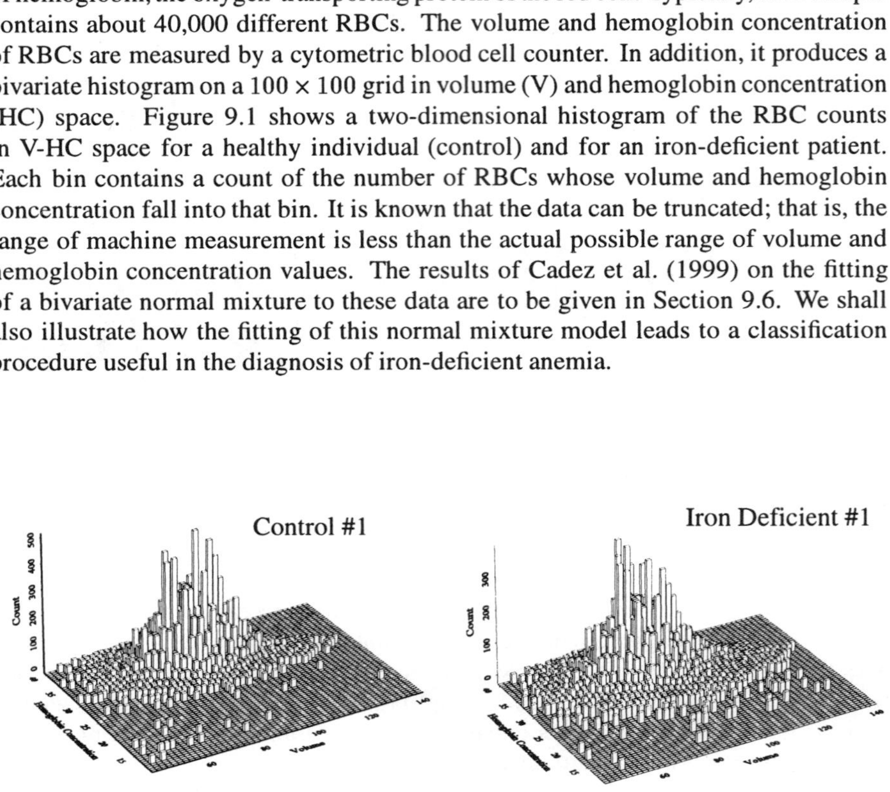

Fig. 9.1 Two-dimensional V-HC histograms of the raw data RBC counts for a control subject and a subject with iron-deficient anemia. From Cadez et al. (1999).

9.2 BINNED AND TRUNCATED DATA

We let $\boldsymbol{Y}_1, \ldots, \boldsymbol{Y}_n$ denote n independent observations, where $\boldsymbol{Y}_j$ has the mixture density (9.1). Suppose that the sample space $\mathcal{Y}$ of $\boldsymbol{Y}_j$ is partitioned into v mutually exclusive p-dimensional regions $\mathcal{Y}_r$ $(r = 1, \ldots, v)$. We consider the case where only the number n_r of the $\boldsymbol{Y}_j$ that fall in $\mathcal{Y}_r$ $(r = 1, \ldots, v_o)$ is recorded, where $v_o \leq v$. That is, individual observations are not recorded but only the regions $\mathcal{Y}_r$ in which they fall are recorded; further, even such observations are made only if the value of $\boldsymbol{Y}_j$ falls in one of the regions $\mathcal{Y}_r$ $(r = 1, \ldots, v_o)$.

For given

$$n = \sum_{r=1}^{v_o} n_r,$$

the observed data vector

$$\boldsymbol{a} = (n_1, \ldots, n_r)^T$$

has a multinomial distribution, consisting of n draws on v_o categories with probabilities $P_r(\boldsymbol{\Psi})/P(\boldsymbol{\Psi})$, $r = 1, \ldots, v_o$, where

$$P_r(\boldsymbol{\Psi}) = \int_{\mathcal{Y}_r} f(\boldsymbol{y}_j; \boldsymbol{\Psi})\, d\boldsymbol{y}_j \tag{9.2}$$

and

$$P(\boldsymbol{\Psi}) = \sum_{r=1}^{v_o} P_r(\boldsymbol{\Psi}).$$

Thus the log likelihood for $\boldsymbol{\Psi}$ that can be formed on the basis of $\boldsymbol{a}$ is given by

$$\log L(\boldsymbol{\Psi}) = \sum_{r=1}^{v_o} n_r \log\{P_r(\boldsymbol{\Psi})/P(\boldsymbol{\Psi})\} + C_1, \tag{9.3}$$

where

$$C_1 = \log\{n!/\prod_{r=1}^{v_o} n_r!\}.$$

9.3 APPLICATION OF EM ALGORITHM

9.3.1 Missing Data

We can solve this problem within the EM framework by introducing the vectors

$$\boldsymbol{u} = (n_{v_o+1}, \ldots, n_v)^T \tag{9.4}$$

and

$$\boldsymbol{y}_{r+} = (\boldsymbol{y}_{r1}^T, \ldots, \boldsymbol{y}_{r,n_r}^T)^T \quad (r = 1, \ldots, v) \tag{9.5}$$

as the missing data. The vector $\boldsymbol{u}$ contains the unobservable frequencies in the case of truncation ($v_o < v$), while $\boldsymbol{y}_{r+}$ contains the n_r unobservable individual observations in the rth region $\mathcal{Y}_r$ $(r = 1, \ldots, v)$. The conditional distribution of $\boldsymbol{u}$ given $\boldsymbol{a}$ is specified to be

$$f(\boldsymbol{u} \mid \boldsymbol{a}; \boldsymbol{\Psi}) = C_2\{P(\boldsymbol{\Psi})\}^n \prod_{r=v_o+1}^{v} \{P_r(\boldsymbol{\Psi})\}^{n_r}, \tag{9.6}$$

where C_2 is the normalizing constant

$$C_2 = \left(\sum_{r=v_o+1}^{v} n_r + n - 1 \right)! / \{(n-1)! \prod_{r=v_o+1}^{v} n_r!\}.$$

Conditional on $\boldsymbol{a}$ and $\boldsymbol{u}$, $\boldsymbol{y}_{rs}$ $(s = 1, \ldots, n_r)$, denote n_r independent observations with density $f(\boldsymbol{y}_j;\, \boldsymbol{\Psi})/P_r(\boldsymbol{\Psi})$ for $r = 1, \ldots, v$.

The EM machinery is then invoked by declaring

$$(\boldsymbol{a}^T,\, \boldsymbol{u}^T,\, \boldsymbol{y}_{1+}^T,\, \ldots,\, \boldsymbol{y}_{v+}^T)^T \tag{9.7}$$

as the complete-data vector. The log likelihood for this complete-data specification is, ignoring an additive combinatorial term not involving $\boldsymbol{\Psi}$, equal to

$$\log L_c(\boldsymbol{\Psi}) = \sum_{r=1}^{v} \sum_{s=1}^{n_r} \log f(\boldsymbol{y}_{rs};\, \boldsymbol{\Psi}), \tag{9.8}$$

and it implies the (incomplete-data) log likelihood $\log L(\boldsymbol{\Psi})$.

However, if we work with the complete-data vector specified by (9.7), then the M-step of the EM algorithm will still require an iterative procedure, since $f(\boldsymbol{y}_j;\, \boldsymbol{\Psi})$ is a mixture density. Consequently, for a mixture density, a further extension of the complete-data vector is needed to include the zero–one component-indicator variables in

$$\boldsymbol{z}_{rs} = (z_{1rs}, \ldots, z_{grs})^T \quad (r = 1, \ldots, v;\ s = 1, \ldots, n_r),$$

where $\sum_{i=1}^{g} z_{irs} = 1$ and where, given the $\boldsymbol{y}_{rs}$, the $\boldsymbol{z}_{rs}$ are conditionally independent with

$$\begin{aligned} \operatorname{pr}\{Z_{irs} = 1 \mid \boldsymbol{y}_{rs}\} &= \pi_i f_i(\boldsymbol{y}_{rs};\, \boldsymbol{\theta}_i)/f(\boldsymbol{y}_{rs};\, \boldsymbol{\Psi}) \\ &= \tau_i(\boldsymbol{y}_{rs};\, \boldsymbol{\Psi}) \end{aligned} \tag{9.9}$$

for $i = 1, \ldots, g$. Note that $\tau_i(\boldsymbol{y}_{rs};\, \boldsymbol{\Psi})$ is the conditional probability that the random variable $\boldsymbol{Y}_{rs}$ with realization $\boldsymbol{y}_{rs}$ belongs to the ith component of the mixture given $\boldsymbol{y}_{rs}$. This posterior probability is of course the same as if $\boldsymbol{Y}_{rs}$ had been obtained by sampling from the entire sample space $\mathcal{Y}$ and not just the subspace $\mathcal{Y}_r$.

With the inclusion of these component-indicator vectors $\boldsymbol{z}_{rs}$ in the complete-data specification, the complete-data log likelihood becomes

$$\log L_c(\boldsymbol{\Psi}) = \sum_{i=1}^{g} \sum_{r=1}^{v} \sum_{s=1}^{n_r} z_{irs} \log\{\pi_i\, f_i(\boldsymbol{y}_{rs};\, \boldsymbol{\theta}_i)\}. \tag{9.10}$$

9.3.2 E-Step

On the $(k+1)$th iteration of the EM algorithm, the E-step requires the calculation of $Q(\boldsymbol{\Psi};\, \boldsymbol{\Psi}^{(k)})$, the conditional expectation of the complete-data log likelihood, $\log L_c(\Psi)$, given the observed data vector $\boldsymbol{a}$ of frequencies, using the current value

$\boldsymbol{\Psi}^{(k)}$ for $\boldsymbol{\Psi}$. On taking the expectation of (9.10) over $\boldsymbol{y}_{1+}, \ldots, \boldsymbol{y}_{v+}$ and finally $\boldsymbol{u}$, it follows that

$$Q(\boldsymbol{\Psi}; \boldsymbol{\Psi}^{(k)}) = \sum_{i=1}^{g} \sum_{r=1}^{v} n_r^{(k)} E_{\boldsymbol{\Psi}^{(k)}}[\tau_i(\boldsymbol{Y}_j; \boldsymbol{\Psi}^{(k)})\{\log f_i(\boldsymbol{Y}_j; \boldsymbol{\Psi}) + \log \pi_i\} \mid \boldsymbol{Y}_j \in \mathcal{Y}_r], \tag{9.11}$$

where

$$\begin{aligned} n_r^{(k)} &= n_r \qquad (r = 1, \ldots, v_o) \\ &= E_{\boldsymbol{\Psi}^{(k)}}(n_r \mid \boldsymbol{a}) \quad (r = v_o + 1, \ldots, v). \end{aligned}$$

With the conditional distribution of $\boldsymbol{u}$ given $\boldsymbol{a}$ specified by (9.6), it can be shown that

$$E_{\boldsymbol{\Psi}^{(k)}}(n_r \mid \boldsymbol{a}) = nP_r(\boldsymbol{\Psi}^{(k)})/P(\boldsymbol{\Psi}^{(k)}) \quad (r = v_o + 1, \ldots, v); \tag{9.12}$$

see Dempster et al. (1977), McLachlan and Jones (1988), and McLachlan and Krishnan (1997, Section 2.8).

9.3.3 M-Step

On the M-step at the $(k + 1)$th iteration, we have on differentiation of $Q(\boldsymbol{\Psi}; \boldsymbol{\Psi}^{(k)})$ with respect to the mixing proportions that their updates are given by

$$\pi_i^{(k+1)} = \frac{\sum_{r=1}^{v} n_r^{(k)} E_{\boldsymbol{\Psi}^{(k)}}\{\tau_i(\boldsymbol{Y}_j; \boldsymbol{\Psi}^{(k)}) \mid \boldsymbol{Y}_j \in \mathcal{Y}_r\}}{\sum_{r=1}^{v} n_r^{(k)}} \tag{9.13}$$

for $i = 1, \ldots, g$.

If we let $\boldsymbol{\xi}$ denote the distinct elements of $\boldsymbol{\theta}_1, \ldots, \boldsymbol{\theta}_g$ known *a priori* to be distinct, then it follows that $\boldsymbol{\xi}^{(k+1)}$ is root of the equation

$$\partial Q(\boldsymbol{\Psi}; \boldsymbol{\Psi}^{(k)})/\partial \boldsymbol{\xi} = \boldsymbol{0}; \tag{9.14}$$

that is, a root of

$$\sum_{i=1}^{g} \sum_{r=1}^{v} n_r^{(k)} E_{\boldsymbol{\Psi}^{(k)}}\{\tau_i(\boldsymbol{Y}_j; \boldsymbol{\Psi}^{(k)})\, \partial \log f_i(\boldsymbol{Y}_j; \boldsymbol{\theta}_i)/\partial \boldsymbol{\xi} \mid \boldsymbol{Y}_j \in \mathcal{Y}_r\} = \boldsymbol{0}, \tag{9.15}$$

on interchanging the operations of differentiation and expectation.

9.3.4 M-Step for Normal Components

We now specialize the implementation of the M-step to the case of fitting a normal mixture density,

$$f(\boldsymbol{y}_j; \boldsymbol{\Psi}) = \sum_{i=1}^{g} \pi_i \phi(\boldsymbol{y}_j; \boldsymbol{\mu}_i, \boldsymbol{\Sigma}_i), \tag{9.16}$$

to binned and truncated data. This has been done in McLachlan and Jones (1988) for a univariate normal mixture; see also McLachlan and Krishnan (1997, Section 2.8). Using the multivariate analogue of their univariate results, it follows that on the $(k+1)$th iteration of the EM algorithm, the estimates of the component means $\boldsymbol{\mu}_i$ and the component-covariance matrices $\boldsymbol{\Sigma}_i$ are given by

$$\boldsymbol{\mu}_i^{(k+1)} = \sum_{r=1}^{v} n_r^{(k)} E_{\boldsymbol{\Psi}^{(k)}}\{\tau_i(\boldsymbol{Y}_j;\, \boldsymbol{\Psi}^{(k)})\, \boldsymbol{Y}_j \mid \boldsymbol{Y}_j \in \mathcal{Y}_r\} / C_i(\boldsymbol{\Psi}^{(k)}) \tag{9.17}$$

and

$$\boldsymbol{\Sigma}_i^{(k+1)} = \frac{\sum_{r=1}^{v} n_j^{(k)} E_{\boldsymbol{\Psi}^{(k)}}\{\tau_i(\boldsymbol{Y}_j;\, \boldsymbol{\Psi}^{(k)})\, (\boldsymbol{Y}_j - \boldsymbol{\mu}_i^{(k+1)})(\boldsymbol{Y}_j - \boldsymbol{\mu}_i^{(k+1)})^T \mid \boldsymbol{Y}_j \in \mathcal{Y}_r\}}{C_i(\boldsymbol{\Psi}^{(k)})}, \tag{9.18}$$

where

$$C_i(\boldsymbol{\Psi}^{(k)}) = \sum_{r=1}^{v} n_r^{(k)}\, E_{\boldsymbol{\Psi}^{(k)}}\{\tau_i(\boldsymbol{Y}_j;\, \boldsymbol{\Psi}^{(k)}) \mid \boldsymbol{Y}_j \in \mathcal{Y}_r\}. \tag{9.19}$$

In the univariate case ($p = 1$), McLachlan and Jones (1988) expressed the conditional expectations in (9.17) and (9.18) in terms of the standard normal distribution and density. Jones and McLachlan (1990a) subsequently gave an algorithm (MGT) for implementing the EM algorithm in the univariate case.

As noted by Cadez et al. (2000), although the multivariate theory is a straightforward extension of the univariate case, the practical implementation of this theory is considerably more complex due to the fact that the calculation of the multidimensional integrals in the conditional expectations in (9.13), (9.17), and (9.18) are more complex than their univariate analogues.

9.4 PRACTICAL IMPLEMENTATION OF EM ALGORITHM

9.4.1 Computational Issues

From the previous section, it can be seen that each EM iteration requires the calculation of integrals that have to be performed numerically at least in the multivariate case. Naive implementation of the procedure can lead to computationally inefficient results. Cadez et al. (2000) have summarized some of the difficulties involved in implementing the EM algorithm for multivariate mixtures. They go on to propose a number of straightforward numerical techniques.

9.4.2 Numerical Integration at Each EM Iteration

One way to effect the E-step is to use a Monte Carlo E-step whereby the calculation of the Q-function is approximated by sampling from the (current) conditional distribution of the complete-data vector given the observed data. However, because of the slow convergence of this Monte Carlo procedure for this problem, Cadez et al. (2000) concentrated on numerical multidimensional techniques using Romberg integration; see, for example, Thisted (1988) for details. They note that an important aspect of Romberg integration is selection of the *order* of integration. Low-order schemes use relatively few function evaluations in the initialization phase, but may converge slowly. Higher-order schemes may take longer at the initialization phase, but converge faster. Thus, order selection can substantially affect the computation time of numerical integration.

9.4.3 Integration over Truncated Regions

For a mixture density that is defined on the whole space we must define bins to cover regions extending from grid boundaries to infinity. In the one-dimensional case it suffices to define two additional bins: one extending from the upper endpoint of the last bin to ∞, and the other extending from $-\infty$ to the lower end point of the first bin. As pointed out by Cadez et al. (2000), in the multivariate case it is more natural to define a single bin

$$\mathcal{Y}_{v_o,v} = \mathcal{Y} \setminus \sum_{r=1}^{v_o} \mathcal{Y}_r$$

that covers everything but the data grid than to explicitly describe the out-of-grid regions. The reason is that all expected values over the subspace $\mathcal{Y}_{v_o,v}$ can be calculated without actually doing any integration. For example, the following integrals over the truncated regions can be calculated as

$$\int_{\mathcal{Y}_{v_o,v}} \phi(\boldsymbol{w};\, \boldsymbol{\mu}_i,\, \boldsymbol{\Sigma}_i)\, d\boldsymbol{w} = 1 - \sum_{r=1}^{v_o} \int_{\mathcal{Y}_r} \phi(\boldsymbol{w};\, \boldsymbol{\mu}_i,\, \boldsymbol{\Sigma}_i)\, d\boldsymbol{w}, \tag{9.20}$$

$$\int_{\mathcal{Y}_{v_o,v}} \boldsymbol{w}\, \phi(\boldsymbol{w};\, \boldsymbol{\mu}_i,\, \boldsymbol{\Sigma}_i)\, d\boldsymbol{w} = \boldsymbol{\mu}_i - \sum_{r=1}^{v_o} \int_{\mathcal{Y}_r} \boldsymbol{w}\, \phi(\boldsymbol{w};\, \boldsymbol{\mu}_i,\, \boldsymbol{\Sigma}_i)\, d\boldsymbol{w}, \tag{9.21}$$

$$\begin{aligned} &\int_{\mathcal{Y}_{v_o,v}} (\boldsymbol{w} - \boldsymbol{\mu}_i)(\boldsymbol{w} - \boldsymbol{\mu}_i)^T\, \phi(\boldsymbol{w};\, \boldsymbol{\mu}_i,\, \boldsymbol{\Sigma}_i)\, d\boldsymbol{w} \\ &\qquad = \boldsymbol{\Sigma}_i - \sum_{r=1}^{v_o} \int_{\mathcal{Y}_r} (\boldsymbol{w} - \boldsymbol{\mu}_i)(\boldsymbol{w} - \boldsymbol{\mu}_i)^T\, \phi(\boldsymbol{w};\, \boldsymbol{\mu}_i,\, \boldsymbol{\Sigma}_i)\, d\boldsymbol{w}. \end{aligned} \tag{9.22}$$

No extra work is required to obtain the integrals on the right-hand side of equations (9.20) to (9.22), since these integrals are effectively required in the calculation of the conditional expectation terms in the EM equations (9.13) and (9.17) to (9.19).

For efficiency, Cadez et al. (2000) take advantage of the sparseness of the bin counts. They optimize the time spent on integrating over the (numerous) empty bins that do not significantly contribute to the integral or the accuracy of the integration.

With this in mind, they use the standard (computationally cheap) EM algorithm on a random sample of the data to provide a good starting point for the more computationally complex binned/truncated version of EM. They take a subset of points within each bin, randomize their coordinates around the bin centers (they use the uniform distribution within each bin), and treat the newly obtained data as nonbinned and nontruncated. The standard EM algorithm is relatively fast, as a closed form solution exists for each EM step (without any integration). Once the standard algorithm converges to a solution in the parameter space, they use these parameters as initial starting points for the full algorithm (for binned, truncated data) which then refines these guesses to a final solution, typically taking just a few iterations. Note that this initialization scheme cannot affect the accuracy of the results, as the full algorithm is used as the final criterion for convergence.

9.4.4 EM Algorithm for Binned Multivariate Data

The algorithm developed by Cadez et al. (2000) for fitting mixture models to multivariate binned, truncated data consists of the following stages:

- Draw a small number of data points from the multivariate histogram, adding some counts to all the bins to prevent occurrences of empty bins.
- Fit the mixture model using the standard EM algorithm (that is, fit the mixture model to the nonbinned, nontruncated data).
- Use the parameter estimates from the standard mixture analysis to refine them with the EM algorithm applied to the binned and truncated data. This consists of iteratively applying the EM equations (9.13) and (9.17) to (9.19) until convergence as measured by the change in the log likelihood defined by (9.3).

9.5 SIMULATIONS

Cadez et al. (2000) performed some simulation experiments to observe the effect of the number of bins v and the sample size n on the estimate $\hat{\boldsymbol{\Psi}}$ of the parameter vector $\boldsymbol{\Psi}$ in a two-component bivariate normal mixture. The component-covariance matrices were set equal to the identity matrix, and the means were specified to be $\boldsymbol{\mu}_1 = (-1.5, 0)^T$ and $\boldsymbol{\mu}_2 = (1.5, 0)^T$. The range of the grid extended from $(-5, -5)$ to $(5, 5)$ so that truncation was relatively rare. The sample size was varied in steps of 10 from $n = 10$ to $n = 1,000$, while the number of bins per dimension was varied in steps of 5 from $b = 5$ to $b = 100$ so that the original unbinned samples were quantized into $v_o = b^2$ bins. Ten random samples were generated for each combination of v_o and n.

To measure the quality of the solution, Cadez et al. (2000) calculated the Kullback–Leibler (KL) distance between each estimated density and the true known density. They calculated the average KL distance over the 10 samples for each value of n and b, for both the binned and the standard EM algorithms. In total, each of the standard and binned algorithms were run 20,000 different times to generate the reported results.

In Figure 9.2, the KL distance on the log scale is plotted as a function of bin size for specific values of n (n = 100, 300, 1000) in a comparison of the standard and binned versions of the EM algorithm. For each of the three values of n, the curves have the same qualitative shape: a rapid improvement in quality as one moves from $b = 5$ to $b = 20$, with relatively flat performance (that is, no sensitivity to b) above $b = 20$. For each of the three values of n, the binned EM "tracks" the performance of the standard EM quite closely: the difference between the two becomes less as n increases. The variability in the curves is due to the variability in the ten randomly sampled data sets for each particular combination of b and n. Note that for $b \geq 20$ the difference between the binned and standard versions of the EM algorithm is smaller than the "natural" variability due to random sampling effects.

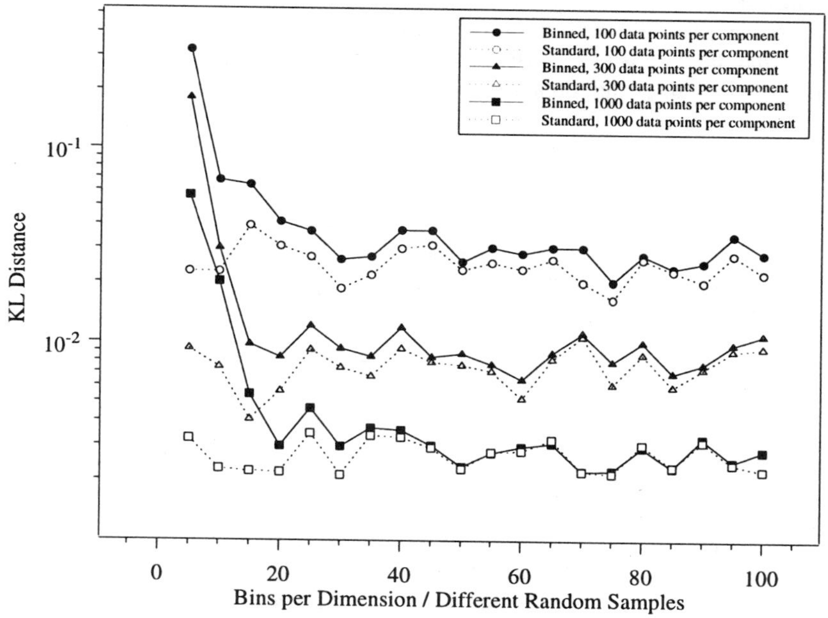

Fig. 9.2 Average KL distance (log scale) between the estimated densities and the true density as a function of the number of bins for different sample sizes n compared to the standard EM on the unbinned data. From Cadez et al. (2000).

9.6 EXAMPLE 9.1: RED BLOOD CELL DATA

As mentioned at the beginning of this chapter, the work of Cadez et al. (2000) was motivated by a real-world application in medical diagnosis based on two-dimensional histograms characterizing RBC and hemoglobin measurements (Figure 9.1). McLaren

(1996) has summarized prior work on this problem; see also McLaren et al. (2000). Mixture models are particularly useful in this context as a model since it is plausible that different components in the model correspond to blood cells in different states. Cadez et al. (1999) generalized the earlier work of McLaren et al. (1991) and McLaren (1996) on one-dimensional volume data to the analysis of two-dimensional volume–hemoglobin histograms. Mixture densities were fitted to histograms from 97 control subjects and 83 subjects with iron deficient anemia, using the binned/truncated EM procedure described in Section 9.3.

Figure 9.3 shows contour probability plots of fitted two-component bivariate mixture densities for three control and three iron-deficient subjects, where only the lower 10% of the density is displayed (since the differences between the two populations are more obvious in the tails). One can clearly see systematic variability within the control and the iron-deficient groups, as well as between the two groups. Since the number of bins is relatively large ($b = 100$ in each dimension), as is the number of data points ($n = 40,000$), the simulation results from the previous section would tend to suggest that these density estimates are likely to be relatively accurate (compared to running the EM algorithm on unbinned data).

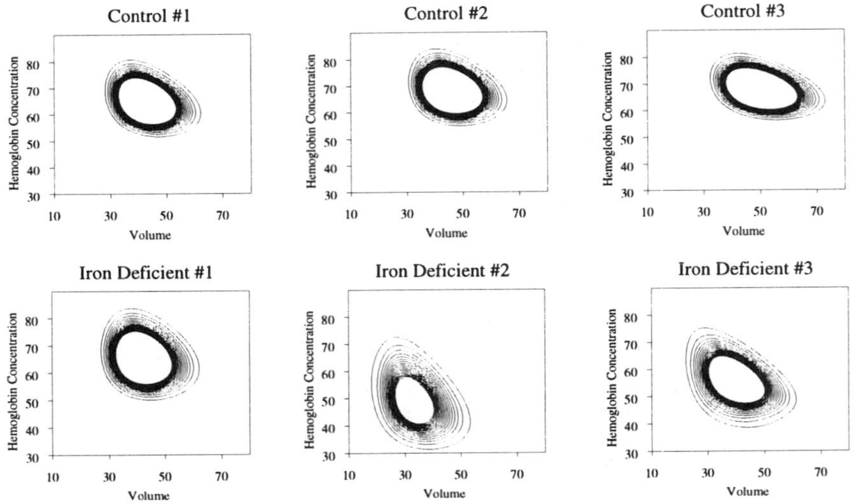

Fig. 9.3 Contour plots from estimated density estimates for three typical Control patients and three typical iron-deficient anemia patients. The lowest 10% of the probability contours are plotted to emphasize the systematic difference between the two groups. From Cadez et al. (1999).

Cadez et al. (1999) investigated how the fitted densities for each of the control and iron-deficient subjects could be used to form a discriminant rule for the diagnosis of a new subject. One approach is to take the fitted parameter vector $\hat{\boldsymbol{\Psi}}$, or elements of it, as the feature vector for the formation of a discriminant rule. In the case where $\boldsymbol{\Psi}$ contains the parameters of a two-component bivariate normal mixture, the feature vector would consist of eleven elements, corresponding to $\hat{\pi}_1$, the two elements in $\hat{\boldsymbol{\mu}}_1$ and in $\hat{\boldsymbol{\mu}}_2$, and the three distinct elements in $\hat{\boldsymbol{\Sigma}}_1$ and in $\hat{\boldsymbol{\Sigma}}_2$. Cadez et al. (1999) considered a number of models for the group-conditional distribution of this eleven-dimensional feature vector, including a two-component normal mixture model. The cross-validated error rate of this rule was found to be 1.5%. This compares with a rate of about 4% on the same subjects using algorithms such as CART (Breiman et al., 1984) or C4.5 (Quinlan, 1993) directly on features from the histogram, such as univariate means and standard deviations (that is, using no mixture modeling). Thus, the ability to fit mixture densities to binned and truncated data can play a significant role in improved classification performance for this particular problem.

10

Mixture Models for Failure-Time Data

10.1 INTRODUCTION

It is only in relatively recent times that the potential of finite mixture models has started to be exploited in survival and reliability analyses. Mixture models can be used to analyze failure-time data in a variety of situations. As a flexible way of modeling data, mixture models have obvious applications in situations where a single parametric family may not suffice. For example, following open-heart surgery, Blackstone, Naftel, and Turner (1986) consider that the risk of death can be characterized by three merging phases: an early phase in which the risk is relatively high, a middle phase of constant risk, and finally a late phase in which the risk starts to increase as the patient ages. These phases overlap each other in time and thus cannot be modeled satisfactorily by attempting to fit a separate parametric model to each discrete time period thought to correspond to the different phases. However, the use of a three-component mixture model with three components corresponding to the three phases of death provides an effective way of modeling survival; see, for example, McGiffin et al. (1993) and McLachlan and McGiffin (1994). These models can be viewed in the mixtures-of-experts framework, as considered in Chapter 5. This was the approach taken by Rosen and Tanner (1999), who proposed a mixture of Cox experts in which the components were taken to follow the Cox proportional hazards model.

The Weibull distribution is perhaps the most widely used lifetime distribution model because of its flexibility and simple expressions for the density, survival, and hazard functions. The Weibull hazard function can be monotone increasing, decreasing, or constant, depending upon the value of the shape parameter for the distribution. It has been found that such flexibility provides a good fit for a wide variety of lifetime data. Hence it is not surprising that mixtures of Weibull distributions are widely used in survival and reliability analyses with heterogeneous data. For example, Patra and Dey (1999) considered a multivariate mixture of Weibull distributions in reliability modeling. Some characteristics of mixtures of Weibulls can be found in Gupta and Gupta (1996).

When the shape parameter is set equal to one in the Weibull distribution, we obtain the exponential. Mixtures of exponential distributions are also widely used in applied science and the social sciences. It is frequently adopted to model the distribution of time to failure in those situations where the hazard function is observed empirically to decline with time. Heckman et al. (1990) have presented nonparametric tests to distinguish mixtures of exponentials from more general models with declining hazards. McLachlan (1995) has provided a review of mixtures of exponential distributions.

Another situation where mixture distributions can play a useful role is in modeling time to failure in the case of competing risks or failures. In this chapter, we shall focus on the latter situation. The results are to be presented mainly in a survival analysis context, but they apply equally well to problems in reliability analysis.

10.2 COMPETING RISKS

10.2.1 Mixtures of Survival Functions

In the analysis of failure-time data, it is often necessary to consider different types of failure. For simplicity of exposition, we shall consider the case of $g = 2$ different types of failure or causes, but the results extend to an arbitrary number g. An item is taken to have failed with the occurrence of the first failure from either cause, and we observe the time T to failure and the type of failure i $(i = 1, 2)$. In the case where the study terminates before failure occurs, T is the censoring time and the censoring indicator δ is set equal to zero to indicate that the failure time is right-censored.

The traditional approach to the modeling of the distribution of failure time in the case of competing risks is to postulate the existence of so-called latent failure times, T_1 and T_2, corresponding to the two causes and to proceed with the modeling of $T = \min(T_1, T_2)$ on the basis that the two causes are independent of each other; see David and Moeschberger (1978) and Kalbfleisch and Prentice (1980).

An alternative approach is to adopt a two-component mixture model, whereby the survival function of T is modeled as

$$S(t;\, \boldsymbol{x}) = \pi_1(\boldsymbol{x})\, S_1(t;\, \boldsymbol{x}) + \pi_2(\boldsymbol{x})\, S_2(t;\, \boldsymbol{x}), \qquad (10.1)$$

where the ith component survival function $S_i(t;\, \boldsymbol{x})$ denotes the conditional survival function, given that failure is due to the ith cause, and $\pi_i(\boldsymbol{x})$ is the probability of failure from the ith cause $(i = 1, 2)$. Here $\boldsymbol{x}$ is a vector of covariates associated with the item. It is common to assume that the mixing proportions $\pi_i(\boldsymbol{x})$ have the logistic form,

$$\pi_1(\boldsymbol{x};\, \boldsymbol{\beta}) = 1 - \pi_2(\boldsymbol{x};\, \boldsymbol{\beta}) = \exp(a + \boldsymbol{b}^T\boldsymbol{x})/\{(1 + \exp(a + \boldsymbol{b}^T\boldsymbol{x})\}, \qquad (10.2)$$

where $\boldsymbol{\beta} = (a, \boldsymbol{b}^T)^T$ is the vector of logistic regression coefficients; see Farewell (1977, 1982, 1986).

The mixture model (10.1) is expressed in terms of the survival functions, but it could be expressed equivalently in terms of the density functions, namely,

$$f(t;\, \boldsymbol{x}) = \pi_1(\boldsymbol{x})\, f_1(t;\, \boldsymbol{x}) + \pi_2(\boldsymbol{x})\, f_2(t;\, \boldsymbol{x}), \tag{10.3}$$

where

$$f_i(t;\, \boldsymbol{x}) = -\partial \log S_i(t;\, \boldsymbol{x})/\partial t \quad (i = 1, \ldots, g).$$

This mixture approach provides an alternative method to that of Prentice et al. (1978), who characterize the joint distribution of the failure time and the type of failure in terms of cause-specific hazard functions.

The mixture approach was adopted by Larson and Dinse (1985), who assumed also that the component survival functions follow a proportional hazards model (Cox, 1972) and that the baseline hazard functions $h_{0i}(t)$ $(i = 1, 2)$ are piecewise constant for simplicity. That is,

$$\begin{aligned} h_i(t;\, \boldsymbol{x}) &= e^{\boldsymbol{\gamma}_i^T \boldsymbol{x}} h_{0i}(t) \\ &= e^{\boldsymbol{\gamma}_i^T \boldsymbol{x} + \alpha_{im}}, \qquad \text{if } t \in J_m, \end{aligned} \tag{10.4}$$

where $\boldsymbol{\gamma}_i$ is a vector of parameters, $J_1, \ldots, J_M$ are M prespecified disjoint intervals that totally exhaust the nonnegative real line, and α_{im} is the parameter representing the log of the ith component baseline hazard on the mth interval J_m $(i = 1, 2;\ m = 1, \ldots, M)$. As an alternative to this specification of $h_{0i}(t)$, we can adopt some parametric form for the ith component baseline hazard function $h_{0i}(t)$ $(i = 1, 2)$. For example, Gordon (1990a) adopted the Gompertz distribution to specify the conditional survival functions in the context of estimating the "cure" rate of breast cancer after a treatment therapy. She also examined the applicability of the mixture model in fitting competing risks data through a simulation study (Gordon, 1990b). Kuk (1992) and Kuk and Chen (1992), on the other hand, considered a semiparametric version of (10.4) by treating the baseline hazard functions $h_{0i}(t)$ as nuisance parameters to be eliminated during the analysis.

10.2.2 Latent Failure-Time Approach

With the traditional approach to the handling of competing risks, consideration is given to the hypothetical latent failure times corresponding to each cause in the absence of the other (Moeschberger and David, 1971). Accordingly, we let T_1 be the latent failure time due to the first cause in the absence of the second; and, likewise, we let T_2 be the latent failure time due to the second cause in the absence of the first. We let $f_{li}(t;\, \boldsymbol{x})$ and $S_{li}(t;\, \boldsymbol{x})$ denote the density and survival functions of T_i $(i = 1, 2)$. With this approach, it is common to assume that the competing risks are independent. Under this assumption, the survival function $S(t;\, \boldsymbol{x})$ for the observable $T = \min(T_1,\, T_2)$ is given by

$$S(t;\, \boldsymbol{x}) = \int_t^\infty f_{l1}(u;\, \boldsymbol{x}) S_{l2}(u;\, \boldsymbol{x})\, du + \int_t^\infty f_{l2}(u;\, \boldsymbol{x}) S_{l1}(u;\, \boldsymbol{x})\, du. \tag{10.5}$$

We can write (10.5) as the two-component mixture model (10.1), where the mixing proportions $\pi_i(\boldsymbol{x})$ are given by

$$\pi_1(\boldsymbol{x}) = 1 - \pi_2(\boldsymbol{x}) = \int_0^\infty f_{l1}(u;\, \boldsymbol{x}) S_{l2}(u;\, \boldsymbol{x})\, du \tag{10.6}$$

and the component survival functions $S_i(t;\, \boldsymbol{x})$ are given by

$$S_1(t;\, \boldsymbol{x}) = \left\{ \int_t^\infty f_{l1}(u;\, \boldsymbol{x}) S_{l2}(u;\, \boldsymbol{x})\, du \right\} / \pi_1(\boldsymbol{x}) \tag{10.7}$$

and

$$S_2(t;\, \boldsymbol{x}) = \left\{ \int_t^\infty f_{l2}(u;\, \boldsymbol{x}) S_{l1}(u;\, \boldsymbol{x})\, du \right\} / \pi_2(\boldsymbol{x}). \tag{10.8}$$

It can be seen from the forms (10.7) and (10.8) for the component survival functions that if the latent failure due to one risk is much greater than the other, then this will affect the component-survival function corresponding to the other risk through the effect of truncation in the integrand of the integrals on the right-hand sides of (10.7) and (10.8). It suggests that in practice there may be a need to have the components of the two-component mixture model (10.1) interrelated or some constraints imposed on them in modeling competing risks. We shall demonstrate this in the case study to follow shortly.

Note that modeling the overall hazard function as a mixture is equivalent to using a latent failure-time approach.

10.2.3 ML Estimation for Mixtures of Survival Functions

We now consider the fitting of the mixture model (10.1) by maximum likelihood. The observed data are of the form $\boldsymbol{y} = (\boldsymbol{y}_1^T, \ldots, \boldsymbol{y}_n^T)^T$, where

$$\boldsymbol{y}_j = (t_j, \boldsymbol{x}_j^T, \delta_j)^T \quad (j = 1, \ldots, n)$$

and where $\delta_j = i$ if the jth entity failed due to cause i $(i = 1, 2)$ and $\delta_j = 0$ if the jth entity had not failed by the end of the study (that is, the failure time is right-censored at time t_j).

We let $S_i(t;\, \boldsymbol{\theta}_i,\, \boldsymbol{x})$ denote some postulated parametric form for $S_i(t;\, \boldsymbol{x})$, where $\boldsymbol{\theta}_i$ is a vector of unknown parameters $(i = 1, 2)$. The vector of all unknown parameters is given by $\boldsymbol{\Psi} = (\boldsymbol{\beta}^T, \boldsymbol{\theta}_1^T, \boldsymbol{\theta}_2^T)^T$. The log likelihood for $\boldsymbol{\Psi}$ that can be formed from the observed data $\boldsymbol{y}$ is given by

$$\begin{aligned} \log L(\boldsymbol{\Psi}) \;=\; & \sum_{j=1}^n [I_{[1]}(\delta_j) \log\{\pi_1(\boldsymbol{x}_j;\, \boldsymbol{\beta}) f_1(t_j;\, \boldsymbol{\theta}_1,\, \boldsymbol{x}_j)\} \\ & + I_{[2]}(\delta_j) \log\{\pi_2(\boldsymbol{x}_j;\, \boldsymbol{\beta}) f_2(t_j;\, \boldsymbol{\theta}_2,\, \boldsymbol{x}_j)\} \\ & + I_{[0]}(\delta_j) \log S(t_j;\, \boldsymbol{\Psi},\, \boldsymbol{x}_j)], \end{aligned} \tag{10.9}$$

where $I_{[h]}(\delta_j)$ is the indicator function that equals one if $\delta_j = h\,(h = 0, 1, 2)$.

If the component-survival functions belong to the exponential family, then we can fit the model (10.1) within the framework of mixtures of GLMs, as discussed in Chapter 5. McLachlan et al. (1997) have developed a program for fitting this model in the case where the component-survival functions are specified by assuming proportional hazards and taking the baseline hazard functions $h_{0i}(t)$ $(i = 1, 2)$ to have the Gompertz form. That is, the ith component hazard function is specified as

$$h_i(t;\, \boldsymbol{\theta}_i,\, \boldsymbol{x}) = \exp(\boldsymbol{\gamma}_i^T\, \boldsymbol{x})\, h_{0i}(t) \quad (i = 1, 2), \tag{10.10}$$

where

$$h_{0i}(t) = \exp(\lambda_i + \kappa_i\, t),$$

with $\kappa_i > 0$, and $\boldsymbol{\theta}_i = (\lambda_i,\, \kappa_i,\, \boldsymbol{\gamma}_i^T)^T$ $(i = 1, 2)$. The identifiability of mixtures of Gompertz distributions has been established by Gordon (1990a) in the case of mixing proportions that do not depend on any covariates. The extension to the case of mixing proportions specified by the logistic model (10.2) is straightforward. It follows that a sufficient condition for identifiability of the Gompertz mixture model is that the matrix $(\boldsymbol{x}_1^+,\, \ldots,\, \boldsymbol{x}_n^+)^T$ be of full rank, where

$$\boldsymbol{x}_j^+ = (1,\, \boldsymbol{x}_j)^T.$$

This model does not admit a closed-form solution on the M-step of the EM algorithm. But the calculation can be simplified by replacing the M-step with two computationally simple conditional maximization (CM) steps, as with the extension of the EM algorithm that Meng and Rubin (1993) termed the expectation–conditional maximization (ECM) algorithm. The two CM-steps here correspond to the subdivision of the parameter vector $\boldsymbol{\Psi}$ into $\boldsymbol{\beta}$ and $(\boldsymbol{\theta}_1^T,\, \boldsymbol{\theta}_2^T)^T$.

10.3 EXAMPLE 10.1: HEART-VALVE DATA

10.3.1 Description of Problem

To illustrate the application of mixture models for competing risks in practice, we consider the problem studied in Ng et al. (1999). They considered the use of the two-component mixture model (10.1) to estimate the probability that a patient aged x years would undergo a rereplacement operation after having his/her native aortic valve replaced by a xenograft prosthesis. At the time of the initial replacement operation, the surgeon has the choice of using either a mechanical valve or a biologic valve, such as a xenograft (made from porcine valve tissue) or an allograft (human donor valve). Modern-day mechanical valves are very reliable, but a patient must take blood-thinning drugs for the rest of his/her life to avoid thromboembolic events. On the other hand, biologic valves have a finite working life, and thus they have to be replaced if the patient were to live for a sufficiently long enough time after the initial replacement operation. Thus inferences about the probability that a patient of a given age will need to undergo a rereplacement operation can be used to assist a heart surgeon in deciding on the type of valve to be used in view of the patient's age.

With respect to this problem, we now let T denote the time to either a rereplacement operation or to death without a rereplacement operation. For brevity, we shall henceforth refer to the event of a rereplacement operation as a reoperation. The data set consisted of $n = 950$ cases of aortic valve replacements that were performed with xenograft prostheses. There were 62 subsequent valve-replacement operations due to either xenograft degeneration or some other reason, while 198 patients died without a reoperation. The remaining 690 survival times were all censored. The proportion of censored observations was as large as 73%.

The observed data are of the form

$$\boldsymbol{y}_1 = (t_1, x_1, \delta_1)^T, \ldots, \boldsymbol{y}_n = (t_n, x_n, \delta_n)^T,$$

where $\delta_j = 1$ if the jth patient undergoes a reoperation, $\delta_j = 2$ if the jth patient dies without a reoperation, and $\delta_j = 0$ if the jth patient is still alive without having undergone a reoperation by the end of the study (that is, the failure time is then right-censored at time t_j).

10.3.2 Mixture Models with Unconstrained Components

For a patient aged x years at the time of the initial replacement operation, we can model the survival function of T by (10.1), where $S_1(t;\, x)$ denotes the conditional survival function given the patient undergoes a reoperation and $S_2(t;\, x)$ denotes the conditional survival function given the patient dies without needing a reoperation. The mixing proportion $\pi_1(x)$ then represents the probability of interest, namely that a patient aged x years at the time of the initial replacement operation will undergo a reoperation in his/her lifetime. The logistic model

$$\pi_1(x;\, \boldsymbol{\beta}) = 1 - \pi_2(x;\, \boldsymbol{\beta}) = \exp(a + bx)/\{1 + \exp(a + bx)\}, \qquad (10.11)$$

where $\boldsymbol{\beta} = (a,\, b)^T$, was adopted for the mixing proportions $\pi_1(x)$ and $\pi_2(x)$, corresponding to reoperation and death without reoperation, respectively, as a function of the single covariate x, being the age of the patient at the time of the initial replacement operation.

With this mixture approach, Ng et al. (1999) found that the specification of typically used parametric forms for the component-survival functions in the mixture model (10.1) gave unsatisfactory results. For example, they initially considered parametric forms for the component-survival functions specified by (10.10). The assumption of the Gompertz model for death in the absence of the competing risk of reoperation (that is, for the latent survival function for death) is often used in the literature to model death in the absence of competing risks. However, it does not follow that this model will be applicable for death without a reoperation. Indeed, Ng et al. (1999) found that the Gompertz model for the baseline hazard function and proportional hazards for the effect of age was inadequate for the second component corresponding to death without a reoperation.

10.3.3 Constrained Mixture Models

Because a xenograft prosthesis will always need replacement if the patient were to live for a sufficiently long enough time after the initial operation, the two mixture model components corresponding to reoperation and death without a reoperation need to be interrelated in their parametric specification. Ng et al. (1999) handled this situation by constraining the components so that the component-hazard function for death without a reoperation was always greater than the hazard for reoperation. They effected this by setting the component-survival function corresponding to death without a reoperation to be equal to the component survival function for reoperation multiplied by some additional survival function. More specifically, they set

$$S_2(t;\, x) = S_a(t;\, x)\, S_1(t;\, x), \tag{10.12}$$

where $S_a(t;\, x)$ denotes some additional survival function.

From (10.12), it follows that

$$\begin{aligned} h_2(t;\, x) &= h_a(t;\, x) + h_1(t;\, x) && (10.13) \\ &> h_1(t;\, x) \qquad \forall t, && (10.14) \end{aligned}$$

as desired for this problem. The model (10.12) for the component survival function $S_2(t;\, x)$ for death without a reoperation has an easy to understand interpretation. Given that a patient will die without a reoperation, his/her conditional hazard function can be viewed as being equal to the conditional hazard function for reoperation plus some an additional hazard denoted here by $h_a(t;\, x)$.

In their initial analysis, Ng et al. (1999) took the same parametric form for $S_a(t;\, x)$; that is, they assumed a proportional hazards function with the Gompertz distribution used to model the baseline hazard function. Consequently, the component-hazard function for death without a reoperation no longer had the proportional hazards form.

However, this assumption of the Gompertz model for both $h_1(t;\, x)$ and $h_a(t;\, x)$ implies that their sum $h_2(t;\, x)$, corresponding to death without reoperation, must also be an increasing function. However, such a monotone hazard function may not be appropriate in some situations; for example, in survival data where the presence of short-term deaths is evident. The occurrence of death very soon after the treatments results in a nonmonotone U-shaped hazard function with three phases (a decreasing hazard, followed by a constant hazard, and finally by an increasing hazard); see; for example, Blackstone et al. (1986) and McGiffin et al. (1993). This led Ng et al. (2000) to specify $S_a(t;\, x)$ by a Weibull model with hazard function of the form

$$h_a(t;\, x) = \lambda_a \kappa_a t^{(\kappa_a - 1)} \exp(\gamma_a x), \tag{10.15}$$

where $\lambda_a,\, \kappa_a > 0$. With $h_1(t;\, x)$ still specified by a Gompertz model, the assumption of the Weibull model for $h_a(t;\, x)$ implies that $h_2(t;\, x)$ is an increasing function if $\kappa_a \geq 1$ and is a U-shaped function if $\kappa_a < 1$.

We shall refer to this constrained mixture model with one component based on the Gompertz distribution and the other on a Weibull times a Gompertz, as the

Weibull*Gompertz–Gompertz mixture model. The FORTRAN program of McLachlan et al. (1997) for the fitting of mixtures of Gompertz components with proportional hazards was modified to handle the constrained form (10.14). In Figure 10.1 we have plotted the estimate of the probability $\pi_1(x; \boldsymbol{\beta})$ that a patient aged x years at the initial replacement operation will undergo a reoperation, as obtained by this constrained mixture model, along with the corresponding estimate of $\pi_1(x)$ obtained by the latent failure-time approach under the assumption of independent competing risks. The 90% confidence limits are given by the dotted curves. They were obtained by applying the nonparametric bootstrap approach with the resampling scheme slightly modified for the competing risks problem. Let N_i be the number of cause i failures $(i = 1, 2)$, and let N_3 be the number of censored observations. The bootstrap data were obtained by sampling separately from each of the three sets, corresponding to failure from cause $i\,(i = 1, 2)$ and to the censored observations, with the sizes of these bootstrap subsamples taken equal to N_1, N_2, and N_3, respectively. A similar resampling scheme has been used by Golbeck (1992) in bootstrapping life-table estimators. In this application, $K = 100$ bootstrap samples were used.

It can be seen from Figure 10.1 that beyond 30 years, the Weibull*Gompertz–Gompertz mixture model gives an increasingly lower estimate of the probability of reoperation with increasing age of the patient. The difference is due to how the censored observations are "treated" during the estimation process. With this mixture model, elderly patients with censored observations are regarded as having a higher chance of dying before a reoperation is needed than with the latent failure-time approach.

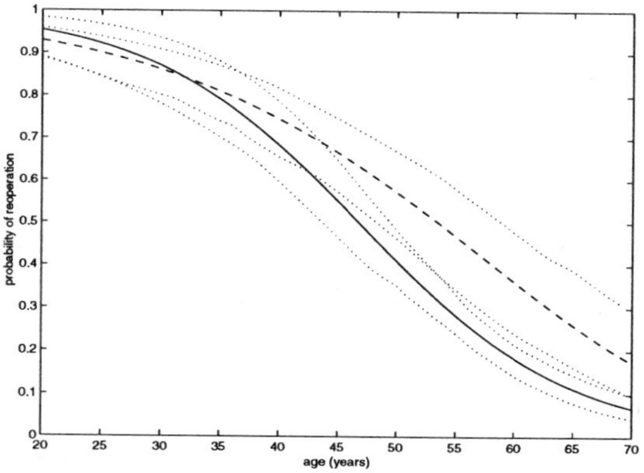

Fig. 10.1 Estimated probability of reoperation at a given age of patient: Weibull*Gompertz–Gompertz mixture model (solid curve); latent failure-time approach (dashed curve); 90% confidence limits (dotted curves). From Ng et al. (2000).

10.3.4 Conditional Probability of a Reoperation

Up to now we have focused on the probability that a patient aged x years at the time of the initial replacement operation of the aortic valve by a xenograft prosthesis will have to undergo a reoperation. As a patient can avoid a reoperation by dying first, it is relevant to consider the conditional probability of a reoperation within a specified time t after the initial operation given that the patient does not die without a reoperation during this period. We shall denote this conditional probability by $CP_R(t;\, x)$. In Figure 10.2 we have plotted the estimate of this conditional probability, as obtained by Ng et al. (2000), versus t for various values of the age x of the patient, along with the corresponding plots given by the latent failure-time approach applied under the assumption of independent competing risks. Although in Figure 10.2 we have plotted the conditional probability of reoperation for t up to 18 years for a patient aged $x = 70$ years at the time of the initial replacement operation, this estimate for large x should only be used for values of t in the practical range of interest. This is because for large values of t, the proportional hazards model (10.10) for the effect of age x provides only a crude approximation to reality for large x.

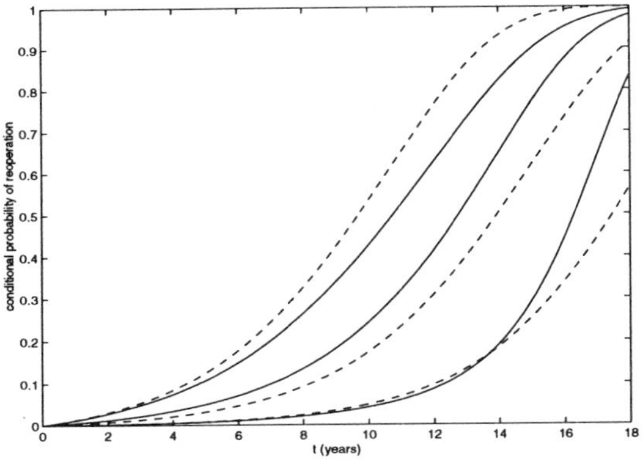

Fig. 10.2 Conditional probability of reoperation (xenograft valve) for specified age of patient: Weibull*Gompertz–Gompertz mixture model (solid curves); latent failure-time approach (dashed curves). With each model, $x = 30$ years (top curve), $x = 50$ years (middle curve), and $x = 70$ years (bottom curve). From Ng et al. (2000).

10.3.5 Advantages of Mixture Model-Based Approach

An attractive feature of the mixture model approach is that it does not have to make assumptions about the independence of the competing risks, as with the latent failure-time approach. As pointed by Lagakos (1979), assumption of independence of risks seems unreasonable and questionable in most real-life situations. But any model that allows for dependence is restricted to have a simple parametric form for the joint distribution of the latent times, for example, a bivariate normal model, as proposed by Nádas (1971). Furthermore, the dependence of the competing risks is nontestable. Cox (1959; 1962, p. 112) and Tsiatis (1975) showed that for any joint distribution for the latent-failure times, there exists a joint distribution with independent latent-failure times that gives the same distribution of the observable failure times. Thus from the observable failure times and causes of failure alone, it is impossible to distinguish between an independent competing-risks model and an infinitude of dependent models. The identifiability of the latent failure-time approach with covariates has been considered by Heckman and Honoré (1989) and Slud (1992). Klein and Moeschberger (1987) determined the effects of an incorrect assumption of independence and concluded that departures from independence were of great consequence on the estimates of the overall and marginal survival functions. In contrast to the latent failure-time approach, the postulated component survival functions and mixing proportions of the mixture model are able to be estimated directly from the observable data.

10.4 LONG-TERM SURVIVOR MODEL

10.4.1 Definition

In some situations where the aim is to estimate the survival distribution for a particular type of failure, a certain fraction of the population, say π_1, may never experience this type of failure. It is characterized by the overall survival curve being leveled at a nonzero probability. In some applications, the surviving fractions are said to be "cured" (Boag, 1949; Berkson and Gage, 1952; Gordon, 1990a). For example, in a clinical trial for assessing the efficacy of a treatment therapy for breast cancer, a patient may be considered cured if she lives an apparently normal life span and dies of causes other than breast cancer and shows no evidence of the latter disease at death. This is the definition of "personal cure" in Haybittle (1983); see also Brinkley and Haybittle (1975). In a criminological study, there is the possibility that some offenders will not reoffend after release (Copas and Heydari, 1997). The estimation of π_1 is of interest in many of these applications.

It is assumed that an entity or individual has probability $\pi_2 = 1-\pi_1$ of failing from the cause of interest and probability π_1 of never experiencing failure from this cause. In the usual framework for this problem, it is assumed further that the entity cannot fail from any other cause during the course of the study (that is, during follow-up). We let T be the random variable denoting the time to failure, where $T = \infty$ denotes the event that the individual will not fail from the cause of interest. The probability of this latter event is π_1.

The unconditional survival function of T can then be expressed as

$$S(t) = \pi_1 + \pi_2 \, S_2(t), \tag{10.16}$$

where $S_2(t)$ denotes the conditional survival function for failure from the cause of interest. The mixture model (10.16) with the first component having mass one at $T = \infty$ can be regarded as a nonstandard mixture model, as discussed in a general context in Chapter 5. Nonstandard mixture models are used in a variety of applications in survival analysis; see, for example, Longini and Halloran (1996) for a class of frailty models with point mass at the origin.

The model (10.16) is sometimes referred to as the long-term survival (LTS) mixture model for the obvious reason that individuals who will never fail due to the cause of interest can be viewed as being long-term survivors of the cause; see Langlands et al. (1979), Farewell (1982), and Greenhouse and Wolfe (1984). The recent monograph by Maller and Zhou (1996) provides a wide range of applications of the LTS mixture model. Here we have not taken $S_2(t)$ nor the π_i to depend on covariates, but this situation can be handled as in Section 10.2.

In the typical situation in which the LTS mixture model is applied, the observed data are of the form $(t_j, \delta_j)^T$ for the jth entity, where $\delta_j = 2$ implies that the jth individual was observed to fail from the cause of interest at time t_j during the follow-up period, and $\delta_j = 0$ implies that the failure time was censored at time t_j. That is, the jth individual had not experienced failure by time t_j. In this context, the model (10.16) can be fitted by maximum likelihood after the specification of a parametric model $S_2(t;\, \boldsymbol{\theta}_2)$ for $S_2(t)$. An individual who fails from the cause of interest at time t_j $(j = 1, \ldots, n)$ contributes a likelihood factor,

$$\pi_2\, f_2(t_j;\, \boldsymbol{\theta}_2), \tag{10.17}$$

where $f_2(t;\, \boldsymbol{\theta}_2)$ is the density of T corresponding to $S_2(t;\, \boldsymbol{\theta}_2)$. An individual who has been followed to time t_j without failure contributes a likelihood factor

$$\pi_1 + \pi_2\, S_2(t_j;\, \boldsymbol{\theta}_2), \tag{10.18}$$

which is the probability that an individual never experiences failure.

The log likelihood, $\log L(\boldsymbol{\Psi})$, for $\boldsymbol{\Psi} = (\pi_1,\, \boldsymbol{\theta}_2^T)^T$ is given by

$$\log L(\boldsymbol{\Psi}) = \sum_{j=1}^{n} [I_{[2]}(\delta_j) \log\{\pi_2\, f_2(t_j;\, \boldsymbol{\theta}_2)\} + I_{[0]}(\delta_j) \log\{\pi_1 + \pi_2\, S_2(t_j;\, \boldsymbol{\theta}_2)\}]. \tag{10.19}$$

It can be seen that this model is simpler to fit than the two-component mixture model (10.1) in which the survival function for time to failure from a cause or causes other than the cause of interest has to be included in the estimation process.

10.4.2 Modified Long-Term Survivor Model

We now consider a generalization of the LTS problem where during the follow-up period an individual may be observed to fail from a cause other than the cause of interest. For example, in a study of breast cancer, some individuals may die from another cause and without any symptoms of breast cancer. The LTS mixture model is not directly applicable to such problems. An obvious way to handle this situation is to define the failure time T to be the time to either failure from the cause of interest or failure from a competing cause. A two-component mixture model with components representing the competing causes and the cause of interest, respectively, can be adopted. This competing-risks model was considered in Section 10.2.

10.4.3 Partial ML Approach for Modified Long-Term Survival Model

The competing-risks model (10.1) requires the specification of a suitable parametric form for the conditional survival function $S_1(t;\ \boldsymbol{\theta}_1)$ for time to failure from causes other than the cause of interest. In situations where only a few failures from competing causes occur during the observational period, there may be insufficient data to estimate $\boldsymbol{\theta}_1$. Moreover, it can be seen from (10.9) that even if we were to ignore those failures from competing causes, we would still have to model $S_1(t;\ \boldsymbol{\theta}_1)$ as it appears in the contribution $S(t_j;\boldsymbol{\Psi})$ from a censored failure time t_j. Ng and McLachlan (1998) considered how the LTS mixture model can still be used to handle the possibility of risk of failure from a competing cause during the observational period. Their proposed modification, which is based on a partial ML approach, circumvents the need to model failure from the competing risks, but still provides a consistent estimator of the conditional survival function for the cause of interest and for the proportion who fail from a competing cause. With this partial ML approach, we do not fully utilize the information that there are failures from competing causes observed during the follow-up period. Rather we form their likelihood contributions based on only a subset of the information, namely that the failure of interest did not occur for this group of patients (those with $\delta_j = 1$) during the follow-up period. Hence we are effectively treating a patient who failed due to a competing cause as having an observation time censored at the end of the follow-up period. That is, if a failure from a competing cause is recorded at time t_j, then this event contributes the term

$$\pi_1 + \pi_2 S_2(t_j^*;\ \boldsymbol{\theta}_2),$$

where $t_j^* = t_j + u_j$, and u_j is the time from the occurrence of failure from a competing cause to the end of the follow-up period.

Because all the available information has not been used in forming the likelihood, Ng and McLachlan (1998) referred to this approach as the partial ML approach, and they referred to the approach based on the competing-risks model (10.1) as the full ML approach. The simplicity of the partial ML approach is somewhat offset by a loss in efficiency of the estimates so obtained. Ng and McLachlan (1998) performed some simulations to compare the efficiency of the partial ML approach relative to the full ML approach for survival in the presence of competing risks. They concluded that the partial ML approach is reasonably efficient, provided that the risk of failure from a competing cause is low compared with the risk of failure from the cause of interest. Practically, for clinical studies, this situation is justified because the risk of failure from the disease under study is usually much higher than that of failure due to other reasons for a cured patient. In these circumstances, the partial ML approach has particularly high relative efficiency if there are limited failures from competing causes; for example, when the follow-up period is only of medium duration and the sample size is not large. In these situations, it is not possible with the full ML approach to estimate reliably the conditional survival function and hence the probability of failure from competing causes.

10.4.4 Interpretation of Cure Rate in Presence of Competing Risks

With model (10.16), the probability $\pi_2 = 1 - \pi_1$ represents the chance of failing from the cause of interest in the presence of other risks. Equivalently, π_1 is the probability that an individual will not fail from the cause of interest, and thus will fail from one of the competing causes. In some applications such as with the estimation of the cure rate of breast cancer, patients who die shortly after the treatment, apparently from a competing cause, might not be regarded as cured even if they have no symptoms of the disease at the time of their death. For with early deaths, it is not clear whether they would have suffered a relapse from the disease under study if they had continued to live for a longer time. Also, some of the early deaths might even be due to the hazards of the treatment, such as postoperative complications. In these applications, π_1 can be adjusted by, for example, excluding those patients with death times smaller than T_0 as being cured. In so doing this, we are taking a patient to be cured if he/she lives an apparently normal life span and dies from a competing cause at time at least T_0 after treatment and without any symptoms of the disease. That is, the cure rate is taken to be

$$\pi_1 \, S_1(T_0). \tag{10.20}$$

With the partial ML approach, the conditional survival function $S_1(t)$ is not estimated. But if the smallest censored time is greater than T_0, then we are still able to estimate $S_1(T_0)$ by

$$1 - \frac{n_{T_0}}{n \pi_1}, \tag{10.21}$$

where n is the total number of patients and n_{T_0} is the number of patients who died from a competing cause before time T_0. For example, with breast cancer, it was reported in Burch (1976, p. 387) that there is a probability of 0.73 for a recurrence within 5 years. A larger value of T_0 could be used if a tighter interpretation of normal life span is assumed. Obviously, T_0 should be chosen before the study so as not to induce biases.

10.4.5 Example 9.2: Breast Cancer Data

Ng and McLachlan (1998) analyzed the breast cancer data of Boag (1949) to illustrate the application of the full ML and the proposed partial ML approaches in the estimation of cure rate. The data were recorded for 121 patients having treatment for breast cancer. For their analysis, the patients were classified as follows:

1. Those who responded favorably to treatment and were free from signs or symptoms of breast cancer at their death from other causes at time t $(\delta_j = 1)$;

2. those who died due to breast cancer at time t_j $(\delta_j = 2)$;

3. those who were still alive at the end of follow-up period $(\delta_j = 0)$.

The number (proportion) in each category was 18(15%), 78(64%), and 25(21%), respectively. These data are given in Table 10.1.

Table 10.1 Breast Cancer Data

Survival Times (in months) of Those								Length of Time (in months) Survivors Were in the Study	
Who Died with Cancer Present						Who Died from Other Causes			
0.3	12.2	17.5	28.2	41	78	0.3	110	111	136
5.0	12.3	17.9	29.1	42	80	4.0	111	112	141
5.6	13.5	19.8	30.0	44	84	7.4	112	113	143
6.2	14.4	20.4	31	46	87	15.5	132	114	167
6.3	14.4	20.9	31	48	89	23.4	162	114	177
6.6	14.8	21.0	32	48	90	46		117	179
6.8	15.7	21.0	35	51	97	46		121	189
7.5	16.2	21.1	35	51	98	51		123	201
8.4	16.3	23.0	38	52	100	65		129	203
8.4	16.5	23.6	39	54	114	68		131	203
10.3	16.8	24.0	40	56	126	83		133	213
11.0	17.2	24.0	40	60	131	88		134	228
11.8	17.3	27.9	41	78	174	96		134	

Source: Boag (1949).

For the partial ML approach, those patients with $\delta_j = 1$ are treated as censored at the end of follow-up period. This information, which would be usually known in practice, was not available to Ng and McLachlan (1998). Hence they randomly chose one of the censored times to be the length of the study for each individual whose death was not due to breast cancer.

For the full ML approach, they adopted the Gompertz model

$$S_1(t;\,\boldsymbol{\theta}_1) = \exp[-\exp(\lambda_1)\{\exp(\kappa_1 t) - 1\}/\kappa_1],$$

for the conditional survival function for death from causes other than breast cancer and the exponential model

$$S_2(t;\,\boldsymbol{\theta}_1) = \exp(-\lambda_2)$$

for the conditional survival function for death from breast cancer. The results are given in Table 10.2. Also included there are the cure rates adjusted with $T_0 = 4$ years ($n_{T_0} = 7$), $T_0 = 5$ years ($n_{T_0} = 8$), and $T_0 = 6$ years ($n_{T_0} = 10$).

From Table 10.2, it can be observed that the partial ML approach gives estimates comparable with the full ML approach. The relatively good performance of the partial ML approach here can be explained by the fact there is a relatively low risk of death

Table 10.2 Estimates (Estimated Asymptotic Variances) Obtained by the Full and Partial ML Approaches

Method	$\hat{\lambda}_1$	$\hat{\kappa}_1$	$\hat{\lambda}_2$	$\hat{\pi}_1$	Cure Rate		
					$T_0 = 4$	$T_0 = 5$	$T_0 = 6$
Full ML	-5.396	-0.002	0.0225	0.32918	0.271	0.263	0.247
	(0.168)	(2E-5)	(1E-5)	(0.00215)	(0.002)	(0.002)	(0.002)
Partial ML			0.0219	0.32194	0.264	0.256	0.239
			(1E-5)	(0.00230)	(0.002)	(0.002)	(0.002)

Source: Adapted from Ng and McLachlan (1998).

from other causes compared to death from breast cancer and the sample size is small. The estimates of the cure rate are comparable to those (0.26 and 0.27, respectively) obtained by Boag (1949) and Berkson and Gage (1952) using different approaches to analyze the breast cancer data.

The conditional survival function for death from breast cancer, $S_2(t; \hat{\boldsymbol{\theta}}_2)$, is plotted in Figure 10.3 for the partial and full ML approaches, along with a nonparametric estimate as proposed by Taylor (1995). It can be seen that there is close agreement between the three estimates of the conditional survival function for death due to breast cancer.

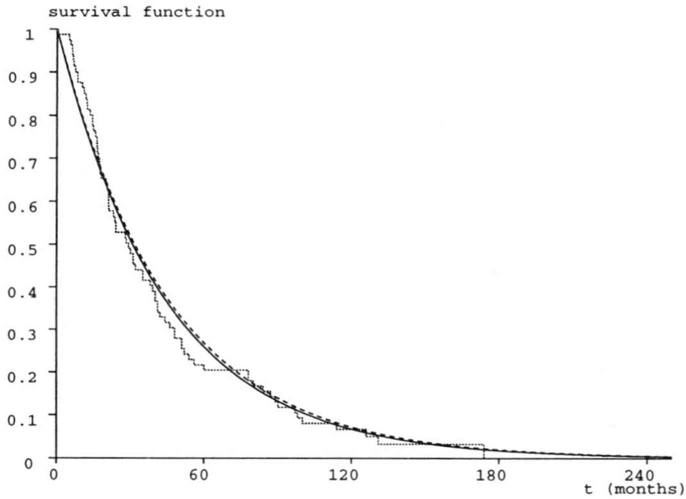

Fig. 10.3 The conditional survival function for death due to competing risks: full ML approach ——— ; partial ML approach - - - - - - ; nonparametric approach ·······. From Ng and McLachlan (1998).

10.5 ANALYSIS OF MASKED SYSTEM-LIFE DATA

10.5.1 Masked Cause of Failure

We conclude this chapter by considering a problem in reliability analysis where mixture models form the basis for the estimation of the life of electronic parts in series systems as used, for example, in the telecommunications industry. It illustrates the wide use of mixture models in the statistical analysis of reliability data. This problem has been considered by Albert and Baxter (1995), among others.

Suppose there is a series system which functions if and only if all its constituents are functioning. The system is assumed to comprise g components or parts. The life of part i is a random variable with the survival function $S(t;\ \boldsymbol{\theta}_i)$ and hazard function $h(t;\ \boldsymbol{\theta}_i)$ $(i = 1, \ldots, g)$, where the functional form of S is known and $\boldsymbol{\theta}_i$ is a vector of unknown parameters to be estimated. We let $\boldsymbol{\xi}$ denote the elements of $\boldsymbol{\theta}_1, \ldots, \boldsymbol{\theta}_g$ known *a priori* to be distinct. Here the vector $\boldsymbol{\Psi}$ of all unknown parameters is equal to $\boldsymbol{\xi}$ in the absence of any mixing proportions to be estimated.

The g parts are grouped into $g^*(< g)$ modules; and to expedite repair, the module containing the failed part is replaced. That is, only the identity of the module containing the failed part is known. This is known as masking (Usher and Guess, 1989; Guess, Usher, and Hodgson, 1990; Usher, 1996). The observed data $\boldsymbol{y}$ consists of n independent observations on the system so that $\boldsymbol{y} = (\boldsymbol{y}_1^T, \ldots, \boldsymbol{y}_n^T)^T$, where $\boldsymbol{y}_j = (t_j,\ M_j)^T$ and where t_j is the observed life length of the jth system and M_j is the set of labels defining the parts in the module that failed in the jth system $(j = 1, \ldots, n)$.

The log likelihood for $\boldsymbol{\Psi}$ that can be formed on the basis of the observed data $\boldsymbol{y}$ is given by

$$\begin{aligned} \log L(\boldsymbol{\Psi}) &= \sum_{j=1}^{n} \log[\{\sum_{i \in M_j} h(t_j;\ \boldsymbol{\theta}_i)\}\{\prod_{i=1}^{g} S(t_j;\ \boldsymbol{\theta}_i)\}] \\ &= \sum_{j=1}^{n} \{\log \sum_{i \in M_j} h(t_j;\ \boldsymbol{\theta}_i) + \sum_{i=1}^{g} \log S(t_j;\ \boldsymbol{\theta}_i)\}. \end{aligned} \tag{10.22}$$

10.5.2 Application of EM Algorithm

To apply the EM algorithm to this problem, we define the complete-data vector $\boldsymbol{y}_c$ to be

$$\boldsymbol{y}_c = (\boldsymbol{y}^T,\ \boldsymbol{z}^T)^T,$$

where

$$\boldsymbol{z} = (\boldsymbol{z}_1^T, \ldots, \boldsymbol{z}_n^T)^T$$

and $z_{ij} = (\boldsymbol{z}_j)_i = 1$ if part i were the part that failed in the jth system, and is zero otherwise.

The complete-data log likelihood, $\log L_c(\boldsymbol{\Psi})$, is given by

$$\log L_c(\boldsymbol{\Psi}) = \sum_{i=1}^{g}\sum_{j=1}^{n} z_{ij}\{\log h(t_j;\, \boldsymbol{\theta}_i) + \log S(t_j;\, \boldsymbol{\theta}_i)\}.$$

Concerning the E-step on the $(k+1)$th iteration of the EM algorithm, we have that the current conditional expectation of the complete-data log likelihood is

$$\begin{aligned} Q(\boldsymbol{\Psi};\, \boldsymbol{\Psi}^{(k)}) &= E_{\boldsymbol{\Psi}^{(k)}}\{\log L_c(\boldsymbol{\Psi}) \mid \boldsymbol{y}\} \\ &= \sum_{i=1}^{g}\sum_{j=1}^{n} \tau_i(t_j;\, \boldsymbol{\Psi}^{(k)})\, \{\log h(t_j;\, \boldsymbol{\theta}_i) + \log S(t_j;\, \boldsymbol{\theta}_i)\}\,, \end{aligned}$$

where

$$\begin{aligned} \tau_i(t_j;\, \boldsymbol{\Psi}^{(k)}) &= E_{\boldsymbol{\Psi}^{(k)}}\{Z_{ij} \mid \boldsymbol{y}_j\} \\ &= \mathrm{pr}_{\boldsymbol{\Psi}^{(k)}}\{Z_{ij} = 1 \mid \boldsymbol{y}_j\} \end{aligned}$$

for $i = 1, \ldots, g;\ j = 1, \ldots, n$. It follows that

$$\mathrm{pr}_{\boldsymbol{\Psi}^{(k)}}\{Z_{ij} = 1 \mid \boldsymbol{y}_j\} = \begin{cases} h(t_j;\, \boldsymbol{\theta}_i) / \sum_{h \in M_j} h(t_j;\, \boldsymbol{\theta}_h) & \text{if } i \in M_j, \\ 0 & \text{otherwise.} \end{cases} \tag{10.23}$$

10.5.3 Exponential Components

As an illustration, we consider the special case of an exponential distribution for the parts,

$$f(y;\, \alpha_i) = \alpha_i \exp(-\alpha_i t_j) I_{(0,\infty)}(t_j),$$

where $\theta_i = \alpha_i\ (i = 1, \ldots, g)$, as considered by Albert and Baxter (1995). Then

$$Q(\boldsymbol{\Psi};\, \boldsymbol{\Psi}^{(k)}) = \sum_{i=1}^{g}\sum_{j=1}^{n} \tau_i(t_j;\, \boldsymbol{\Psi}^{(k)})\{\log \alpha_i - \alpha_i t_j\}. \tag{10.24}$$

On differentiation of the right-hand side of (10.24) with respect to α_i, we have that $\alpha_i^{(k+1)}$ is given by

$$\alpha_i^{(k+1)} = \{\sum_{j=1}^{n} t_j / \sum_{j=1}^{n} \tau_i(t_j;\, \boldsymbol{\Psi}^{(k)})\}^{-1} \quad (i = 1, \ldots, g), \tag{10.25}$$

where

$$\tau_i(t_j;\, \boldsymbol{\Psi}^{(k)}) = \begin{cases} \alpha_i^{(k)} / \sum_{h \in M_j} \alpha_h^{(k)} & \text{if } i \in M_j, \\ 0 & \text{otherwise.} \end{cases} \tag{10.26}$$

10.5.4 Weibull Components

Albert and Baxter (1995) also considered the case where the density of the life of each part is Weibull with survival function

$$S(t;\, \boldsymbol{\theta}_i) = \exp\{-(\alpha_i\, t)^{\kappa_i}\} I_{(0,\,\infty)}(t),$$

where $\boldsymbol{\theta}_i = (\alpha_i,\, \kappa_i)^T$ $(i = 1, \ldots, g)$.

The E-step is still easily effected with $\tau_i(t_j;\, \boldsymbol{\Psi}^{(k)})$ being given by (10.23), but with

$$h(y_i;\, \boldsymbol{\theta}_i) = \alpha_i^{\kappa_i} \kappa_i\, t_j^{\kappa_i - 1} \quad (i = 1,\, \ldots,\, g).$$

However, the solution to the M-step does not exist in closed form. On differentiation of $Q(\boldsymbol{\Psi};\, \boldsymbol{\Psi}^{(k)})$ with respect to α_i and κ_i, it follows that $\alpha_i^{(k+1)}$ and $\kappa_i^{(k+1)}$ satisfy the equations

$$\alpha_i^{(k+1)} = \left\{ \sum_{j=1}^{n} t_j^{\kappa_i} \Big/ \sum_{j=1}^{n} \tau_i(t_j;\, \boldsymbol{\Psi}^{(k)}) \right\}^{1/\kappa_i} \tag{10.27}$$

and

$$\sum_{j=1}^{n} \Big[\tau_i(t_j;\, \boldsymbol{\Psi}^{(k)}) \{1/\kappa_i + \log(\alpha_i\, t_j)\} - (\alpha_i\, t_j)^{\kappa_i} \log(\alpha_i\, t_j) \Big] = 0 \tag{10.28}$$

for $i = 1,\, \ldots,\, g$.

These equations do not admit a closed-form solution. But they suggest that the solution to the M-step on the $(k+1)$th iteration of the EM algorithm be replaced by two computationally simple CM steps, thus using an ECM algorithm.

We partition $\boldsymbol{\Psi}$ as

$$\boldsymbol{\Psi} = (\boldsymbol{\Psi}_1^T,\, \boldsymbol{\Psi}_2^T)^T,$$

where

$$\boldsymbol{\Psi}_1 = (\alpha_1,\, \ldots,\, \alpha_g)^T$$

is the vector of Weibull scale parameters and

$$\boldsymbol{\Psi}_2 = (\kappa_1,\, \ldots,\, \kappa_g)^T$$

is the vector of Weibull shape parameters.

With the first CM-step, we choose $\boldsymbol{\Psi}_1^{(k+1)}$ to be the value of $\boldsymbol{\Psi}_1$ that maximizes the Q-function with $\boldsymbol{\Psi}_2$ fixed at $\boldsymbol{\Psi}_2^{(k)}$; that is, $\boldsymbol{\Psi}_1^{(k+1)}$ is the value of $\boldsymbol{\Psi}_1$ that maximizes

$$Q(\boldsymbol{\Psi}_1,\, \boldsymbol{\Psi}_2^{(k)};\, \boldsymbol{\Psi}^{(k)})$$

with respect to $\boldsymbol{\Psi}_1$. We then choose $\boldsymbol{\Psi}_2^{(k+1)}$ to be the value of $\boldsymbol{\Psi}_2$ that maximizes $Q(\boldsymbol{\Psi}_1^{(k+1)},\, \boldsymbol{\Psi}_2;\, \boldsymbol{\Psi}^{(k)})$ with respect to $\boldsymbol{\Psi}_2$.

This leads to $\alpha_i^{(k+1)}$ being given by (10.27) where κ_i is fixed at $\kappa_i^{(k)}$, and $\kappa_i^{(k+1)}$ is then given as a root of (10.28) with α_i fixed at $\alpha_i^{(k+1)}$.

Albert and Baxter (1995) suggested a similar approach to the computation of $\boldsymbol{\Psi}^{(k+1)}$, which they called pseudo-alternating EM (PAEM) algorithm. With this approach, $\boldsymbol{\Psi}_2^{(k+1)}$ is chosen to maximize the actual (incomplete-data) log likelihood, $\log L(\boldsymbol{\Psi})$, and not the Q-function, subject to $\boldsymbol{\Psi} = \boldsymbol{\Psi}_1^{(k+1)}$. This corresponds to using the ECME (expectation–conditional maximization either) algorithm of Liu and Rubin (1994). The ECM and ECME algorithms preserve the appealing convergence properties of the EM algorithm, such as its monotone convergence. The reader is referred to McLachlan and Krishnan (1997) for a detailed account of the ECM and ECME algorithms.

11

Mixture Analysis of Directional Data

11.1 INTRODUCTION

To illustrate the use of finite mixture models in the analysis of directional data, we present an example on the use of mixtures of Kent distributions as an aid in joint set identification. When examining a rock mass, joint sets and their orientations can play a significant role with regard to how the rock mass will behave. To identify joint sets present in the rock mass, measurements of the orientation of individual fractures can be measured on exposed rock faces and the resulting data examined for heterogeneity. In this chapter, we focus on the recent case study of Peel, Whiten, and McLachlan (2001), who used the EM algorithm to fit mixtures of Kent component distributions to the fracture data to aid in the identification of joint sets. An extra uniform component is also included in the model to accommodate the noise present in the data.

11.2 JOINT SETS

We introduce the concept of directional data through the example considered by Peel et al. (2001) on joint sets. In nature a rock mass is almost never a single block of uniform solid rock, instead it usually contains many fractures, or discontinuities. Since these fractures arise from a force acting on the rock mass, they will generally form in a nearly parallel fashion. These families of parallel fractures are termed joint sets. Figure 11.1 shows a hypothetical cross-section of a rock mass with a single joint set present.

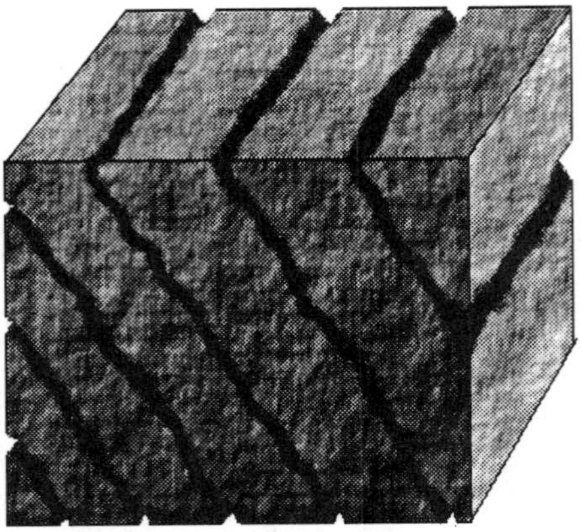

Fig. 11.1 Hypothetical cross-section of rock mass with single joint set present.

Since, over time, forces from more than one direction often act on a rock mass, a number of joint sets may be present (see Figure 11.2). How the rock mass will react to new external forces is greatly affected by the orientation of these joint sets, since the rock will most likely separate at its weakest point, being the fractures.

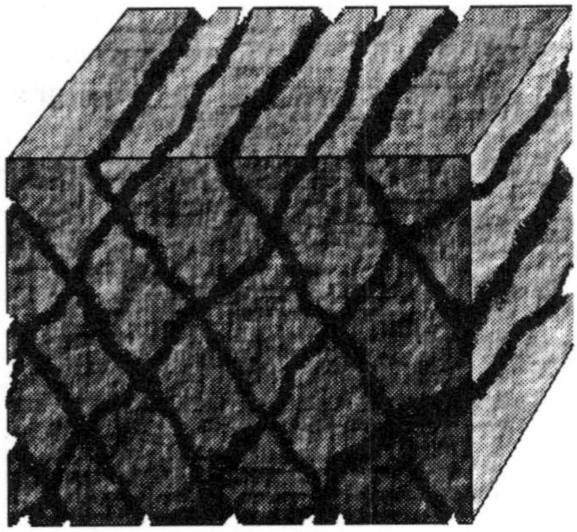

Fig. 11.2 Hypothetical cross-section of rock mass with two joint sets present.

In the mining industry it is advantageous to be able to predict how a rock mass will react. Obviously, when modeling a mine tunnel, the rock structure of the roof of a tunnel is extremely important. Other applications are in the design of ore extraction systems and the stability of open pit walls. When more than one joint set is present, blocks are formed (see Figure 11.3). The shape and location of these blocks is very significant. With certain configurations the roof will be very unstable, with the blocks falling causing a cave in. This is the case in the block caving method of mining, where the roof is purposely caved in and the collapsed rock is progressively removed. In other cases, this situation can be very dangerous, at worst costing lives, at best costing extra time and resources to clear the debris. In both situations, it is paramount to be able to predict, with some confidence, the rock mass structure.

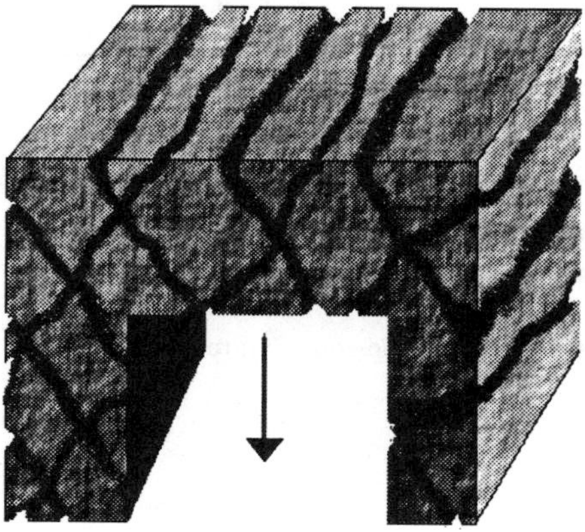

Fig. 11.3 Hypothetical cross-section of rock mass with two joint sets present forming blocks.

To determine the joint sets present, measurements are taken on site to determine what fractures are evident on an exposed rock face. This is accomplished by taking one or more lines across an exposed rock surface, and any fractures that intersect this line are described by two angles, dip direction and dip angle (as shown in Figure 11.4), corresponding to the direction of the normal to the fracture plane. Other measurements are also taken such as the distance along the measurement line and the nature of the fracture; see Peel et al. (2001) for further details.

The fracture measurements are used to discern clusters of fractures corresponding to joint sets. The current approach is to plot the dip direction and dip angle of each fracture on a polar plot. Then contours of the density of the data are added to enable the user to manually distinguish between joint sets (see Figure 11.5).

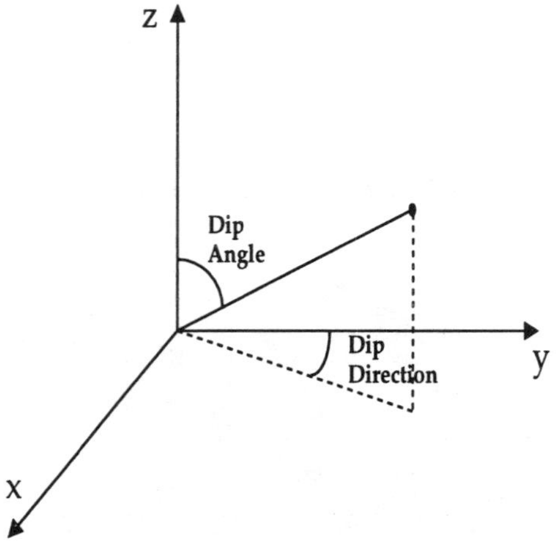

Fig. 11.4 Representation of dip angle and dip direction.

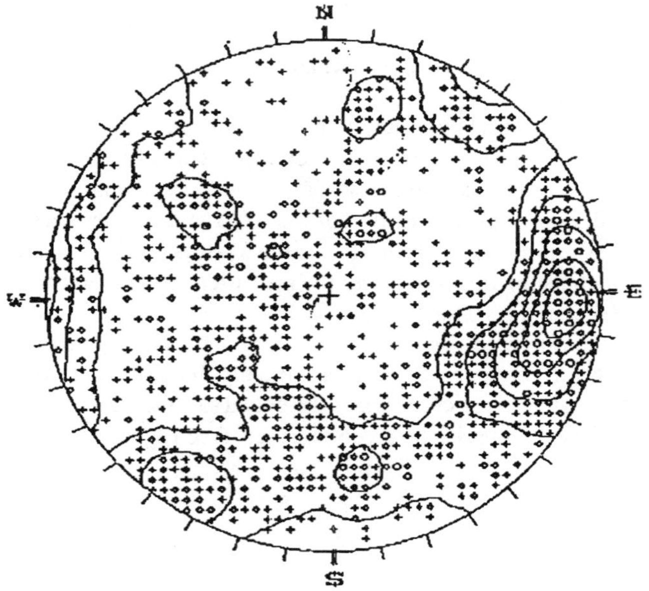

Fig. 11.5 Example of a contour plot.

Often since a large dense joint set can overshadow the other joint sets, points belonging to the first joint set located are removed and the data recontoured. This procedure is repeated at the user's discretion. This method is quite effective, but can be ambiguous with different users producing very different interpretations of the data. Also, the amount of user time and input needed is quite high. Hence there is a need for an automated method of clustering the joint sets, as proposed in Peel et al. (2001).

11.3 DIRECTIONAL DATA

The dip direction and dip angle measured are obviously directional vectors, so the measured samples are two-dimensional directional data. A two-dimensional directional sample can simply be thought of as a sample of points that lie on the surface of a unit sphere. Hence the sample points can be represented in either polar or Cartesian coordinates. Denoting the polar coordinates by (γ_1, γ_2) $(0 \leq \gamma_1 \leq 2\pi, 0 \leq \gamma_2 \leq \frac{\pi}{2})$, then γ_1 would be the dip direction and γ_2 would be the dip angle, as described in Figure 11.4. The Cartesian coordinates of a point are given by the directional cosines,

$$\begin{aligned} x_1 &= \cos(\gamma_2), \\ x_2 &= \sin(\gamma_2)\cos(\gamma_1), \\ x_3 &= \sin(\gamma_2)\sin(\gamma_1), \end{aligned}$$

where we have followed the notation of Kent (1982) in using x_1, x_2, and x_3 to refer to the z-, y-, and x-coordinates, respectively.

The polar coordinates (γ_1, γ_2) can be obtained from the directional cosines by

$$\gamma_1 = \tan^{-1}(x_3/x_2), \; \gamma_2 = \cos^{-1}(x_1).$$

It is also useful to note that the angle, θ, between two vectors $\boldsymbol{x}$ and $\boldsymbol{y}$ with directional cosines (x_1, x_2, x_3) and (y_1, y_2, y_3), respectively, is given by the inverse cosine of the dot product of the two vectors; that is,

$$\theta = \cos^{-1}(x_1 y_1 + x_2 y_2 + x_3 y_3).$$

In the sequel both the polar and Cartesian representations will be used.

Concerning the use of the mixture model

$$f(\boldsymbol{y}_j; \boldsymbol{\Psi}) = \sum_{i=1}^{g} \pi_i f_i(\boldsymbol{y}_j; \boldsymbol{\theta}_i) \tag{11.1}$$

for the analysis of directional data, an obvious choice for the component densities in (11.1) would be the Fisher distribution (due to R.A. Fisher), which can be thought of as an extension of the von Mises distribution from a circle to a sphere. The Fisher distribution is analogous to a circular bivariate normal, which corresponds to a normal distribution with a diagonal covariance matrix with equal eigenvalues. This allows for distributions of circular shape of varying size on the surface of the sphere.

11.4 INITIAL WORK ON CLUSTERING OF DIRECTIONAL DATA

We briefly consider work that existed prior to the study of Peel et al. (2001) for analyzing heterogeneous directional data. Fisher, Lewis, and Embleton (1987) state that there are three possible approaches to "multimodal" directional data: Partition the data into groups visually using contour plots; use a clustering method; or use a probability model (such as a mixture of Fisher distributions). Fisher et al. (1987) go on to point out with regard to clustering methods for directional data, that there is "little currently available in the literature which seems to be of practical use." One example is given of a clustering method based on a nonparametric density approach in Schaeben (1984).

With regard to fitting mixtures of Fisher distributions, Fisher et al. (1987) refer to Stephens (1969), who investigated fitting mixtures of Fisher distributions and found the estimation of the parameters to be tedious. However, it must be stated that the comment of Stephens (1969) was made when the available computer power was considerably less than that available today and the EM algorithm was not available.

Hsu, Walker, and Ogren (1986) fitted a mixture of bivariate von Mises distributions; that is, the dip angle and direction were assumed to be distributed independently, each with a von Mises distribution. This assumption does not allow for any correlation between dip direction and dip angle, which can certainly occur. If the cluster's elliptical contours are visualized on the surface of the sphere, then the use of bivariate von Mises component distributions constrains the axis of the cluster ellipses to be parallel to the longitude and latitude lines of the sphere. Other work in the literature includes Lund (1999), which examines the cluster analysis of circular data.

Nonmixture based cluster analysis of joint set data is discussed in Pal Roy (1995), Aler, Du Mouza, and Arnould (1996), and Hammah and Curran (1998).

11.5 MIXTURE OF KENT DISTRIBUTIONS

Peel et al. (2001) used the Kent, rather than the Fisher distribution for the mixture model components, so as to provide greater flexibility. The Kent distribution can be thought of as a generalization of the Fisher distribution, or as a special case of the more general 8-parameter family of distributions known as the Fisher–Bingham family; see Kent (1982). In the same way that the Fisher distribution is comparable to a bivariate normal with a constrained covariance matrix, the Kent distribution is comparable to a bivariate normal where the covariance matrices are unconstrained. This allows for distributions of any elliptical shape, size, and orientation on the surface of the sphere.

Hence the use of Kent distributions provides a more flexible alternative to model the data.

The Kent density is given in terms of the directional cosines $\boldsymbol{y} = (y_1, y_2, y_3)^T$ and is given by

$$f_K(\boldsymbol{y}; \boldsymbol{\theta}) = C_K \exp[\kappa(\boldsymbol{y}^T \boldsymbol{\omega}_1) + \beta\{(\boldsymbol{y}^T \boldsymbol{\omega}_2)^2 - (\boldsymbol{y}^T \boldsymbol{\omega}_3)^2\}], \tag{11.2}$$

where

$$C_K = (2\pi)^{-1} \exp(-\kappa)(\kappa^2 - 4\beta^2)^{1/2}$$

and $\boldsymbol{\theta} = (\kappa, \beta, \boldsymbol{\omega}_1^T, \boldsymbol{\omega}_2^T, \boldsymbol{\omega}_3^T)^T$ is the parameter vector. The parameter $\boldsymbol{\omega}_1$ is the vector of the directional cosines that define the mean or center of the distribution. The parameters $\boldsymbol{\omega}_2$ and $\boldsymbol{\omega}_3$ relate to the orientation of the distribution. If we visualize the elliptical contours of the distribution on the surface of the sphere, then $\boldsymbol{\omega}_2$ and $\boldsymbol{\omega}_3$ define the directions of the major and minor axes, respectively, of the ellipses. The parameter $\boldsymbol{\omega}_2$ is a vector containing the directional cosines corresponding to a point on the major axis of the ellipse, while $\boldsymbol{\omega}_3$ contains the directional cosines corresponding to a point on the minor axis. Hence, in conjunction with the $\boldsymbol{\omega}_1$ parameter, the direction of the major and minor axes are defined.

The physical interpretation of κ is as a measure of the spread of the distribution: The smaller the value of κ, the more dispersed the distribution will be and, as κ increases, the distribution compacts to the point on the sphere defined by $\boldsymbol{\omega}_1$. In addition to κ, there is the parameter β which can be thought of as a circularity measure. The smaller the value of β the closer the distribution is to circular; the larger the value of β the more elliptical the distribution becomes. When $\beta = 0$, corresponding to maximum circularity, the Kent distribution reduces to the Fisher distribution.

For a mixture of g Kent distributions, the density is given by

$$f(\boldsymbol{y}_j; \boldsymbol{\Psi}) = \sum_{i=1}^{g} \pi_i f_K(\boldsymbol{y}_j; \boldsymbol{\theta}_i), \tag{11.3}$$

where $f_K(\boldsymbol{y}_j, \boldsymbol{\theta}_i)$ is the Kent density with parameter vector

$$\boldsymbol{\theta}_i = (\kappa_i, \beta_i, \boldsymbol{\omega}_{i1}^T, \boldsymbol{\omega}_{i2}^T, \boldsymbol{\omega}_{i3}^T)^T \qquad (i = 1, \ldots, g).$$

The method to be described in the next section for the fitting of a mixture of Kent distributions involves approximating the MLEs of the Kent component distributions by their estimates obtained by the method of moments.

11.6 MOMENT ESTIMATION OF KENT DISTRIBUTION

The following steps are proposed in Kent (1982) to estimate the parameters of a single Kent distribution from a sample $(\gamma_{11}, \gamma_{21})^T, \ldots, (\gamma_{1n}, \gamma_{2n})^T$. Let $(y_{11}, y_{21}, y_{31})^T$,

$\ldots, (y_{1n}, y_{2n}, y_{3n})^T$ denote the respective directional cosines. Then the moment estimates are calculated as follows.

Step 1. Calculate the sample mean direction

$$\overline{\gamma}_1 = \sum_{j=1}^{n} \gamma_{1j}/n, \quad \overline{\gamma}_2 = \sum_{j=1}^{n} \gamma_{2j}/n,$$

and

$$R^2 = S_{y_1}^2 + S_{y_2}^2 + S_{y_3}^2,$$

where $S_{y_1} = \sum_{j=1}^{n} y_{1j}$, $S_{y_2} = \sum_{j=1}^{n} y_{2j}$, and $S_{y_3} = \sum_{j=1}^{n} y_{3j}$. Next calculate the mean resultant length, $\bar{R} = R/n$, and the matrix $\mathbf{S}$, given by

$$\mathbf{S} = \begin{pmatrix} \sum y_{1j}^2 & \sum y_{1j}y_{2j} & \sum y_{1j}y_{3j} \\ \sum y_{1j}y_{2j} & \sum y_{2j}^2 & \sum y_{2j}y_{3j} \\ \sum y_{1j}y_{3j} & \sum y_{2j}y_{3j} & \sum y_{3j}^2 \end{pmatrix}.$$

Step 2. Compute the matrix

$$\mathbf{H} = \begin{pmatrix} \cos\overline{\gamma}_2 & -\sin\overline{\gamma}_2 & 0 \\ \sin\overline{\gamma}_2\cos\overline{\gamma}_1 & \cos\overline{\gamma}_2\cos\overline{\gamma}_1 & -\sin\overline{\gamma}_1 \\ \sin\overline{\gamma}_2\sin\overline{\gamma}_1 & \cos\overline{\gamma}_2\sin\overline{\gamma}_1 & \cos\overline{\gamma}_1, \end{pmatrix},$$

and then compute the matrix $\mathbf{B}$ given by

$$\mathbf{B} = \mathbf{H}^T\mathbf{S}\mathbf{H}.$$

Then $\hat{\alpha}$ is defined by

$$\hat{\alpha} = \tfrac{1}{2}\tan^{-1}\{2b_{23}/(b_{22} - b_{33})\}.$$

Step 3. Compute the matrix $\mathbf{K}$, where

$$\mathbf{K} = \begin{pmatrix} 1 & 0 & 0 \\ 0 & \cos\hat{\alpha} & -\sin\hat{\alpha} \\ 0 & \sin\hat{\alpha} & \cos\hat{\alpha} \end{pmatrix}.$$

Put

$$\hat{\mathbf{G}} = \boldsymbol{HK} = (\hat{\omega}_1, \hat{\omega}_2, \hat{\omega}_3),$$

where $\hat{\omega}_1, \hat{\omega}_2$ and $\hat{\omega}_3$ are 3×1 column vectors. Then calculate

$$V = \hat{\mathbf{G}}^T S \hat{\mathbf{G}}$$

and

$$W = v_{22} - v_{33},$$

where $v_{ij} = (V)_{ij}$.

Step 4. When κ is large, the parameter estimates of $\hat{\kappa}$ and $\hat{\beta}$ are given approximately by

$$\hat{\kappa} = (2 - 2\bar{R} - W)^{-1} + (2 - 2\bar{R} + W)^{-1} \tag{11.4}$$

and

$$\hat{\beta} = \tfrac{1}{2}[(2 - 2\bar{R} - W)^{-1} - (2 - 2\bar{R} + W)^{-1}], \tag{11.5}$$

and the mean direction $(\bar{y}_1, \bar{y}_2, \bar{y}_3)^T$ is given by $\hat{\omega}_1$.

11.7 UNIFORM COMPONENT FOR BACKGROUND NOISE

If the data contain a significant amount of noise, as in the example of Peel et al. (2001), an extra component can be included in the mixture model, corresponding to Poisson noise, as proposed in Banfield and Raftery (1993); see also Campbell et al. (1997). In the present context, a uniform distribution on the unit sphere is appropriate, defined by

$$f_o(\boldsymbol{y}_j) = 1/(4\pi) \qquad \boldsymbol{y}_j \in \mathcal{Y}, \tag{11.6}$$

where $\mathcal{Y}$ denotes the surface on the unit sphere. There is no problem defining the range of the uniform distribution because the whole data space is clearly defined in this application, that is, the surface of the sphere. We let π_0 be the mixing proportion associated with the noise component so that now $\sum_{i=0}^{g} \pi_i = 1$. The mixture model can now be written as

$$f(\boldsymbol{y}_j; \boldsymbol{\Psi}) = \sum_{i=1}^{g} \pi_i f_K(\boldsymbol{y}_j; \boldsymbol{\theta}_i) + \pi_0 f_0(\boldsymbol{y}_j). \tag{11.7}$$

This formulation allows the data to determine the amount of noise present via the mixing proportion π_0, rather than the user setting the level in some *ad hoc* manner.

In the example of Peel et al. (2001), the data are axial (antipodally symmetric); that is, every data point is represented by two points on the surface of the sphere, a pole and a dipole. They handled the presence of antipodal symmetry by representing each data point by the pole closest to the center of the cluster in question (using angular distance, as defined in Section 11.3). In effect, the data space is restricted to a hemisphere, and so if a uniform component distribution is used, it must be suitably changed to $f_o(\boldsymbol{y}_j) = 1/(2\pi)$.

Another choice of distribution in place of the Kent for the component distributions under antipodal symmetry is the Bingham distribution. It applies only to antipodally symmetric data, and so it avoids having to adjust the data to be in a form amenable to Kent component distributions, which can handle data not necessarily antipodally symmetric.

11.8 APPLICATION OF EM ALGORITHM

Peel et al. (2001) used the EM algorithm to fit the mixture model (11.7). As with the fitting of a finite mixture model to independent data, the E-step requires on the $(k+1)$th iteration the calculation of the current values $\tau_i(\boldsymbol{y}_j;\ \boldsymbol{\Psi}^{(k)})$ of the posterior probabilities of component membership. The M-step effectively requires the calculation of the MLEs of the parameters of the component Kent distributions considered separately. On the $(k+1)$th iteration of the EM algorithm, it follows that the updated estimate $\boldsymbol{\theta}_i^{(k+1)}$ for $\boldsymbol{\theta}_i$ is obtained by solving

$$\sum_{j=1}^{n} \tau_i(\boldsymbol{y}_j;\ \boldsymbol{\Psi}^{(k)}) \partial \log f_K(\boldsymbol{y}_j;\ \boldsymbol{\theta}_i)/\partial \boldsymbol{\theta}_i = \boldsymbol{0} \tag{11.8}$$

for each i $(i = 1, \ldots, g)$. However, this solution does not exist in closed form, and so it has to be computed iteratively. A more convenient approach, as outlined by Kent (1982), is to use the moment estimates. Hence the moment estimates were used here in place of the MLEs. If $\boldsymbol{\Psi}^{(k+1)}$ denotes the updated estimate of $\boldsymbol{\Psi}$ so obtained on the $(k+1)$th iteration, it does not follow now that the objective function $Q(\boldsymbol{\Psi};\ \boldsymbol{\Psi}^{(k)})$ is globally maximized at $\boldsymbol{\Psi} = \boldsymbol{\Psi}^{(k+1)}$, since the moment rather than the ML estimates of the component parameters $\boldsymbol{\theta}_i$ are being used. However, the use of the moment estimates should make little difference, as Kent (1982) noted that if the eccentricity $2\beta/\kappa$ is small or if κ is large, then the moment estimates are close to the MLEs. Peel et al. (2001) noted in their work that the likelihood function was not decreased after each such EM iteration.

Of course one could check for each $\boldsymbol{\Psi}^{(k+1)}$ whether the inequality

$$Q(\boldsymbol{\Psi}^{(k+1)};\ \boldsymbol{\Psi}^{(k)}) \geq Q(\boldsymbol{\Psi}^{(k)};\ \boldsymbol{\Psi}^{(k)}) \tag{11.9}$$

holds. This inequality is sufficient to ensure that the likelihood is not decreased. Choosing $\boldsymbol{\Psi}^{(k+1)}$ so that (11.9) holds corresponds to using a generalized EM algorithm to compute the MLE of $\boldsymbol{\Psi}$.

The algorithm can be initialized by fitting a mixture without the noise component and allocating all points beyond a specified threshold from all group means to the noise component and refitting the mixture with a noise component. To provide starting values for the EM algorithm, a small number of random starts, modified versions of various hierarchical methods, and k-means using the angle between vectors as a distance measure can be used. This provides an automated approach with the user simply providing the sample and specifying which of the above methods is to be utilized to provide starting values.

11.9 EXAMPLE 11.1: TWO MINING SAMPLES

To demonstrate their mixture model-based approach, Peel et al. (2001) analyzed two mining samples, referred to as Site 1 and Site 2 (supplied by the Julius Kruttschnitt Mineral Research Centre (JKMRC) at the University of Queensland). Site 1 consists of 860 measurements of dip angle and direction, and Site 2 consists of 531 measurements. Also available for these two sites were the user interpretations based on the use of contour plots, as described in Section 11.2. These user-defined groupings are given in Figures 11.6 and 11.7 for Site 1 and Site 2, respectively, with the groupings denoted by rectangular regions bounded by the solid lines and labeled $J1$, $J2$, $\ldots$, Jg. The contour graphs are superimposed on the plots, where the sample points are indicated by various symbols, depending on the number of values recorded at the point (+, ◇, or o is used if one, two, or three values, respectively, were recorded at a point).

To initialize the EM algorithm, Peel et al. (2001) used ten random starts, ten k-means, and seven hierarchical methods. The number of components fitted was taken to be the same as the user interpretation. The results produced are shown in Figures 11.6 and 11.7 for Sites 1 and 2, respectively. In these and subsequent figures in this chapter, the points determined as noise are denoted by dots.

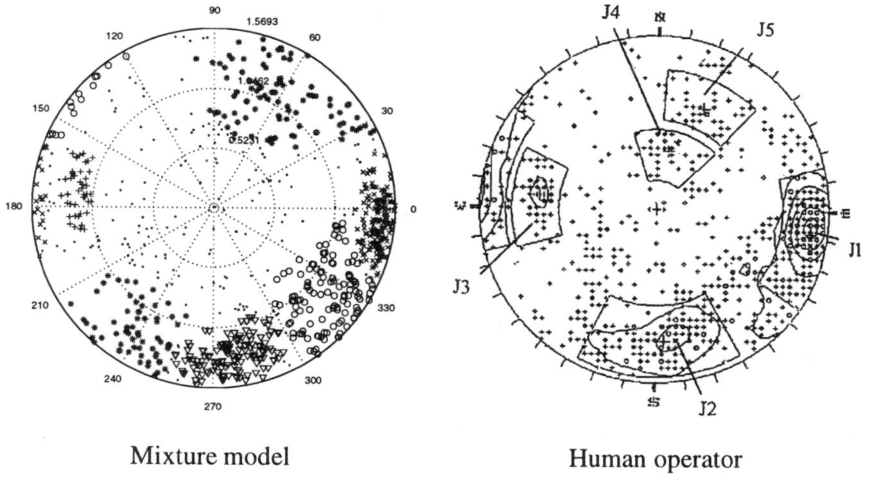

Fig. 11.6 Polar plot of the results for Site 1. From Peel et al. (2001).

Firstly, we compare the result obtained by fitting a mixture model to the user interpretation for Site 1 (Figure 11.6). There is some agreement between the solutions, with the groups designated as J1, J2, and J3 by the user roughly corresponding to the

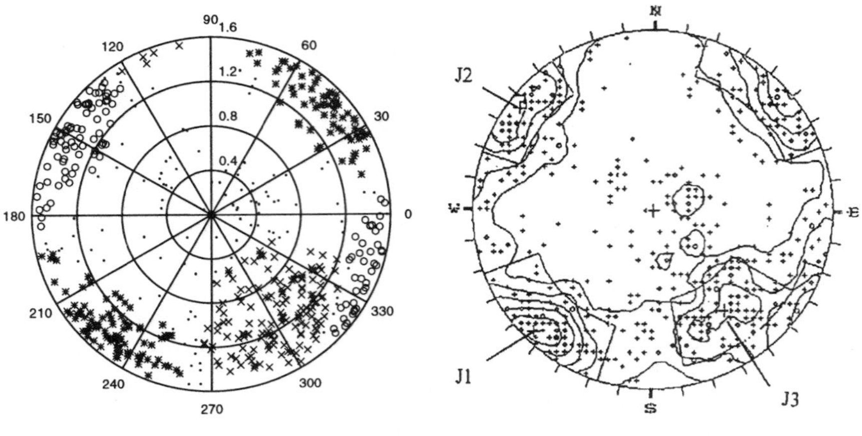

Mixture model   Human operator

Fig. 11.7 Polar plot of the results for Site 2. From Peel et al. (2001).

groups denoted by the x, inverted Δ, and + symbols in the mixture model solution. However, the user-defined groups J4 and J5 are modeled by a single group in the mixture model solution (denoted by the symbol *), and the extra component (denoted by the symbol o) is used to fit the points to the right of J2.

The results produced when fitting a mixture model for Site 2 were much closer to the user-defined groups (see Figure 11.7). The user-defined groups J1 and J2 matched very well to the mixture model solution (denoted by * and o symbols, respectively), whereas group J3 was shifted slightly in the mixture model solution (denoted by the symbol x). Overall, the computed results correspond very nicely to what the human user determined. It should be noted that, as stated in Section 11.2, the user's solutions are not necessarily the only solutions, but are possible solutions. It is, however, encouraging that the two results are similar.

11.10 DETERMINING THE NUMBER OF JOINT SETS

An important question that needs to be addressed is how many joint sets are present or, in the mixture model framework, the value of g. This question has previously been examined for directional data by Hsu et al. (1986), who looked at a stepwise method for determining the number of components in a mixture with example applications to joint set data. Their stepwise procedure uses a bootstrap method to determine the null distribution of Watson's U^2 statistic. An alternative would be to utilize the likelihood ratio test statistic, as described in Chapter 6.

For their example, Peel et al. (2001) felt that a quick, crude approximation would suffice due to the need for reasonably fast analysis. For this reason, AIC and BIC were employed. To examine the use of the AIC and BIC criteria to provide a guide to the number of joint sets, Peel et al. (2001) conducted some simulations. Eight simulated samples, with various parameter configurations, were examined. The mixture model (11.7) was fitted to each sample for $g = 0$ to $g = 10$, and the AIC and BIC criteria were used to determine the choice of g. Here $g = 0$ corresponds to fitting just the uniform distribution. Each model was fitted using ten random starts, ten k-means starts, and seven hierarchical methods to provide partitions of the sample to initialize the EM algorithm.

The results are given in Table 11.1, which reports the number of groups estimated by AIC and BIC for the various samples. The number of sample and noise points are also given in Table 11.1. A comparison between the true grouping and the grouping obtained by fitting the mixture model for Sample 1 is given in Figure 11.8. In all cases, the true number of components lies between the estimates provided by AIC and BIC. In this way, AIC and BIC provided a useful interval or range for the number of components. The estimate given by BIC was correct in six simulations out of eight, while AIC was correct in three of the simulations.

Table 11.1 Number of Groups Estimated by AIC and BIC for Simulated Samples

Sample	$\sum_{i=1}^{g} n_i$	n_0	True g	AIC	BIC
1	400	300	4	4	4
2	200	500	3	3	2
3	0	700	0	1	0
4	300	400	3	4	2
5	600	100	1	2	1
6	200	500	2	2	2
7	500	200	5	6	5
8	600	100	6	7	6
Site 1	860	—	5*	6	2
Site 2	531	—	3*	4	2

*Number of groups estimated by user from contour plots.
Source: Adapted from Peel et al. (2001).

In the literature, AIC and BIC have been found to be very sensitive to the number of sample points and parameters (see Cutler and Windham, 1994), with the methods failing for small sample size relative to the number of parameters. To assess this sensitivity of these two criteria for this application, Peel et al. (2001) performed some further simulations to investigate the effect of sample size on the accuracy of the criteria. Samples of various sizes n were generated from four well-separated groups, each consisting of $n/7$ points. The remaining $3n/7$ points were noise points generated from a uniform distribution. The AIC and BIC criteria were applied to each of these samples. The experiment was repeated for three less separated groups. Their results are reported in Table 11.2. As expected, the criteria fail when the sample size is small. However, for reasonably large sample sizes, it can be seen that the results are excellent, with both criteria repeatedly determining values for g that correspond to the true number of groups.

Table 11.2 Number of Groups Estimated by AIC and BIC for a Range of Samples Sizes n

n	True $g = 4$ AIC	True $g = 4$ BIC	True $g = 3$ AIC	True $g = 3$ BIC
100	0	0	1	1
200	4	0	3	1
300	4	0	3	1
400	4	0	3	2
500	4	4	3	2
600	4	4	3	3
700	4	4	3	3
800	4	4	4	3
900	4	4	3	3
1000	4	4	3	3
1500	4	4	3	3
2000	4	4	4	3

Source: Adapted from Peel et al. (2001).

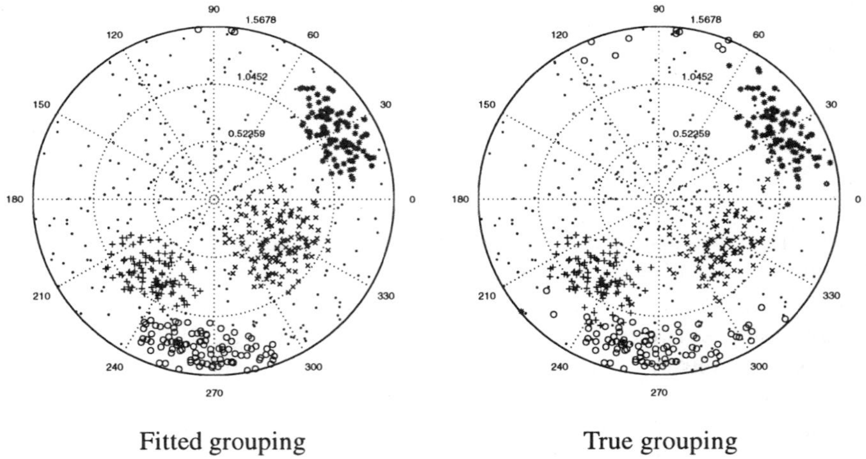

Fig. 11.8 Comparison of true grouping to that obtained fitting a mixture model for simulated Sample 1. From Peel et al. (2001).

11.11 DISCUSSION

The mixture model approach described in the previous sections offers a practical method of analyzing joint sets in two ways: firstly, to identify joint sets in noisy data and, secondly, to determine a range for the number of joint sets suggested by the data. In the case of identifying joint sets, the mixture approach using Kent component distributions (plus a uniformly distributed component) performed very well on actual data. The results approximately matched those independently obtained by an operator using contour plots. An exact match was not expected since even two experienced operators can produce differing results. However, the results produced by fitting a mixture of Kent/uniform distributions provided a reasonable interpretation of the data. An important feature of the model is that it is robust with respect to noise due to the inclusion of the extra uniform component in the model. With regard to determining the number of joint sets, in the simulation experiment discussed in Section 1.10, the AIC and BIC criteria were found to give useful and accurate bounds on the true number of simulated joint sets.

This mixture model-based method of Peel et al. (2001) provides a fully automated method to identify how many joint sets are present and their location and shape, in the form of model parameters, a partition of the sample, and/or posterior probabilities. The manual interpretation of joint sets using the progressive contouring has several disadvantages. Firstly, the resulting interpretation is subjective and often varies according to who made the interpretation. Secondly, the presentation of the contours on a planar circle distorts the original hemisphere and may result in a biased interpretation. Thirdly, it is difficult to remove accurately a particular joint set so that the remaining data can be examined, particularly in the case of overlapping joint sets. Fourthly, it is time-consuming and requires an experienced user. On the other hand, the mixture model approach is based on a sound statistical model of the joint set distribution that overcomes these problems as it provides a quantitative method not dependent on the user's interpretation, and it places no bias on the interpretation of joints at different angles. Moreover, it is capable of describing elongated clusters that often occur in practice as well as circular clusters, and it is able to handle overlapping clusters in a satisfactory manner. Further, it requires only a limited amount of the user's time, this being mainly to read the computer output, and does not require any special experience.

12

Variants of the EM Algorithm for Large Databases

12.1 INTRODUCTION

As set out in some detail in McLachlan and Krishnan (1997, Section 1.7), the EM algorithm has a number of desirable properties, including reliable global convergence. That is, starting from an arbitrary point $\boldsymbol{\Psi}^{(0)}$ in the parameter space, convergence is nearly always to a local maximizer, barring very bad luck in the choice of $\boldsymbol{\Psi}^{(0)}$ or some local pathology in the log likelihood function. In particular, the likelihood is not decreased after each EM iteration. We have seen in the preceding chapters that the ML fitting of finite mixture models is straightforward using the EM algorithm, at least for independent data. The E-step is simply equivalent to updating the posterior probabilities of component membership, and the M-step exists in closed form for component distributions belonging to commonly used parametric families such as the normal.

In considering methods for speeding up the convergence of the EM algorithm, it is highly desirable if the simplicity and stability of the EM algorithm can be preserved. Against this background in the context of mixture models, we considered in Chapter 2 some methods that have been suggested for speeding up the convergence of the EM algorithm. In this chapter we shall focus on methods for improving the speed of the EM algorithm for the ML fitting of mixture models to large databases that preserve the simplicity of implementation of the EM in its standard form. One of these methods is the incremental version of the EM algorithm as proposed recently by Neal and Hinton (1998), whereby an M-step is performed after a single or block of observations have had their component posterior probabilities updated. With this version, the likelihood is still increased after each scan. We shall also consider how the EM algorithm can be scaled to handle very large databases with a limited memory buffer, as with the scalable version proposed recently by Bradley et al. (1999); see also Bradley and Fayyad (1998), Bradley, Fayyad, and Reina (1998), and Fayyad, Reina, and Bradley (1998).

In other recent work on speeding up the EM algorithm for mixtures, Celeux et al. (1999) have proposed a componentwise EM (CEM) algorithm, which is similar in construction to a version of the general-purpose space-alternating generalized EM (SAGE) algorithm of Fessler and Hero (1994, 1995); see also Hero and Fessler (1995).

12.2 INCREMENTAL EM ALGORITHM

12.2.1 Introduction

We consider here the Incremental EM (IEM) algorithm proposed by Neal and Hinton (1998) for improving the rate of convergence of the EM algorithm. With the standard EM algorithm in the context of a finite mixture model being fitted to independent data, the E-step updates the posterior probabilities of component membership of each observation before the M-step is performed. With the IEM algorithm, only a partial E-step is undertaken before the next M-step is performed. More specifically, suppose the observed data $\boldsymbol{y}_1, \ldots, \boldsymbol{y}_n$ are divided into B blocks. If r equals the integer part of n/B, then we let each block contain r observations apart from, say, the Bth, which will have more than r when n is not a multiple of B. The IEM algorithm proceeds by implementing the E-step for only a block of observations at a time before performing a M-step. In this way, each data point $\boldsymbol{y}_j$ is visited after B partial E-steps and B (full) M-steps have been performed. That is, a "pass" or scan of the IEM algorithm consists of B partial E-steps and B full M-steps.

12.2.2 Definition of Partial E-Step

We let $\boldsymbol{\Psi}^{(k)}$ denote the value of $\boldsymbol{\Psi}$ after the kth scan, and we let $\boldsymbol{\Psi}^{(k+b/B)}$ denote the value of $\boldsymbol{\Psi}$ after the bth iteration on the $(k+1)$th scan $(b = 1, \ldots, B)$. In the context of a g-component mixture model, the IEM algorithm is implemented on the $(b+1)$th iteration of the $(k+1)$th scan as follows for $b = 0, \ldots, B-1$:

E-step For $i = 1, \ldots, g$, replace the unobservable component-indicator variable z_{ij} in the complete-data log likelihood by $\tau_i(\boldsymbol{y}_j; \boldsymbol{\Psi}^{(k+b/B)})$ for those $\boldsymbol{y}_j$ in the $(b+1)$th block.

12.2.3 Block Updating of Sufficient Statistics

If the component distributions belong to the exponential family as, for example, in the case of normal components, then it is computationally advantageous to work in terms of the current conditional expectations of the sufficient statistics. We now illustrate this for normal components, but firstly we need the following notation.

For the bth block ($b = 1, \ldots, B$) and current value $\boldsymbol{\Psi}^{(k)}$ for $\boldsymbol{\Psi}$, let

$$T_{i1,b}^{(k)} = \sum_{j \in S_b} \tau_i(\boldsymbol{y}_j;\, \boldsymbol{\Psi}^{(k)}), \tag{12.1}$$

$$\boldsymbol{T}_{i2,b}^{(k)} = \sum_{j \in S_b} \tau_i(\boldsymbol{y}_j;\, \boldsymbol{\Psi}^{(k)})\, \boldsymbol{y}_j, \tag{12.2}$$

and

$$\boldsymbol{T}_{i3,b}^{(k)} = \sum_{j \in S_b} \tau_i(\boldsymbol{y}_j;\, \boldsymbol{\Psi}^{(k)})\, \boldsymbol{y}_j \boldsymbol{y}_j^T \tag{12.3}$$

for $i = 1, \ldots, g$, where S_b is a subset of $\{1, \ldots, n\}$, containing the subscripts of those $\boldsymbol{y}_j$ that belong to the bth block ($b = 1, \ldots, B$). The (current) conditional expectations of the sufficient statistics can then be expressed in terms of sums over the B blocks of the terms (12.1) to (12.3) to give

$$\boldsymbol{T}_{iq}^{(k)} = \sum_{b=1}^{B} \boldsymbol{T}_{iq,b}^{(k)} \qquad (i = 1, \ldots, g;\; q = 1, 2, 3). \tag{12.4}$$

In the definitions (12.1) to (12.4), the superscript k denotes a general iteration number and not necessarily the scan number of the IEM algorithm.

From (3.2) to (3.8), it follows that on the M-step on the $(b+1)$th iteration of the $(k+1)$th scan of the IEM algorithm, the estimates of π_i, $\boldsymbol{\mu}_i$, and $\boldsymbol{\Sigma}_i$ are updated as follows:

$$\pi_i^{(k+\,(b+1)/B)} = T_{i1}^{(k+\,b/B)}/n, \tag{12.5}$$

$$\boldsymbol{\mu}_i^{(k+\,(b+1)/B)} = \boldsymbol{T}_{i2}^{(k+\,b/B)}/T_{i1}^{(k+\,b/B)}, \tag{12.6}$$

and

$$\boldsymbol{\Sigma}_i^{(k+\,(b+1)/B)} = \left\{\boldsymbol{T}_{i3}^{(k+\,b/B)} - T_{i1}^{(k+\,b/B)^{-1}} \boldsymbol{T}_{i2}^{(k+\,b/B)} \boldsymbol{T}_{i2}^{(k+\,b/B)^T}\right\} / T_{i1}^{(k+\,b/B)} \tag{12.7}$$

for $i = 1, \ldots, g$.

The conditional expectations $\boldsymbol{T}_{iq}^{(k+\,b/B)}$ of the sufficient statistics on the right-hand side of (12.5) to (12.7), which are calculated effectively on the E-step prior to this M-step, can be expressed in terms of their values on the previous iteration, using the result

$$\boldsymbol{T}_{iq}^{(k+\,b/B)} = \boldsymbol{T}_{iq}^{(k+\,(b-1)/B)} - \boldsymbol{T}_{iq,b+1}^{(k-1+\,b/B)} + \boldsymbol{T}_{iq,b+1}^{(k+\,b/B)} \quad (i = 1, \ldots, g;\; q = 1, 2, 3). \tag{12.8}$$

That is, on the partial E-step on the $(b+1)$th iteration of the $(k+1)$th scan, only the $(b+1)$th block terms

$$\boldsymbol{T}_{iq,b+1}^{(k+\,b/B)} \quad (i = 1, \ldots, g;\; q = 1, 2, 3)$$

have to be computed, since the first and second terms on the right-hand side of (12.8) are already available from the previous iteration and the last iteration of the previous scan, respectively.

12.2.4 Justification of IEM Algorithm

The argument for improved rate of convergence is that the IEM algorithm exploits new information more quickly rather than waiting for a complete scan of the data before parameters are updated by an M-step. The theoretical justification for the IEM algorithm has been provided by Neal and Hinton (1998).

As pointed out by Thiesson, Meek, and Heckerman (1999), the use of a partial E-step raises two issues not encountered with the standard EM algorithm. One is how to assess convergence when using the IEM algorithm with the E-step implemented partially through blocks of the data. After the $(b+1)$th iteration on the $(k+1)$th scan, the log likelihood can be approximated as

$$\log L(\boldsymbol{\Psi}^{(k+(b+1)/B)}) \approx \log L(\boldsymbol{\Psi}^{(k+b/B)}) \\ + \sum_{j \in S_b} \{\log f(\boldsymbol{y}_j;\, \boldsymbol{\Psi}^{(k+(b+1)/B)}) - \log f(\boldsymbol{y}_j;\, \boldsymbol{\Psi}^{(k+b/B)})\}.$$

The second issue is related to the initial scan through the data. The first few blocks may contain few observations with a high probability of membership of a component of the mixture, which may result in that component having a negligible estimate for its mixing proportion. As a consequence, observations in the subsequent blocks may have practically zero estimates for their posterior probabilities of membership of that component, even if they had actually arisen from that component. This problem of "premature component starvation" can be avoided by running the standard EM algorithm for the first few scans or by at least waiting until the E-step is performed for several blocks before performing the first M-step.

Of course this would be regarded as being overly conservative in most data mining applications involving very large databases; see, for example, Fayyad and Smyth (1999) and the references therein.

12.2.5 Gain in Convergence Time

Concerning the time taken to perform the IEM algorithm for one scan, the B partial E-steps take more time to implement than the one full E-step of the standard EM algorithm, the additional time involving the subtraction of the second term on the right-hand side of (12.8) and the inversion of the component-covariances matrices, which have to be performed after each of the B partial steps. Also, one scan of the IEM algorithm requires $(B-1)$ additional M-steps in updating the estimates using (12.5) to (12.7). McLachlan and Ng (2000a) have investigated the tradeoff between this additional computation on one scan of the IEM algorithm and the fewer number of scans. In Sections 12.3 we present some of their simulation results for normal mixtures, in which the performance of the IEM version is compared with the standard EM algorithm for various number of blocks B. It shall be seen that the time to convergence is reduced by up to a factor of 41%.

The choice of the number of blocks B so as to optimize the convergence time of the IEM algorithm is an interesting problem. The initial work of McLachlan and Ng (2000a) suggests using $B \approx n^{2/5}$ as a simple guide. The optimal choice will depend on the number of unknown parameters, and McLachlan and Ng (2000a) suggest modifying this guide to $B \approx n^{1/3}$ in the case of component-covariance matrices specified to be diagonal.

12.2.6 IEM Algorithm for Singleton Blocks

We will see in the simulations in Section 12.3 that the time to convergence starts to increase as the number of blocks B becomes sufficiently large. This is because of the additional computation time required in having to perform more M-steps and having to invert the component-covariance matrices in the more frequent updating of the posterior probabilities of component membership on each scan of the whole data set.

McLachlan and Ng (2000a) looked at the extreme case of $B = n$ blocks; that is, one observation per block. In order for the updating of the posterior probabilities of component membership to be feasible in practice for multivariate data, some further computational shortcuts are needed beyond those that work in terms of block updates to the sufficient statistics, as given in Section 12.2.3. In particular, one needs to avoid having to invert the component-covariance matrices after each update of the component posterior probabilities for a single observation.

There are formulas that express the inverse of a sample covariance matrix in terms of the inverse of the sample covariance matrix based on one more observation; see, for example, Friedman (1989) and the references therein. McLachlan and Ng (2000a) have modified these formulas to suit the present situation where the weight of a single observation in the full sample does not change from 1 to 0 (that is, is deleted) but is merely updated to another value between 0 and 1.

12.2.7 Efficient Updating Formulas

We assume without loss of generality that the jth block consists of the jth observation $\boldsymbol{y}_j$ $(j = 1, \ldots, n)$. It can be seen from Section 12.2.3 that in order to update the posterior probabilities from

$$\tau_i(\boldsymbol{y}_j;\, \boldsymbol{\Psi}^{(k+(j-1)/n)}) \quad \text{to} \quad \tau_i(\boldsymbol{y}_j;\, \boldsymbol{\Psi}^{(k+j/n)})$$

on the $(j+1)$th iteration of the $(k+1)$th scan of the IEM algorithm, we need to update the current values of π_i, $\boldsymbol{\mu}_i$, and $\boldsymbol{\Sigma}_i^{-1}$ for $i = 1, \ldots, g$.

For convenience, in the remainder of this section we write $\tau_i(\boldsymbol{y}_j;\, \boldsymbol{\Psi}^{(k+(j-1)/n)})$, $\pi_i^{(k+(j-1)/n)}$, $\boldsymbol{\mu}_i^{(k+(j-1)/n)}$, and $\boldsymbol{\Sigma}_i^{(k+(j-1)/n)}$ as τ_{ij}, π_i, $\boldsymbol{\mu}_i$, and $\boldsymbol{\Sigma}_i$, respectively. The corresponding quantities with $\boldsymbol{\Psi}^{(k+(j-1)/n)}$ replaced by $\boldsymbol{\Psi}^{(k+j/n)}$ are denoted by τ_{ij}^*, π_i^*, $\boldsymbol{\mu}_i^*$, and $\boldsymbol{\Sigma}_i^*$, respectively.

McLachlan and Ng (2000a) showed that when $\boldsymbol{\Psi}^{(k+(j-1)/n)}$ is updated to $\boldsymbol{\Psi}^{(k+j/n)}$ in $\tau_i(\boldsymbol{y}_j;\, \boldsymbol{\Psi})$ $(i = 1, \ldots, g)$ (that is, τ_{ij} is updated to τ_{ij}^* in the present notation), then π_i, $\boldsymbol{\mu}_i$, $\boldsymbol{\Sigma}_i^{-1}$, and $|\boldsymbol{\Sigma}_i|$ can be updated as follows:

$$\pi_i^* = (n\pi_i - \tau_{ij} + \tau_{ij}^*)/n, \tag{12.9}$$

$$\boldsymbol{\mu}_i^* = \boldsymbol{\mu}_i - (\tau_{ij} - \tau_{ij}^*)(\boldsymbol{y}_j - \boldsymbol{\mu}_i)/(n\pi_i^*), \tag{12.10}$$

$$\boldsymbol{\Sigma}_i^{*^{-1}} = \frac{\pi_i^*}{\pi_i} \left[\boldsymbol{\Sigma}_i^{-1} + \frac{(\tau_{ij} - \tau_{ij}^*) \boldsymbol{\Sigma}_i^{-1} (\boldsymbol{y}_j - \boldsymbol{\mu}_i)(\boldsymbol{y}_j - \boldsymbol{\mu}_i)^T \boldsymbol{\Sigma}_i^{-1}}{n\pi_i^* - (\tau_{ij} - \tau_{ij}^*)(\boldsymbol{y}_j - \boldsymbol{\mu}_i)^T \boldsymbol{\Sigma}_i^{-1} (\boldsymbol{y}_j - \boldsymbol{\mu}_i)} \right], \tag{12.11}$$

and

$$|\boldsymbol{\Sigma}_i^*| = \left(\frac{\pi_i}{\pi_i^*} \right)^p |\boldsymbol{\Sigma}_i| \left[1 - \frac{\tau_{ij} - \tau_{ij}^*}{n\pi_i^*} (\boldsymbol{y}_j - \boldsymbol{\mu}_i)^T \boldsymbol{\Sigma}_i^{-1} (\boldsymbol{y}_j - \boldsymbol{\mu}_i) \right] \tag{12.12}$$

for $i = 1, \ldots, g$.

The use of (12.9) to (12.12) considerably reduces the amount of computation time in the updating of the g component posterior probabilities for $\boldsymbol{y}_j$. In particular, the use of (12.11) and (12.12) avoids having to calculate directly the inverses and determinants of the g component-covariance matrices. Unfortunately, it is not possible to generalize these updating formulas to the case of blocks consisting of more than one observation.

12.3 SIMULATIONS FOR IEM ALGORITHM

In this section we report some recent simulations of McLachlan and Ng (2000a) on the performance of the IEM algorithm with different number of blocks B relative to the standard EM algorithm.

12.3.1 Simulation 1

A sample of size $n = 256 \times 256$ was generated from a seven-component trivariate normal mixture. The estimates obtained in Liang, MacFall, and Harrington (1994) were used as the values of our population parameters. These seven components correspond to seven tissue types in the segmentation of a two-dimensional magnetic resonance image (MRI) of the human brain. The standard EM algorithm and the IEM version for various number of blocks B were fitted to these simulated data, using the same random starts. In order to avoid the problem of premature component starvation with the IEM algorithm, a full E-step was performed before running the initial M-step.

With this implementation, the EM and IEM algorithms converged to the same local maximum of the log likelihood. The overall CPU time (in seconds) and the number of scans are displayed in Table 12.1. In addition, the average CPU times of the E-step (T_E) and the M-step (T_M) for each scan are displayed in parentheses. It can be seen that when the data are partitioned into blocks with the IEM algorithm, the average times T_E and T_M increase, but the number of scans to convergence decreases, relative to the standard EM algorithm. The overall CPU time is a tradeoff between the additional time of computation per scan and the fewer number of scans required because of the more frequent updating due to B partial E-steps instead of one full E-step ($B = 1$) between each M-step. Eventually, as the data are partitioned into more and more blocks, the time to convergence eventually starts to rise. Similar results have been reported by Thiesson et al. (1999) with the IEM algorithm used to fit mixtures to two real-world databases. Their first data set, MSNBC, is a sparse data

Table 12.1 CPU Times (in Seconds) and the Number of Scans for the Standard EM Algorithm and the IEM Version for B Blocks (Simulation 1)

Algorithm	CPU Times Overall (T_E, T_M)	No. of Scans
Standard EM	601 (4.76, 1.07)	101
Incremental EM		
$B = 4$	458 (4.89, 1.26)	72
$B = 8$	427 (4.89, 1.26)	67
$B = 16$	414 (4.89, 1.26)	65
$B = 32$	408 (4.90, 1.25)	64
$B = 64$	405 (4.90, 1.28)	63
$B = 128$	407 (4.92, 1.28)	63
$B = 256$	411 (4.94, 1.32)	63
$B = 256^2 = n$	2352 (28.40, 6.87)	63
$B = 256^2$ with updating formulas	1143 (10.34, 6.80)	63

Source: Adapted from McLachlan and Ng (2000a).

set, which encoded the stories that people read on the MSNBC web site on December 22, 1998. The second data set holds data records for a single particular subphonetic event observed for continuous speech. For the extreme case of $B = n$, the use of the updating formulas (12.9) to (12.12) considerably reduces the amount of computation time.

In Figure 12.1, the log likelihood is plotted against the number of scans. It shows how the log likelihood changes in the initial fifty scans. It can be seen that the IEM algorithm converges with fewer scans than the standard EM algorithm. In Figure 12.2,

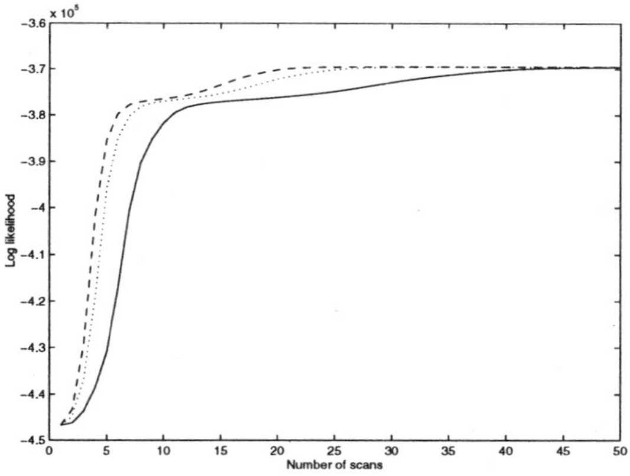

Fig. 12.1 Log likelihood versus number of scans. Standard EM (solid curve); Incremental EM: $B = 4$ (dotted curve); $B = 64$ (dashed curve). From McLachlan and Ng (2000a).

the log likelihood is plotted against the elapsed time. It shows how the likelihood changes during the initial 400 seconds. It can be seen that the fastest time to convergence with the IEM algorithm was obtained with $B = 64$ blocks. For this value of B, the time to convergence is reduced by a factor of 33%.

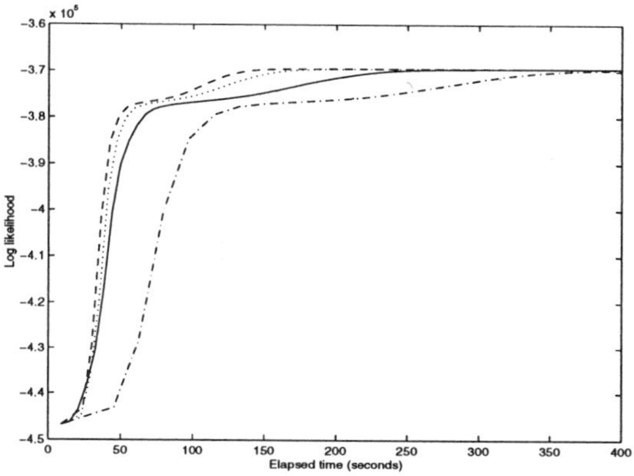

Fig. 12.2 Log likelihood versus elapsed time. Standard EM (solid curve); Incremental EM: $B = 4$ (dotted curve); $B = 64$ (dashed curve); $B = 256 \times 256$ with updating formulas (dashed-dotted curve). From McLachlan and Ng (2000a).

12.3.2 Simulation 2

In the second simulated data set considered by McLachlan and Ng (2000a), a sample of size $n = 2000$ was generated from a four-component eight-dimensional normal mixture. The population parameters are those described in Fukunaga (1990, p. 46). The standard EM algorithm and the IEM version with various number of blocks B were applied to this simulated data set, using the same random starts. For this simulated data, both the EM and its IEM variant converge to the same local maximum of the log likelihood. The overall CPU time, the average times T_E and T_M for the E- and M-steps respectively, and the number of scans are displayed in Table 12.2. Similar to the results of the first simulation, it can be seen that when the data are partitioned into blocks, the overall CPU time decreases, but eventually starts to rise as the data are partitioned into more and more blocks. It can be seen that the fastest time to convergence with the IEM algorithm was obtained with $B = 20$ blocks. For this value of B, convergence time is reduced by a factor of 41%.

As demonstrated in these two simulation examples, the decrease in the number of scans eventually slows down as the data are partitioned into more and more blocks, while at the same time the average times of the E-step (T_E) and the M-step (T_M) for each scan continue to rise. Subsequently, the overall CPU time stops decreasing and commences to increase. Thus it is more beneficial to choose a relatively small B.

Table 12.2 CPU Times (in Seconds) and the Number of Scans for the Standard EM Algorithm and the IEM Version for B Blocks (Simulation 2)

Algorithm	CPU Times Overall (T_E, T_M)	No. of Scans
Standard EM	186 (0.34, 0.07)	446
Incremental EM		
$B = 4$	137 (0.37, 0.10)	287
$B = 8$	119 (0.37, 0.10)	249
$B = 16$	112 (0.38, 0.10)	230
$B = 20$	109 (0.39, 0.10)	218
$B = 40$	113 (0.40, 0.11)	215
$B = 80$	118 (0.42, 0.12)	213
$B = 200$	144 (0.52, 0.14)	210
$B = 400$	196 (0.68, 0.20)	210
$B = 2000 = n$	510 (1.82, 0.47)	210
$B = 2000$ with updating formulas	249 (0.63, 0.49)	210

Source: Adapted from McLachlan and Ng (2000a).

12.4 LAZY EM ALGORITHM

In fitting a mixture model to a data set by maximum likelihood via the EM algorithm, the posterior probability of ith component membership of the mixture, $\tau_i(\boldsymbol{y}_j; \boldsymbol{\Psi}^{(k)})$, for some $\boldsymbol{y}_j$ are often observed to be close to one for some i; that is,

$$\max_i \tau_i(\boldsymbol{y}_j; \boldsymbol{\Psi}^{(k)}) \geq C \tag{12.13}$$

for some $\boldsymbol{y}_j$ after the kth iteration, where C is some threshold specified to be very close to one, say $C \geq 0.95$. If (12.13) continues to hold beyond the kth iteration, then there would be no need to update these posterior probabilities when performing the E-step on subsequent EM iterations. To this end, suppose that the n sample observations $\boldsymbol{y}_1, \ldots, \boldsymbol{y}_n$ are divided into blocks as before with the IEM algorithm, but where now there are only $B = 2$ blocks and the $\boldsymbol{y}_j$ belonging to the first block satisfy (12.13) with the remaining observations put in the second block. Then the lazy EM (LEM) algorithm would be implemented in the same manner as the standard EM algorithm, except that on the E-step, only the posterior probabilities of component membership of those observations $\boldsymbol{y}_j$ in the second block would be updated.

The empirical justification for the LEM algorithm is that if an observation $\boldsymbol{y}_j$ has a very strong posterior probability of belonging to one of the components of the mixture, then there is a good chance that it will not change its probabilistic component membership in subsequent iterations; and if it does, then the change will occur gradually rather than suddenly. The idea therefore is to run the LEM algorithm for a number of iterations k_1 (say, $k_1 \leq 8$) between full E-steps. Initially, the EM algorithm would be run for k_0 iterations before switching to the LEM algorithm, which is then run for k_1 iterations in between full EM steps.

The LEM algorithm is justified theoretically by noticing that the formulation of Neal and Hinton (1998) is applicable for an arbitrary and not necessarily an exhaustive set of data blocks, as long as the data are visited regularly. Hence the LEM algorithm is guaranteed to converge to a local maximum (or saddle point).

12.5 SPARSE EM ALGORITHM

We have seen with the LEM algorithm that for some observations $\boldsymbol{y}_j$, their posterior probabilities of component membership are not updated on every E-step. With the so-called Sparse EM (SPEM) algorithm, proposed by Neal and Hinton (1998), some of these posterior probabilities for a given data point $\boldsymbol{y}_j$ are held fixed for some components while the posterior probabilities of $\boldsymbol{y}_j$ for the remaining components in the mixture are updated. For example, if the current values of the posterior probabilities of component membership for a given $\boldsymbol{y}_j$ were very small (for example, $\tau_i(\boldsymbol{y}_j; \boldsymbol{\Psi}^{(k)}) < C = 0.01$) for, say, the first three components of a six-component mixture being fitted, then with the SPEM algorithm we would fix the posterior probabilities for membership of $\boldsymbol{y}_j$ with respect to the first three components at their current values and only update the posterior probabilities for the last three components. To examine this more closely, let $\boldsymbol{\Psi}^{(k)}$ be the value of $\boldsymbol{\Psi}$ after the kth iteration of the EM algorithm and suppose that the SPEM version is to be implemented on the next iteration, where the posterior probabilities of component membership of $\boldsymbol{y}_j$ are to be held fixed for components i, where $i \in A_j$, with A_j being a subset of $\{1, \ldots, g\}$. We let A_j^c be the complement of A_j. Then on the E-step of the SPEM algorithm implemented on the $(k+1)$th iteration, if A_j is the null set for observation $\boldsymbol{y}_j$, update the posterior probabilities of component membership to $\tau_i(\boldsymbol{y}_j; \boldsymbol{\Psi}^{(k)})$ $(i = 1, \ldots, g)$. If A_j is not the null set, then update the posterior probabilities of component membership to

$$\sum_{h \in A_j^c} \tau_h(\boldsymbol{y}_j; \boldsymbol{\Psi}^{(k-1)}) \frac{\tau_i(\boldsymbol{y}_j; \boldsymbol{\Psi}^{(k)})}{\sum_{h \in A_j^c} \tau_h(\boldsymbol{y}_j; \boldsymbol{\Psi}^{(k)})} \tag{12.14}$$

for those components i which do not belong to A_j; otherwise do not update the posterior probabilities $\tau_i(\boldsymbol{y}_j; \boldsymbol{\Psi}^{(k-1)})$. As (12.14) reduces to

$$\sum_{h \in A_j^c} \tau_h(\boldsymbol{y}_j; \boldsymbol{\Psi}^{(k-1)}) \frac{\pi_i^{(k)} f_i(\boldsymbol{y}_j; \boldsymbol{\theta}_i^{(k)})}{\sum_{h \in A_j^c} \pi_h^{(k)} f_h(\boldsymbol{y}_j; \boldsymbol{\theta}_h^{(k)})}, \tag{12.15}$$

only the updates of the prior-weighted component densities, $\pi_i^{(k)} f_i(\boldsymbol{y}_j; \boldsymbol{\theta}_i^{(k)})$, have to be calculated for those components $i \in A_j^c$. Therefore, this sparse E-step will take time proportional only to the number of components $i \in A_j^c$ $(j = 1, \ldots, n)$. The M-step can also be done efficiently by updating only the contribution to the sufficient statistics for those components $i \in A_j^c$. For example,

$$T_{i1}^{(k)} = \sum_{j=1}^{n} \left[I_{A_j}(i) \tau_i(\boldsymbol{y}_j;\, \boldsymbol{\Psi}^{(k-1)}) + I_{A_j^c}(i) \tau_i(\boldsymbol{y}_j;\, \boldsymbol{\Psi}^{(k)}) \right], \tag{12.16}$$

where $I_{A_j}(i)$ is the indicator function for the set A_j. The first term on the right-hand side of (12.16) can be saved for use in the subsequent SPEM iterations. Similar arguments apply to $\boldsymbol{T}_{i2}^{(k)}$ and $\boldsymbol{T}_{i3}^{(k)}$.

After running the SPEM version a number of iterations k_1, a full EM step is then performed, and a new set A_j $(j = 1, \ldots, n)$ is selected. Note that if we replace $\boldsymbol{\Psi}^{(k-1)}$ by $\boldsymbol{\Psi}^{(k)}$ in the first term in (12.14) or (12.15), then we obtain the updated posterior probabilities of component membership as on the E-step of the standard EM algorithm. Hence we can switch on any E-step to the standard EM algorithm by using the current value of $\boldsymbol{\Psi}$ in this first term in (12.15) for all values of i $(i = 1, \ldots, g)$.

12.6 SPARSE IEM ALGORITHM

A sparse version of the IEM algorithm (SPIEM) can be formulated by combining the sparse E-step of SPEM algorithm and the partial E-step of the IEM algorithm. Let $\boldsymbol{\Psi}^{(k)}$ be the value of $\boldsymbol{\Psi}$ after the kth scan and suppose that the SPEM algorithm is to be implemented on the subsequent bth iteration of the $(k+1)$th scan $(b = 1, \ldots, B)$. Then on the E-step on the $(b+1)$th iteration for $b = 0, \ldots, B-1$, the posterior probabilities of component membership for all $j \in S_{(b+1)}$ are updated to

$$\sum_{h \in A_j^c} \tau_h(\boldsymbol{y}_j;\, \boldsymbol{\Psi}^{(k-1)}) \, \frac{\tau_i(\boldsymbol{y}_j;\, \boldsymbol{\Psi}^{(k+b/B)})}{\sum_{h \in A_j^c} \tau_h(\boldsymbol{y}_j;\, \boldsymbol{\Psi}^{(k+b/B)})}$$

for those components i which do not belong to A_j; otherwise leave the posterior probabilities unchanged as $\tau_i(\boldsymbol{y}_j;\, \boldsymbol{\Psi}^{(k-1)})$.

To avoid the problem of premature component starvation, we suggest running a full EM step at the first scan and then performing the IEM for five scans before running the SPIEM algorithm. After running the SPIEM algorithm for a number of scans k_1, a full EM step (B iterations of IEM) is then performed, and a new set A_j $(j = 1, \ldots, n)$ is selected. The efficient M-step described in the previous section can also be applied in the SPIEM algorithm.

12.6.1 Some Simulation Results

McLachlan and Ng (2000a) performed two simulations to compare the relative performances of the SPIEM, IEM, and standard EM algorithms. Their first simulation used the same simulated data set used in Section 12.3.1. The number of blocks B was set to be 64, and they considered the combinations of $k_1 = 2, 3, 4, 5, 6$ and $C = 0.05, 0.01, 0.005$. For these simulated data, all the algorithms converged to the same local maximum of the log likelihood. The overall CPU time, the average times for the E-step (T_E) and M-step (T_M), and the number of scans are displayed in Table 12.3.

Table 12.3 CPU Times (in Seconds) and the Number of Scans for the SPIEM Algorithm for $B = 64$ Blocks (Simulation 1)

k_1	C	CPU Times Overall (T_E, T_M)	No. of Scans
2	0.05	323 (2.88, 0.75)	83
2	0.01	319 (3.31, 0.88)	71
2	0.005	325 (3.39, 0.89)	71
3	0.05	316 (2.62, 0.69)	89
3	0.01	300 (3.03, 0.80)	73
3	0.005	308 (3.12, 0.82)	73
4	0.05	343 (2.42, 0.64)	105
4	0.01	318 (2.84, 0.74)	83
4	0.005	313 (2.97, 0.78)	78
5	0.05	325 (2.25, 0.59)	107
5	0.01	304 (2.82, 0.71)	80
5	0.005	291 (2.84, 0.72)	77
6	0.05	369 (2.22, 0.57)	124
6	0.01	312 (2.73, 0.69)	85
6	0.005	315 (2.82, 0.71)	83

Source: Adapted from McLachlan and Ng (2000a).

On comparing Table 12.3 with Table 12.1 for $B = 64$, it can be seen that the number of scans to convergence for each combination of C and k_1 with the SPIEM algorithm is larger than that for the IEM algorithm. However, as to be anticipated, T_E and T_M for the SPIEM algorithm are smaller than that for the IEM algorithm, which allows the SPIEM algorithm for the present combinations of C and k_1 to converge faster than the IEM algorithm. Concerning the optimal choice of C and k_1, the overall CPU time is a tradeoff between the computation time per scan and the number of scans required. It can be seen from Table 12.3 that the SPIEM algorithm converges fastest for $C = 0.005$ and $k_1 = 5$.

In Figure 12.3, the log likelihood is plotted against the number of scans for the various algorithms. It shows how the log likelihood changes in the initial fifty scans for these algorithms. It is noted that for the implementation of the SPIEM algorithm, the IEM algorithm is performed for the first five scans before running the first SPIEM step. Thus in Figure 12.3, the curves corresponding to the IEM and the SPIEM algorithms coincide up to the fifth scan. From Figure 12.3, it is observed that the SPIEM algorithm takes more scans to converge compared with the IEM algorithm. But, the SPIEM algorithm takes less scans when compared to the standard EM algorithm. In Figure 12.4, the log likelihood is plotted against the elapsed time for the various algorithms. It shows how the log likelihood changes during the initial 250 seconds.

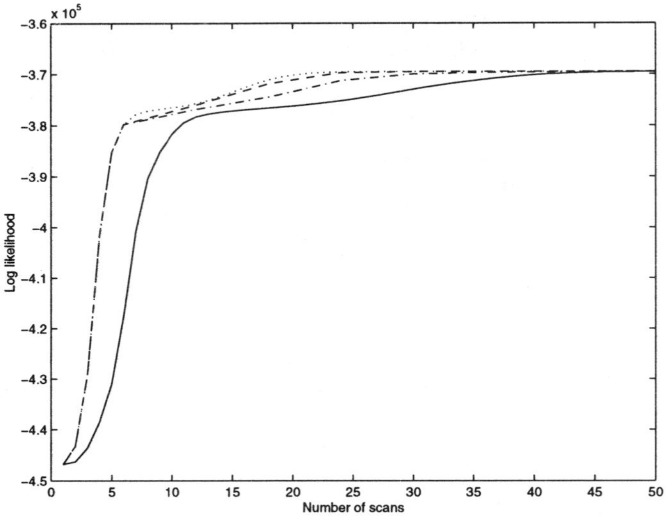

Fig. 12.3 Log likelihood versus number of scans. Standard EM (solid curve); Incremental EM with $B = 64$ (dotted curve); SPIEM with $B = 64$, $k_1 = 5$, and $C = 0.005$ (dashed curve); SPIEM with $B = 64$, $k_1 = 5$, and $C = 0.05$ (dashed-dotted curve). From McLachlan and Ng (2000a).

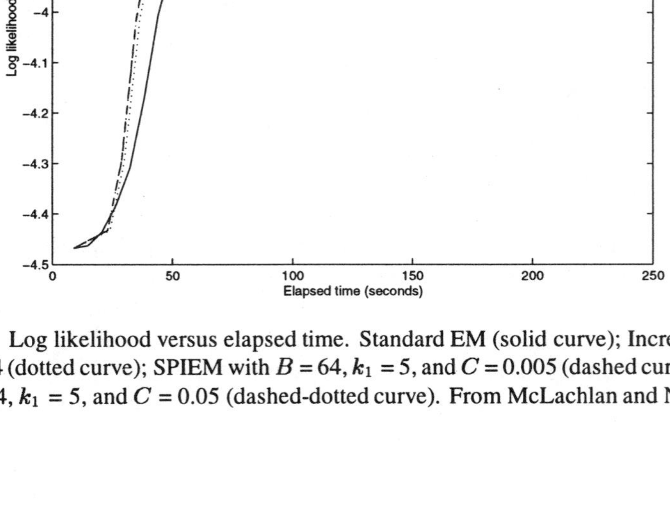

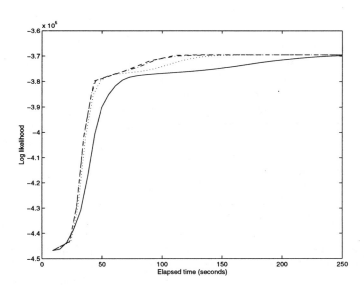

Fig. 12.4 Log likelihood versus elapsed time. Standard EM (solid curve); Incremental EM with $B = 64$ (dotted curve); SPIEM with $B = 64$, $k_1 = 5$, and $C = 0.005$ (dashed curve); SPIEM with $B = 64$, $k_1 = 5$, and $C = 0.05$ (dashed-dotted curve). From McLachlan and Ng (2000a).

The second simulation of McLachlan and Ng (2000a) used the simulated data set adopted in Section 12.3.2. The number of blocks B was set to be 20, and the combinations of $k_1 = 2, 3, 4, 5, 6$ and $C = 0.05, 0.01, 0.005$ were considered. For these simulated data, some runs of the SPIEM algorithm with $C = 0.05$ and $C = 0.01$ converged to a slightly smaller local maximum of the log likelihood. The overall CPU time, the number of scans, and the values of the local maxima of the log likelihood are displayed in Table 12.4.

Table 12.4 CPU Times (in Seconds) and the Number of Scans for the SPIEM Algorithm for $B = 20$ Blocks (Simulation 2)

k_1	C	CPU Times Overall (T_E, T_M)	No. of Scans	Log Likelihood
2	0.05	96 (0.30, 0.07)	254	−30384.7
2	0.01	86 (0.34, 0.08)	203	−30381.6
2	0.005	91 (0.35, 0.09)	203	−30381.6
3	0.05	95 (0.27, 0.07)	273	−30384.7
3	0.01	84 (0.31, 0.08)	209	−30381.6
3	0.005	86 (0.34, 0.08)	201	−30381.6
4	0.05	100 (0.26, 0.07)	300	−30384.7
4	0.01	84 (0.30, 0.07)	215	−30381.6
4	0.005	85 (0.32, 0.08)	205	−30381.6
5	0.05	100 (0.25, 0.06)	317	−30384.7
5	0.01	83 (0.30, 0.07)	221	−30381.6
5	0.005	81 (0.31, 0.08)	203	−30381.6
6	0.05	104 (0.24, 0.06)	341	−30384.7
6	0.01	90 (0.29, 0.07)	243	−30384.7
6	0.005	82 (0.31, 0.07)	208	−30381.6

Source: Adapted from McLachlan and Ng (2000a).

On comparing Table 12.4 with Table 12.2 for $B = 20$, it can be seen that T_E and T_M for the SPIEM algorithm are smaller for each combination of C and k_1 considered than that for the IEM algorithm, which is to be expected. For some of these combinations of C and k_1, the SPIEM algorithm required a fewer number of scans to convergence, while for all combinations its time to convergence is smaller than that of the IEM algorithm. As in the first simulation, the convergence time of the SPIEM algorithm is smallest for $C = 0.005$ and $k_1 = 5$.

12.6.2 Summary of Results for the IEM and SPIEM Algorithms

From the results obtained for the IEM and SPIEM algorithms applied to these two simulated data sets, we see that the overall CPU time is smallest for the SPIEM algorithm (comparing Table 12.1 to 12.3 and Table 12.2 to 12.4). For the implementation of the SPIEM algorithm, we suggest taking $C = 0.005$ and $k_1 = 5$. For these values of C and k_1, the time to convergence is reduced by a factor of 52% for the first simulated data set, compared with the standard EM algorithm, and a factor of 28%, compared with the IEM algorithm for $B = 64$. For the second simulation, the time to convergence is reduced by a factor of 56%, compared with the standard EM algorithm, and a factor of 26%, compared with the IEM algorithm for $B = 20$. It can be seen that the reduction in the time to convergence between the IEM and the SPIEM algorithms is smaller for the second simulated data set, the apparent reason being that there are only $g = 4$ components being fitted to the second set.

12.7 A SCALABLE EM ALGORITHM

12.7.1 Introduction

The standard EM algorithm and its variants require multiple scans of the data to converge. For very large databases, these scans become prohibitively expensive. Bradley et al. (1999) have developed a scalable EM algorithm requiring at most one scan of the database. Their algorithm works within the confines of a limited memory (RAM) buffer allocated by the user, and it utilizes a forward-only cursor. We now describe a scalable EM algorithm proposed by Bradley et al. (1999) for fitting a g-component normal mixture model. It is based on identifying regions of the data that are compressible and regions that must be maintained in memory. The algorithm of Bradley et al. (1999) uses primary and secondary compression of the data. Their primary compression purges sample points that are unlikely to change component membership during the iterative fitting process. Their secondary compression identifies dense regions of data that are not near the current values of the component means of the mixture model. We firstly describe their algorithm without secondary data compression.

12.7.2 Primary Compression of the Data

The scalable EM algorithm is designed for use in the case of a fixed maximum memory buffer size. It proceeds by selecting a random sample of size $n^{(0)}$ from the available data to fill up the memory buffer. The standard EM algorithm is implemented from some initial value $\boldsymbol{\Psi}^{(0)}$ to produce $\boldsymbol{\Psi}^{(1)}$. After completion of this first standard EM iteration, primary compression is performed by locating those observations that are strongly identified with a component of the mixture model. To this end, the blocks $B_1^{(1)}, \ldots, B_g^{(1)}$ are formed, where

$$B_b^{(1)} = \{\boldsymbol{y}_j : (\boldsymbol{y}_j - \boldsymbol{\mu}_b^{(1)})^T \boldsymbol{\Sigma}_b^{(1)^{-1}} (\boldsymbol{y}_j - \boldsymbol{\mu}_b^{(1)}) \leq C_b\} \quad (b = 1, \ldots, g), \quad (12.17)$$

where $C_1, \ldots, C_g$ are chosen so that the proportion α_{PC} of the existing observations in the memory buffer are in $B_1^{(1)}, \ldots, B_g^{(1)}$. Figure 12.5 illustrates the idea of primary compression; the elliptical regions correspond to the blocks $B_1^{(1)}, \ldots, B_g^{(1)}$.

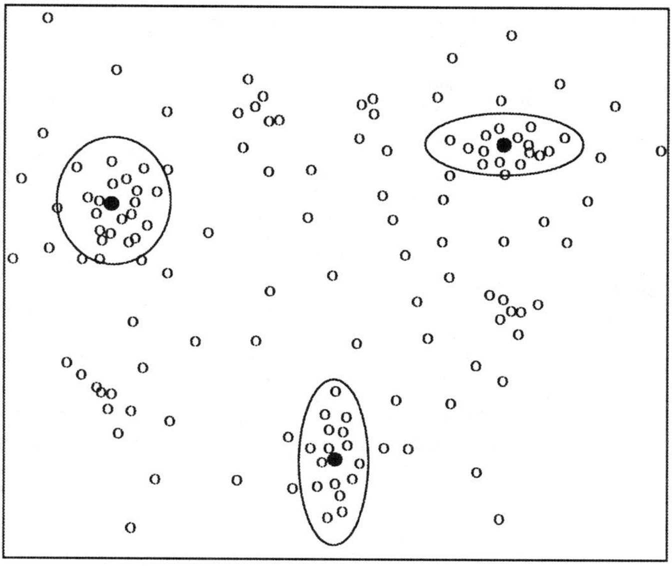

Fig. 12.5 Illustration of the idea of primary compression; the elliptical regions correspond to the blocks $B_1^{(1)}, \ldots, B_g^{(1)}$.

If an observation $\boldsymbol{y}_j$ belongs to more than one block according to (12.17), then it is given a unique assignment to block B_h, where

$$h = \arg \min_b (\boldsymbol{y}_j - \boldsymbol{\mu}_b^{(1)})^T \boldsymbol{\Sigma}_b^{(1)^{-1}} (\boldsymbol{y}_j - \boldsymbol{\mu}_b^{(1)}) \qquad (b = 1, \ldots, g).$$

Bradley et al. (1999) set $\alpha_{PC} = 0.5$ in their experiments.

The observations in $B_1^{(1)}, \ldots, B_g^{(1)}$ are purged from memory and replaced by the scalar $T_{1,b}^{(1)}$, the vector $\boldsymbol{T}_{2,b}^{(1)}$, and the distinct elements of the symmetric matrix $\boldsymbol{T}_{3,b}^{(1)}$ for $b = 1, \ldots, g$, where $T_{1,b}^{(1)}$ is the number of observations in $B_b^{(1)}$; that is,

$$T_{1,b}^{(1)} = \sum_{j=1}^{n^{(0)}} \delta_{bj}^{(1)} \qquad (12.18)$$

and where

$$\boldsymbol{T}_{2,b}^{(1)} = \sum_{j=1}^{n^{(0)}} \delta_{bj}^{(1)} \boldsymbol{y}_j \qquad (12.19)$$

and

$$\boldsymbol{T}_{3,b}^{(1)} = \sum_{j=1}^{n^{(0)}} \delta_{bj}^{(1)} \boldsymbol{y}_j \boldsymbol{y}_j^T \qquad (12.20)$$

for $b = 1, \ldots, g$. Here $\delta_{bj}^{(1)}$ is one or zero, according to whether $\boldsymbol{y}_j$ is in or not in the bth block. We put

$$\overline{\boldsymbol{y}}_b^{(1)} = \boldsymbol{T}_{2,b}^{(1)} / T_{1,b}^{(1)} \qquad (b = 1, \ldots, g) \tag{12.21}$$

and

$$n_{PC}^{(1)} = \sum_{b=1}^{g} T_{1,b}^{(1)}. \tag{12.22}$$

Thus $n_{PC}^{(1)}$ denotes the number of observations that have undergone primary compression after the first EM iteration.

Before the second iteration is undertaken, the available memory space, as a consequence of the primary data compression, is filled by a random sample from the balance of observations in the original data set. We let $n_s^{(1)}$ denote the number of new and retained singleton observations now in the memory, and we relabel them $\boldsymbol{y}_j (j = 1, \ldots, n_s^{(1)})$. We let

$$n^{(1)} = n_{PC}^{(1)} + n_s^{(1)}.$$

12.7.3 Updating of Parameter Estimates

After the buffer refill, the parameter estimates are updated on the basis of the new sample of observations, the set of observations retained in the buffer, and the compressed representations of the discarded data as follows:

$$\pi_i^{(2)} = n^{(1)^{-1}} \{ \sum_{j=1}^{n_s^{(1)}} \tau_i(\boldsymbol{y}_j; \boldsymbol{\Psi}^{(1)}) + \sum_{b=1}^{g} \tau_i(\overline{\boldsymbol{y}}_b; \boldsymbol{\Psi}^{(1)}) \, T_{1,b}^{(1)} \}, \tag{12.23}$$

$$\boldsymbol{\mu}_i^{(2)} = \frac{\sum_{j=1}^{n_s^{(1)}} \tau_i(\boldsymbol{y}_j; \boldsymbol{\Psi}^{(1)}) \, \boldsymbol{y}_j + \sum_{b=1}^{g} \tau_i(\overline{\boldsymbol{y}}_b; \boldsymbol{\Psi}^{(1)}) \, \boldsymbol{T}_{2,b}^{(1)}}{n^{(1)} \pi_i^{(2)}}, \tag{12.24}$$

and

$$\begin{aligned} \boldsymbol{\Sigma}_i^{(2)} = \; & (n^{(1)} \pi_i^{(2)})^{-1} \{ \sum_{j=1}^{n_s^{(1)}} \tau_i(\boldsymbol{y}_j; \boldsymbol{\Psi}^{(1)}) \boldsymbol{y}_j \boldsymbol{y}_j^T + \sum_{b=1}^{g} \tau_i(\overline{\boldsymbol{y}}_b; \boldsymbol{\Psi}^{(1)}) \, \boldsymbol{T}_{3,b}^{(1)} \} \\ & - \boldsymbol{\mu}_i^{(2)} \boldsymbol{\mu}_i^{(2)^T} \end{aligned} \tag{12.25}$$

for $i = 1, \ldots, g$.

Provided that the current mixing proportions $\pi_1^{(1)}, \ldots, \pi_g^{(1)}$ or the current component-covariance matrices $\boldsymbol{\Sigma}_1^{(1)}, \ldots, \boldsymbol{\Sigma}_g^{(1)}$ are not too disparate, then for a given block sample mean $\overline{\boldsymbol{y}}_b^{(1)}$, $\tau_i(\overline{\boldsymbol{y}}_b^{(1)}; \boldsymbol{\Psi}^{(1)})$ should be greatest for $i = b$ and, in many situations, close to one for $i = b$ and hence close to zero for $i \neq b$.

12.7.4 Merging of Sufficient Statistics

Primary data compression is then undertaken on the singleton observations in the memory as on the first iteration but with $\boldsymbol{\Psi}^{(1)}$ replaced by $\boldsymbol{\Psi}^{(2)}$. We let $B_1^{(2)}, \ldots, B_g^{(2)}$ be the blocks of observations so obtained. The compressed quantities (sufficient statistics) for these new g blocks are merged with those statistics from the previous iteration to give

$$T_{1,b}^{(2)} = T_{1,b}^{(1)} + \sum_{j=1}^{n^{(1)}} \delta_{bj}^{(2)}, \tag{12.26}$$

$$\boldsymbol{T}_{2,b}^{(2)} = \boldsymbol{T}_{2,b}^{(1)} + \sum_{j=1}^{n^{(1)}} \delta_{bj}^{(2)} \boldsymbol{y}_j, \tag{12.27}$$

and

$$\boldsymbol{T}_{3,b}^{(2)} = \boldsymbol{T}_{3,b}^{(1)} + \sum_{j=1}^{n^{(1)}} \delta_{bj}^{(2)} \boldsymbol{y}_j \boldsymbol{y}_j^T \tag{12.28}$$

for $b = 1, \ldots, g$, where $\delta_{bj}^{(2)}$ is one or zero, according to whether $\boldsymbol{y}_j$ is in or not in block $B_b^{(2)}$.

With the two sets of sufficient statistics so merged, a third random sample is selected from the original data set to fill up the memory before proceeding to the third iteration, and so on. The process is terminated when the change in the log likelihood after an iteration is arbitrarily small. The algorithm has the provision to handle components that have small estimates for their mixing proportions.

12.7.5 Secondary Data Compression

The scalable algorithm of Bradley et al. (1999) also has the provision for secondary compression of the data, which we now describe. As explained by Bradley et al. (1999), the motivation for secondary data compression is based on the observation that for a dense region of data not near the normal component means, the posterior probabilities will be approximately the same for all $\boldsymbol{y}_j$ constituting the dense region. This secondary compression phase has three stages. The first consists of locating candidate dense regions (clusters) in the data in the buffer. The second consists of applying a criterion to see if the candidate clusters are sufficiently dense. The third state merges acceptably dense clusters with those obtained on the previous iteration provided the merged clusters remain dense according to the criterion (12.29) below. After the kth iteration, the k-means algorithm is used to cluster those observations that are not in one of the g blocks, $B_1^{(k)}, \ldots, B_g^{(k)}$, into $\alpha_k = 2g$ blocks. The data are normalized first to have a global sample variance of one for each variable in the feature vector. We let $\boldsymbol{S}_b$ denote the sample covariance matrix of the observations in the bth block so formed $(b = 1, \ldots, 2g)$. If

$$\max_v (\boldsymbol{S}_b)_{vv} \leq \alpha_{SC}, \tag{12.29}$$

then the observations in this block are purged and replaced by the sufficient statistics $T_{1,b}^{(k)}$, $\boldsymbol{T}_{2,b}^{(k)}$, and $\boldsymbol{T}_{3,b}^{(k)}$ computed from (12.18) to (12.20) with $\delta_{bj}^{(1)}$ replaced by

$\delta_{bj}^{(k)}$, which is one or zero, according to whether $\boldsymbol{y}_j$ is in or not in the bth block for $b = 1, \ldots, 2g$; see Figure 12.6 in which the elliptical (almost spherical) regions correspond to the dense regions. Bradley et al. (1999) set the threshold α_{SC} for secondary compression to be equal to 0.5. Suppose that $g_1^{(k)}$ of these $2g$ blocks satisfy (12.29). We relabel the sufficient statistics corresponding to these $g_1^{(k)}$ blocks so that they are given by $T_{1,b}^{(k)}$, $\boldsymbol{T}_{2,b}^{(k)}$, and $\boldsymbol{T}_{3,b}^{(k)}$ for $b = g+1, \ldots, g+g_1^{(k)}$. Using equations (12.26) to (12.28), a given set is merged with a set obtained from secondary compression on the previous iteration if their means are nearest (in Euclidean distance) and the resulting merged block satisfies criterion (12.29). Now let $g_1^{(k)}$ denote the number of distinct sets of sufficient statistics after this merging process. Then these $g_1^{(k)}$ sets of sufficient statistics are appended to the g sets of sufficient statistics obtained by primary compression after the kth M-step to give $g + g_1^{(k)}$ sets of $T_{1,b}^{(k)}$, $\boldsymbol{T}_{2,b}^{(k)}$, and $\boldsymbol{T}_{3,b}^{(k)}$ $(b = 1, \ldots, g + g_1^{(k)})$ before proceeding to the $(k+1)$th iteration.

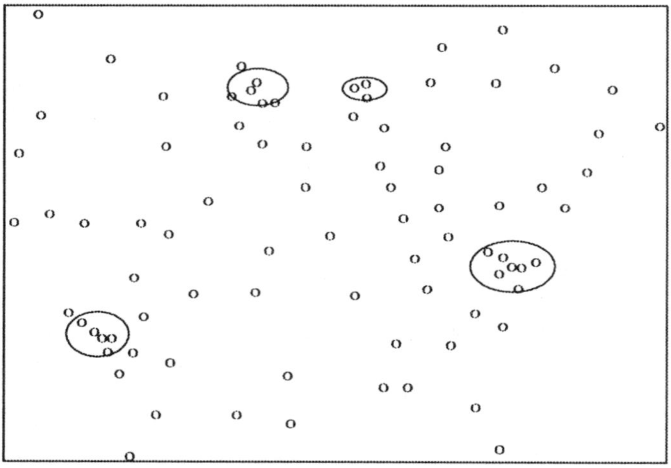

Fig. 12.6 Illustration of the idea of secondary compression.

12.7.6 Tuning Constants

As noted by Bradley et al. (1999), their algorithm can handle the simultaneous fitting of multiple mixture models. The scalable algorithm of Bradley et al. (1999) has three major tuning constants: the compression factor α_{PC}, the denseness tolerance α_{SC} in the secondary compression, and the number of secondary clusters α_k in the application of the k-means algorithm. In order to assess the sensitivity of their scalable EM algorithm to these tuning parameters, Bradley et al. (1999) used it to fit (multivariate) normal mixture models to synthetic data generated from a mixture of $g = 10$ normal components whose means were chosen uniformly on the interval $[0.0, 10.0]$ and diagonal covariance matrices whose diagonal elements were chosen uniformly on $[0.8, 1.2]$.

For values of the primary compression factor α_{PC} near zero, there is little compression, and so the memory buffer fills and the scalable EM algorithm terminates before the entire database is able to be processed. Hence poor results are obtained as measured by the size of the log likelihood at the solution. For large values of α_{PC}, almost all of the data are compressed in the primary compression and so data resolution is lost with poor results being obtained for the fit of the mixture model. Optimal values for α_{PC} occur between these two extremes. On the basis of some simulation results for various α_{PC}, Bradley et al. (1999) noted that the scalable algorithm was fairly insensitive to α_{PC} away from the extrema; the optimal value occurred at $\alpha_{PC} = 0.5$. Concerning the tuning constant α_{SC} in the secondary compression of the data, there is little compression for small α_{SC} and so data resolution is preserved. As α_{SC} increases from zero, the loss of data resolution starts to increase. But Bradley et al. (1999) found that there is practically no falloff in the resulting fit of the mixture model until α_{SC} goes past 0.6.

12.7.7 Simulation Results

On the basis of some simulated data and four real-world data sets, Bradley et al. (1999) compared their scalable EM (SCEM) algorithm with the main *de facto* practices for dealing with large databases: sampling-based and online methods. In reporting their results below, we let ONEM denote their online version of the EM algorithm whereby a single observation is read at a time and the mixture fit updated before purging the observation from memory. The sampling-based version of the EM algorithm, which is applied to a randomly selected sample of the available set is denoted by SAEM. In its applications to the real-world data sets, the sampling rate used by Bradley et al. (1999) varied from 1% to 10% for smaller data sets and varied from 0.1% to 1% for larger data sets to keep the buffer sizes small in the move to the large databases.

Bradley et al. (1999) applied the scalable EM algorithm to four real-world databases: (1) Reuter's information retrieval database, consisting of $n = 12,902$ news articles with the observation vector $\boldsymbol{y}_j$ containing the word count for $p = 302$ keywords for an article; (2) a subset of the U.S. Bureau Census "adult," containing $n = 299,285$ numeric observations of $p = 11$ dimensions; (3) the astronomy database from the Second Palomar Sky Survey at Caltech with $n = 648,291$ observations, each representing objects on a photographic plate with $p = 29$ numeric attributes; (4) the REV digits recognition database with $n = 13,711$ observations of $p = 64$ dimensions, where each observation represents the gray-scale level of an 8×8 image of a handwritten digit. Normal mixture models were fitted to each database, using random starts. The results for the first three of these databases are summarized in Tables 12.5 to 12.7, which gives the average (over several runs) of the log likelihood at the solution found by the three variants of the EM algorithm minus the value of the log likelihood for the best solution for the scalable version. The results given are averages; for example, for the first database, they are averages of 10 solutions for the scalable version, 50 solutions for SAEM (10% and 5%), and 100 solutions for ONEM and SAEM (1%). It can be seen that for each of these three databases, the scalable version outperforms the sampling-based version. Of course the sampling-based version would generally be the best given enough memory. Rather surprisingly, the online version is the best for the astronomy database. For the fourth database (REV digits), the scalable and sampling-based versions gave comparable results superior to the online version for 10% and 5% buffer size, but at the 1% level the performance of the sampling-based version deteriorated to that of the online version.

Table 12.5 Log Likelihood Difference at Solution for Variants of the EM Algorithm for Reuter's Information Retrieval Database

Version of EM	Buffer Size as Percentage of Database Size		
	10%	5%	1%
Scalable	0	−51.8	−228.8
Sampling	-63.9×10^4	-114.8×10^4	-1568.9×10^4
	100%		
Online	−408.4		

Source: Adapted from Bradley et al. (1999).

Table 12.6 Log Likelihood Difference at Solution for Variants of the EM Algorithm for Census Database

Version of EM	Buffer Size as Percentage of Database Size			
	5%	1%	0.5%	0.1%
Scalable	−0.0178	0	−1.8706	−3.0047
Sampling	−8.119	−14204.2	−1486.35	−85241.84
	100%			
Online	−33.6861			

Source: Adapted from Bradley et al. (1999).

Table 12.7 Log Likelihood Difference at Solution for Variants of the EM Algorithm for Astronomy Database

Version of EM	Buffer Size as Percentage of Database Size		
	10%	5%	1%
Scalable	−197.826	−86.737	0
Sampling	−216.901	−225.36	−892.29
	100%		
Online	43.490		

Source: Adapted from Bradley et al. (1999).

12.8 MULTIRESOLUTION *K*D-TREES

12.8.1 Introduction

Moore (1999) has made use of multiresolution kd-trees ($mrkd$-trees) to speed up the fitting process of the EM algorithm. Here kd stands for k-dimensional where, in our notation, $k = p$, the dimension of a feature vector $\boldsymbol{y}_j$. His approach builds a multiresolution data structure to summarize the database at all resolutions of interest simultaneously. Precisely, the $mrkd$-tree is a binary tree that recursively splits the whole set of data points into partitions. Each node in the $mrkd$-tree is associated with a certain partition (subset) of the data points, and each non-leaf-node has two children nodes, which divide their parents' data points between them. Hence the root node owns all the data points, while the leaf-nodes are the smallest possible partitions this $mrkd$-tree offers.

The $mrkd$-tree is constructed top-down, and the partition of the current node (a non-leaf-node) is implemented using the hyper-rectangle of the node, which is defined as the smallest bounding box that contains all the data points owned by the node. Based on this p-dimensional hyper-rectangle, the splitting dimension v_s is identified as the dimension of the hyper-rectangle for which the range of the points is greatest. The current node is split at the midrange of this splitting dimension. That is, the left child of the current node contains those data points owned by the node whose values in the v_s dimension are less than the midrange in the v_s dimension of the hyper-rectangle. The right child then contains the remainder of the parents' data points. The splitting procedure continues until the range of data points in the v_s dimension of the hyper-rectangle of the node is smaller than some threshold γ. This node is then declared to be a leaf-node and is left unsplit. Hence if $\gamma = 0$, then all leaf nodes denote singleton or coincident points. Moore (1999) took γ to be 1% of the range in the v_s dimension.

12.8.2 EM Algorithm Based on Multiresolution *K*D-Trees

Moore (1999) proposed to scale the EM algorithm for handling large databases, based on the $mrkd$-tree. He illustrated his approach for the normal mixture model, but noted that his approach is not necessarily confined to normal components.

With the help of the multiresolution data structure built up by the mrk-d tree, the computation of the sufficient statistics can be restructured as follows on the $(k+1)$th iteration of the EM algorithm. For the non-leaf-node NLN and current value $\boldsymbol{\Psi}^{(k)}$ for $\boldsymbol{\Psi}$, let

$$T_{i1,NLN}^{(k)} = \sum_{j \in NLN} \tau_i(\boldsymbol{y}_j; \boldsymbol{\Psi}^{(k)}), \tag{12.30}$$

$$T_{i2,NLN}^{(k)} = \sum_{j \in NLN} \tau_i(\boldsymbol{y}_j; \boldsymbol{\Psi}^{(k)})\boldsymbol{y}_j, \tag{12.31}$$

and

$$T_{i3,NLN}^{(k)} = \sum_{j \in NLN} \tau_i(\boldsymbol{y}_j; \boldsymbol{\Psi}^{(k)})\boldsymbol{y}_j\boldsymbol{y}_j^T \tag{12.32}$$

for $i = 1, \ldots, g$.

For a leaf node LN, the (current) conditional expectations of the sufficient statistics are simplified by treating all the data points in the node to have the same posterior probabilities $\tau_i(\overline{\boldsymbol{y}}_{LN}; \boldsymbol{\Psi}^{(k)})$, where

$$\tau_i(\overline{\boldsymbol{y}}_{LN}; \boldsymbol{\Psi}^{(k)}) = \frac{\pi_i^{(k)} f_i(\overline{\boldsymbol{y}}_{LN}; \boldsymbol{\theta}_i^{(k)})}{\sum_{h=1}^{g} \pi_h^{(k)} f_h(\overline{\boldsymbol{y}}_{LN}; \boldsymbol{\theta}_h^{(k)})} \quad (i = 1, \ldots, g), \tag{12.33}$$

where

$$\overline{\boldsymbol{y}}_{LN} = \sum_{j \in LN} \boldsymbol{y}_j / n_{LN}$$

and where n_{LN} is the number of data points belonging to the leaf node LN. Thus equations (12.30) to (12.32) are approximated as

$$T_{i1,LN}^{(k)} = \tau_i(\overline{\boldsymbol{y}}_{LN}; \boldsymbol{\Psi}^{(k)}) n_{LN}, \tag{12.34}$$

$$\boldsymbol{T}_{i2,LN}^{(k)} = \tau_i(\overline{\boldsymbol{y}}_{LN}; \boldsymbol{\Psi}^{(k)}) n_{LN} \overline{\boldsymbol{y}}_{LN}, \tag{12.35}$$

and

$$\boldsymbol{T}_{i3,LN}^{(k)} = \tau_i(\overline{\boldsymbol{y}}_{LN}; \boldsymbol{\Psi}^{(k)}) \sum_{j \in LN} \boldsymbol{y}_j \boldsymbol{y}_j^T \tag{12.36}$$

for $i = 1, \ldots, g$.

To update the estimates of the component mixing proportions, means, and covariance matrices on the M-step of the EM algorithm, it is first required to perform the E-step, which, for the normal mixture model, involves computing the (current) conditional expectations of the sufficient statistics, $T_{i1}^{(k)}$, $\boldsymbol{T}_{i2}^{(k)}$, and $\boldsymbol{T}_{i3}^{(k)}$, on the basis of all the data points; that is, to compute equations (12.30) to (12.32) for the root node. As the data points of the root node are divided among its children nodes, it follows that

$$\boldsymbol{T}_{iq}^{(k)} = \sum_{LN \in S} \boldsymbol{T}_{iq,LN}^{(k)} \quad (i = 1, \ldots, g;\ q = 1, 2, 3),$$

where S denotes the set of all leaf nodes LN. In practice, the leaf nodes should be very small (or γ small) in order that the simplified equations (12.34) to (12.36) be applicable. However, in this situation, the number of leaf nodes will be close to the number of data points n, and hence there is very little computational gain over the standard EM algorithm.

Thus Moore (1999) introduced a further step to reduce the computation time as follows. For each $i = 1, \ldots, g$ at a given node, compute the minimum and maximum values that any point $\boldsymbol{y}_j$ in the node can have for its current posterior probabilities $\tau_i(\boldsymbol{y}_j; \boldsymbol{\Psi}^{(k)})$. Denote these minimum and maximum values by $\tau_{i,\min}^{(k)}$ and $\tau_{i,\max}^{(k)}$. If the differences between these bounds satisfies

$$\tau_{i,\max}^{(k)} - \tau_{i,\min}^{(k)} < \beta \tau_{i,\text{total}}^{(k)} \tag{12.37}$$

for all $i = 1, \ldots, g$, where β is a small threshold (say, 0.1), the node is treated as if it is a leaf node. Hence its descendents need not be searched at this iteration. In (12.37), $\tau_{i,\text{total}}^{(k)}$ is the sum of the posterior probabilities of ith component membership for all data points. It is not known in advance, but an approximation to it can be given by $\tau_{i,\text{total}}^{(k-1)}$ from the previous iteration.

It is very difficult to calculate $\tau_{i,\min}^{(k)}$ and $\tau_{i,\max}^{(k)}$ for all $\boldsymbol{y}_j$ belonging to the node because they depend not only on the mean and covariance matrix of the ith component, but also on all of the other components. However, it is much easier to obtain the bounds on the density at the points belonging to the node by computing the closest and farthest point from $\boldsymbol{\mu}_i^{(k)}$ within the hyper-rectangle for each $i = 1, \ldots, g$. Precisely, let $\Delta_{ij}^{(k)^2}$ denote the Mahalanobis squared distance between $\boldsymbol{\mu}_i^{(k)}$ and any points within the hyper-rectangle. Let $\Delta_{i,\max}^{(k)^2}$ and $\Delta_{i,\min}^{(k)^2}$ be the farthest and shortest squared distances, respectively. Then a lower bound on the ith component density at points $\boldsymbol{y}_j$ in the node is given by

$$f_{i,\min}^{(k)} = (2\pi)^{-p/2} \mid \boldsymbol{\Sigma}_i^{(k)} \mid^{-1/2} \exp(-\tfrac{1}{2}\Delta_{i,\max}^{(k)^2}).$$

Similarly, an upper bound $f_{i,\max}^{(k)}$ is obtained for this density. It follows that a lower bound on the $\tau_{ij}^{(k)}$ is given by

$$\tau_{i,\min}^{(k)} = \pi_i^{(k)} f_{i,\min}^{(k)} / (\pi_i^{(k)} f_{i,\min}^{(k)} + \sum_{h \neq i} \pi_h^{(k)} f_{h,\max}^{(k)}).$$

Similarly, an upper bound of the $\tau_{ij}^{(k)}$ can be obtained.

Moore (1999) presented some simulations from a normal mixture for various combinations of the parameter values. He reported that speedups vary from 8-fold to 1000-fold. For example, for a simulated data set of size $n = 160,000$ from a $g = 20$ bivariate normal mixture, the number of nodes in the mrk-d tree was 22,737 with the default value of 0.1 for β, yielding a speedup of about 200-fold. The speedup was defined to be the ratio of the time for the final iteration of the standard EM to that using the mrkd-tree. As Moore (1999) notes, the cost of building the tree is not included in the speedup calculation, but it is negligible in all the cases he considered. He also notes that the quality of the clustering produced by the mrkd-tree version of the EM is indistinguishable from that obtained with the standard EM algorithm when viewed visually and when measured in terms of the log likelihood.

Moore (1999) also reported encouraging results in preliminary experiments in applying the mrkd-tree-based EM algorithm to some large real data sets. For example, on the clustering of 800,000 galaxies into 1,000 clusters on the basis of some three-dimensional data, the standard EM needed 35 minutes per iteration, while the mrkd-tree-based EM required only 13 seconds. With 1.6 million galaxies, the standard EM needed 70 minutes per iteration, while the mrkd-tree version required only 14 seconds.

The sparse EM algorithm may be applied under the mrkd-tree data structure. That is, instead of considering all g components at all nodes, it is possible to freeze those $\tau_i(\boldsymbol{y}_j; \boldsymbol{\Psi}^{(k)})$ that are close to zero in some nodes and hence all descendents of those nodes. In particular, near the tree's leaves, the posterior probabilities of component membership need to be computed only for a small fraction of the number of components g. This leads to a further reduction in computation time.

13

Hidden Markov Models

13.1 INTRODUCTION

The hidden Markov model (HMM) is increasingly being adopted in applications, since it provides a convenient way of formulating an extension of a mixture model to allow for dependent data. To see this, consider the mixture model,

$$f(\boldsymbol{y}_j;\, \boldsymbol{\Psi}) = \sum_{i=1}^{g} \pi_i\, f_i(\boldsymbol{y}_j;\, \boldsymbol{\theta}_i) \tag{13.1}$$

for the density of an observation $\boldsymbol{Y}_j$. In our treatment of mixture models in the previous chapters for independent data, we have introduced the component-indicator vector $\boldsymbol{z}_j$, where $z_{ij} = (\boldsymbol{z}_j)_i$ is one or zero, according to whether $\boldsymbol{y}_j$ is viewed as belonging, or not belonging, to the ith component of the mixture $(i = 1, \ldots, g)$. These vectors $\boldsymbol{z}_1, \ldots, \boldsymbol{z}_n$ are taken to be i.i.d. according to a multinomial distribution consisting of one draw on g categories with probabilities $\pi_1, \ldots, \pi_g$; that is,

$$\boldsymbol{Z}_1, \ldots, \boldsymbol{Z}_n \overset{\text{i.i.d.}}{\sim} \text{Mult}_g(1, \boldsymbol{\pi}). \tag{13.2}$$

The feature vectors $\boldsymbol{y}_1, \ldots, \boldsymbol{y}_n$ are assumed to be conditionally independent given $\boldsymbol{z}_1, \ldots, \boldsymbol{z}_n$; that is,

$$f(\boldsymbol{y}_1, \ldots, \boldsymbol{y}_n \mid \boldsymbol{z}_1, \ldots, \boldsymbol{z}_n;\, \boldsymbol{\xi}) = \prod_{j=1}^{n} f(\boldsymbol{y}_j \mid \boldsymbol{z}_j;\, \boldsymbol{\xi}), \tag{13.3}$$

where

$$f(\boldsymbol{y}_j \mid \boldsymbol{z}_j; \boldsymbol{\xi}) = \prod_{i=1}^{g} f_i(\boldsymbol{y}_j; \boldsymbol{\theta}_i)^{z_{ij}} \tag{13.4}$$

and $\boldsymbol{\xi}$ denotes the unknown parameters in $\boldsymbol{\theta}_1, \ldots, \boldsymbol{\theta}_g$ known *a priori* to be distinct. From (13.2) and (13.3), the $\boldsymbol{Y}_j$ will be i.i.d. according to the mixture density (13.1). Titterington (1990) contrived the nomenclature *hidden multinomial* for the mixture model.

The hidden Markov extension relaxes the independence assumption on the $\boldsymbol{Y}_j$ by taking successive observations to be correlated through their component of origin. With this approach, the independence assumption (13.2) on the component-indicator vectors $\boldsymbol{z}_j$ is relaxed. Usually, a stationary Markovian model is formulated for the distribution of the hidden vectors $\boldsymbol{Z}_1, \ldots, \boldsymbol{Z}_n$. In one dimension, this Markovian model is a Markov chain (see, for example, Holst and Lindgren (1991)), and in two and higher dimensions it is a Markov random field (MRF); see Besag (1986, 1989). The conditional distribution of the observed vector $\boldsymbol{Y}_j$ is formulated as before to depend only on the value of $\boldsymbol{Z}_j$, the component of origin (state of the Markov process), and to be conditionally independent as in (13.3). With the relaxation of (13.2), the marginal density of the feature vector $\boldsymbol{Y}_j$ will not have its simple representation (13.1) of a mixture density as in the independence case.

Hidden Markov models are closely related to state-space models, in which unobserved state variables determine the distribution of the observations. In many applications, the goal is reconstruction of the state variable based on an observation set. This is achieved by the Kalman filter in Gaussian linear state-space models; see Leroux and Puterman (1992).

For general HMMs, Lindgren (1978) constructed consistent and asymptotically normal estimators of the component distributions, but he did not consider estimation of the transition probabilities. Leroux (1992b) established the consistency of the MLE for general HMMs under mild conditions, while local asymptotic normality was proved by Bickel and Ritov (1996). Recently, Bickel, Ritov, and Rydén (1998) showed that under mild conditions the MLE is asymptotically normal and that the observed information matrix is a consistent estimator of the expected information. The relation between HMMs and graphical models has been reviewed recently in Smyth, Heckerman, and Jordan (1997). Ghahramani and Jordan (1997) have proposed a generalization of HMMs in which each hidden component (state) is factored into multiple components and is therefore represented in a distributed manner.

The hidden Markov chain is often a realistic model when the observations $\boldsymbol{y}_j$ appear sequentially in time and tend to cluster or to alternate between different possible components (subpopulations). It is finding widespread application in many areas, including econometrics (Hamilton, 1989; Chib, 1996), biology (Albert, 1991; Leroux and Puterman, 1992; Wang and Puterman, 1999), computational molecular biology (Krogh et al., 1994; Churchill, 1995), finance (Rydén, Teräsvirta, and Asbrink, 1998), neurophysiology (Fredkin and Rice, 1992), and speech and character recognition (Rabiner and Juang, 1986; Kundu and He, 1991; Digalakis, 1999); see also the monograph of MacDonald and Zucchini (1997). In speech recognition applications, the indexing subscript j generally represents speech frames and $\boldsymbol{Y}_j$ is taken to be a finitely many-valued vector, representing random functions of the underlying (hidden) prototypical spectra; see Rabiner (1989) and Juang and Rabiner (1991).

In image-processing applications, the subscript j indexes pixel sites, but we use the term "pixel" liberally as in Besag, York, and Mollie (1991), allowing for pixel arrays that have no direct connection with "picture elements." In some applications, the hidden variable $\boldsymbol{Z}_j$ may represent some discretized characteristic of the pixel, about which inference is required to be made and $\boldsymbol{Y}_j$ is an observable feature of a pixel statistically related to the hidden characteristic of the pixel. For instance, in the example to be considered in Section 13.5 on the tissue-segmentation of a region of the human brain into the component regions of white matter, gray matter, and cerebrospinal fluid, $\boldsymbol{z}_j$ is a three-dimensional indicator vector denoting the component membership of the jth pixel on which the feature vector $\boldsymbol{y}_j$ has been recorded. The latter contains the multispectral magnetic resonance images obtained for the jth pixel. In this example, the $\boldsymbol{z}_j$ will not be independently distributed because of the spatial correlation between neighboring pixels. In some other applications, the hidden variable $\boldsymbol{Z}_j$ may take values over a continuum representing the gray levels or a similar characteristic of the true image and $\boldsymbol{Y}_j$ is a blurred version of the hidden true image $\boldsymbol{Z}_j$; see Qian and Titterington (1991).

Parameter estimation in hidden Markov models usually relies on maximum likelihood or Bayesian methods, moments methods being intractable in this setting. The dependency structure can only exacerbate the difficulties met in mixture estimation for i.i.d. data; see, for example, Robert, Celeux, and Diebolt (1993) and Archer and Titterington (1997).

13.2 HIDDEN MARKOV CHAIN

13.2.1 Definition

We first consider the case where the dependence between the unobservable (hidden) component-indicator vectors $\boldsymbol{Z}_j$ is specified by a stationary Markov chain with transition probability matrix $\boldsymbol{A} = ((\pi_{hi})),\, h, i = 1, \ldots, g$. Thus

$$\text{pr}\{Z_{i,j+1} = 1 \mid Z_{hj} = 1)\} = \pi_{hi} \quad (h, i = 1, \ldots, g) \tag{13.5}$$

for each j $(j = 1, \ldots, n-1)$. The initial distribution of the Markov chain is defined by π_{0i} $(i = 1, \ldots, g)$.

We can therefore write the distribution of $\boldsymbol{Z}$ as

$$p(\boldsymbol{z}; \boldsymbol{\beta}) = p(\boldsymbol{z}_1; \boldsymbol{\beta}) \prod_{j=2}^{n} p(\boldsymbol{z}_j \mid \boldsymbol{z}_{j-1}; \boldsymbol{\beta}), \tag{13.6}$$

where

$$p(\boldsymbol{z}_1; \boldsymbol{\beta}) = \prod_{i=1}^{g} \pi_{0i}^{z_{i1}} \tag{13.7}$$

and

$$p(\boldsymbol{z}_j \mid \boldsymbol{z}_{j-1};\, \boldsymbol{\beta}) = \prod_{h=1}^{g} \prod_{i=1}^{g} \pi_{hi}^{z_{h,j-1} z_{ij}}, \tag{13.8}$$

where $\boldsymbol{\beta}$ contains the initial probabilities π_{0i} and the transition probabilities π_{ij}. Here we are using $p(\cdot)$ as a generic notation for a probability function.

13.2.2 Some Examples

Bickel et al. (1998) contains some examples of hidden Markov chains. In one, there are $g = 2$ hidden components with the ith component distribution of $\boldsymbol{Y}_j$ being Poisson with mean μ_i $(i = 1, 2)$. Albert (1991) proposed this hidden Markov chain as a model for a series of daily counts of epileptic seizures in one patient; see also Le et al. (1992). Leroux and Puterman (1992) applied this model to fetal lamb movements.

In another example, there are again $g = 2$ hidden components, where the ith component density of $\boldsymbol{Y}_j$ is normal with mean μ_i and variance σ^2 $(i = 1, 2)$. This model has been used to model electric current through channels in ion membranes; see Guttorp (1995, p. 109) and Fredkin and Rice (1992) for further details.

13.3 APPLYING EM ALGORITHM TO HIDDEN MARKOV CHAIN MODEL

13.3.1 EM Framework

We now describe the application of the EM algorithm to this problem, known in the HMM literature as the Baum–Welch algorithm. Baum and his collaborators formulated this algorithm before the appearance of the EM algorithm in Dempster et al. (1977) and established the convergence properties for this algorithm; see Baum and Petrie (1966), Baum and Eagon (1967), and Baum et al. (1970). The E-step can be implemented exactly, but it does require a forward and backward recursion through the data, which is time-consuming and numerically sensitive, even though modified algorithms have been designed (Devijver, 1985).

The M-step can be implemented in closed form, provided that the MLEs of the component densities $f_i(\boldsymbol{y}_j;\, \boldsymbol{\theta}_i)$ are available in closed form. In our description here of the EM process, we shall take $\boldsymbol{Y}_j$ to be discrete, where now $f_i(\boldsymbol{y}_j)$ denotes the probability that $\boldsymbol{Y}_j = \boldsymbol{y}_j$, given its membership of the ith component of the chain; that is,

$$f_i(\boldsymbol{y}_j) = \operatorname{pr}\{\boldsymbol{Y}_j = \boldsymbol{y}_j \mid Z_{ij} = 1\} \quad (i = 1, \ldots, g;\; j = 1, \ldots, n). \tag{13.9}$$

The vector of $\boldsymbol{\Psi}$ of unknown parameters thus consists of $\boldsymbol{\beta}$ and the component probabilities for the distinct values taken on by the $\boldsymbol{Y}_j$.

The complete data are declared to be the observed data vector $\boldsymbol{y}$ and the hidden vector, $\boldsymbol{z} = (\boldsymbol{z}_1^T, \ldots, \boldsymbol{z}_n^T)^T$, containing the component labels. The complete-data log likelihood is given by

$$\begin{aligned} \log L_c(\boldsymbol{\Psi}) &= \log p(\boldsymbol{z}) + \log f(\boldsymbol{y} \mid \boldsymbol{z}) \\ &= \log p(\boldsymbol{z}) + \sum_{i=1}^{g} \sum_{j=1}^{n} \log f_i(\boldsymbol{y}_j;\, \boldsymbol{\theta}_i), \end{aligned} \tag{13.10}$$

where

$$\log p(\boldsymbol{z}) = \sum_{i=1}^{g} z_{i1} \log \pi_i + \sum_{h=1}^{g} \sum_{i=1}^{g} \sum_{j=1}^{n-1} z_{h,j} z_{i,j+1} \log \pi_{hi}. \tag{13.11}$$

13.3.2 E-Step

The E-step requires the calculation of the conditional expectation of (13.10), given the observed data $\boldsymbol{y}$. On taking this expectation, we have on the $(k+1)$th iteration that

$$\begin{aligned} Q(\boldsymbol{\Psi}; \boldsymbol{\Psi}^{(k)}) &= \sum_{i=1} \tau_{i1}^{(k)} \log \pi_{0i} + \sum_{h=1}^{g} \sum_{i=1}^{g} \sum_{j=1}^{n-1} \tau_{hij}^{(k)} \log \pi_{hi} \\ &\quad + \sum_{i=1}^{g} \sum_{j=1}^{n} \tau_{ij}^{(k)} \log f_i(\boldsymbol{y}_j; \boldsymbol{\theta}_i), \end{aligned} \tag{13.12}$$

where $\tau_{hij}^{(k)}$ and $\tau_{ij}^{(k)}$ denote the current values of the conditional probabilities defined as

$$\tau_{hij} = \text{pr}\{Z_{hj} = 1, Z_{i,j+1} = 1 \mid \boldsymbol{y}\} \quad (j = 1, \ldots, n-1) \tag{13.13}$$

and

$$\tau_{ij} = \text{pr}\{Z_{ij} = 1 \mid \boldsymbol{y}\} \quad (j = 1, \ldots, n). \tag{13.14}$$

From (13.13), we have that

$$\tau_{ij} = \sum_{h=1}^{g} \tau_{hi,j-1} \quad (j = 2, \ldots, n) \tag{13.15}$$

and

$$\tau_{i1} = \pi_{0i} f_i(\boldsymbol{y}_1) / \sum_{h=1}^{g} \pi_{0h} f_h(\boldsymbol{y}_1).$$

13.3.3 Forward–Backward Recursions on E-Step

The posterior probabilities τ_{hij} and τ_{ij} can be expressed in terms of the following probabilities:

$$a_{ij} = \text{pr}\{\boldsymbol{Y}_1 = \boldsymbol{y}_1, \ldots, \boldsymbol{Y}_j = \boldsymbol{y}_j, Z_{ij} = 1\} \quad (j = 1, \ldots, n) \tag{13.16}$$

and

$$b_{ij} = \text{pr}\{\boldsymbol{Y}_{j+1} = \boldsymbol{y}_{j+1}, \ldots, \boldsymbol{Y}_n = \boldsymbol{y}_n \mid Z_{ij} = 1\} \quad (j = n-1, n-2, \ldots, 1). \tag{13.17}$$

Rabiner (1989) refers to the a_{ij} as the "forward" probabilities and to the b_{ij} as the "backward" probabilities. It follows that τ_{hij} can be expressed as

$$\tau_{hij} = \frac{a_{hj}\,\pi_{hi} f_i(\boldsymbol{y}_{j+1}) b_{i,j+1}}{\sum_{h=1}^{g}\sum_{i=1}^{g} a_{hj}\,\pi_{hi} f_i(\boldsymbol{y}_{j+1}) b_{i,j+1}}, \tag{13.18}$$

since the numerator is $\mathrm{pr}\{Z_{hj} = 1,\, Z_{i,j+1} = 1,\, \boldsymbol{Y} = \boldsymbol{y}\}$ and the denominator is $\mathrm{pr}\{\boldsymbol{Y} = \boldsymbol{y}\}$.

The values of $a_{ij}^{(k)}$ are calculated by forward recursion as follows on the $(k+1)$th iteration.

Initialization:

$$a_{i1}^{(k)} = \pi_{0i}^{(k)} f_i^{(k)}(\boldsymbol{y}_1) \quad (i = 1, \ldots, g).$$

Induction:

$$a_{i,j+1}^{(k)} = [\sum_{h=1}^{g} a_{hj}^{(k)} \pi_{hi}^{(k)}] f_i^{(k)}(\boldsymbol{y}_{j+1}) \quad (j = 1, \ldots, n-1).$$

Termination:

$$\mathrm{pr}_{\boldsymbol{\Psi}^{(k)}}(\boldsymbol{Y}_1 = \boldsymbol{y}_1, \ldots, \boldsymbol{Y}_n = \boldsymbol{y}_n) = \sum_{i=1}^{g} a_{in}^{(k)},$$

where $\mathrm{pr}_{\boldsymbol{\Psi}^{(k)}}$ denotes the probability operator with $\boldsymbol{\Psi}$ replaced by $\boldsymbol{\Psi}^{(k)}$.

The values of $b_{hj}^{(k)}$ are calculated by backward recursion as follows on the $(k+1)$th iteration.

Initialization:

$$b_{hn}^{(k)} = 1 \quad (h = 1, \ldots, g).$$

Induction:

$$b_{hj}^{(k)} = \sum_{i=1}^{g} \pi_{hi}^{(k)} f_i^{(k)}(\boldsymbol{y}_{j+1}) b_{i,j+1}^{(k)} \quad (j = n-1, \ldots, 1;\ h = 1, \ldots, g).$$

The final computation in the E-step consists of plugging these values and the current parameter values into equation (13.18) as follows:

$$\tau_{hij}^{(k)} = \frac{a_{hj}^{(k)}\,\pi_{hi}^{(k)} f_i^{(k)}(\boldsymbol{y}_{j+1}) b_{i,j+1}^{(k)}}{\sum_{h=1}^{g}\sum_{i=1}^{g} a_{hj}^{(k)}\,\pi_{hi}^{(k)} f_i^{(k)}(\boldsymbol{y}_{j+1}) b_{i,j+1}^{(k)}} \quad (j = 1, \ldots, n-1). \tag{13.19}$$

13.3.4 M-Step

The M-step consists in finding the updated estimates of the parameters from the Q-function (13.12). They are a combination of the MLEs for the multinomial parameters and Markov chain transition probabilities. The updated parameters are computed as

$$\pi_{0i}^{(k+1)} = \tau_{i1}^{(k)},$$

$$\pi_{hi}^{(k+1)} = \sum_{j=1}^{n-1} \tau_{hij}^{(k)} / \sum_{j=1}^{n-1} \tau_{hj}^{(k)},$$

and

$$f_i^{(k+1)}(\boldsymbol{y}_j) = \sum_{m=1}^{n-1} \tau_{im}^{(k)} \delta(\boldsymbol{y}_m - \boldsymbol{y}_j) / \sum_{m=1}^{n-1} \tau_{im}^{(k)},$$

where $\delta(\boldsymbol{u} - \boldsymbol{v})$ is one if $\boldsymbol{u} = \boldsymbol{v}$ and zero otherwise.

Recently, Dunmur and Titterington (1998a) studied the influence of initial conditions on the ML estimation for a homogeneous binary Markov chain corrupted by binary channel noise. They concluded that the MLEs are dependent on the initial estimates. In some cases, the algorithm converges not to a sensible solution, but to the fixed point of the forward–backward algorithm used on the E-step.

13.3.5 Numerical Instabilities

As explained in Leroux and Puterman (1992), the forward–backward algorithm for the calculation of the a_{ij} and the b_{ij} is numerically unstable in many situations. This is because a_{ij} converges rapidly to zero or diverges to infinity as j increases, thus making it impossible to calculate and store long sequences. Various methods have been proposed for avoiding this problem in speech recognition (Devijver, 1985). In general, Leroux and Puterman (1992) suggest determining for each j the value of r for which $10^{-r} \sum_i a_{ij}$ lies between 0.1 and 1.0 and multiplying a_{ij} by 10^{-r} $(i = 1, \ldots, g)$. Then $a_{i,j+1}$ $(i = 1, \ldots, g)$ is computed. A similar procedure is applied to the b_{ij}. Afterward, the a_{ij} and the b_{ij} can be reconstructed for the purpose of computing τ_{ij} and τ_{hij}; see also Turner, Cameron, and Thomson (1998).

13.4 HIDDEN MARKOV RANDOM FIELD

Concerning more general Markov models, Qian and Titterington (1990) considered parameter estimation for the Gibbs chain, which is a more general model than the Markov chain model. We consider now estimation for a hidden MRF, which is perhaps the most familiar of more general HMMs. Here now the subscript j indexes a pixel site.

13.4.1 Specification of Markov Random Field

The distribution of $\boldsymbol{Z}$ is specified by that of a Gibbs distribution,

$$p(\boldsymbol{z}) = \exp\{-U(\boldsymbol{z};\ \boldsymbol{\beta})\}/C(\boldsymbol{\beta}), \tag{13.20}$$

where $C(\boldsymbol{\beta})$ is a normalizing constant (the partition function), and the energy function $U(\boldsymbol{z};\ \boldsymbol{\beta})$ takes the form

$$U(\boldsymbol{z};\ \boldsymbol{\beta}) = \sum_{s \in S} V_s(\boldsymbol{z};\ \boldsymbol{\beta}),$$

where S is a class of subsets of the sites, and V_s is the potential function associated with subset s. In the context of images, the structure of the Gibbs function is intended to reflect plausible local, spatial correlation in the true scene; see Geman and Geman (1984) and Besag (1986). As in the hidden Markov chain case, the marginal density $f(\boldsymbol{y}_j;\ \boldsymbol{\xi})$ of $\boldsymbol{Y}_j$ does not take a simple form.

We need the following notation. Let

$$\partial j = \{j_1, \ldots, j_r\}$$

contain the labels of r pixels in some prescribed neighborhood of pixel j, and put

$$\boldsymbol{z}_{\partial j} = (\boldsymbol{z}_{j_1}^T, \ldots, \boldsymbol{z}_{j_r}^T)^T.$$

The simplest departure from independence is a first-order MRF, in which the neighbors of each pixel j comprise its available N, S, E, and W adjacencies. On the boundary, where pixels have less neighbors, assumptions are made for convenience. A first-order assumption is viewed by Besag (1986) to be unrealistic for most practical purposes. For a second-order field, the available pixels which are diagonally adjacent to pixel j are included also. Thus each interior pixel j has eight neighbors.

Regarding possible choices for $p(\boldsymbol{z})$, Besag (1986) has found that his ICM algorithm has performed well for the (second-order) locally dependent MRF model for which

$$\operatorname{pr}\{Z_{ij} = 1 \mid \boldsymbol{z}_{\partial j}\} \propto \exp(\beta u_{ij}), \tag{13.21}$$

where u_{ij} is the number of the prescribed neighbors of pixel j belonging to the ith component $(i = 1, \ldots, g)$.

13.4.2 Application of EM Algorithm

To examine the application of the EM algorithm to this more general HMM, we shall continue to assume that the feature vectors $\boldsymbol{y}_j$ are conditionally independent as specified by (13.3). But we no longer assume that each $\boldsymbol{Y}_j$ is discrete.

From (13.3) and (13.20), the complete-data log likelihood is given by

$$\begin{aligned}\log L_c(\boldsymbol{\Psi}) &= \log p(\boldsymbol{z}) + \sum_{j=1}^{n} \log f(\boldsymbol{y}_j \mid \boldsymbol{z}_j; \boldsymbol{\xi}) \\ &= -U(\boldsymbol{z};\ \boldsymbol{\beta}) - \log C(\boldsymbol{\beta}) + \sum_{i=1}^{g}\sum_{j=1}^{n} z_{ij} \log f_i(\boldsymbol{y}_j;\ \boldsymbol{\theta}_i).\end{aligned} \tag{13.22}$$

The EM algorithm for the hidden MRF is considerably more difficult. As explained by Qian and Titterington (1991), it is impossible to obtain an explicit solution via the EM algorithm where "explicit" means that an explicit formula exists, not requiring numerical integration or summation on the E-step nor iterative solution on the M-step. To see this, the E-step on the $(k+1)$th iteration requires the computation of the conditional expectation of (13.22), using the current fit $\boldsymbol{\Psi}^{(k)}$ for $\boldsymbol{\Psi}$, which is given by

$$Q(\boldsymbol{\Psi};\, \boldsymbol{\Psi}^{(k)}) = -\sum_{s} W_s^{(k)}(\boldsymbol{\beta}) - \log C(\boldsymbol{\beta}) + \sum_{i=1}^{g}\sum_{j=1}^{n} \tau_{ij}^{(k)} \log f_i(\boldsymbol{y}_j;\, \boldsymbol{\theta}_i), \quad (13.23)$$

where

$$W_s^{(k)}(\boldsymbol{\beta}) = E_{\boldsymbol{\Psi}^{(k)}}\{V_i(\boldsymbol{Z};\, \boldsymbol{\beta}) \mid \boldsymbol{y})\} \quad (13.24)$$

and

$$\tau_{ij}^{(k)} = \text{pr}_{\boldsymbol{\Psi}^{(k)}}\{Z_{ij} = 1 \mid \boldsymbol{y}\}. \quad (13.25)$$

Exact computation of both (13.24) and (13.25) is typically impossible (Kay and Titterington, 1986). Concerning the M-step, we have to calculate the updated value $\boldsymbol{\xi}^{(k+1)}$ for $\boldsymbol{\xi}$ containing the distinct elements of $\boldsymbol{\theta}_1, \ldots, \boldsymbol{\theta}_g$ known *a priori* to be distinct. It can be seen from (13.23) that the computation of $\boldsymbol{\xi}^{(k+1)}$ involves the maximization of the same type of function as in the fitting of a mixture model to i.i.d. data. But maximization of (13.23) with respect to $\boldsymbol{\beta}$ is not easy, in general, largely because the computation of the partition function $C(\boldsymbol{\beta})$ is hardly ever feasible. Thus, even if we have a realization from a MRF, maximum likelihood is not a practical proposition.

13.4.3 Restoration Step

Qian and Titterington (1991) have suggested an approximation that replaces the E-step by a stochastic restoration step. For the E-step on the $(k+1)$th iteration, the idea is to replace $\boldsymbol{z}$ in the complete-data log likelihood by a current estimate of $\boldsymbol{z}$, $\boldsymbol{z}^{(k)}$. The latter might be the segmentation obtained by running the ICM algorithm of Besag (1986) or the *maximum a posteriori* (MAP) estimate of Geman and Geman (1984), based on the observed data $\boldsymbol{y}$ and the current estimate $\boldsymbol{\Psi}^{(k)}$. In the special case of binary images, the MAP estimate can be computed exactly (Greig, Porteous, and Seheult, 1989).

The M-step can then be carried out by taking $\boldsymbol{\Psi}^{(k+1)}$ to be the value of $\boldsymbol{\Psi}$ that maximizes, $\log L_c(\boldsymbol{\Psi};\, \boldsymbol{z}^{(k)})$, the complete-data log likelihood with $\boldsymbol{z}$ replaced by $\boldsymbol{z}^{(k)}$. However, this is analogous to using the classification ML approach as described in Section 2.21 to estimate the parameters in the component densities. As explained there, it leads to biased parameter estimates; see also Qian and Titterington (1991) on this point in the present context.

13.4.4 An Improved Approximation to EM Solution

As discussed in Qian and Titterington (1991), an improved estimator of $\boldsymbol{\xi}$ can be obtained using the idea of the pseudo likelihood. The latter was proposed by Besag (1975) as a method for estimating $\boldsymbol{\beta}$ when ML estimation of $\boldsymbol{\beta}$ is intractable. It is given by

$$p_{PL}(\boldsymbol{z};\,\boldsymbol{\beta}) = \prod_{j=1}^{n} p(\boldsymbol{z}_j \mid \boldsymbol{z}_{\partial j};\,\boldsymbol{\beta}). \tag{13.26}$$

The maximizer of the pseudo likelihood is often consistent (Geman and Graffigne, 1987); see Cadez and Smyth (1998) for a recent application of it as an approximation to the MLE. Also, Dunmur and Titterington (1998b) have noted the link of the pseudo likelihood with the mean-field approximation as considered by Zhang (1992, 1993).

Kay (1986), Kay and Titterington (1986), and Qian and Titterington (1991, 1992) have considered approximations to the EM algorithm for this problem. As can be seen from (13.23) and (13.25), we can carry out the maximization of the Q-function for updating the $\boldsymbol{\theta}_i$ as for i.i.d. data. This is provided that we can calculate the posterior probabilities $\tau_{ij}^{(k)}$. We can do this approximately if, analogous to the idea of the pseudo likelihood, we approximate $p(\boldsymbol{z})$ as

$$p(\boldsymbol{z};\,\boldsymbol{\beta}^{(k)}) = \prod_{j=1}^{n} p(\boldsymbol{z}_j \mid \boldsymbol{z}_{\partial j}^{(k)}). \tag{13.27}$$

Then

$$\begin{aligned} \tau_{ij}^{(k)} &= \mathrm{pr}_{\boldsymbol{\Psi}^{(k)}}\{Z_{ij} = 1 \mid \boldsymbol{y},\, \boldsymbol{z}_{\partial j}^{(k)}\} \\ &= \pi_{ij}^{(k)} f_i(\boldsymbol{y}_j;\,\boldsymbol{\theta}_i^{(k)}) / \sum_{h=1}^{g} \pi_{hj}^{(k)} f_h(\boldsymbol{y}_j;\,\boldsymbol{\theta}_h^{(k)}), \end{aligned} \tag{13.28}$$

where

$$\begin{aligned} \pi_{ij}^{(k)} &= \mathrm{pr}_{\boldsymbol{\Psi}^{(k)}}\{Z_{ij} = 1 \mid \boldsymbol{z}_{\partial j}^{(k)}\} \\ &= \exp(\beta^{(k)} u_{ij}^{(k)}) / \sum_{h=1}^{g} \exp(\beta^{(k)} u_{hj}^{(k)}), \end{aligned} \tag{13.29}$$

if we adopt model (13.21). Here

$$u_{ij}^{(k)} = \sum_{m \in \partial j} z_{im}^{(k)} \quad (i = 1, \ldots, g).$$

Analogous to the ICE algorithm as suggested by Owen (1986), we could replace $u_{ij}^{(k)}$ by

$$\sum_{m \in \partial j} \tau_{im}^{(k)}.$$

It follows that if the $\boldsymbol{\theta}_i$ have no elements known *a priori* to be in common, then the updated estimate $\boldsymbol{\theta}_i^{(k+1)}$ is a solution of

$$\sum_{j=1}^{n} \tau_{ij}^{(k)} \partial \log f_i(\boldsymbol{y}_j;\, \boldsymbol{\theta}_i)/\partial \boldsymbol{\theta}_i = \boldsymbol{0} \quad (i = 1,\, \ldots,\, g). \tag{13.30}$$

If $\boldsymbol{\beta}$ is not specified, but has to be estimated from the observed data, then $\boldsymbol{\beta}^{(k+1)}$ can be computed by maximization of the pseudo likelihood

$$\prod_{j=1}^{n} p(\boldsymbol{z}_j^{(k)} \mid \boldsymbol{z}_{\partial j}^{(k)};\, \boldsymbol{\beta}).$$

13.4.5 Approximate M-Step for Normal Components

To illustrate the above approximation to the EM solution, suppose that the ith component density is normal with mean $\boldsymbol{\mu}_i$ and covariance matrix $\boldsymbol{\Sigma}_i$ $(i = 1,\, \ldots,\, g)$; that is,

$$f_i(\boldsymbol{y}_j;\, \boldsymbol{\theta}_i) = \phi(\boldsymbol{y}_j;\, \boldsymbol{\mu}_i,\, \boldsymbol{\Sigma}_i).$$

Then using (13.30), we have that the updated estimates $\boldsymbol{\mu}_i^{(k+1)}$ and $\boldsymbol{\Sigma}_i^{(k+1)}$ for $\boldsymbol{\mu}_i$ and $\boldsymbol{\Sigma}_i$ on the $(k+1)$th iteration of the EM algorithm are given by

$$\boldsymbol{\mu}_i^{(k)} = \sum_{j=1}^{n} \tau_{ij}^{(k)}\, \boldsymbol{y}_j \Big/ \sum_{j=1}^{n} \tau_{ij}^{(k)} \tag{13.31}$$

and

$$\boldsymbol{\Sigma}_i^{(k+1)} = \sum_{j=1}^{n} \tau_{ij}^{(k)}\, (\boldsymbol{y}_j - \boldsymbol{\mu}_i^{(k+1)})(\boldsymbol{y}_j - \boldsymbol{\mu}_i^{(k+1)})^T \Big/ \sum_{j=1}^{n} \tau_{ij}^{(k)}, \tag{13.32}$$

where

$$\tau_{ij}^{(k)} = \frac{\pi_{ij}^{(k)} \phi(\boldsymbol{y}_j;\, \boldsymbol{\mu}_i^{(k)},\, \boldsymbol{\Sigma}_i^{(k)})}{\sum_{h=1}^{g} \pi_{hj}^{(k)} \phi(\boldsymbol{y}_j;\, \boldsymbol{\mu}_h^{(k)},\, \boldsymbol{\Sigma}_h^{(k)})}. \tag{13.33}$$

13.5 EXAMPLE 13.1: SEGMENTATION OF MR IMAGES

Tissue segmentation of multispectral magnetic resonance (MR) images of the human brain has a large potential to facilitate an imaging-based medical diagnosis. The segmentation of MR images provides a way to quantify the volume sizes of different tissue types of the human brain and then to display the tissue structures in three dimensions (3D). In particular, the existence of tumors and their volume sizes provide useful information in the diagnosis of patients. On the other hand, accurate estimation of the tissue parameters may help to monitor changes in brain hemodynamics and metabolism resulting from neuronal activity.

Ng et al. (1995) considered the segmentation of two-dimensional (2D) MR images, while McLachlan et al. (1996) considered the extension to the segmentation of 3D images where $\boldsymbol{y}_j$ was recorded for a scene consisting of 256^3 voxels. The vector $\boldsymbol{y}_j$ contains the image intensities corresponding to T_1 (the spin-lattice relaxation time), T_2 (the spin-spin relaxation time), and the proton density (ρ_D).

In the example reported here from McLachlan et al. (1996), 3D MR images were acquired from the head of a volunteer with a 27-cm field of view. All images were acquired with a resolution of $256 \times 128 \times 96$ and were zero-filled to $256 \times 256 \times 256$ during Fourier transformation. The imaging matrix was rotated so that the read and slow-phase directions were perpendicular to the principal axis of the hippocampus.

The contextual segmentation of these 3D images was carried out (with $\beta = 1$) in the MRF specified by (13.21). The component densities were taken to be trivariate normal and their parameters estimated by (13.31) and (13.32) using the approximate version of the EM algorithm. The noncontextual segmentation was used to define initially the component memberships of the voxels in the images. Three groups were specified, corresponding to the three tissue types — brain-white matter, brain-gray matter, and cerebrospinal fluid (CSF) — after removal of the regions from the image corresponding to bone, fat, skull, and muscle. In Figure 13.1 we depict for a given slice of one of the images, the T_1-weighted image, along with the segmented regions corresponding to CSF and the gray and white matter.

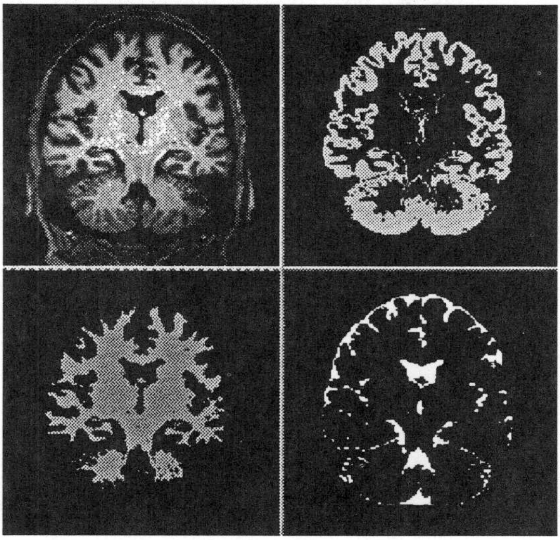

Fig. 13.1 From top left-hand corner clockwise: T_1-weighted image and segmented regions corresponding to gray matter, CSF, and white matter. From McLachlan et al. (1996).

13.6 BAYESIAN APPROACH

Robert et al. (1993) and Shephard (1994) have demonstrated that the hidden Markov chain model is amenable to Bayesian estimation based on Gibbs sampling. There are some difficulties for Bayesian estimation for the more general hidden MRF model, as considered by Qian and Titterington (1989, 1992), Robert and Titterington (1998), and Rydén and Titterington (1998). Robert, Rydén, and Titterington (1999) have proposed a series of on-line convergence controls for MCMC methods applied to hidden Markov chains.

The posterior density for the HMM is given by

$$p(\boldsymbol{y},\, \boldsymbol{z}, \boldsymbol{\xi},\, \boldsymbol{\beta}) = \{\prod_{j=1}^{n} f(\boldsymbol{y}_j \mid \boldsymbol{z}_j;\, \boldsymbol{\xi})\}\, p(\boldsymbol{z};\, \boldsymbol{\beta})\, p(\boldsymbol{\beta})\, p(\boldsymbol{\xi}), \tag{13.34}$$

where $p(\boldsymbol{\xi})$ and $p(\boldsymbol{\beta})$ denote the prior densities for $\boldsymbol{\xi}$ and $\boldsymbol{\beta}$, respectively, assuming that $\boldsymbol{\xi}$ and $\boldsymbol{\beta}$ are *a priori* independent. Bayesian estimation for this model can be implemented by posterior simulation along the lines described in Chapter 4 for a mixture model fitted to i.i.d. data, that is, for the hidden multinomial model. As for the latter model, Gibbs sampling can be used to generate realizations of the parameter vector, given here by $\boldsymbol{\Psi} = (\boldsymbol{\xi}^T,\, \boldsymbol{\beta}^T)^T$ with $\boldsymbol{\Psi}$ augmented by the hidden component-indicator vector $\boldsymbol{z}$. For this to be practical, simulation from all the complete conditional distributions has to be easy. Rydén and Titterington (1998) have investigated this for HMMs and have provided modifications to the procedure to combat any intractability.

The generation of the parameter vector $\boldsymbol{\beta}$ in the distribution of $\boldsymbol{Z}$ involves simulating $\boldsymbol{\beta}$ from

$$p(\boldsymbol{\beta} \mid \boldsymbol{\xi},\, \boldsymbol{z},\, \boldsymbol{y}) \propto p(\boldsymbol{z} \mid \boldsymbol{\beta})\, p(\boldsymbol{\beta}). \tag{13.35}$$

The generation of $\boldsymbol{\xi}$ can be undertaken in a manner similar to that for the hidden multinomial model. It involves simulating $\boldsymbol{\xi}$ from

$$p(\boldsymbol{\xi} \mid \boldsymbol{\beta},\, \boldsymbol{z},\, \boldsymbol{y}) \propto p(\boldsymbol{\xi}) \prod_{i=1}^{g} \prod_{j=1}^{n} f_i(\boldsymbol{y}_j;\, \boldsymbol{\theta}_i)^{z_{ij}}. \tag{13.36}$$

Finally, the hidden component-indicator vector $\boldsymbol{z}$ can be undertaken by generating each $\boldsymbol{z}_j$ from the conditional distribution

$$p(\boldsymbol{z}_j \mid \boldsymbol{z}_{\partial j},\, \boldsymbol{y},\, \boldsymbol{\beta},\, \boldsymbol{\xi}) \propto p(\boldsymbol{z}_j \mid \boldsymbol{z}_{\partial j};\, \boldsymbol{\beta}) \tag{13.37}$$

for $j = 1, \ldots, n$. This typically involves simulation from a binomial or multinomial distribution.

As explained in Rydén and Titterington (1998), simulation from (13.35) is easy for a hidden Markov chain model if $p(\boldsymbol{\beta})$ is suitably chosen, since it involves simulating from Beta or Dirichlet distributions; see Robert et al. (1993). But in the case of the hidden MRF model, simulation from (13.35) runs into difficulties because $p(\boldsymbol{z} \mid \boldsymbol{\beta})$ typically involves the partition function $C(\boldsymbol{\beta})$.

Heikkinen and Högmander (1994) and Rydén and Titterington (1998) suggest replacing $p(\boldsymbol{z} \mid \boldsymbol{\beta})$ by the pseudo likelihood $p_{PL}(\boldsymbol{z}; \boldsymbol{\beta})$ defined by (13.26). The modified version of (13.35) is therefore

$$p(\boldsymbol{\beta} \mid \boldsymbol{\xi}, \boldsymbol{z}, \boldsymbol{y}) \propto p_{PL}(\boldsymbol{z} \mid \boldsymbol{\beta})\, p(\boldsymbol{\beta}). \tag{13.38}$$

Rydén and Titterington (1998) call the estimate of $\boldsymbol{\Psi}$ obtained by using Gibbs sampling based on (13.35) to (13.37) the BL solution, $\boldsymbol{\Psi}_{BL}$, and that based on (13.36) to (13.38) the BP estimate, $\boldsymbol{\Psi}_{BP}$. Rydén and Titterington (1998) establish some asymptotic properties of the MCMC procedures denoted by BL and BP.

They also consider further versions of the BP solution obtained by using the pseudo likelihood formed from coarser partitions of the index set $\{1, 2, \ldots, n\}$. For the general partition of this set into R blocks, $\{\delta_1, \ldots, \delta_r\}$, they use the pseudo likelihood

$$p_{PL(\delta)}(\boldsymbol{z} \mid \boldsymbol{\beta}) = \prod_{r=1}^{R} p(\boldsymbol{z}_{\delta_r} \mid \boldsymbol{z}_{\partial(\delta_r)}, \boldsymbol{\beta}), \tag{13.39}$$

where $\partial(\delta_r)$ denotes the combined sets of neighbors of all points in the rth block δ_r, but excluding any points in δ_r from the combined neighborhood. The idea is that the block-pseudo likelihood might be more efficient than the point-pseudo likelihood. This was found to be the case in the examples considered by Rydén and Titterington (1998).

In other Bayesian work on HMMs, Dunmur and Titterington (1997) have derived versions of the Gibbs sampler suitable for hidden Markov mesh random fields.

13.7 EXAMPLES OF GIBBS SAMPLING WITH HIDDEN MARKOV CHAINS

In Section 4.8.4 on the Bayesian approach to mixture estimation, we described the approach of Robert and Titterington (1998) that enabled Gibbs sampling to be carried out without the need of a proper prior for the fitting of univariate normal mixtures. Their results were presented for the more general hidden Markov chain model rather for independent data. Hence their methods can be used for Bayesian estimation for a hidden Markov chain. Concerning the prior distribution for the transition probabilities, they modeled each row of the transition matrix by a Dirichlet distribution $\mathcal{D}(1, \ldots, 1)$. The priors and the simulation of the component parameters are essentially the same as discussed in Section 4.8.4 for the independent data case (that is, the hidden multinomial model). In particular, they showed that their choice of a partially proper prior led to a proper posterior distribution for $\boldsymbol{\Psi}$.

For the same model as above, Robert and Titterington (1998) also applied the prior feedback method for the computation of the MLE. This method was discussed in Section 4.10 for the hidden multinomial model. As the component densities were normal with unequal densities, they showed how to modify the prior density for the variances to ensure that the posterior density is bounded. They also considered the case of Poisson component distributions.

In other recent work, Robert et al. (2000) have considered Bayesian inference in hidden Markov chains, using reversible jump MCMC methods to enable the number of components g to be unspecified. Their work is therefore an extension of the approach of Richardson and Green (1997) for i.i.d. data (the hidden multinomial mixture model) discussed in Chapter 4. As Robert et al. (2000) note, the Markovian structure together with the reversibility constraint lead to a higher level of complexity than in Richardson and Green (1997).

We report here one of the examples in Robert and Titterington (1998) on some $n = 341$ univariate observations simulated from a three-component hidden Markov chain model with transition matrix

$$\boldsymbol{A} = \begin{pmatrix} 0.62 & 0.25 & 0.13 \\ 0.09 & 0.18 & 0.73 \\ 0.21 & 0.62 & 0.13 \end{pmatrix}, \tag{13.40}$$

and corresponding normal components with means $\mu_1 = 0.25$, $\mu_2 = 3.37$, and $\mu_3 = 2.45$, and variances $\sigma_1^2 = 5.20$, $\sigma_2^2 = 0.372$, and $\sigma_3^2 = 0.314$. The Bayes estimate of the transition matrix $\boldsymbol{A}$ based on 5,000 iterations of the Gibbs sampler was

$$\tilde{\boldsymbol{A}} = \begin{pmatrix} 0.51 & 0.34 & 0.15 \\ 0.20 & 0.10 & 0.70 \\ 0.19 & 0.73 & 0.08 \end{pmatrix}. \tag{13.41}$$

The estimated component means and variances were $\hat{\mu}_1 = 0.207$, $\hat{\mu}_2 = 3.34$, and $\hat{\mu}_3 = 2.44$ and $\hat{\sigma}_1^2 = 8.38$, $\hat{\sigma}_2^2 = 0.571$, and $\hat{\sigma}_3^2 = 0.357$.

Robert and Titterington (1998) then applied the prior feedback method to this data set, using 30 iterations each with 5000 iterations of the Gibbs sampler. The posterior expectations of the component mean and variances so obtained are displayed in Figure 13.2. It can be seen that some of the prior feedback estimates are not yet stable after 30 iterations, although close to the true values of the parameters. At the end of the 30 iterations, the estimated transition matrix $\hat{\boldsymbol{A}}$ was

$$\hat{\boldsymbol{A}} = \begin{pmatrix} 0.65 & 0.27 & 0.09 \\ 0.20 & 0.07 & 0.73 \\ 0.16 & 0.82 & 0.02 \end{pmatrix}.$$

The estimated component means and variances were $\hat{\mu}_1 = 0.61$, $\hat{\mu}_2 = 3.34$, and $\hat{\mu}_3 = 2.42$, and $\hat{\sigma}_1^2 = 8.18$, $\hat{\sigma}_2^2 = 0.490$, and $\hat{\sigma}_3^2 = 0.336$, which Robert and

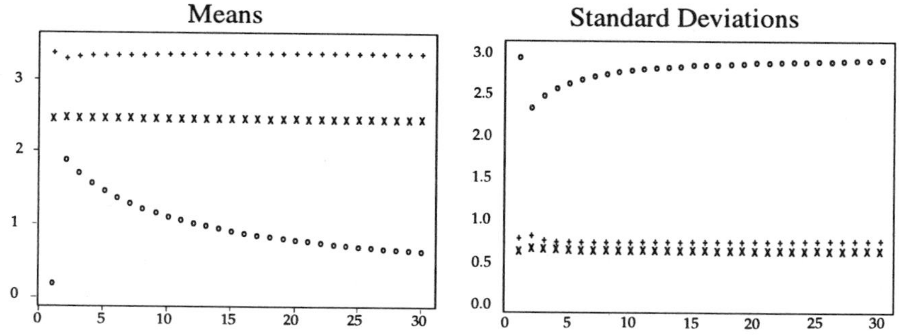

Fig. 13.2 Convergence of the prior feedback estimates of the component means and variances of the normal hidden Markov chain model. From Robert and Titterington (1998).

Titterington (1998) noted does not differ considerably from the corresponding Bayes estimate (13.41), while being closer to the true matrix (13.40).

Appendix
Mixture Software

A.1 EMMIX

We very briefly describe some available software for the fitting of mixture models, commencing with the EMMIX program (previously known as MIXFIT (McLachlan and Peel (1998b)) that has been developed jointly by the authors (McLachlan et al., 1999). Since its publication, the EMMIX program has been developed further and is being extended to handle mixed continuous and categorical data. The current version is available from the World Wide Web address given below.

Authors: McLachlan, Peel, Adams, and Basford (1999).

Language/Platform: FORTRAN 77 and PC executable

Availability:

`http://www.maths.uq.edu.au/~gjm/emmix/emmix.html`

Description

The program EMMIX was developed as a general tool to fit mixtures of normal or t components by maximum likelihood via the EM algorithm to continuous multivariate data. If the user does not supply a starting value for the vector of unknown parameters or an initial grouping of the data for the application of the EM algorithm, it automatically provides starting values by considering a selection obtained from three sources:

(1) random starts;

(2) hierarchical clustering-based starts;

(3) k-means clustering-based starts.

Concerning (1), there is an additional option whereby the user can first subsample the data before using a random start based on the subsample each time. This is to limit the effect of the central limit theorem, which would have the randomly selected starts being similar for each component in large samples.

Concerning (2), the user has the option of using, in either standardized or unstandardized form, the results from seven hierarchical methods (nearest neighbor, farthest neighbor, group average, median, centroid, flexible sorting, and Ward's method). There are several algorithm parameters that the user can optionally specify; alternatively, default values are used. All these computations are automatically carried out by the program. The user only has to provide the data set and the form of the component-covariance matrices (equal, unequal, or diagonal), and to specify the starting procedures and the number of components that are to be fitted. Summary information is automatically given as output for the final fit. The default final fit is taken to be the one corresponding to the largest of the local maxima located. However, the summary information can be recovered for any distinct fit. Indeed, it is not suggested that the mixture analysis of a data set should always be based solely on a single solution of the likelihood equation. It may be informative to consider the various solutions collectively, particularly with clustering applications. Also, with the fitting of a mixture of normal components with unrestricted component-covariance matrices, there is a need to monitor the relative sizes of the component variances (and generalized variances for multivariate data) of a solution as an aid in the detection of spurious local maximizers, as discussed in Section 3.10. To this end, the EMMIX program outputs the generalized component variances for each solution.

As well as the options pertaining to the automatic provision of starting values covered above, several other options are available, including:

- the provision of standard errors for the fitted parameters in the mixture model via various methods (Section 2.16);
- the bootstrapping of the likelihood ratio statistic λ for testing $g = g_0$ versus $g = g_0 + 1$ components in the mixture model, where the value g_0 is specified by the user (Section 6.6);
- the fitting of mixture models to partially classified data.

More precise information on the EMMIX program, including its implementation, are given in the "User's Guide to EMMIX," which is supplied with the program.

A.2 SOME OTHER MIXTURE SOFTWARE

We list below summary information on some other available mixture software. A review of some of these programs may be found in Haughton (1997). It should be noted that the list below is not intended to be comprehensive, as there are other programs available for fitting mixtures, especially in specialized cases; for example, mixtures of Weibulls in reliability applications.

AUTOCLASS

Authors: Cheeseman and Stutz (1996)

Language/Platform: C

Availability: `http://ic-www.arc.nasa.gov/ic/projects/bayes-group/people/cheeseman/`

Description: AUTOCLASS adopts a Bayesian approach to fit mixtures of normal or uniform distributions to continuous multivariate variables, and mixtures of Bernoulli distributions to discrete data. The program is also able to handle missing data and the case of an unspecified number of components.

BINOMIX

Author: Erdfelder (1993)

Language/Platform: BASIC

Availability: `http://www.psychologie.uni-bonn.de/~erdfel_e/comp/binomix.htm`

Description: BINOMIX fits a mixture of binomial or beta-binomial distributions using the EM algorithm.

C.A.MAN

Authors: Böhning, Schlattmann, and Lindsay (1992, 1998)

Language/Platform: FORTRAN 77 and PC executable

Availability: `http://ftp.ukbf.fu-berlin.de/sozmed/caman.html`

Description: C.A.MAN stands for Computer-Assisted Mixture Analysis.
C.A.MAN will fit mixtures of normal (with equal or unequal variances), Poisson, geometric, binomial, exponential, or Laplace univariate distributions, using one of four fitting methods, including the EM algorithm. It also has a semiparametric method to estimate an appropriate number of components. The maximum number of data points that the PC version can analyze is 500. Haughton (1997) refers to a personal correspondence from Professor Böhning, in which he points out that in practice it is feasible to group the data into 500 bins of histogram like frequencies to allow analysis; see also Section 5.8.5.

MCLUST/EMCLUST

Authors: Fraley and Raftery (1999)

Language/Platform: FORTRAN with interfaces to the S-PLUS software package and the R language.

Availability:
`http://www.stat.washington.edu/fraley/software.html`

Description: MCLUST is a software package for hierarchical clustering on the basis of mixtures of normal components under various parameterizations of the component-covariance matrices. The EM algorithm is used in the fitting process, and BIC is used for the determination of the number of components. It has the option to include an additional component in the model for background (Poisson) noise.

MGT

Authors: Jones and McLachlan (1990b)

Language/Platform: FORTRAN 77

Availability: `AS 254 StatLib`
`http://www.stat.cmu.edu/apstat/`

Description: MGT is a subroutine for fitting a mixture of univariate normal distributions to binned and truncated data. The subroutine also provides the standard errors of the estimates.

MIX

Authors: Macdonald and Pitcher (1979)

Language/Platform: Macintosh, DOS, and Windows executables

Availability: MIX is commercially available with a demo obtainable from

`http://icarus.math.mcmaster.ca/peter/mix/mix.html`

Description: MIX works with univariate binned data, with a maximum of 80 bins and 15 components. The program will fit mixtures of normal, log normal, gamma, exponential, or Weibull components and provides standard errors of the estimated parameters.

MIXBIN

Authors: J. Uebersax

Language/Platform: QuickBASIC and PC executable

Availability: `http://members.xoom.com/jsuebersax/papers.html`

Description: MIXBIN fits mixtures of binomial distributions via an EM-type algorithm and also gives the asymptotic standard errors by inverting the observed information matrix. The likelihood ratio test statistic, AIC, and BIC are also computed. The program restricts the minimum trial size to 3 outcomes and the maximum trial size to 12 outcomes.

Program for Gompertz Mixtures

Authors: McLachlan et al. (1997)

Language/Platform: FORTRAN 77

Availability: Journal of Statistical Software

`http://www.stat.ucla.edu/journals/jss/v02.i07`

Description: McLachlan et al. (1997) have provided an algorithm for fitting mixtures of two Gompertz distributions to censored survival data.

MPLUS

Author: B. Muthén and L. Muthén

Language/Platform: Windows 95

Availability: MPLUS is commercially available with a demo obtainable from

`http://www.statmodel.com/`

Description: M*plus* is a statistical modeling program that includes tools to fit latent class models. The criteria AIC and BIC are used for model selection, and standard errors of the estimates are supplied.

MULTIMIX

Authors: Jorgensen and Hunt (1996)

Language/Platform: FORTRAN 77

Availability: `ftp://ftp.math.waikato.ac.nz/pub/maj/`

Description: MULTIMIX adopts the location model (Section 5.2.1) to fit mixtures to mixed continuous and categorical variables. There is also a version of MULTIMIX that implements the approach of Little and Rubin (1987) to missing data. An introduction to the program is given by Jorgensen and Hunt (1996) and Hunt and Jorgensen (1999); see also Section 5.2.2.

NORMIX

Author: Wolfe (1965, 1967, 1970)

Language/Platform: FORTRAN 77 and PC executable

Availability: `http://alumnus.caltech.edu/~wolfe/`

Description: NORMIX fits mixtures of normals or Bernoulli distributions from specified initial values of the parameters or from initial partitions obtained by various hierarchical clustering methods.

SNOB

Authors: Wallace and Dowe (1994)

Language/Platform: FORTRAN 77

Availability: `http://www.cs.monash.edu.au/~dld/Snob.html`

Description: SNOB is based on the Minimum Message Length (MML) approach of Wallace and Boulton (1968) and Wallace and Freeman (1987). It allows the fitting of mixtures of discrete distributions (multistate Bernoulli or categorical), normal (with diagonal covariance matrices), Poisson, and von Mises distributions. The input data can contain missing values and the number of components can be estimated.

Software for Flexible Bayesian Modeling and Markov Chain Sampling

Author Neal (1999)

Language/Platform: C

Availability:

`http://www.cs.toronto.edu/~radford/fbm.software.html`

Description: This software allows the fitting of mixtures via a Bayesian approach.

References

Abramowitz, M. and Stegun, I.A. (Eds.). (1964). *Handbook of Mathematical Functions.* Washington, D.C.: National Bureau of Standards.

Aitchison, J. and Dunsmore, I.R. (1975). *Statistical Predication Analysis.* Cambridge: Cambridge University Press.

Aitkin, M. (1980). Mixture applications of the EM algorithm in GLIM. *Compstat 1980, Proceedings Computational Statistics.* Vienna: Physica-Verlag, pp. 537–541.

Aitkin, M. (1991). Posterior Bayes factors (with discussion). *Journal of the Royal Statistical Society B* **53**, 111-142.

Aitkin, M. (1996). A general maximum likelihood analysis of overdispersion in generalized linear models. *Statistics and Computing* **6**, 251–262.

Aitkin, M. (1997). Contribution to the discussion of paper by S. Richardson and P.J. Green. *Journal of the Royal Statistical Society B* **59**, 766.

Aitkin, M. (1999a). A general maximum likelihood analysis of variance components in generalized linear models. *Biometrics* **55**, 117–128.

Aitkin, M. (1999b). Meta-analysis by random effect modelling in generalized linear models. *Statistics in Medicine* **18**, 2343–2351.

Aitkin, M. and Aitkin, I. (1994). Efficient computation of maximum likelihood estimates in mixture distributions, with reference to overdispersion and variance components. In *Proceedings XVIIth International Biometric Conference.* Alexandria, Virginia: Biometric Society, pp. 123–138.

Aitkin, M. and Aitkin, I. (1996). A hybrid EM/Gauss-Newton algorithm for maximum likelihood in mixture distributions. *Statistics and Computing* **6**, 127–130.

Aitkin, M., Anderson, D., Francis, B.J., and Hinde, J. (1989). *Statistical Modelling in GLIM*. Oxford: Oxford University Press.

Aitkin, M., Anderson, D., and Hinde, J. (1981). Statistical modelling of data on teaching styles (with discussion). *Journal of the Royal Statistical Society B* **144**, 419–461.

Aitkin, M., Finch, S.J., Mendell, N.R., and Thode, H.C. (1996). A new test for the presence of a normal mixture distribution based on the posterior Bayes factor. *Statistics and Computing* **6**, 121–126.

Aitkin, M. and Rubin, D.B. (1985). Estimation and hypothesis testing in finite mixture models. *Journal of the Royal Statistical Society B* **47**, 67–75.

Aitkin, M. and Tunnicliffe Wilson, G. (1980). Mixture models, outliers, and the EM algorithm. *Technometrics* **22**, 325–331.

Akaike, H. (1973). Information theory and an extension of the maximum likelihood principle. In *Second International Symposium on Information Theory*, B.N. Petrov and F. Csaki (Eds.). Budapest: Akadémiai Kiadó, pp. 267–281. (Reproduced in (1992) in *Breakthroughs in Statistics* **1**, S. Kotz and N.L. Johnson (Eds.). New York: Springer-Verlag, pp. 610–624.)

Akaike, H. (1974). A new look at the statistical model identification. *IEEE Transactions on Automatic Control* **19**, 716–723.

Albert, J.R.G. and Baxter, L.A. (1995). Applications of the EM algorithm to the analysis of life length data. *Applied Statistics* **44**, 323–341.

Albert, P.S. (1991). A two-state Markov mixture model for a time series of epileptic seizure counts. *Biometrics* **47**, 1371–1381.

Aler, J., Du Mouza, J., and Arnould, M. (1996). Measurement of the fragmentation efficiency of rock mass blasting and its mining applications. *International Journal of Rock Mechanics and Mining Sciences & Geomechanics Abstracts* **33**, 125–139.

Anderson, D.A. (1988). Some models for overdispersed binomial data. *Australian Journal of Statistics* **30**, 125–148.

Anderson, D.A. and Hinde, J.P. (1988). Random effects in generalized linear models and the EM algorithm. *Communications in Statistics—Theory and Methods* **17**, 3847–3856.

Anderson, E. (1935). The irises of the Gaspé Peninsula. *Bulletin of the American Iris Society* **59**, 2–5.

Anderson, T.W. (1963). Asymptotic theory for principal component analysis. *Annals of Mathematical Statistics* **34**, 122–148.

Anderson, T.W. and Rubin, H. (1956). Statistical inference in factor analysis. In *Proceedings of the Third Berkeley Symposium*, Vol. 5. Berkeley: University of California, pp. 122–148.

Andrews, D.F. and Herzberg, A.M. (1985). *Data: A Collection of Problems from Many Fields for the Student and Research Worker.* New York: Springer-Verlag.

Archer, G.E.B. and Titterington, D.M. (1997). Parameter estimation for hidden Markov chains. *Technical Report*. Glasgow: Department of Statistics, University of Glasgow.

Ashman, K.M. and Bird, C.M. (1993). Globular cluster clustering in M31. *The Astrophysical Journal* **106**, 2281–2290.

Ashman, K.M. and Bird, C.M. (1994). Detecting bimodality in astronomical data sets. *The Astronomical Journal* **108**, 2348–2361.

Atkinson, S.E. (1992). The performance of standard and hybrid EM algorithms for ML estimates of the normal mixture model with censoring. *Journal of Statistical Computation and Simulation* **44**, 105–115.

Attias H. (1999a). Inferring parameters and structure of latent variable models by variational Bayes. In *Proceedings of the 15th Conference on Uncertainty in Artificial Intelligence.*

Attias H. (1999b). Independent factor analysis. *Neural Networks* **11**, 803–851.

Attias, H. (2000). Independent factor analysis with temporally structured factors. In *Neural Information Processing Systems 12*, S. A. Solla, T. K. Leen, and K.-R. Múller (Eds.). Cambridge, Massachusetts: MIT Press. (to appear).

Atwood, L.D., Wilson, A.F., Bailey-Wilson, J.E., Carruth, J.N., and Elston, R.C. (1996). On the distribution of the likelihood ratio test statistic for a mixture of two normal distributions. *Communications in Statistics—Simulation and Computation* **25**, 733–740.

Banfield, J.D. and Raftery, A.E. (1993). Model-based Gaussian and non-Gaussian clustering. *Biometrics* **49**, 803–821.

Banjeree, T. and Paul, S.R. (1999). An extension of Morel-Nagaraj's finite mixture distribution for modelling multinomial clustered data. *Biometrika* **86**, 723–727.

Barnard, G.A. (1963). Contribution to the discussion of paper by M.S. Bartlett. *Journal of the Royal Statistical Society B* **25**, 294.

Barndorff-Nielsen, O. (1965). Identifiability of mixtures of exponential families. *Journal of Mathematical Analysis and Applications* **12**, 115–121.

Bartlett, M.S. (1938). Further aspects of multiple regression. *Proceedings of the Cambridge Philosophical Society* **34**, 33–40.

Basford, K.E., Greenway, D.R., McLachlan, G.J., and Peel, D. (1997a). Standard errors of fitted means under normal mixture models. *Computational Statistics* **12**, 1–17.

Basford, K.E. and McLachlan, G.J. (1985a). Estimation of allocation rates in a cluster analysis context. *Journal of the American Statistical Association* **80**, 286–293.

Basford, K.E. and McLachlan, G.J. (1985b). Cluster analysis in a randomized complete block design. *Communications in Statistics—Theory and Methods* **14**, 451–463.

Basford, K.E. and McLachlan, G.J. (1985c). The mixture method of clustering applied to three-way data. *Journal of Classification* **2**, 109–125.

Basford, K.E. and McLachlan, G.J. (1985d). Likelihood estimation for normal mixture models. *Applied Statistics* **34**, 282–289.

Basford, K.E., McLachlan, G.J., and York, M.G. (1997b). Modelling the distribution of stamp paper thickness via finite normal mixtures: the 1872 stamp issue of Mexico revisited. *Journal of Applied Statistics* **24**, 169–179.

Basilevsky, A. (1994). *Statistical Factor Analysis and Related Methods.* New York: Wiley.

Baum, L.E. and Eagon, J.A. (1967). An inequality with applications to statistical estimation for probabilistic functions of Markov processes and to a model for ecology. *Bulletin of the American Mathematical Society* **73**, 360–363.

Baum, L.E. and Petrie, T. (1966). Statistical inference for probabilistic functions of finite Markov chains. *Annals of Mathematical Statistics* **37**, 1554–1563.

Baum, L.E., Petrie, T., Soules, G., and Weiss, N. (1970). A maximization technique occurring in the statistical analysis of probabilistic functions of Markov chains. *Annals of Mathematical Statistics* **41**, 164–171.

Bautista, M.G., Smith, D.W., and Steiner, R.L. (1997). A cluster-based approach to means separation. *Journal of Agricultural, Biological, and Environmental Statistics* **2**, 179–197.

Baxter, R.A. and Oliver, J.J. (2000). Finding overlapping components with MML. *Statistics and Computing* **10**, 5–16.

Bechtel, Y.C., Bonaiti-Pellieé, C., Poisson, N., Magnette, J., and Bechtel, P.R. (1993). A population and family study of N-acetyltransferase using caffeine urinary metabolites. *Clinical Pharmacology and Therapeutics* **54**, 134–141.

Becker, R.A., Chambers, J.M., and Wilks, A.R. (1988). *The New S Language: A Programming Environment for Data Analysis and Graphics*. Pacific Grove, California: Wadsworth and Brooks/Cole.

Beers, T.C. and Sommer-Larsen, J. (1995). Kinematics of metal-poor stars in the galaxy. *The Astrophysical Journal Supplement Series* **96**, 175–221.

Behboodian, J. (1970). On a mixture of normal distribution. *Biometrika* **57**, 215–217.

Belbin, T.R. and Rubin, D.B. (1995). The analysis of repeated-measures data on schizophrenic reaction times using mixture models. *Statistics in Medicine* **14**, 747–768.

Bensmail, H. and Celeux, G. (1996). Regularized Gaussian discriminant analysis through eigenvalue decomposition. *Journal of the American Statistical Association* **91**, 1743–1748.

Bensmail, H., Celeux, G., Raftery, A.E., and Robert, C.P. (1997). Inference in model-based cluster analysis. *Statistics and Computing* **7**, 1–10.

Berdai, A. and Garel, B. (1996). Detecting a univariate normal mixture with two components. *Statistics and Decisions* **16**, 35–51.

Berkson, J. and Gage, R.P. (1952). Survival curve for cancer patients following treatment. *Journal of the American Statistical Association* **47**, 501 – 515.

Bernardo, J.M. and Girón, J. (1988). A Bayesian analysis of simple mixture problems. In *Bayesian Statistics 3*, J.M. Bernardo, M.H. DeGroot, D.V. Lindley, and A.F.M. Smith (Eds.). Oxford: Oxford University Press, pp. 67–78.

Besag, J. (1975). Statistical analysis of non-lattice data. *The Statistician* **24**, 179–195.

Besag, J. (1986). On the statistical analysis of dirty pictures (with discussion). *Journal of the Royal Statistical Society B* **48**, 259–302.

Besag, J. (1989). Towards Bayesian image analysis. *Journal of Applied Statistics* **16**, 395–407.

Besag, J., York, J. and Mollie, A. (1991). Bayesian image restoration with two applications in spatial statistics (with discussion). *Annals of the Institute of Statistical Mathematics* **43**, 1–59.

Bezdek, J.C. (1981). *Pattern recognition with Fuzzy Objective Function Algorithms*. New York: Plenum.

Bezdek, J.C., Hathaway, R.M., and Huggins, V.J. (1985). Parametric estimation for normal mixtures. *Pattern Recognition* **3**, 79–84.

Bhattacharya, C.G. (1967). A simple method for resolution of a distribution into its Gaussian components. *Biometrics* **23**, 115–135.

Bickel, P.J. and Chernoff, H. (1993). Asymptotic distribution of the likelihood ratio statistic in a prototypical non regular problem. In *Statistics and Probability*, J.K. Ghosh, S.K. Mira, K.R. Parthasarthy, and B.L.S. Prakasa Rao (Eds.). New York: Wiley, pp. 83–96.

Bickel, P.J. and Ritov, Y. (1996). Inference in hidden Markov models I: local asymptotic normality in the stationary case. *Bernoulli* **2**, 199–228.

Bickel, P.J., Ritov, Y., and Rydén, T. (1998). Asymptotic normality of the maximum likelihood estimator for general hidden Markov models. *Annals of Statistics* **26**, 1614–1635.

Biernacki, C., Celeux, G., and Govaert, G. (1998). Assessing a mixture model for clustering with the integrated classification likelihood. *Technical Report No. 3521*. Rhône-Alpes: INRIA.

Biernacki, C., Celeux, G., and Govaert, G. (1999). An improvement of the NEC criterion for assessing the number of clusters in a mixture model. *Pattern Recognition Letters* **20**, 267-272.

Biernacki, C. and Govaert, G. (1997). Using the classification likelihood to choose the number of clusters. *Computing Science and Statistics* **29**, 451–457.

Biernacki C. and Govaert, G. (1999). Choosing models in model-based clustering and discriminant analysis. *Journal of Statistical Computation and Simulation* **64**, 49-71.

Billard, L. (1997). A voyage of discovery. *Journal of the American Statistical Association* **92**, 1–12.

Billio, M., Montfort, A., and Robert, C.P. (1999). Bayesian estimation of switching ARMA models. *Journal of Econometrics* **93**, 229–255.

Bird, C.M. (1994a). Substructure and cd peculiar velocities: A2670. *The Astrophysical Journal* **422**, 480–485.

Bird, C.M. (1994b). Substructure in clusters and central galaxy peculiar velocities. *The Astronomical Journal* **107**, 1637–1648.

Bird, C.M. (1995). The effects of substructure on galaxy cluster mass determination. *The Astrophysical Journal* **445**, L81–L84.

Bird, C.M., Davis, D.S., and Beers, T.C. (1995). Subcluster mergers and galaxy infall in A2151. *The Astronomical Journal* **109**, 920–927.

Bishop, C.M. (1995). *Neural Networks for Pattern Recognition*. Oxford: Oxford University Press.

Bishop, C.M. (1998). Latent variable models. In *Learning in Graphical Models*, M.I. Jordan (Ed.). Dordrecht: Kluwer, pp. 371–403.

Bishop, C.M., Svensén, M., and Williams, C.K.I. (1999). GTM: the generative topographic mapping. *Neural Computation* **10**, 215–234.

Bishop, C.M. and Tipping, M.E. (1998). A hierarchical latent variable model for data visualization. *IEEE Transactions on Pattern Analysis and Machine Intelligence* **20**, 281–293.

Blackstone, E.H., Naftel, D.C., and Turner, M.E. (1986). The decomposition of time-varying hazard into phases, each incorporating a separate stream of concomitant information. *Journal of the American Statistical Association* **81**, 615–24.

Blischke, W.R. (1978). Mixtures of distributions. In *International Encyclopedia of Statistics*, Vol. 1, W.H. Kruskal and J.M. Tanur (Eds.). New York: The Free Press, pp. 174–180.

Blum, J.R. and Susarla, V. (1977). Estimation of a mixing distribution function. *Annals of Probability* **5**, 200–209.

Boag, J.W. (1949). Maximum likelihood estimates of the proportion of patients cured by cancer therapy. *Journal of the Royal Statistical Society B* **11**, 15–53.

Bogardus, C., Lillioja, S., Nyomba, B.L., Zurlo, F., Swinburn, B., Puente, A.E.-D., Knowler, W.C., Ravussin, E., Mott, D.M., and Bennett, P.H. (1989). Distribution of in vivo insulin action in Pima-Indians as mixture of 3 normal-distributions. *Diabetes* **38**, 1423-1432.

Böhning, D. (1995). A review of reliable maximum likelihood algorithms for semi-parametric mixture models. *Journal of Statistical Planning and Inference* **47**, 5–28.

Böhning, D. (1999). *Computer-Assisted Analysis of Mixtures and Applications: Meta-Analysis, Disease Mapping and Others*. New York: Chapman & Hall/CRC.

Böhning, D., Dietz, E., Schaub, R., Schlattmann, P., and Lindsay, B. (1994). The distribution of the likelihood ratio for mixtures of densities from the one-parameter exponential family. *Annals of the Institute of Statistical Mathematics* **46**, 373–388.

Böhning, D., Schlattmann, P., and Lindsay, B. (1992). Computer-assisted analysis of mixtures (C.A.MAN): statistical algorithms. *Biometrics* **48**, 283–303.

Böhning, D., Schlattmann, P., and Lindsay, B. (1998). Recent developments in computer-assisted analysis of mixtures. *Biometrics* **54**, 525–536.

Booth, J.G. and Sarkar, S. (1998). Monte Carlo approximations of bootstrap variances. *The American Statistician* **52**, 354–357.

Box, G.E.P. and Cox, D.R. (1964). An analysis of transformations. *Journal of the Royal Statistical Society B* **26**, 211–252.

Bozdogan, H. (1990). On the information-based measure of covariance complexity and its applications to the evaluation of multivariate linear models. *Communications in Statistics—Theory and Methods* **19**, 221–278.

Bozdogan, H. (1993). Choosing the number of component clusters in the mixture-model using a new informational complexity criterion of the inverse-Fisher information matrix. In *Information and Classification*, O. Opitz, B. Lausen, and R. Klar (Eds.). Heidelberg: Springer-Verlag, pp. 40–54.

Bozdogan, H. and Sclove, S.L. (1984). Multi-sample cluster analysis using Akaike's information criterion. *Annals of the Institute of Statistical Mathematics* **36**, 163–180.

Bradley, P.S. and Fayyad, U.M. (1998). Refining initial points for k-means clustering. In *Proceedings of the Fifteenth International Conference on Machine Learning (ICML98)*. San Francisco: Morgan Kaufmann, pp. 91–99.

Bradley, P.S., Fayyad, U.M., and Reina, C.A. (1998). Scaling clustering algorithms to large databases. In *Proceedings of the Fourth International Conference on Knowledge Discovery and Data Mining*. Menlo Park, California: AAAI Press, pp. 9–15.

Bradley, P.S., Fayyad, U.M., and Reina, C.A. (1999). Scaling EM (expectation-maximization) clustering to large databases. *Technical Report No. MSR-TR-98-35*. Seattle: Microsoft Research.

Breiman, L., Friedman, J.H., Olshen, R.A., and Stone, C.J. (1984). *Classification and Regression Trees*. Belmont, California: Wadsworth.

Breslow, N.E. (1984). Extra-Poisson variation in log-linear models. *Applied Statistics* **33**, 38–44.

Breslow, N.E. (1987). Tests of hypotheses in overdispersed Poisson regression and other quasi-likelihood models. *Journal of the American Statistical Association* **85**, 565–571.

Breslow, N.E. and Clayton, D.G. (1993). Approximate inference in generalized linear mixed models. *Journal of the American Statistical Association* **88**, 9–25.

Bridges, T.J., Ashman, K.M., Zepf, S.E., Carter, D., Hanes, D.A., Sharples, R.M., and Kavelaars, J.J. (1997). Kinematics and metallicities of globular clusters in M104. *Monthly Notices of the Royal Astronomical Society* **284**, 376–384.

Brillinger, D.R. (1986). The natural variability of vital rates and associated statistics (with discussion). *Biometrics* **42**, 693–734.

Brinkley, D. and Haybittle, J.L. (1975). The curability of breast cancer. *Lancet* **2**, 95–97.

Broniatowski, M., Celeux, G., and Diebolt, J. (1983). Reconaissance de densités par un algorithme d'apprentissage probabiliste. In *Data Analysis and Informatics*, Vol. 3. Amsterdam: North–Holland, pp. 359–374.

Brooks, S.P. (1998). Markov chain Monte Carlo method and its application. *The Statistician* **47**, 69–100.

Brooks, S.P. and Morgan, B.J.T. (1995). Optimisation using simulated annealing. *The Statistician* **44**, 241–257.

Brooks, S.P., Morgan, B.J.T., Ridout, M.S., and Pack, S.E. (1998). Finite mixture models for proportions. *Biometrics* **53**, 1097–1115.

Brooks, S.P. and Roberts, G.O. (1998). Convergence assessment techniques for Markov chain Monte Carlo. *Statistics and Computing* **8**, 319–335.

Bryant, P.G. (1991). Large-sample results for optimization-based clustering. *Journal of Classification* **8**, 31–44.

Bryant, P.G. and Williamson, J.A. (1978). Asymptotic behaviour of classification maximum likelihood estimates. *Biometrika* **65**, 273–281.

Bryant, P.G. and Williamson, J.A. (1986). Maximum likelihood and classification: a comparison of three approaches. In *Classification as a Tool of Research*, W. Gaul and M. Schafer (Eds.). Amsterdam: North-Holland, pp. 33–45.

Burch, P.R.J. (1976). *The Biology of Cancer. A New Approach.* Edinburgh: MTP Press, pp. 370 – 392.

Butler, R.W. (1986). Predictive likelihood inference with applications (with discussion). *Journal of the Royal Statistical Society B* **48**, 1–38.

Byar, D.P. and Green, S.B. (1980). The choice of treatment for cancer patients based on covariate information: application to prostate cancer. *Bulletin du Cancer (Paris)* **67**, 477–490.

Byers, S. and Raftery, A.E. (1998). Nearest-neighbor clutter removal for estimating features in spatial point processes. *Journal of the American Statistical Association* **93**, 577–584.

Cadez, I.V., McLaren, C.E., Smyth, P., and McLachlan, G.J. (1999). Hierarchical models for screening of iron deficiency anemia. In *Proceedings of the Sixteenth International Conference on Machine Learning*. San Francisco: Morgan Kaufmann, pp. 77–86.

Cadez, I.V. and Smyth, P. (1998). Modeling of inhomogeneous Markov random fields with applications to cloud screening. *Technical Report No. 98-2.* Irvine: Department of Information Science, University of California Irvine.

Cadez, I.V., Smyth, P., McLachlan, G.J., and McLaren C.E. (2000). Maximum likelihood estimation of mixture densities for binned and truncated multivariate data. *Machine Learning* (to appear).

Campbell, J.G., Fraley, C., Murtagh, F., and Raftery, A.E. (1997). Linear flaw detection in woven textiles using model-based clustering. *Pattern Recognition Letters* **18**, 1539–1548.

Campbell, N.A. (1984). Mixture models and atypical values. *Mathematical Geology* **16**, 465–477.

Campbell, N.A. and Mahon, R.J. (1974). A multivariate study of variation in two species of rock crab of genus *Leptograpsus*. *Australian Journal of Zoology* **22**, 417–425.

Cao, G. and West, M. (1996). Practical Bayesian inference using mixtures of mixtures. *Biometrics* **52**, 1334–1341.

Cao, R., Cuevas, A., and Fraiman, R. (1995). Minimum distance density-based estimation. *Computational Statistics and Data Analysis* **20**, 611–631.

Carlin, B. and Chib, S. (1995). Bayes model choice via Markov chain Monte Carlo methods. *Journal of the Royal Statistical Society B* **57**, 473–484.

Carroll, R.J., Roeder, K., and Wasserman, L. (1999). Flexible parametric measurement error models. *Biometrics* **55**, 44–54.

Casella, G., Mengersen, K.L., Robert, C.P., and Titterington, D.M. (2000). Perfect slice samplers for mixtures of distributions. *Technical Report BU-1453-M*. Ithaca, New York: Department of Biometrics, Cornell University.

Casella, G., Robert, C.P., and Wells, M.T. (1999). Exact Monte Carlo analysis of mixture models. *Technical Report*. Ithaca, New York: Department of Biometrics, Cornell University.

Cassie, R.M. (1954). Some uses of probability paper in the analysis of size frequency distributions. *Australian Journal of Marine and Freshwater Research* **5**, 513–522.

Castelli, V. and Cover, T. (1996). The relative value of labeled and unlabeled samples in pattern recognition with an unknown mixing parameter. *IEEE Transactions on Information Theory* **42**, 2102–2117.

Celeux, G. (1999). Bayesian inference for mixture: the label switching problem. *Technical Report*. Rhone-Alpes: INRIA.

Celeux, G., Chauveau, D., and Diebolt, J. (1996). Stochastic versions of the EM algorithm: an experimental study in the mixture case. *Journal of Statistical Computation and Simulation* **55**, 287–314.

Celeux, G., Chrétien, S., Forbes, F., and Mkhadri, A. (1999). A component-wise EM algorithm for mixtures. *Technical Report No. 3746*. Rhone-Alpes: INRIA.

Celeux, G. and Diebolt, J. (1985). The SEM algorithm: a probabilistic teacher algorithm derived from the EM algorithm for the mixture problem. *Computational Statistics Quarterly* **2**, 73–82.

Celeux, G. and Diebolt, J. (1986a). The SEM and EM algorithms for mixtures: Numerical and statistical aspects. In *Proceedings of the* 7th *Franco-Belgium Meeting of Statistics*. Bruxelles: Publication des Facultés Universitaries St. Louis.

Celeux, G. and Diebolt, J. (1986b). L'algorithme SEM: un algorithme d'apprentissage probabiliste pour la reconnaissance de mélanges de densités. *Random Structures and Algorithms* **34**, 35–52.

Celeux, G. and Govaert, G. (1991). Clustering criteria for discrete data and latent class models. *Journal of Classification* **8**, 157–176.

Celeux, G. and Govaert, G. (1993). Comparison of the mixture and the classification maximum likelihood in cluster analysis. *Journal of Statistical Computation and Simulation* **47**, 127–146.

Celeux, G. and Govaert, G. (1995). Gaussian parsimonious clustering model. *Pattern Recognition* **28**, 781–793.

Celeux, G., Hurn, M., and Robert, C.P. (2000). Computational and inferential difficulties with mixture posterior distributions. *Journal of the American Statistical Association* (to appear).

Celeux, G. and Soromenho, G. (1996). An entropy criterion for assessing the number of clusters in a mixture model. *Classification Journal* **13**, 195–212.

Chandra, S. (1977). On the mixtures of probability distributions. *Scandinavian Journal of Statistics: Theory and Applications* **39**, 105–112.

Chang, P.C. and Afifi, A.A. (1974). Classification based on dichotomous and continuous variables. *Journal of the American Statistical Association* **69**, 336–339.

Chang, W.C. (1983). On using principal components before separating a mixture of two multivariate normal distributions. *Applied Statistics* **32**, 267–275.

Charlier, C.V.L. (1906). Researchers into the theory of probability. *Lunds Universities Årskrift, Ny foljd* **2.1**, *No. 5*.

Charlier, C.V.L. and Wicksell, S.D. (1924). On the dissection of frequency functions. *Arkiv för Mathematik Astronomi och Fysik* **18**, *No. 6*.

Cheeseman, P. and Heckerman, D. (1997). Bayesian classification: efficient approximations for the marginal likelihood of Bayesian networks with hidden variables. *Machine Learning* **29**, 181–212.

Cheeseman, P. and Stutz, J. (1996). Bayesian classification (AutoClass): theory and results. In *Advances in Knowledge Discovery and Data Mining*, U.M. Fayyad, G. Piatetsky-Shapiro, P. Smyth, and R. Uthurusamy (Eds.). Menlo Park, California: The AAAI Press, pp. 61–83.

Chen, J. (1992). Optimal rate of convergence in finite mixture models. *Annals of Statistics* **23**, 221–234.

Chen, J. (1994). Generalized likelihood ratio test of the number of components in finite mixture models. *Canadian Journal of Statistics* **22**, 387–400.

Chen, J. (1998). Penalized likelihood-ratio test for finite mixture models with multinomial observations. *Canadian Journal of Statistics* **26**, 583–599.

Chen, J. and Cheng, P. (1997). On testing the number of components in finite mixture models with known relevant component distributions. *Canadian Journal of Statistics* **25**, 389–400.

Chen, J. and Kalbfleisch, J.D. (1996). Penalized minimum-distance estimates in finite mixture models. *Canadian Journal of Statistics* **24**, 167–175.

Chen, K., Xu, L., and Chi, H. (1999). Improved learning algorithms for mixture of experts in multiclass classification. *Neural Networks* **12**, 1229–1252.

Cheng, R.C.H. and Traylor, L. (1995). Non-regular maximum likelihood problems (with discussion). *Journal of the Royal Statistical Society B* **57**, 3–44.

Chernick, M.R. (1999). *Bootstrap Methods: A Practitioner's Guide.* New York: Wiley.

Chernoff, H. and Lander, E. (1995). Asymptotic distribution of the likelihood ratio test that a mixture of two binomials is a single binomial. *Journal of Statistical Planning and Inference* **43**, 19–40.

Chhikara, R.S. and Register, D.T. (1979). A numerical classification method for partitioning of a large dimensional mixed data set. *Technometrics* **21**, 531–538.

Chib, S. (1995). Marginal likelihood from the Gibbs output. *Journal of the American Statistical Association* **90**, 1313–1321.

Chib, S. (1996). Calculating posterior distributions and modal estimates in Markov mixture models. *Journal of Econometrics* **75**, 79–97.

Choi, K. (1969). Estimators for the parameters of a finite mixture of distributions. *Annals of the Institute of Statistical Mathematics* **21**, 107–116.

Choi, K. and Bulgren, W.G. (1968). An estimation procedure for mixtures of distributions. *Journal of the Royal Statistical Society B* **30**, 444–460.

Chuang, R. and Mendell, N.R. (1997). The approximate null distribution of the likelihood ratio test for a mixture of two bivariate normal distributions with equal covariance. *Communications in Statistics—Simulation and Computation* **26**, 631–648.

Churchill, G.A. (1995). Accurate restoration of DNA sequences (with discussion). In *Case Studies in Bayesian Statistics*, Vol. 2, C. Gatsonis, J.S. Hodges, R.E. Kass, and N.D. Singpurwalla (Eds.). New York: Springer, pp. 90–148.

Clark, V.A., Chapman, L.M., Coulson, A.H., and Hasselblad, V. (1968). Dividing the blood pressures from the Los Angeles heart study into two normal distributions. *John Hopkins Medical Journal* **122**, 77–83.

Clarke, B.R. (1989). An unbiased minimum distance estimator of the proportion parameter in a mixture of two normal distributions. *Statistics & Probability Letters* **7**, 275–281.

Clarke, B.R. and Heathcote, C.R. (1994). Robust estimation of k-component univariate normal mixtures. *Annals of the Institute of Statistical Mathematics* **46**, 83–93.

Clayton, D.G. and Kaldor, J. (1987). Empirical Bayes estimates for age-standardized relative risks. *Biometrics* **43**, 671–681.

Clogg, C.C. (1977). Unrestricted and restricted maximum likelihood latent structure analysis: a manual for users. *Working Paper No. 09*, Pittsburgh: Population Issues Research Center, Pennsylvania State University.

Clogg, C.C. (1995). Latent class models. In *Handbook of Statistical Modeling for the Social and Behavioral Sciences*, G. Arminger, C.C. Clogg, and M.E. Sobel (Eds.). New York: Plenum, pp. 311–359.

Cohen, A.C. (1967). Estimation in mixtures of two normal distributions. *Technometrics* **9**, 15–28.

Conway, J.H. and Sloane, N.J.A. (1988). *Sphere Packings, Lattices and Groups*. New York: Springer-Verlag.

Cooley, C.A. and MacEachern, S.N. (1999). Prior elicitation in the classification problem. *Canadian Journal of Statistics* **27**, 299–313.

Copas, J.A. and Heydari, F. (1997). Estimating the risk of reoffending by using exponential mixture models. *Journal of the Royal Statistical Society A* **160**, 237–252.

Cowles, M.K. and Carlin, B.P. (1996). Markov chain Monte Carlo convergence diagnostics: a comparative review. *Journal of the American Statistical Association* **91**, 883–904.

Cox, D. R. (1959). The analysis of exponentially distributed life-times with two types of failure. *Journal of the Royal Statistical Society B* **21**, 411–421.

Cox, D. R. (1962). *Renewal Theory*. London: Methuen.

Cox, D. R. (1972). Regression models and life-tables (with discussion). *Journal of the Royal Statistical Society B* **34**, 187–220.

Cox, D.R. and Wermuth, N. (1992). Response models for mixed binary and quantitative variables. *Biometrika* **79**, 441–461.

Craigmile, P.F. and Titterington, D.M. (1998). Parameter estimation for finite mixtures of uniform distributions. *Communications in Statistics—Theory and Methods* **26**, 1981–1995.

Cramér, H. (1946). *Mathematical Methods of Statistics*. Princeton, New Jersey: Princeton University Press.

Crawford, S. (1994). An application of the Laplace method to finite mixture distributions. *Journal of the American Statistical Association* **89**, 259–267.

Crawford, S.L., DeGroot, M.H., Kadane, J.B., and Small, M.J. (1992). Modeling lake-chemistry distributions: approximate Bayesian methods for estimating a finite-mixture model. *Technometrics* **34**, 441–453.

Crouch, E.A.C. and Spiegelman, D. (1990). The evaluation of integrals of the form $\int_{-\infty}^{\infty} f(t) \exp(-t^2)dt$: an application to logistic-normal models. *Journal of the American Statistical Association* **85**, 464–469.

Cutler, A. and Cordiero-Braña, O.I. (1996). Minimum Hellinger distance estimation for finite mixture models. *Journal of the American Statistical Association* **91**, 1716–1723.

Cutler, A. and Windham, M.P. (1994). Information-based validity functionals for mixture analysis. In *Proceedings of the First US/Japan Conference on the Frontiers of Statistical Modeling in Informational Approach*. Amsterdam: Kluwer, pp. 149–170.

Ćwik, J. and Koronacki, J. (1997). A combined adaptive-mixtures/plug-in estimator of multivariate probability densities. *Computational Statistics and Data Analysis* **26**, 199–218.

Dacunha-Castelle, D. and Gassiat, E. (1997a). Testing in locally conic models. *ESAIM: Probability and Statistics* **1**, 285–317.

Dacunha-Castelle, D. and Gassiat, E. (1997b). The estimation of the order of a mixture model. *Bernoulli* **3**, 279–299.

Dasgupta, A. and Raftery, A.E. (1998). Detecting features in spatial point processes with clutter via model-based clustering. *Journal of the American Statistical Association* **93**, 294–302.

DasGupta, S. (1999). Learning mixtures of Gaussians. *Technical Report No. UCB/CSD 99-1047*. Berkeley: Computer Science Division, University of California.

Davé, R.N. and Krishnapuram, R. (1997). Robust clustering methods: a unified view. *IEEE Transactions on Fuzzy Systems* **5**, 270–293.

David, H.A. and Moeschberger, M.L. (1978). *The Theory of Competing Risks*. London: Griffin.

Davies, R.B. (1977). Hypothesis testing when a nuisance parameter is present only under the alternative. *Biometrika* **64**, 247–254.

Davis, D.S., Bird, C.M., Mushotzky, R.F., and Odewahn, S.C. (1995). Abell-548—an X-ray and optical analysis of substructure. *The Astrophysical Journal* **440**, 48–59.

Davison, A.C. and Hinkley, D.V. (1997). *Bootstrap Methods and Their Application*. Cambridge: Cambridge University Press.

Dawid, A.P. (1976). Properties of diagnostic data distributions. *Biometrics* **32**, 647–658.

Day, N.E. (1969). Estimating the components of a mixture of two normal distributions. *Biometrika* **56**, 463–474.

Dean C., Lawless, J.F., and Willmot, E. (1989). A mixed Poisson-inverse-Gaussian regression model. *Canadian Journal of Statistics* **17**, 171–181.

Deely, J.J. and Kruse, R.L. (1968). Construction of sequences estimating the mixture distribution. *Annals of Mathematical Statistics* **39**, 286–288.

Dellaportas, P. (1998). Bayesian classification of Neolithic tools. *Applied Statistics* **47**, 279–297.

Dempster, A.P., Laird, N.M., and Rubin, D.B. (1977). Maximum likelihood from incomplete data via the EM algorithm (with discussion). *Journal of the Royal Statistical Society B* **39**, 1–38.

DeSarbo, W.S. and Cron, W.L. (1988). A maximum likelihood methodology for clusterwise linear regression. *Journal of Classification* **5**, 249–282.

De Veaux, R.D. (1989). Mixtures of linear regressions. *Computational Statistics and Data Analysis* **8**, 227–245.

De Veaux, R.D. and Kreiger, A.M. (1990). Robust estimation of a normal mixture. *Statistics & Probability Letters* **10**, 1–7.

Devijver, P.A. (1985). Baum's forward-backward algorithm revisited. *Pattern Recognition Letters* **3**, 369–373.

Devroye, L.P. and Györfi, L. (1985). *Nonparametric Density Estimation: the L_1 View*. New York: Wiley.

Diebolt, J. and Ip, E.H.S. (1996). Stochastic EM: method and application. In *Markov Chain Monte Carlo in Practice*, W.R. Gilks, S. Richardson, and D.J. Spiegelhalter (Eds.). London: Chapman & Hall, pp. 259–273.

Diebolt, J. and Robert, C.P. (1990). Bayesian estimation of finite mixture distributions: Part II, Sampling implementation. *Technical Report III*. Paris: Laboratoire de Statistique Théorique et Appliquée, Université Paris VI.

Diebolt, J. and Robert, C.P. (1994). Estimation of finite mixture distributions through Bayesian sampling. *Journal of the Royal Statistical Society B* **56**, 363–375.

Dietz, E. (1992). Estimation of heterogeneity—a GLM-approach. In *Advances in GLIM and Statistical Modelling*, L. Fahrmeir, F. Francis, R. Gilchrist, and G. Tutz (Eds.). Berlin: Springer Verlag, pp. 66–72.

Dietz, E. and Böhning, D. (1996). Statistical inference based on a general model of unobserved heterogeneity. In *Advances in GLIM and Statistical Modelling*, L. Fahrmeir, F. Francis, R. Gilchrist, and G. Tutz (Eds.). Berlin: Springer-Verlag, pp. 75–82.

Digalakis, V.V. (1999). Online adaptation of hidden Markov models using incremental estimation algorithms. *IEEE Transactions on Speech and Audio Processing* **7**, 253–261.

Do, K.-A. and McLachlan, G.J. (1984). Estimation of mixing proportions: a case study. *Applied Statistics* **33**, 134–140.

Doerge, R.W., Zeng, Z.-B., and Weir, B.S. (1997). Statistical issues in the search for gene affecting quantitative traits in experimental populations. *Statistical Science* **12**, 195–219.

Doetsch, G. (1928). Die elimination des dopplereffekts auf spektroskopische feinstrukturen und exakte bestimmung der komponenten. *Zeitschrift für Physik* **49**, 705–730.

Doetsch, G. (1936). Zerlegung einer function in Gauss'sche fehlerkurven. *Mathematische Zeitschrift* **41**, 283–318.

Donoho, D.L. (1988). One-sided inference about functionals of a density. *Annals of Statistics* **16**, 1390–1420.

Dressler, A. (1980). Catalog of morphological types in 55 rich clusters of galaxies. *The Astrophysical Journal Supplement Series* **42**, 565–609.

Duda, R.O. and Hart, P.E. (1973). *Pattern Classification and Scene Analysis*. New York: Wiley.

Dunmur, A.P. and Titterington, D.M. (1997). Computational Bayesian analysis of hidden Markov mesh models. *IEEE Transactions on Pattern Analysis and Machine Intelligence* **19**, 1296–1300.

Dunmur, A.P. and Titterington, D.M. (1998a). The influence of initial conditions on maximum likelihood estimation of the parameters of a binary hidden Markov model. *Statistics & Probability Letters* **40**, 67–73.

Dunmur, A.P. and Titterington, D.M. (1998b). Parameter estimation in latent profile models. *Computational Statistics and Data Analysis* **27**, 371–388.

Durairajan, T.M. and Kale, B.K. (1979). Locally most powerful similar test for the mixing proportion. *Sankhyā B* **41**, 91–100.

Durairajan, T.M. and Kale, B.K. (1982). Locally most powerful similar test for mixing proportions. *Sankhyā A* **44**, 153–161.

Efron, B. (1979). Bootstrap methods: another look at the jackknife. *Annals of Statistics* **7**, 1–26.

Efron, B. (1982). *The Jackknife, the Bootstrap and Other Resampling Plans*. Philadelphia: SIAM.

Efron, B. (1986). Double exponential families and their use in generalized linear regression. *Journal of the American Statistical Association* **81**, 709–721.

Efron, B. (1992). Poisson overdispersion estimates based on the method of asymmetric maximum likelihood. *Journal of the American Statistical Association* **87**, 98–107.

Efron, B. (1994). Missing data, imputation and the bootstrap (with discussion). *Journal of the American Statistical Association* **89**, 463–479.

Efron, B. and Hinkley, D.V. (1978). Assessing the accuracy of the maximum likelihood estimator: Observed versus expected Fisher information (with discussion). *Biometrika* **65**, 457-487.

Efron, B. and Tibshirani, R. (1993). *An Introduction to the Bootstrap*. London: Chapman & Hall.

Eisenberger, I. (1964) Genesis of bimodal distributions. *Technometrics* **6**, 357–363.

Erdfelder, E. (1993). BINOMIX, a BASIC program for maximum likelihood analyses of finite and beta–binomial mixture distributions. *Behaviour Research Methods, Instruments, & Computers* **25**, 416–418.

Escobar, M.D. and West, M. (1995). Bayesian density estimation and inference using mixtures. *Journal of the American Statistical Association* **90**, 577–588.

Evans, M., Guttman, I., and Olkin, I. (1992). Numerical aspects in estimating the parameters of a mixture of normal distributions. *Journal of Computational and Graphical Statistics* **1**, 351-365.

Everitt, B.S. (1981). A Monte Carlo investigation of the likelihood ratio test for the number of components in a mixture of normal distributions. *Multivariate Behavioral Research* **16**, 171–180.

Everitt, B.S. (1984). A note on parameter estimation for Lazarsfeld's latent class model using the EM algorithm. *Multivariate Behavioral Research* **19**, 79–89.

Everitt, B.S. (1985). Mixture distributions. In *Encyclopedia of Statistical Sciences*, Vol. 5, S. Kotz and N.L. Johnson (Eds.). New York: Wiley, pp. 559–569.

Everitt, B.S. (1988a). A Monte Carlo investigation of the likelihood-ratio test for number of classes in latent class analysis. *Multivariate Behavioral Research* **23**, 531–538.

Everitt, B.S. (1988b). A finite mixture model for the clustering of mixed mode data. *Statistics & Probability Letters* **6**, 305–309.

Everitt, B.S. (1996). An introduction to finite mixture distributions. *Statistical Methods in Medical Research* **5**, 107–127.

Everitt, B.S. and Hand, D.J. (1981). *Finite Mixture Distributions*. London: Chapman & Hall.

Everitt, B.S and Merette, C. (1990). The clustering of mixed-mode data: a comparison of possible approaches. *Journal of Applied Statistics* **17**, 283–297.

Faddy, M.J. (1994). On variation in Poisson processes. *Mathematical Scientist* **19**, 47–51.

Farewell, V.T. (1977). A model for a binary variable with time-censored observations. *Biometrika* **64**, 43–46.

Farewell, V.T. (1982). The use of mixture models for the analysis of survival data with long-term survivors. *Biometrics* **38**, 1041–1046.

Farewell, V.T. (1986). Mixture models in survival analysis: are they worth the risk? *Canadian Journal of Statistics* **14**, 257–262.

Farewell, V.T. and Sprott, D. (1988). The use of a mixture model in the analysis of count data. *Biometrics* **44**, 1191–1194.

Fayyad, U.M., Reina, C.A., and Bradley, P.S.(1998). Initialization of iterative refinement clustering algorithms. In *Proceedings of the Fourth International Conference on Knowledge Discovery and Data Mining*. Menlo Park, California: AAAI Press, pp. 194–198.

Fayyad, U.M. and Smyth, P. (1999). Cataloging and mining massive datasets for science data analysis. *Journal of Computational and Graphical Statistics* **8**, 589–610.

Feng, Z.D. and McCulloch, C.E. (1992). Statistical inference using maximum likelihood estimation and the generalized likelihood ratio when the true parameter is on the boundary of the parameter space. *Statistics & Probability Letters* **13**, 325–332.

Feng, Z.D. and McCulloch, C.E. (1994). On the likelihood ratio test statistic for the number of components in a normal mixture with unequal variances. *Biometrics* **50**, 1158–1162.

Feng, Z.D. and McCulloch, C.E. (1996). Using bootstrap likelihood ratio in finite mixture models. *Journal of the Royal Statistical Society B* **58**, 609–617.

Fessler, J.A. and Hero, A.O. (1994). Space-alternating generalized expectation-maximization algorithm. *IEEE Transactions on Signal Processing* **42**, 2664–2677.

Fessler, J.A. and Hero, A.O. (1995). Penalized maximum-likelihood image reconstruction using space-alternating generalized EM algorithms. *IEEE Transactions on Image Processing* **4**, 1417–1429.

Fisher, N.I., Lewis, T., and Embleton, B.J.J. (1987). *Statistical Analysis of Spherical Data*. Cambridge: Cambridge University Press.

Fisher, N.I., Mammen, E., and Marron, J.S. (1994). Testing for multimodality. *Computational Statistics and Data Analysis* **18**, 499–512.

Fisher, R.A. (1936). The use of multiple measurements in taxonomic problems. *Annals of Eugenics* **7**, 179–188.

Flury, B.D., Airoldi, J.-P., and Biber, J.-P. (1992). Gender identification of water pipits *(Anthus-spinoletta)*, using mixtures of distributions. *Journal of Theoretical Biology* **158**, 465–480.

Folks, J.L. and Chhikara, R.S. (1978). The inverse Gaussian distribution and its statistical application—a review. *Journal of the Royal Statistical Society B* **40**, 263–289.

Follmann, D.A. and Lambert, D. (1989). Generalizing logistic regression by non-parametric mixing. *Journal of the American Statistical Association* **84**, 295–300.

Follmann, D.A. and Lambert, D. (1991). Identifiability for nonparametric mixtures of logistic regressions. *Journal of Statistical Planning and Inference* **27**, 375–381.

Fong, D.Y.T. and Yip, P. (1993). An EM algorithm for a mixture model of count data. *Statistics & Probability Letters* **17**, 53–60.

Fong, D.Y.T. and Yip, P.S.F. (1995). A note on information loss in analyzing a mixture model for count data. *Communications in Statistics—Theory and Methods* **24**, 3197–3209.

Formann, A.K. (1994). Measuring change in latent subgroups using dichotomous data: unconditional, conditional, and semiparametric maximum likelihood estimation. *Journal of the American Statistical Association* **89**, 1027–1034.

Formann, A.K. and Kohlmann, T. (1996). Latent class analysis in medical research. *Statistical Methods in Medical Research* **5**, 179–211.

Fowlkes, E.B. (1979). Some methods for studying the mixture of two normal (lognormal) distributions. *Journal of the American Statistical Association* **74**, 561–575.

Fraley, C. and Raftery, A.E. (1998). How many clusters? Which clustering method? Answers via model-based cluster analysis. *Computer Journal* **41**, 578–588.

Fraley, C. and Raftery, A.E. (1999). MCLUST: Software for model-based cluster analysis. *Journal of Classification* **16**, 297-306.

Fredkin, D.R. and Rice, J.A. (1992). Maximum likelihood estimation and identification directly from single-channel recordings. *Proceedings of the Royal Society of London B* **249**, 125–132.

Friedman, H.P. and Rubin, J. (1967). On some invariant criteria for grouping data. *Journal of the American Statistical Association* **62**, 1159–1178.

Friedman, J.H. (1987). Exploratory projection pursuit. *Journal of the American Statistical Association* **82**, 249–266.

Friedman, J.H. (1989). Regularized discriminant analysis. *Journal of the American Statistical Association* **84**, 165–175.

Friedman, J.H. (1991). Multivariate adaptive regression splines. *Annals of Statistics* **19**, 1–141.

Frigui, H. and Krishnapuram, R. (1996). A robust algorithm for automatic extraction of an unknown number of clusters from noisy data. *Pattern Recognition Letters* **17**, 1223–1232.

Fryer, J.G. and Robertson, C.A. (1972). A comparison of some methods for estimating mixed normal distributions. *Biometrika* **59**, 639–648.

Fukunaga, K.S. (1990). *Introduction to Statistical Pattern Recognition.* Boston: Academic Press.

Furman, W.D. and Lindsay, B.G. (1994a). Testing for the number of components in a mixture of normal distributions using moment estimators. *Computational Statistics and Data Analysis* **17**, 473–492.

Furman, W.D. and Lindsay, B.G. (1994b). Measuring the effectiveness of moment estimators as starting values in maximizing mixture likelihoods. *Computational Statistics and Data Analysis* **17**, 493–507.

Galton, F. (1869). *Heredity Genius: An Inquiry into Its Laws and Consequences.* London: Macmillan.

Gan, L. and Jiang, J. (1999). A test for a global maximum. *Journal of the American Statistical Association* **94**, 847–854.

Ganesalingam, S. and McLachlan, G.J. (1978). The efficiency of a linear discriminant function based on unclassified initial samples. *Biometrika* **65**, 658–662.

Ganesalingam, S. and McLachlan, G.J. (1979a). Small sample results for a linear discriminant function estimated from a mixture of normal populations. *Journal of Statistical Computation and Simulation* **9**, 151–158.

Ganesalingam, S. and McLachlan G.J. (1979b). A case study of two clustering methods based on maximum likelihood. *Statistica Neerlandica* **33**, 81–90.

Ganesalingam, S. and McLachlan, G.J. (1980a). A comparison of the mixture and classification approaches to cluster analysis. *Communications in Statistics—Theory and Methods* **9**, 923–933.

Ganesalingam, S. and McLachlan, G.J. (1980b). Error rate estimation on the basis of posterior probabilities. *Pattern Recognition* **11**, 405–413.

Ganesalingam, S. and McLachlan, G.J. (1981). Some efficiency results on the estimation of the mixing proportion in a mixture of two normal distributions. *Biometrics* **37**, 23–33.

García-Escudero, L.A. and Gordaliza, A. (1999). Robustness properties of k means and trimmed means. *Journal of the American Statistical Association* **94**, 956–969.

Garel, B. (1998). Asymptotic theory of the likelihood ratio test for the identification of a mixture. *Technical Report No. 05-98.* Toulouse: Université Paul Sabatier.

Gelfand, A.E. and Smith, A.F.M. (1990). Sampling-based approaches to calculating marginal densities. *Journal of the American Statistical Association* **85**, 398–409.

Gelman, A., Carlin, J., Stern, H., and Rubin, D. (1995). *Bayesian Data Analysis.* London: Chapman & Hall.

Gelman, A. and King, G. (1990). Estimating the electoral consequences of legislative redirecting. *Journal of the American Statistical Association* **85**, 274–282.

Gelman, A. and Rubin, D.B. (1996). Markov chain Monte Carlo methods in biostatistics. *Statistical Methods in Medical Research* **5**, 339–355.

Geman, S. and Geman, D. (1984). Stochastic relaxation, Gibbs distributions, and the Bayesian restoration of images. *IEEE Transactions on Pattern Analysis and Machine Intelligence* **6**, 721–741.

Geman, S. and Graffigne, C. (1987). Markov random field image models and their applications to computer vision. In *Proceedings of the International Congress of Mathematicians.* Berkeley: American Mathematical Society, pp. 1498–1517.

Geman, S. and Hwang, C.-R. (1982). Nonparametric maximum likelihood estimation by the method of sieves. *Annals of Statistics* **10**, 401–414.

Genovese, C. and Wasserman, L. (1999). Rates of convergence for the Gaussian mixture sieve. *Technical Report No. 685*. Pittsburgh: Department of Statistics, Carnegie-Mallon University.

Geyer, C.J. and Thompson, E.A. (1992). Constrained Monte Carlo maximum likelihood for dependent data (with discussion). *Journal of the Royal Statistical Society B* **54**, 657–699.

Ghahramani, Z. (1995). Factorial learning and the EM algorithm. In *Advances in Neural Information Processing Systems 17*, G. Tesauro, D.S. Touretzky, and T.K. Leen. (Eds.). Cambridge, Massachusetts: MIT Press, pp. 1125–1132.

Ghahramani, Z. and Beal, M.J. (2000). Variational inference for Bayesian mixtures of factor analysers. In *Neural Information Processing Systems 12* S. A. Solla, T. K. Leen, and K.-R. Múller (Eds.). Cambridge, Massachusetts: MIT Press (to appear).

Ghahramani, Z. and Hinton, G.E. (1997). The EM algorithm for factor analyzers. *Technical Report No. CRG-TR-96-1*. Toronto: The University of Toronto.

Ghahramani, Z. and Hinton, G.E. (1998). Hierarchical nonlinear factor analysis and topographic maps. In *Advances in Neural Information Processing Systems 10*, M.I. Jordan, M.J. Kearns, and S.A. Solla (Eds.). Cambridge, Massachusetts: MIT Press, pp. 486–492.

Ghahramani, Z. and Jordan, M.J. (1997). Factorial hidden Markov models. *Machine Learning* **29**, 245–275.

Ghosh, J.H. and Sen, P.K. (1985). On the asymptotic performance of the log likelihood ratio statistic for the mixture model and related results. In *Proceedings of the Berkeley Conference in Honor of Jerzy Neyman and Jack Kiefer*, Vol. 2. Monterey: Wadsworth, pp. 789–806.

Gilks, W.R., Oldfield, L., and Rutherford, A. (1989). In *Leucoctye Typing IV*, W. Knapp, B. Dörken, W.R. GIlks, S.F. Schlossman, L. Boumsell, J.M. Harlan, T. Kishimoto, C. Morimoto, J. Ritz, S. Shaw, R. Silverstein, T. Springer, T.F. Tedder, and R.F. Todd. (Eds.). Oxford: Oxford University Press, pp. 6–12.

Gilks, W.R., Richardson, S., and Spiegelhalter, D.J. (Eds.) (1996). *Markov Chain Monte Carlo in Practice*. London: Chapman & Hall.

Gnanadesikan, R., Harvey, J.W., and Kettenring, J.R. (1993). Mahalanobis metrics for cluster analysis. *Sankhyā A* **55**, 494–505.

Goffinet, B., Loisel, P., and Laurent, B. (1992). Testing in normal mixture models when the proportions are known. *Biometrika* **79**, 842–846.

Golbeck, A.L. (1992). Bootstrapping current life table estimators. In *Bootstrapping and Related Techniques*, K.H. Jöckel, G. Rothe, G., and W. Sendler (Eds.). Heidelberg: Springer-Verlag, pp. 197–201.

Goodman, L.A. (1974a). The analysis of qualitative variables when some of the variables are unobservable. Part I—a modified latent structure approach. *Journal of the American Statistical Association* **79**, 1179–1259.

Goodman, L. (1974b). Exploratory latent structure analysis using both identifiable and unidentifiable models. *Biometrika* **61**, 215–231.

Gordon, N.H. (1990a). Application of the theory of finite mixtures for the estimation of 'cure' rates of treated cancer patients. *Statistics in Medicine* **9**, 397–407.

Gordon, N.H. (1990b). Maximum likelihood estimation for mixtures of two Gompertz distributions when censoring occurs. *Communications in Statistics—Simulation and Computation* **19**, 737–747.

Govaert, G. and Nadif, M. (1996). Comparison of the mixture and the classification maximum likelihood in cluster analysis with binary data. *Computational Statistics and Data Analysis* **23**, 65–81.

Green, P.J. (1984). Iteratively reweighted least squares for maximum likelihood estimation, and some robust and resistant alternatives. *Journal of the Royal Statistical Society B* **46**, 149–192.

Green, P.J. (1994). Contribution to the discussion on paper by U. Grenader and M.I. Miller. *Journal of the Royal Statistical Society B* **56**, 589–590.

Green, P.J. (1995). Reversible jump Markov chain Monte Carlo computation and Bayesian model determination. *Biometrika* **82**, 711–732.

Greenhouse, J.B. and Wolfe, R.A. (1984). A competing risk derivation of a mixture model for the analysis of survival data. *Communications in Statistics—Theory and Methods* **13**, 3133–3154.

Greig, D.M., Porteous, B.T., and Seheult, A.H. (1989). Exact maximum *a posteriori* estimation for binary images. *Journal of the Royal Statistical Society B* **51**, 271–279.

Grenader, U. and Miller, M.I. (1994). Representations of knowledge in complex systems (with discussion). *Journal of the Royal Statistical Society B* **56**, 549–603.

Gruet, M.A., Philippe, A., and Robert, C.P. (1999). MCMC control spreadsheets for exponential mixture estimation. *Journal of Computational and Graphical Statistics* **8**, 298-317.

Guess, F.M., Usher, J.S., and Hodgson, T.J. (1990). Estimating system and component reliabilities via the EM algorithm (with discussion). *Journal of Statistical Planning and Inference* **29**, 75–85.

Gupta, A.K. and Miyawaki, T. (1978). On a uniform mixture model. *Biometrical Journal* **20**, 631–637.

Gupta, P.L. and Gupta, R.C. (1996). Ageing characteristics of the Weibull mixtures. *Probability in the Engineering and Informational Sciences* **10**, 591–600.

Gupta, S.S. and Huang, W.T. (1981). On mixtures of distributions: a survey and some new results on ranking and selection. *Journal of the Indian Statistical Association B* **43**, 245–290.

Gutierrez, R.G., Carroll, R.J., Wang, N., Lee, G.-H., and Taylor, B.H. (1995). Analysis of tomato root initiation using a normal mixture distribution. *Biometrics* **51**, 1461–1468.

Guttorp, P. (1995). *Stochastic Modeling of Scientific Data.* London: Chapman & Hall.

Habbema, J.D.F., Hermans, J., and van den Broek, K. (1974). A stepwise discriminant analysis program using density estimation. *Compstat 1974, Proceedings Computational Statistics.* Vienna: Physica-Verlag, pp. 101–110.

Hamilton, J.D. (1989). A new approach to the economic analysis of nonstationary time series and the business cycle. *Econometrica* **57**, 357–384.

Hammah, R.E. and Curran, J.H. (1998). Fuzzy cluster algorithm for the automatic identification of joint sets. *International Journal of Rock Mechanics and Mining Sciences* **35**, 889-905.

Hampel, F.R. (1973). Robust estimation: a condensed partial survey. *Z. Wahrscheinlickeitstheorie verw. Gebiete* **27**, 87–104

Harding, J.P. (1948). The use of probability paper for the graphical analysis of polymodal frequency distributions. *Journal of Marine Biological Association* **28**, 141–153.

Harrison, J. and Stevens, C.F. (1976). Bayesian forecasting (with discussion). *Journal of the Royal Statistical Society B* **38**, 205–247.

Hartigan, J.A. (1985a). Statistical theory in clustering. *Journal of Classification* **2**, 63–76.

Hartigan, J.A. (1985b). A failure of likelihood asymptotics for normal mixtures. In *Proceedings of the Berkeley Conference in Honor of Jerzy Neyman and Jack Kiefer*, Vol. 2, Monterey: Wadsworth, pp. 807–810.

Hartigan, J. (1988). The SPAN test for unimodality. In *Classification and Related Methods of Data Analysis*, H.H. Bock (Ed.). Amsterdam: North-Holland.

Hartigan, J. and Hartigan, P. (1985). The dip test for unimodality. *Annals of Statistics* **13**, 70–84.

Hartigan, J. and Mohanty, S. (1992). The RUNT test for multimodality. *Journal of Classification* **9**, 63–70.

Hasselblad, V. (1966). Estimation of parameters for a mixture of normal distributions. *Technometrics* **8**, 431–444.

Hasselblad, V. (1969). Estimation of finite mixtures of distributions from the exponential family. *Journal of the American Statistical Association* **64**, 1459–1471.

Hastie, T. and Tibshirani, R.J. (1996). Discriminant analysis by Gaussian mixtures. *Journal of the Royal Statistical Society B* **58**, 155–176.

Hathaway, R.J. (1983). Constrained maximum likelihood estimation for a mixture of m univariate normal distributions. *Technical Report No. 92*. Columbia, South Carolina: University of South Carolina.

Hathaway, R.J. (1985). A constrained formulation of maximum-likelihood estimation for normal mixture distributions. *Annals of Statistics* **13**, 795–800.

Hathaway, R.J. (1986a). Another interpretation of the EM algorithm for mixture distributions. *Statistics & Probability Letters* **4**, 53–56.

Hathaway, R.J. (1986b). A constrained EM algorithm for univariate mixtures. *Journal of Statistical Computation and Simulation* **23**, 211–230.

Haughton, D. (1997). Packages for estimating finite mixtures: a review. *The American Statistician* **51**, 194–205.

Hawkins, D.M. (1981). A new test for multivariate normality and homoscedasticity. *Technometrics* **23**, 105–110.

Hawkins, D.M., Muller, M.W., and ten Krooden, J.A. (1982). Cluster analysis. In *Topics in Applied Multivariate Analysis*, D.M. Hawkins (Ed.). Cambridge: Cambridge University Press, pp. 303–356.

Haybittle, J.L. (1983). Is breast cancer ever cured? *Reviews on Endocrine-Related Cancer* **14**, 13–18.

Heckman, J.J. and Honoré, B.E. (1989). The identifiability of the competing risks model. *Biometrika* **76**, 325–330.

Heckman, J.J., Robb, R., and Walker, J.R. (1990). Testing the mixture of exponentials hypothesis and estimating the mixing distribution by the method of moments. *Journal of the American Statistical Association* **85**, 582–589.

Heckman, J.J. and Singer, B. (1982). The identification problem in econometric models for duration data. In *Advances in Econometrics*, W. Hildenbrand (Ed.). Cambridge: Cambridge University Press, pp. 39–77.

Heckman, J.J. and Singer, B. (1984). A method for minimizing the impact of distributional assumptions in econometric models of duration. *Econometrica* **52**, 271–320.

Heikkinen, H. and Högmander, H. (1994). Fully Bayesian approach to image restoration with an application in biogeography. *Applied Statistics* **43**, 569–582.

Henna, J. (1985). On estimating the number of constituents of a finite mixture of continuous distributions. *Annals of the Institute of Statistical Mathematics* **37**, 235–240.

Hero, A.O. and Fessler, J.A. (1995). Convergence in norm for EM-type algorithms. *Statistica Sinica* **5**, 41–54.

Hinde, J.P. (1982). Compound Poisson regression models. In *GLIM 82*, R. Gilchrist (Ed.). New York: Springer, pp. 109–121.

Hinton, G.E., Dayan, P., and Revow, M. (1997). Modeling the manifolds of images of handwritten digits. *IEEE Transactions on Neural Networks* **8**, 65–73.

Hinton, G.E. and Zeal, R.S. (1994). Autoencoders, minimum description length and Helmholtz free energy. In *Advances in Neural Information Processing Systems 6*, J.D. Cowan, G. Tesauro, and J. Alspector (Eds.). San Mateo, California: Morgan Kaufmann, pp. 3–10.

Hjort, N.L. (1986). Contribution to the discussion of paper by P. Dianconis and D. Freedman. *Annals of Statistics* **14**, 49–55.

Hoaglin, D.C. (1985). Using quantiles to study shape. In *Explaining Data Tables, Trends, and Shapes*, D.C. Hoaglin, F. Mosteller, and J.W. Tukey (Eds.). New York: Wiley, pp. 417–460.

Hobert, J.P., Robert, C.P., and Titterington, D.M. (1999). On perfect simulation for some mixtures of distributions. *Statistics and Computing* **9**, 287–298.

Holgersson, M. and Jorner, U. (1978). Decomposition of a mixture into normal components: a review. *International Journal of Biomedical Computing* **9**, 367–392.

Holmes, G.K. (1892). Measures of distribution. *Journal of the American Statistical Association* **3**, 141–157.

Holst, U. and Lindgren, G. (1991). Recursive estimation in mixture models with Markov regime. *IEEE Transactions on Information Theory* **37**, 1683–1690.

Hope, A.C.A. (1968). A simplified Monte Carlo significance test procedure. *Journal of the Royal Statistical Society A* **30**, 582-598.

Hosmer, D.W. (1973a). On MLE of the parameters of a mixture of two normal distributions when the sample size is small. *Computational Statistics* **1**, 217–227.

Hosmer, D.W. (1973b). A comparison of iterative maximum likelihood estimates of the parameters of a mixture of two normal distributions under three different types of sample. *Biometrics* **29**, 761–770.

Hosmer, D.W. and Dick, N.P. (1977). Information and mixtures of two normal distributions. *Journal of Statistical Computation and Simulation* **6**, 137–148.

Hsu, Y.-S., Walker, J.J., and Ogren, D.E. (1986). A stepwise method for determining the number of component distributions in a mixture. *Mathematical Geology* **18**, 153–160.

Huber, P.J. (1964). Robust estimation of a location parameter. *Annals of Mathematical Statistics* **35**, 73–101.

Huber, P.J. (1985). Projection pursuit (with discussion). *Annals of Statistics* **13**, 435–475.

Hunt, L.A. and Basford. K.E. (1999) Fitting a mixture model to three-mode three-way data with categorical and continuous variables. *Journal of Classification* **16**, 283-296.

Hunt, L.A. and Jorgensen, M.A. (1999). Mixture model clustering: a brief introduction to the MULTIMIX program. *Australian & New Zealand Journal of Statistics* **40**, 153–171.

Huzurbazar, V.S. (1948). The likelihood equation, consistency, and the maxima of the likelihood function. *Annals of Eugenics* **14**, 185–200.

Irwin, J.O. (1963). The place of mathematics in medical and biological statistics. *Journal of the Royal Statistical Society A* **126**, 1–45.

Ishiguro, M., Sakamoto, Y., and Kitagawa, G. (1997). Bootstrapping log-likelihood and EIC, an extension of AIC. *Annals of the Institute of Statistical Mathematics* **49**, 411-434.

Izenman, A.J. and Sommer, C.J. (1988). Philatelic mixtures and multimodal densities. *Journal of the American Statistical Association* **83**, 941-953.

Jaakkola, T. and Jordan, M.I. (2000). Bayesian logistic regression: a variational approach. *Statistics and Computing* **10**, 25–37.

Jacobs, R.A. (1997). Bias/variance analyses of mixtures-of-experts architectures. *Neural Computation* **9**, 369–383.

Jacobs, R.A., Jordan, M.I., Nowlan, S.J., and Hinton, G.E. (1991). Adaptive mixtures of local experts. *Neural Computation* **3**, 79–87.

Jacobs, R.A., Peng, F., and Tanner, M.A. (1997). A Bayesian approach to model selection in hierarchical mixtures-of-experts architectures. *Neural Networks* **10**, 231–241.

Jacobs, R.A., Tanner, M.A., and Peng, F.. (1996). Bayesian inference for hierarchical mixtures-of-experts with applications to regression and classification. *Statistical Methods in Medical Research* **5**, 375–390.

Jalali, A. and Pemberton, J. (1995). Mixture models for time series. *Journal of Applied Probability* **32**, 123–138.

Jamshidian, M. and Jennrich, R.I. (1993). Conjugate gradient acceleration of the EM algorithm. *Journal of the American Statistical Association* **88**, 221–228.

Jamshidian, M. and Jennrich, R.L. (1997). Acceleration of the EM algorithm by using quasi-Newton methods. *Journal of the Royal Statistical Society B* **59**, 569–587.

Jansen, R.C. (1993). Maximum likelihood in a generalized linear finite mixture model by using the EM algorithm. *Biometrics* **49**, 227–231.

Jedidi, K., Jagpai, H.S., and DeSarbo, W.S. (1997). STEMM: A general finite mixture structural equation model. *Journal of Classification* **14**, 23–50.

Jeffreys, S.H. (1932). An alternative to the rejection of observations. *Proceedings of the Royal Society of London A* **137**, 78–87.

Jeffreys, S.H. (1961). *Theory of Probability*, third edition. Oxford: Claredon Press.

Jewell, N.P. (1982). Mixtures of exponential distributions. *Applied Statistics* **10**, 479–484.

Jiang, W. and Tanner, M.A. (1999a). Hierarchical mixtures-of-experts for exponential family regression models: approximation and maximum likelihood estimation. *Annals of Statistics* **27**, 987–1011.

Jiang, W. and Tanner, M.A. (1999b). On the approximation rate of hierarchical mixtures-of-experts for generalized linear models. *Machine Learning* **11**, 1183–1198.

Jiang, W. and Tanner, M.A. (1999c). On the identifiability of mixtures-of-experts. *Neural Networks* **12**, 1253–1258.

Johnson, M.E. (1987). *Multivariate Statistical Simulation*. New York: Wiley.

Jolion, J.-M., Meer, P., and Bataouche, S. (1991). Robust clustering with applications in computer vision. *IEEE Transactions on Pattern Analysis and Machine Intelligence* **13**, 791–802.

Jones, M.C. and Sibson, R. (1987). What is projection pursuit? (with discussion). *Journal of the Royal Statistical Society B* **150**, 1–36.

Jones, P.N. and McLachlan, G.J. (1989). Modelling mass-size particle data by finite mixtures. *Communications in Statistics—Theory and Methods* **18**, 2629–2646

Jones, P.N. and McLachlan, G.J. (1990a). Laplace-normal mixtures fitted to wind shear data. *Journal of Applied Statistics* **17**, 271–276.

Jones, P.N. and McLachlan, G.J. (1990b). Algorithm AS 254. Maximum likelihood estimation from grouped and truncated data with finite normal mixture models. *Applied Statistics* **39**, 273–282.

Jones, P.N. and McLachlan, G.J. (1991). Fitting mixture distributions to phenylthiocarbamide (PTC) sensitivity. *American Journal of Human Genetics* **48**, 117–120.

Jones, P.N. and McLachlan, G.J. (1992a). Improving the convergence rate of the EM algorithm for a mixture model fitted to grouped truncated data. *Journal of Statistical Computation and Simulation* **43**, 31–44.

Jones, P.N. and McLachlan, G.J. (1992b). Fitting finite mixture models in a regression context. *Australian Journal of Statistics* **34**, 233–240

Jordan, M.I. and Jacobs, R.A. (1992). Hierarchies of adaptive experts. In *Advances in Neural Information Processing Systems 4*, J. Moody, S. Hanson, and R. Lippmann (Eds.). San Mateo, California: Morgan Kaufmann, pp. 985–993.

Jordan, M.I. and Jacobs, R.A. (1994). Hierarchical mixtures of experts and the EM algorithm. *Neural Computation* **6**, 181–214.

Jordan, M.I. and Xu, L. (1995). Convergence results for the EM approach to mixtures of experts architectures. *Neural Networks* **8**, 1409–1431.

Jorgensen, M.A. (1990). Influence-based diagnostics for finite mixture models. *Biometrics* **46**, 1047–1058.

Jorgensen, M.A. (1999). A dynamic EM algorithm for estimating mixture proportions. *Statistics and Computing* **9**, 299–302.

Jorgensen, M.A. and Hunt, L.A. (1996). Mixture modelling clustering of data sets with categorical and continuous variables. In *ISIS: Information, Statistics and Induction in Science*, D.L. Dowe, K.B. Korb, and J.J. Oliver (Eds.). Singapore: World Scientific Publishing, pp. 375–384.

Juang, B.H. and Rabiner, L.R. (1991). Hidden Markov model for speech recognition. *Technometrics* **33**, 251–272.

Kadane, J.B. (1974). The role of identification in Bayesian theory. In *Studies in Bayesian Econometrics and Statistics*, S. Fienberg and A. Zellner (Eds.). New York: American Elsevier, pp. 175–191.

Kalbfleisch, J.D. and Prentice, R.L. (1980). *The Statistical Analysis of Failure Time Data*. New York: Wiley.

Kalmus, H. and Maynard Smith, S. (1965). The antimode and lines of optimal separation in a genetically determined bimodal distribution, with particular reference to phenylthiocarbamide sensitivity. *Annals of Human Genetics* **29**, 127–139.

Kanji, G.K. (1985). A mixture for wind shear data. *Journal of Applied Statistics* **12**, 49–58.

Kao, C.-H. and Zeng, Z.-B. (1997). General formulas for obtaining the MLEs and the asymptotic variance-covariance matrix in mapping quantitative trait loci when using the EM algorithm. *Biometrics* **53**, 653–665.

Karlis, D. and Xekalaki, E. (1998). Minimum Hellinger distance estimation for finite Poisson mixtures. *Computational Statistics and Data Analysis* **29**, 81–103.

Karlis, D. and Xekalaki, E. (1999). On testing for the number of components in finite Poisson mixtures. *Annals of the Institute of Statistical Mathematics* **5**, 149-162.

Kass, R. and Raftery, A.E. (1995). Bayes factors. *Journal of the American Statistical Association* **90**, 773–795.

Kay, J. (1986). Contribution to the discussion of paper by J. Besag. *Journal of the Royal Statistical Society B* **48**, 293.

Kay, J. and Titterington, D.M. (1986). Image labelling and the statistical analysis of incomplete data. *Proceedings of the Second International Conference on Image Processing and Applications*. London: Institute of Electrical Engineers, pp. 44–48.

Kent, J.T. (1982). The Fisher–Bingham distribution on the sphere. *Journal of the Royal Statistical Society B* **44**, 71–80.

Kent, J.T., Tyler, D.E., and Vardi, Y. (1994). A curious likelihood identity for the multivariate t-distribution. *Communications in Statistics—Simulation and Computation* **23**, 441–453.

Keribin, C. (1998). Consistent estimation of the order of mixture models. *Technical Report*. Evry: Laboratoire Analyse et Probabilité, Université d'Evry-Val d'Essonne.

Kharin, Y. (1996). *Robustness in Statistical Pattern Recognition*. Dordrecht: Kluwer.

Kiefer, J. and Wolfowitz, J. (1956). Consistency of the maximum likelihood estimates in the presence of infinitely many incidental parameters. *Annals of Mathematical Statistics* **27**, 887–906.

Kiefer, N.M. (1978). Discrete parameter variation: efficient estimation of a switching regression model. *Econometrica* **46**, 427–434.

Kim, B.S. (1994). Dispersion statistics in overdispersed mixture models. *Communications in Statistics—Theory and Methods* **23**, 27–45.

Kim, B.S. and Margolin, B.H. (1992). Testing goodness of fit of a multinomial against overdispersed alternatives. *Biometrics* **48**, 711–719.

Kitagawa, G. (1989). Non-Gaussian seasonal adjustment. *Computers & Mathematics with Applications* **18**, 503–514.

Klein, J. P. and Moeschberger, M.L. (1987). Independent or dependent competing risks: Does it make a difference? *Communications in Statistics—Simulation and Computation* **16**, 507–533.

Koehler, A.B. and Murphee, E.H. (1988). A comparison of the Akaike and Schwarz criteria for selecting model order. *Applied Statistics* **37**, 187–195.

Konishi, S. and Kitagawa, G. (1996). Generalised information criteria in model selection. *Biometrika* **83**, 875–890.

Kosinski, A. (1999). A procedure for the detection of multivariate outliers. *Computational Statistics and Data Analysis* **29**, 145–161.

Kowalski, J., Tu, X.M., Day, R.S., Mendoza-Blanco, J.R. (1997). On the rate of convergence of the ECME algorithm for multiple regression models with t-distributed errors. *Biometrika* **84**, 269–281.

Kriessler, J.R. and Beers, T.C. (1997). Substructure in galaxy clusters: a two-dimensional approach. *The Astronomical Journal* **113**, 80–100.

Krogh, A., Brown, M., Mian, L.S., Sjölander, K., and Haussler, D. (1994). Hidden Markov models in computational biology: applications to protein monitoring. *Journal of Molecular Biology* **235**, 1501–1531.

Krzanowski, W.J. (1975). Discrimination and classification using both binary and continuous variables. *Journal of the American Statistical Association* **70**, 782–790.

Kuk, A.Y.C. (1992). A semiparametric mixture model for the analysis of competing risks data. *Australian Journal of Statistics* **34**, 169–180.

Kuk, A.Y.C. and Chen, C.-H. (1992). A mixture model combining logistic regression with proportional hazards regression. *Biometrika* **79**, 531–541.

Kullback, S. and Leibler, R.A. (1951). On information and sufficiency. *Annals of Mathematical Statistics* **22**, 79–86.

Kundu, A. and He, Y. (1991). On optimal order in modeling sequence of letters in words of common language as a Markov chain. *Pattern Recognition* **24**, 603–608.

Lagakos, S. W. (1979). General right censoring and its impact on the analysis of survival data. *Biometrics* **35**, 139–156.

Laird, N.M. (1978). Nonparametric maximum likelihood estimation of a mixing distribution. *Journal of the American Statistical Association* **73**, 805–811.

Laird, N.M. (1982). Empirical Bayes estimates using the nonparametric estimate of the prior. *Journal of Statistical Computation and Simulation* **73**, 805–811.

Lambert, D. (1992). Zero-inflated Poisson regression, with an application to defects in manufacturing. *Technometrics* **34**, 1–14.

Lange, K. (1995a). A gradient algorithm locally equivalent to the EM algorithm. *Journal of the Royal Statistical Society B* **57**, 425–437.

Lange, K. (1995b). A quasi-Newton acceleration of the EM algorithm. *Statistica Sinica* **5**, 1–18.

Lange, K., Little, R.J.A., and Taylor, J.M.G. (1989). Robust statistical modeling using the t distribution. *Journal of the American Statistical Association* **84**, 881–896.

Langlands, A.O., Pocock, S.J., Kerr, G.R., and Gore, S.M. (1979). Long-term survival of patients with breast cancer: a study of the curability of the disease. *British Medical Journal* **2**, 1247–1251.

Larson, M.G. and Dinse, G.E. (1985). A mixture model for the regression analysis of competing risks data. *Applied Statistics* **34**, 201–211.

Lavine, M. and West, M. (1992). A Bayesian method of classification and discrimination. *Canadian Journal of Statistics* **20**, 451–461.

Lawless, J.F. (1987). Negative binomial and mixed Poisson regression. *Canadian Journal of Statistics* **15**, 209–225.

Lawley, D.N. (1953). A modified method of estimation in factor analysis and some large sample results. In *Nordisk Psykologi's Monograph Series No. 3*. Copenhagen: Ejnar Mundsgaards, pp. 35–42.

Lawley, D.N. and Maxwell, A.E. (1971), second edition. *Factor Analysis as a Statistical Method*. London: Butterworths.

Lawrence, C.J. and Krzanowski, W.J. (1996). Mixture separation for mixed-mode data. *Statistics and Computing* **6**, 85–92.

Lazarsfeld, P.F. (1950). The logical and mathematical foundation of latent structure analysis. In *Studies in Social Psychology in World War II Vol. IV: Measurement and Prediction)*, S.A. Stouffer, L. Guttman, E.A. Suchman, P.F. Lazarsfeld, S.A. Star, and J.A. Clausen (Eds.). Princeton, New Jersey: Princeton University Press, pp. 362–412.

Lazarsfeld, P.F. and Henry, N.W. (1968). *Latent Structure Analysis*. New York: Houghton Mifflin.

Le, N.D., Leroux, B.G., and Puterman, M.L. (1992). Exact likelihood evaluation in a Markov mixture model for time series of seizure counts. *Biometrics* **48**, 317–323.

Le, N.D., Martin, R.D., and Raftery, A.E. (1996). Modeling flat stretches, bursts, and outliers in time series using mixture transition distribution models. *Journal of the American Statistical Association* **91**, 1504–1514.

Lee, Y. and Nelder, J.A. (1996). Hierarchical generalized linear models (with discussion). *Journal of the Royal Statistical Society B* **58**, 619–678.

Lehmann, E.L. (1980). Efficient likelihood estimators. *The American Statistician* **34**, 233–235.

Lehmann, E.L. (1983). *Theory of Point Estimation*. New York: Wiley.

Lemdani, M. and Pons, O. (1995). Tests for genetic linkage and homogeneity. *Biometrics* **51**, 1033-1041.

Lemdani, M. and Pons, O. (1997). Likelihood ratio tests for genetic linkage. *Statistics & Probability Letters* **33**, 15–22.

Lemdani, M. and Pons, O. (1999). Likelihood ratio tests in contamination models. *Bernoulli* **5**, 705-719.

Leroux, B.G. (1992a). Consistent estimation of a mixing distribution. *Annals of Statistics* **20**, 1350–1360.

Leroux, B.G. (1992b). Maximum likelihood estimation for hidden Markov models. *Stochastic Processes and Their Applications* **40**, 127–1143.

Leroux, B.G. and Puterman, M.L. (1992). Maximum-penalized-likelihood estimation for independent and Markov-dependent mixture models. *Biometrics* **48**, 545–558.

Li, L.A. and Sedransk, N. (1988). Mixtures of distributions: a topological approach. *Annals of Statistics* **16**, 1623–1634.

Liang, K.-Y. and Rathouz, P.J. (1999). Hypothesis testing under mixture models: application to genetic linkage analysis. *Biometrics* **55**, 65–74.

Liang, Z., MacFall, J.R., and Harrington, D.P. (1994). Parameter estimation and tissue segmentation from multispectral MR images. *IEEE Transactions on Medical Imaging* **13**, 441–449.

Lindgren, G. (1978). Markov regime models for mixed distributions and switching regressions. *Scandinavian Journal of Statistics: Theory and Applications* **5**, 81–91.

Lindsay, B.G. (1983). The geometry of likelihoods: a general theory. *Annals of Statistics* **11**, 86–94.

Lindsay, B.G. (1994). Efficiency versus robustness: the case for minimum Hellinger distance estimation and related methods. *Annals of Statistics* **22**, 1081–1114.

Lindsay, B.G. (1995). *Mixture Models: Theory, Geometry and Applications, NSF-CBMS Regional Conference Series in Probability and Statistics*, Vol. 5. Alexandria, Virginia: Institute of Mathematical Statistics and the American Statistical Association.

Lindsay, B.G. and Basak, P. (1993). Multivariate normal mixtures: a fast, consistent method of moments. *Journal of the American Statistical Association* **88**, 468–476.

Lindsay, B.G., Clogg, C.C., and Grego, J. (1991). Semiparametric estimation in the Rasch model and related exponential response models, including a simple latent class model for item analysis. *Journal of the American Statistical Association* **86**, 96–107.

Lindsay, B.G. and Lesperance, M.L. (1995). A review of semiparametric mixture models. *Journal of Statistical Planning and Inference* **47**, 29–99.

Lindsay, B.G. and Roeder, K. (1992a). Residual diagnostics for mixture models. *Journal of the American Statistical Association* **87**, 785–794.

Lindsay, B.G. and Roeder, K. (1992b). Uniqueness of estimation and identifiability in mixture models. *Canadian Journal of Statistics* **21**, 139–147.

Lindstrom, M.J. and Bates, D.M. (1988). Newton-Raphson and EM algorithms for linear mixed-effects models for repeated-measures data. *Journal of the American Statistical Association* **83**, 1014–1022.

Little, R.J.A. and Rubin, D.B. (1987). *Statistical Analysis with Missing Data.* New York: Wiley.

Liu, C. (1997). ML estimation of the multivariate t distribution and the EM algorithm. *Journal of Multivariate Analysis* **63**, 296–312.

Liu, C. (1998). Information matrix computation from conditional information via normal approximation. *Biometrika* **85**, 973–979.

Liu, C. and Rubin, D.B. (1994). The ECME algorithm: a simple extension of EM and ECM with faster monotone convergence. *Biometrika* **81**, 633–648.

Liu, C. and Rubin, D.B. (1995). ML estimation of the t distribution using EM and its extensions, ECM and ECME. *Statistica Sinica* **5**, 19–39.

Liu, C. and Rubin, D.B. (1998). Maximum likelihood estimation of factor analysis using the ECME algorithm with complete and incomplete data. *Statistica Sinica* **8**, 729–747.

Liu, C., Rubin, D.B., and Wu, Y.N. (1998). Parameter expansion to accelerate EM: the PX-EM Algorithm. *Biometrika* **85**, 755–770.

Liu, C. and Sun, D.X. (1997). Acceleration of EM algorithm for mixture models using ECME. *Proceedings of the American Statistical Association (Statistical Computing Section)*. Alexandria, Virginia: American Statistical Association, pp. 109–114.

Longini, I.M. and Halloran, E. (1996). A frailty mixture model for estimating vaccine efficacy. *Applied Statistics* **45**, 165–173.

Louis, T.A. (1982). Finding the observed information matrix when using the EM algorithm. *Journal of the Royal Statistical Society B* **44**, 226–233.

Lund, U. (1999). Cluster analysis for directional data. *Communications in Statistics—Simulation and Computation* **28**, 1001–1009.

MacDonald, I.L. and Zucchini, W. (1997). *Hidden Markov and Other Models for Discrete-Valued Time Series*. London: Chapman & Hall.

Macdonald, P.D.M. (1971). Comment on a paper by K. Choi and W.G. Bulgren. *Journal of the Royal Statistical Society B* **33**, 326–329.

Macdonald, P.D.M. and Pitcher T.J. (1979). Age groups from size frequency data: a versatile and efficient method for analyzing distribution mixtures. *Journal of the Fisheries Research Board of Canada*, **36**, 987–1001.

Maclean, C.J., Morton, N.E., Elston, R.C., and Yee, S. (1976). Skewness in commingled distributions. *Biometrics* **32**, 695–699.

Maller, R.A. and Zhou, X. (1996). *Survival Analysis with Long-Term Survivors*. New York: Wiley.

Mangin, B. and Goffinet, B. (1995). Local asymptotic power of maximum likelihood ratio tests when the information matrix is singular, with an application in genetics. *Scandinavian Journal of Statistics: Theory and Applications* **22**, 465–475.

Mangin, B., Goffinet, B., and Elsen, J.M. (1993). Testing in normal mixture models with some information on the parameters. *Biometrical Journal* **35**, 771–783.

Manton, K.G., Woodbury, M.A., and Stallard, E. (1981). A variance components approach to categorical data models with heterogeneous cell populations: analysis of spatial gradients in lung cancer mortality rates in North Carolina counties. *Biometrics* **37**, 259–269.

Margolin, B.H., Kim, B.S., and Risko, K.J. (1989). The Ames *salmonella*/microsome mutagenicity assay: issues of inference and validation. *Journal of the American Statistical Association* **84**, 651–661.

Markatou, M. (1998). Mixture models, robustness and the weighted likelihood methodology. *Technical Report No. 1998-9*. Stanford: Department of Statistics, Stanford University.

Markatou, M., Basu, A., and Lindsay, B.G. (1998). Weighted likelihood equations with bootstrap root search. *Journal of the American Statistical Association* **93**, 740–750.

Marriott, F.H.C. (1974). *The Interpretation of Multiple Observations*. London: Academic Press.

Marron, J.S. and Wand, M.P. (1992). Exact mean integrated squared error. *Annals of Statistics* **20**, 712–736.

McCullagh, P.A. and Nelder, J. (1989). *Generalised Linear Models*, second edition. London: Chapman & Hall.

McCutcheon, A.C. (1987). *Latent Class Analysis*. Beverly Hills: Sage Publications.

McGiffin, D.C., O'Brien, M.F., Galbraith, A.J., McLachlan, G.J., Stafford, E.G., Gardiner, M.A.H., Pohlner, P.G., Early, L., and Kear, L. (1993). An analysis of risk factors for death and mode-specific death following aortic valve replacement using allograft, xenograft and mechanical valves. *Journal of Thoracic and Cardiovascular Surgery* **106**, 895–911.

McHugh, R.B. (1956). Efficient estimation and local estimation in latent class analysis. *Psychometrika* **21**, 331–347.

McKendrick, A.G. (1926). Applications of mathematics to medical problems. *Proceedings of the Edinburgh Mathematical Society* **44**, 98–130.

McLachlan, G.J. (1975). Iterative reclassification procedure for constructing an asymptotically optimal rule of allocation in discriminant analysis. *Journal of the American Statistical Association* **70**, 365–369.

McLachlan, G.J. (1982a). The classification and mixture maximum likelihood approaches to cluster analysis. In *Handbook of Statistics*, Vol. 2, P.R. Krishnaiah and L. Kanal (Eds.). Amsterdam: North–Holland, pp. 199–208.

McLachlan, G.J. (1982b). On the bias and variance of some proportion estimators. *Communications in Statistics—Simulation and Computation* **11**, 715–726.

McLachlan, G.J. (1987). On bootstrapping the likelihood ratio test statistic for the number of components in a normal mixture. *Applied Statistics* **36**, 318–324.

McLachlan, G.J. (1992). *Discriminant Analysis and Statistical Pattern Recognition*. New York: Wiley.

McLachlan, G.J. (1994). One hundred years of mixtures. *Stats No. 12*, 6–12.

McLachlan, G.J. (1995). Mixtures–models and applications. In *the Exponential Distribution: Theory, Methods, and Applications*, N. Balakrishnan and A.P. Basu (Eds.). Basel: Gordon & Breach, pp. 307–315

McLachlan, G.J. (1997). On the EM algorithm for overdispersed count data. *Statistical Methods in Medical Research* **6**, 76–98.

McLachlan, G.J. (2000). On the grouping of treatments following an ANOVA. (Letter to the Editor). *Journal of Agricultural, Biological, and Environmental Statistics* (to appear).

McLachlan, G.J. and Basford, K.E. (1988). *Mixture Models: Inference and Applications to Clustering*. New York: Marcel Dekker.

McLachlan, G.J. and Gordon, R.D. (1989). Mixture models for partially unclassified data: a case study of renal venous renin levels in essential hypertension. *Statistics in Medicine* **8**, 1291–1300.

McLachlan, G.J. and Jones, P.N. (1988). Fitting mixture models to grouped and truncated data via the EM algorithm. *Biometrics* **44**, 571–578.

McLachlan, G.J. and Krishnan, T. (1997). *The EM Algorithm and Extensions*. New York: Wiley.

McLachlan, G.J. and McGiffin, D.C. (1994). On the role of finite mixture models in survival analysis. *Statistical Methods in Medical Research* **3**, 211-226.

McLachlan, G.J., McLaren, C.E., and Matthews, D. (1995). An algorithm for the likelihood ratio test of one versus two components in a mixture model fitted to grouped and truncated data. *Communications in Statistics—Simulation and Computation* **24**, 965–985.

McLachlan, G.J. and Ng, S.K. (2000a). A sparse version of the incremental EM algorithm for large databases. *Technical Report*. Brisbane: Department of Mathematics, University of Queensland.

McLachlan, G.J. and Ng, S.K. (2000b). A comparison of some information criteria for the number of components in a mixture model. *Technical Report*. Brisbane: Department of Mathematics, University of Queensland.

McLachlan, G.J., Ng, S.K., Adams, P., McGiffin, D.C., and Galbraith, A.J. (1997). An algorithm for fitting mixtures of Gompertz distributions to censored survival data. *Journal of Statistical Software* **2**, *No.* 7.

McLachlan, G.J., Ng, S.K., Galloway, G., and Wang, D. (1996). Clustering of magnetic resonance images. In *Proceedings of the American Statistical Association (Statistical Computing Section)*. Alexandria, Virginia: American Statistical Association, pp. 12–17.

McLachlan, G.J. and Peel, D. (1996). An algorithm for unsupervised learning via normal mixture models. In *ISIS: Information, Statistics and Induction in Science*, D.L. Dowe, K.B. Korb, and J.J. Oliver (Eds.). Singapore: World Scientific Publishing, pp. 354–363.

McLachlan, G.J. and Peel, D. (1997a). On a resampling approach to choosing the number of components in normal mixture models. In *Computing Science and Statistics*, Vol. 28, L. Billard and N.I. Fisher (Eds.). Fairfax Station, Virginia: Interface Foundation of North America, pp. 260–266.

McLachlan, G.J. and Peel, D. (1997b). Contribution to the discussion of paper by S. Richardson and P.J. Green. *Journal of the Royal Statistical Society B* **59**, 779-780.

McLachlan, G.J. and Peel, D. (1998a). Robust cluster analysis via mixtures of multivariate t-distributions. In *Lecture Notes in Computer Science*, Vol. 1451, A. Amin, D. Dori, P. Pudil, and H. Freeman (Eds.). Berlin: Springer-Verlag, pp. 658–666.

McLachlan, G.J. and Peel, D. (1998b). MIXFIT: an algorithm for the automatic fitting and testing of normal mixture models. In *Proceedings of the 14th International Conference on Pattern Recognition*, Vol. I. Los Alamitos, California: IEEE Computer Society, pp. 553–557.

McLachlan, G.J. and Peel, D. (2000). Mixtures of factor analyzers. In *Proceedings of the Seventeenth International Conference on Machine Learning*. San Francisco: Morgan Kaufmann, pp. 599–606.

McLachlan, G.J., Peel, D., Basford, K.E., and Adams, P. (1999). Fitting of mixtures of normal and t-components. *Journal of Statistical Software* **4**, *No.* 2.

McLachlan, G.J. and Scot, D. (1995). On the asymptotic relative efficiency of the linear discriminant function under partial nonrandom classification of the training data. *Journal of Statistical Computation and Simulation* **52**, 415–426.

McLaren, C.E. (1996). Mixture models in haematology: a series of case studies. *Statistical Methods in Medical Research* **5**, 129–153.

McLaren, C.E., Kambour, E.L., McLachlan, G.J., Lukaski, H.C., Li, X., Brittenham, G.M., and McLaren, G.D. (2000). Patient-specific analysis of sequential hematological data by multiple linear regression and mixture distribution modeling. *Statistics in Medicine* **19**, 83–98.

McLaren, C.E., McLachlan, G.J., Halliday, J.W., Webb, S.I., Leggett, B.A., Jazwinska, E.C., Crawford, D.H.G., Gordeuk, V.R., McLaren, G.D., and Powell, L.W. (1998). The distribution of transferrin saturation and hereditary haemochromatosis in Australians. *Gastroenterology* **114**, 543–549.

McLaren, C.E., Wagstaff, M., Brittenham, G.M., and Jacobs, A. (1991). Detection of two component mixtures of lognormal distributions in grouped doubly-truncated data: analysis of red blood cell volume distributions. *Biometrics* **47**, 607–622.

McManus, I.C. (1983). Bimodality of blood pressure levels. *Statistics in Medicine* **2**, 253–258.

Medgyessy, P. (1961). *Decomposition of Superpositions of Functions*. Budapest: Publishing House of the Hungarian Academy of Sciences.

Meilijson, I. (1989). A fast improvement to the EM algorithm on its own terms. *Journal of the Royal Statistical Society B* **51**, 127–138.

Mendell, N.R., Finch, S.J., and Thode, H.C. (1993). Where is the likelihood ratio test powerful for detecting two component normal mixtures? *Biometrics* **49**, 907–915.

Mendell, N.R., Thode, H.C., and Finch, S.J. (1991). The likelihood ratio test for the normal mixture problem: power and sample size analysis. *Biometrics* **47**, 1143–1148. Correction (1992). *Biometrics* **48**, 661.

Meng, X.L. (1997). The EM algorithm and medical studies: a historical link. *Statistical Methods in Medical Research* **6**, 3–23.

Meng, X.L. and Rubin, D.B. (1989). Obtaining asymptotic variance-covariance matrices for missing-data problems using EM. *Proceedings of the American Statistical Association (Statistical Computing Section)*. Alexandria, Virginia: American Statistical Association, pp. 140–144.

Meng, X.L. and Rubin, D.B. (1991). Using EM to obtain asymptotic variance-covariance matrices: The SEM algorithm. *Journal of the American Statistical Association* **86**, 899–909.

Meng, X.L. and Rubin, D.B. (1993). Maximum likelihood estimation via the ECM algorithm: a general framework. *Biometrika* **80**, 267–278.

Meng, X.L. and van Dyk, D. (1997). The EM algorithm—an old folk song sung to a fast new tune (with discussion). *Journal of the Royal Statistical Society B* **59**, 511–567.

Mengersen, K.L. and Robert, C.P. (1996). Testing for mixtures: a Bayesian entropic approach. In *Bayesian Statistics 5*, J.M. Bernardo, J. O. Berger, A. P. Dawid, and A. F. M. Smith (Eds.). Oxford: Clarendon Press, pp. 255–276.

Mengersen, K. L., Robert, C.P. and Guihenneuc-Jouyaux, C. (1999). MCMC convergence diagnostics: a "reviewww". In *Bayesian Statistics 6*, J. Berger, J. Bernardo, A.P. Dawid, and A.F.M. Smith (Eds.). Oxford: Oxford Sciences Publications, pp. 415-440.

Merz, C., Murphy, P., and Aha, D.W. (1997). *UCI Repository of Machine Learning Databases*. Irvine, CA: University of California, Department of Information and Computer Science.

Meyn, S.P. and Tweedie, R.L. (1993). *Markov Chains and Stochastic Stability*. London: Springer-Verlag.

Mira, A. and Roberts, G.O. (1999). Perfect slice samplers. *Technical Report*. Varese: Universita degli Studi dell'Insurbia, Italy.

Moeschberger, M.L. and David, H.A. (1971). Life tests under competing causes of failure and the theory of competing risks. *Biometrics* **27**, 909–933.

Mollie, A. and Richardson, S. (1991). Empirical Bayes estimates of cancer mortality rates using spatial models. *Statistics in Medicine* **10**, 95–112.

Moore, A.W. (1999). Very fast EM-based mixture model clustering using multiresolution kd-trees. In *Advances in Neural Information Processing Systems 11*, M.S. Kearns, S.A. Solla, and D.A. Cohn (Eds.). Cambridge, Massachusetts: MIT Press, pp. 543–549.

Morel, J.G. and Nagaraj, N.K. (1993). A finite mixture distribution for modelling multinomial extra variation. *Biometrika* **80**, 363–371.

Mosimann, J.E. (1962). On the compound multinomial distribution, the multivariate β-distribution, and correlations among proportions. *Biometrika* **49**, 65–82.

Muthén, B. and Shedden, K. (1999). Finite mixture modelling with mixture outcomes using the EM algorithm. *Biometrics* **55**, 463–469.

Nádas, A. (1971). The distribution of the identified minimum of a normal pair determines the distribution of the pair. *Technometrics* **13**, 201–202.

Neal, R. M. (1999). Regression and classification using Gaussian process priors (with discussion). In *Bayesian Statistics 6*, J.M. Bernardo, J. O. Berger, A. P. Dawid, and A. F. M. Smith, (Eds.). Oxford: Oxford University Press, pp. 475-501.

Neal, R.M. and Hinton, G.E. (1998). A view of the EM algorithm that justifies incremental, sparse, and other variants. In *Learning in Graphical Models*, M.I. Jordan (Ed.). Dordrecht: Kluwer, pp. 355–368.

Nelder, J.A. (1971). Contribution to the discussion of paper by R. O'Neill and G.B. Wetherill. *Journal of the Royal Statistical Society B* **33**, 244–246.

Nelder, J.A. (1985). Quasi-likelihood and GLIM. In *Generalized Linear Models*, R. Gilchrist, B. Francis, and J. Whittaker (Eds.). Berlin: Springer-Verlag.

Nelder, J.A. and Lee, Y. (1992). Likelihood, quasi-likelihood and pseudolikelihood: some comparisons. *Journal of the Royal Statistical Society B* **54**, 273–284.

Nelder, J.A. and Wedderburn, R.W.M. (1972). Generalized linear models. *Journal of the Royal Statistical Society A* **135**, 370–384.

Nettleton, D. (1999). Order-restricted hypothesis testing in a variation of the normal mixture model. *Canadian Journal of Statistics* **27**, 383–394.

Newcomb, S. (1886). A generalized theory of the combination of observations so as to obtain the best result. *American Journal of Mathematics* **8**, 343–366.

Newton, M.A. and Raftery, A.E. (1994). Approximate Bayesian inference with the weighted likelihood bootstrap (with discussion). *Journal of the Royal Statistical Society B* **56**, 3–48.

Ng, S.K. and McLachlan, G.J. (1998). On modifications to the long-term survival mixture model in the presence of competing risks. *Journal of Statistical Computation and Simulation* **61**, 77–96.

Ng. S.K., McLachlan, G.J., Galloway, G., and Rose, S.E. (1995). A mixture model approach to segmentation of magnetic resonance images. In *Proceedings of DICTA 95, 3rd Conference of Digital Image Computing: Techniques and Applications.* Brisbane: Australian Pattern Recognition Society, pp. 583–593.

Ng, S.K., McLachlan, G.J., McGiffin, D.C., and O'Brien, M.F. (1999). Constrained mixture models in competing risks problems. *Environmetrics* **10**, 753–767.

Ng, S.K., McLachlan, G.J., McGiffin, D.C., and O'Brien, M.F. (2000). Constrained mixture models with bathtub-shaped component hazard functions. *Technical Report*. Brisbane: Department of Mathematics, University of Queensland.

Nigam, K., McCallum, A., Thrun, S., and Mitchell, T. (2000). Using EM to classify text from labeled and unlabeled documents. *Machine Learning* (to appear).

Ning, Y. and Finch, S.J. (1996). The null distribution of the likelihood ratio test for a mixture of two normals after a restricted Box–Cox transformation. Unpublished manuscript.

Noble, A. (1994). *Bayesian analysis of finite mixture distributions.* Unpublished Ph.D. Thesis, Carnegie Mallon University.

Oliver, J.J., Baxter, R.A., and Wallace, C.S. (1996). Unsupervised learning using MML. In *Proceedings of the Thirteenth International Conference on Machine Learning (ICML96).* San Francisco: Morgan Kaufmann, pp. 364–372.

Olkin, I. (1995). Meta-analysis: reconciling the results of independent studies. *Statistics in Medicine* **14**, 457–472.

Olkin, I. and Tate, R.F. (1961). Multivariate correlation models with mixed discrete and continuous variables. *Annals of Mathematical Statistics* **22**, 92–96.

O'Neill, R. and Wetherill, G.B. (1971). The present state of multiple comparison methods (with discussion). *Journal of the Royal Statistical Society B* **33**, 218–250.

O'Neill, T.J. (1978). Normal discrimination with unclassified observations. *Journal of the American Statistical Association* **73**, 821–826.

Oskrochi, G.R. and Davies, R.B. (1997). An EM-type algorithm for multivariate mixture models. *Statistics and Computing* **7**, 145–151.

Owen, A. (1986). Contribution to the discussion of paper by B.D. Ripley. *Canadian Journal of Statistics* **14**, 106–110.

Pal Roy, P. (1995). Breakage assessment through cluster analysis of joint set orientations of exposed benches of opencast mines. *Geotechnical and Geological Engineering* **13**, 79–92.

Pan, W. (1999). Bootstrapping likelihood for model selection with small samples. *Journal of Computational and Graphical Statistics* **8**, 687–698.

Panel on Nonstandard Mixtures of Distributions. (1989). Statistical models and analysis in auditing. *Statistical Science* **4**, 2–33.

Patra, K. and Dey, D.K. (1999). A multivariate mixture of Weibull distributions in reliability modeling. *Statistics & Probability Letters* **45**, 225–235.

Paul, S.R., Liang, K.-Y., and Self, S. (1989). On testing departure from the binomial and multinomial assumptions. *Biometrics* **45**, 231–236.

Pauler, D.K., Escobar, M.D., Sweeney, J.A., and Greenhouse, J.B. (1996). Mixture models for eye-tracking data: a case study. *Statistics in Medicine* **15**, 1365–1376.

Pearson, K. (1894). Contributions to the theory of mathematical evolution. *Philosophical Transactions of the Royal Society of London A* **185**, 71–110.

Pearson, K. (1895). Contributions to the theory of mathematical evolution, II: skew variation. *Philosophical Transactions of the Royal Society of London A* **186**, 343–414.

Pearson, K. (1906). Walter Frank Raphael Galton, 1860–1906: a memoir. *Biometrika* **5**, 1–52.

Peel, D. (1998). *Mixture Model Clustering and Related Topics*. Unpublished Ph.D. Thesis, University of Queensland, Brisbane.

Peel, D. and McLachlan. G.J. (2000). Robust mixture modelling using the t distribution. *Statistics and Computing* **10**, 335–344.

Peel, D., Whiten, W., and McLachlan. G.J. (2001). Fitting mixtures of Kent distributions to aid in joint set identification. *Journal of the American Statistical Association* **96** (to appear).

Pẽna, D. and Guttman, L. (1988). Bayesian approach to robustifying the Kalman filter. In *Bayesian Analysis of Time Series and Dynamic Models*, J.C. Spall (Ed.). Cambridge, Massachusetts: Marcel Dekker, pp. 227–254.

Peng, F., Jacobs, R.A., and Tanner, M.A. (1996). Bayesian inference in mixtures-of-experts and hierarchical mixtures-of-experts models with an application to speech recognition. *Journal of the American Statistical Association* **91**, 953–960.

Perlman, M.D. (1972). On the strong consistency of approximate maximum likelihood estimators. In *Proceedings of the 6th Berkeley Symposium*, Vol. 6. Berkeley: University of California Press, pp. 263–281.

Perlman, M.D. (1983). The limiting behavior of multiple roots of the likelihood equation. In *Recent Advances in Statistics*, M.H. Rizvi, J.S. Rustagi, and D. Siegmund (Eds.). New York: Academic Press, pp. 339–370.

Peters, B.C. and Coberly, W.A. (1976). The numerical evaluation of the maximum-likelihood estimate of mixture proportions. *Communications in Statistics—Theory and Methods* **5**, 1127–1135.

Peters, B.C. and Walker, H.F. (1978). An iterative procedure for obtaining maximum-likelihood estimators of the parameters for a mixture of normal distributions. *SIAM Journal on Applied Mathematics* **35**, 362–378.

Phillips, D.B. and Smith, A.F.M. (1996). Bayesian model comparison via jump diffusions. In *Markov Chain Monte Carlo in Practice*, W.R. Gilks, S. Richardson, and D.J. Spiegelhalter (Eds.). London: Chapman & Hall, pp. 215–239.

Pickering, G. (1968). *High Blood Pressure.* London: Churchill Ltd.

Pincus, M. (1968). A closed form solution of certain programming problems. *Operations Research* **16**, 690–694.

Platt, R. (1963). Heredity in hypertension. *Lancet* **April**, 899-904.

Polymenis, A. and Titterington, D.M. (1998). On the determination of the number of components in a mixture. *Statistics & Probability Letters* **38**, 295-298.

Polymenis, A. and Titterington, D.M. (1999). A note on the distribution of the likelihood ratio statistic for normal mixture models with known proportions. *Journal of Statistical Computation and Simulation* **64**, 167–175.

Prentice, R.L., Kalbfleisch, J.D., Petersen, A.V., Flournoy, N., Farewell, V.T., and Breslow, N.E. (1978). The analysis of failure times in the presence of competing risks. *Biometrics* **34**, 541–54.

Preston, E.J. (1953). A graphical method for the analysis of statistical distributions into two normal distributions. *Biometrika* **40**, 460–464.

Priebe, C.E. (1994). Adaptive mixtures. *Journal of the American Statistical Association* **89**, 796–806.

Priebe, C.E. and Marchette, D.J. (1993). Adaptive mixture density estimation. *Pattern Recognition* **26**, 771–785.

Propp, J.G. and Wilson, D.B. (1996). Exact sampling with coupled Markov chains and applications to statistical mechanics. *Random Structures and Algorithms* **9**, 223-252.

Qian, W. and Titterington, D.M. (1989). On the use of Gibbs Markov chain models in the analysis of images based on second-order pairwise-interaction distributions. *Journal of Applied Statistics* **16**, 267–281.

Qian, W. and Titterington, D.M. (1990). Parameters estimation for hidden Gibbs chains. *Statistics & Probability Letters* **10**, 49–58.

Qian, W. and Titterington, D.M. (1991). Estimation of parameters in hidden Markov models. *Philosophical Transactions of the Royal Society of London A* **337**, 407–428.

Qian, W. and Titterington, D.M (1992). Stochastic relaxations and EM algorithms for Markov random fields. *Journal of Statistical Computation and Simulation* **40**, 55–69.

Qu, Y., Tan, M., and Kutner, M.H. (1996). Random effects models in latent class analysis for evaluating accuracy of diagnostic tests. *Biometrics* **52**, 797–810.

Quandt, R.E. and Ramsey, J.B. (1978). Estimating mixtures of normal distributions and switching regressions (with discussion). *Journal of the American Statistical Association* **73**, 730–752.

Quetelet, A. (1846). *Lettres à S.A.R. le Duc Régnant de Saxe-Cobourg and Gotha, sur la théorie des probabilité, appliquée aux sciences morales et politiques.* Brussels: Hayez.

Quetelet, A. (1852). Sur quelques propritiétés curieuses que présentent les résultats d'une serie d'observations, faites dans la vue de déterminer une constante, lorsque les chances de rencontrer des écarts en plus et en moins sont égales et indépendantés les unes des autres. *Bulletins de l'Académie royale des sciences, des lettres et des beaux-arts de Belgique* **19**, 303–317,

Quinlan, R. (1993). C4.5: *Programs for Machine Learning.* San Mateo, California: Morgan Kaufmann.

Quinn, B.G., McLachlan, G.J., and Hjort, N.L. (1987). A note on the Aitkin-Rubin approach to hypothesis testing in mixture models. *Journal of the Royal Statistical Society B* **49**, 311–314.

Rabiner, L.R. (1989). A tutorial on hidden Markov models and selected applications in speech recognition. *Proceedings of the IEEE* **77**, 257–286.

Rabiner, L.R. and Juang, B.H. (1986). An introduction to hidden Markov models. *IEEE Acoustics, Speech and Signal Processing Magazine* **3**, 4–16.

Raftery, A.E. (1985). A model for high-order Markov chains. *Journal of the Royal Statistical Society B* **47**, 528–539.

Raftery, A.E. (1996). Hypothesis testing and model selection. In *Markov Chain Monte Carlo in Practice*, W.R. Gilks, S. Richardson, and D.J. Spiegelhalter (Eds.). London: Chapman & Hall, pp. 115–130.

Raftery, A.E. and Tavaré, S. (1994). Estimation and modelling repeated patterns in high order Markov chains with the mixture transition distribution model. *Applied Statistics* **43**, 179–199.

Rao, C.R. (1948). The utilization of multiple measurements in problems of biological classification. *Journal of the Royal Statistical Society B* **10**, 159–203.

Rao, C.R. (1952). *Advanced Statistical Methods in Biometric Research.* New York: Wiley.

Rao, C.R. (1955). Estimation and tests of significance in factor analysis. *Psychometrika* **47**, 69–76.

Reaven, G.M. and Miller, R.G. (1979). An attempt to define the nature of chemical diabetes using a multidimensional analysis. *Diabetologia* **16**, 17–24.

Redner, R.A. (1981). Note on the consistency of the maximum likelihood estimate for nonidentifiable distributions. *Annals of Statistics* **9**, 225–228.

Redner, R.A., and Walker, H.F. (1984). Mixture densities, maximum likelihood and the EM algorithm. *SIAM Review* **26**, 195–239.

Richardson, S. and Green, P.J. (1997). On Bayesian analysis of mixtures with an unknown number of components (with discussion). *Journal of the Royal Statistical Society B* **59**, 731–792. Correction (1998). *Journal of the Royal Statistical Society B* **60**, 661.

Ripley, B.D. (1996). *Pattern Recognition and Neural Networks.* Cambridge: Cambridge University Press.

Rissanen, J. (1986). Stochastic complexity. *Annals of Statistics* **14**, 1080–1100.

Rissanen, J. (1989). *Stochastic Complexity.* Singapore: World Scientific Publishing.

Robbins, H. (1964). An empirical Bayes approach to statistical decision problems. *Annals of Mathematical Statistics* **35**, 1–20.

Robbins, H. (1983). Some thoughts on empirical Bayes estimation. *Annals of Statistics* **11**, 713–723.

Robert, C.P. (1993). Prior feedback: Bayesian tools for maximum likelihood estimation. *Computational Statistics* **8**, 279–294.

Robert, C.P. (1996). Mixtures of distributions: inference and estimation. In *Markov Chain Monte Carlo in Practice*, W.R. Gilks, S. Richardson, and D.J. Spiegelhalter (Eds.). London: Chapman & Hall, pp. 441–464.

Robert, C.P. (1998). *Discretization and MCMC Convergence Assessment*. Lecture Notes In Statistics, Vol. 135. New York: Springer-Verlag.

Robert, C.P. and Casella, G. (1999). *Monte Carlo Statistical Methods.* New York: Springer-Verlag.

Robert, C.P., Celeux, G., and Diebolt, J. (1993). Bayesian estimation of hidden Markov chains: a stochastic implementation. *Statistics & Probability Letters* **16**, 77–83.

Robert, C.P. and Mengersen, K.L. (1995). Reparameterization issues in mixture modelling and their bearings on the Gibbs sampler. *Technical Report No. 9538.* Paris: CREST, INSEE.

Robert, C.P. and Mengersen, K.L. (1999). Reparameterisation issues in mixture modelling and their bearing on MCMC algorithms. *Computational Statistics and Data Analysis* **29**, 325–343.

Robert, C.P., Rydén, T., and Titterington, D.M. (1999). Convergence controls for MCMC algorithms, with applications to hidden Markov chains. *Journal of Statistical Computation and Simulation* **64**, 327–355.

Robert, C.P., Rydén, T., and Titterington, D.M. (2000). Bayesian inference in hidden Markov models through the reversible jump Markov chain Monte Carlo method. *Journal of the Royal Statistical Society B* **62**, 57–75.

Robert, C.P. and Soubiran, C. (1993). Estimation of a mixture model through Bayesian sampling and prior feedback. *Test* **2**, 125–146.

Robert, C.P. and Titterington, D.M. (1998). Reparameterization strategies for hidden Markov models and Bayesian approaches to maximum likelihood estimation. *Statistics and Computing* **8**, 145–158.

Roberts, G.O. and Rosenthal, J.S. (1998). Markov chain Monte Carlo: some practical implications of theoretical results (with discussion). *Canadian Journal of Statistics* **26**, 4–31.

Roberts, S.J., Husmeier, D., Rezek, I., and Penny, W. (1998). Bayesian approaches to Gaussian modeling. *IEEE Transactions on Pattern Analysis and Machine Intelligence* **20**, 1133–1142.

Rocke, D.M. and Woodruff, D.L. (1997). Robust estimation of multivariate location and shape. *Journal of Statistical Planning and Inference* **57**, 245–255.

Roeder, K. (1990). Density estimation with confidence sets exemplified by superclusters and voids in the galaxies. *Journal of the American Statistical Association* **85**, 617–624.

Roeder, K. (1992). Semiparametric estimation of normal mixture densities. *Annals of Statistics* **20**, 929–943.

Roeder, K. (1994). A graphical technique for determining the number of components in a mixture of normals. *Journal of the American Statistical Association* **89**, 487–495.

Roeder, K. and Wasserman, L. (1997). Practical density estimation using mixtures of normals. *Journal of the American Statistical Association* **92**, 894–902.

Rosen, O. and Tanner, M. (1999). Mixtures of proportional hazards regression models. *Statistics in Medicine* **18**, 1119–1131.

Rousseeuw, P.J., Kaufman, L., and Trauwaert, E. (1996). Fuzzy clustering using scatter matrices. *Computational Statistics and Data Analysis* **23**, 135–151.

Rubin, D.B. (1983). Iteratively reweighted least squares. In *Encyclopedia of Statistical Sciences*, Vol. 4, S. Kotz, N.L. Johnson, and C.B. Read (Eds.). New York: Wiley, pp. 272–275.

Rubin, D.B. and Thayer, D.T. (1982). EM algorithms for ML factor analysis. *Psychometrika* **47**, 69–76.

Rubin, D.B. and Wu, Y.N. (1997). Modeling schizophrenic behavior using general mixture components. *Biometrics* **53**, 243–261.

Rubinstein, R.Y. (1981). *Simulation and the Monte Carlo method*. New York: Wiley.

Rydén, T., Teräsvirta, T., and Asbrink, S. (1998). Stylized facts of daily return series and the hidden Markov model. *Journal of Applied Economics* **13**, 217–244.

Rydén, T. and Titterington, D.M. (1998). Computational Bayesian analysis of hidden Markov models. *Journal of Computational and Graphical Statistics* **7**, 194–211.

Sahu, S.K. and Roberts, G.O. (1999). On convergence of the EM algorithm and the Gibbs sampler. *Statistics and Computing* **9**, 55–64.

Sain, S.R., Gray, H.L, Woodward, W.A., and Fisk, M.D. (1999). Outlier detection from a mixture distribution when training data are unlabeled. *Bulletin of the Seismological Society of America* **89**, 294–304.

SAS Institute, Inc. (1993). SAS/ETS User's Guide. Cary, North Carolina: SAS Institute.

Saul, L.K., Jaakkola, T., and Jordan, M.I. (1996). Mean field theory of sigmoid belief networks. *Journal of Artificial Intelligence* **4**, 61–76.

Scallan, A.J. (1999). Fitting a mixture distribution to complex censored survival data using generalized linear models. *Journal of Applied Statistics* **26**, 747–753.

Schaeben, H. (1984). A new cluster algorithm for orientation data. *Mathematical Geology* **16**, 139–154.

Schlattmann, P. and Böhning, D. (1993). Mixture models and disease mapping. *Statistics in Medicine* **12**, 1943–1950.

Schlattmann, P., Dietz, E., and Böhning, D. (1996). Covariate adjusted mixture models and disease mapping with the program DismapWin. *Statistics in Medicine* **15**, 919–929.

Schork, N.J., Allison, D.B., and Thiel, B. (1996). Mixture distributions in human genetics. *Statistical Methods in Medical Research* **5**, 155–178.

Schork, N.J. and Schork, M.A. (1988). Skewness and mixtures of normal distributions. *Communications in Statistics—Theory and Methods* **17**, 3951–3969.

Schork, N.J., Weder, A.B., and Schork, M.A. (1990). On the asymmetry of biological frequency distributions. *Genetic Epidemiology* **7**, 427–446.

Schroeter, P., Vesin, J.-M., Langenberger, T., and Meuli, R. (1998). Robust parameter estimation of intensity distributions for brain magnetic resonance images. *IEEE Transactions on Medical Imaging* **17**, 172-186.

Schwarz, G. (1978). Estimating the dimension of a model. *Annals of Statistics* **6**, 461–464.

Sclove, S.L. (1987). Application of model-selection criteria to some problems in multivariate analysis. *Psychometrika* **52**, 333–343.

Scott, A.J. and Symons, M.J. (1971). Clustering methods based on likelihood ratio criteria. *Biometrics* **27**, 387–397.

Scott, D.W. (1992). *Multivariate Density Estimation*. New York: Wiley.

Seidel, W., Mosler, K., and Alker, M. (2000a). A cautionary note on likelihood ratio tests in mixture models. *Annals of the Institute of Statistical Mathematics* (to appear).

Seidel, W., Mosler, K., and Alker, M. (2000b). Likelihood ratio tests based on subglobal optimization: a power comparison in exponential mixture models. *Statistical Papers* **41**, 85–98.

Self, S.G. and Liang, K.-Y. (1987). Asymptotic properties of maximum likelihood estimators and likelihood ratio tests under nonstandard conditions. *Journal of the American Statistical Association* **82**, 605–610.

Shephard, N. (1994). Partial non-Gaussian state space. *Biometrika* **81**, 115–131.

Shi, J.-Q. and Lee, S.-Y. (2000). Latent variables with mixed continuous and polytomous data. *Journal of the Royal Statistical Society B* **62**, 77–87.

Shoukri, M.M. and Lathrop, G.M. (1993). Statistical testing of genetic linkage under heterogeneity. *Biometrics* **49**, 151–161.

Shoukri, M.M. and McLachlan, G.J. (1994). Parametric estimation in a genetic mixture model with application to nuclear family data. *Biometrics* **50**, 128–139.

Silverman, B.W. (1981). Using kernel density estimates to investigate multimodality. *Journal of the Royal Statistical Society B* **43**, 97–99.

Silverman, B.W. (1986). *Density Estimation for Statistics and Data Analysis*. London: Chapman and Hall.

Simar, L. (1976). Maximum likelihood estimation of a compound Poisson process. *Annals of Statistics* **4**, 1200–1209.

Slud, E. (1992). Nonparametric identifiability of marginal survival distributions in the. presence of dependent competing risks and a prognostic covariate. In *Survival Analysis: State of the Art*, J.P. Klein and P.K. Goel (Eds.). Dordrecht: Kluwer, pp. 355-368. *Journal of the Royal Statistical Society B* **55**, 3–23. Discussion: 53–102.

Smith, A.F.M. and Makov, U.E. (1978). A quasi-Bayes sequential procedure for mixtures. *Journal of the Royal Statistical Society B* **40**, 106–112.

Smith, A.F.M. and Roberts, G.O. (1993). Bayesian computation via the Gibbs sampler and related Markov chain Monte Carlo methods (with discussion). *Journal of the Royal Statistical Society B* **55**, 3–23. Discussion: 53–102.

Smith, D.J., Bailey, T.C., and Munford, G. (1993). Robust classification of high-dimensional data using artificial neural networks. *Statistics and Computing* **3**, 71–81.

Smyth, P. (2000). Model selection for probabilistic clustering using cross-validated likelihood. *Statistics and Computing* **10**, 63–72.

Smyth, P., Heckerman, D., and Jordan, M.I. (1997). Probabilistic independence networks for hidden Markov probability models. *Neural Computation* **9**, 227–269.

Smyth, P., Ide, K., and Ghil, M. (1999). Multiple regimes in Northern Hemisphere height fields via mixture model clustering. *Journal of the Atmospheric Sciences* **56**, 3704–3723.

Solka, J.L., Wegman, E.J., Priebe, C.E., Poston, W.L., and Rogers, W. (1998). Mixture structure analysis using the Akaike criterion and the bootstrap. *Statistics and Computing* **8**, 177–188.

Soromenho, G. (1993). Comparing approaches for testing the number of components in a finite mixture model. *Computational Statistics* **9**, 65–78.

Stephens, M.A. (1969). Techniques for Directional Data. *Technical Report No. 150*. Stanford, California: Department of Statistics, Stanford University.

Stephens, M.A. (1999). Dealing with multimodal posteriors and non-identifiability in mixture models. *Technical Report*. Oxford: Department of Statistics, Oxford University.

Stigler, S.M. (1986). *The History of Statistics*. Cambridge, Massachusetts: Belknap Press of Harvard University Press.

Streit, R.L. and Luginbuhl, T.E. (1994). Maximum likelihood training of probabilistic neural networks. *IEEE Transactions on Neural Networks* **5**, 764–783.

Strömgren, B. (1934). Tables and diagrams for dissecting a frequency curve into components by the half-invariant method. *Skandinavian Aktuarietidskr* **17**, 7–54.

Sundberg, R. (1974). Maximum likelihood theory for incomplete data from an exponential family. *Scandinavian Journal of Statistics: Theory and Applications*, **1**, 49–58.

Sutradhar, B.C. and Ali, M.M. (1986). Estimation of the parameters of a regression model with a multivariate t error variable. *Communications in Statistics—Theory and Methods* **15**, 429–450.

Swales, J.D. (Ed.). (1985). *Platt Vs. Pickering: An Episode in Recent Medical History.* Cambridge: The Keynes Press.

Symons, M.J., Grimson, R.C., and Yuan, Y.C. (1983). Clustering of rare events. *Biometrics* **39**, 193–205.

Tan, W.Y. and Chang, W.C. (1972). Some comparisons of the method of moments and the method of maximum likelihood in estimating parameters of a mixture of two normal densities. *Journal of the American Statistical Association* **67**, 702–708.

Tanner, M.A. and Wong, W.H. (1987). The calculation of posterior distributions by data augmentation (with discussion). *Journal of the American Statistical Association* **82**, 528–550.

Tarter, M.E. and Lock, M.D. (1993). *Model-Free Curve Estimation*. London: Chapman & Hall.

Tarter, M.E. and Silvers, A. (1975). Implementation and application of bivariate Gaussian mixture decomposition. *Journal of the American Statistical Association* **70**, 47–55.

Taylor, J.M.G. (1995). Semi-parametric estimation in failure time mixture models. *Biometrics* **51**, 899–907.

Teicher, H. (1960). On the mixture of distributions. *Annals of Mathematical Statistics* **31**, 55–73.

Teicher, H. (1963). Identifiability of finite mixtures. *Annals of Mathematical Statistics* **34**, 1265–1269.

Thiesson, B., Meek, C., and Heckerman, D. (1999). Accelerating EM for large databases. *Technical Report No. MSR-TR-99-31*. Seattle: Microsoft Research.

Thisted, R.A. (1988). *Elements of Statistical Computing: Numerical Computation.* London: Chapman & Hall.

Thode, H.C., Finch, S.J., and Mendell, N.R. (1988). Simulated percentage points for the null distribution of the likelihood ratio for a mixture of two normals. *Biometrics* **44**, 1195–1201.

Thompson, T.J., Smith, P.J., and Boyle, J.P. (1998). Finite mixture models with concomitant information: assessing diagnostic criteria for diabetes. *Applied Statistics* **47**, 393–404.

Tibshirani, R. and Knight, K. (1999). Model search by bootstrap "bumping." *Journal of Computational and Graphical Statistics* **8**, 671–686.

Tierney, L. (1994). Markov chains for exploring posterior distributions (with discussion). *Annals of Statistics* **22**, 1701–1762.

Tipping, M.E. and Bishop, C.M. (1997). Mixtures of probabilistic principal component analysers. *Technical Report No. NCRG/97/003*. Birmingham: Neural Computing Research Group, Aston University.

Tipping, M.E. and Bishop, C.M. (1999a). Mixtures of probabilistic principal component analysers. *Neural Computation* **11**, 443–482.

Tipping, M.E. and Bishop, C.M. (1999b). Probabilistic principal components. *Journal of the Royal Statistical Society B* **61**, 611–622.

Titterington, D.M. (1981). Contribution to the discussion of paper by M. Aitkin, D. Anderson and J. Hinde. *Journal of the Royal Statistical Society A* **144**, 459.

Titterington, D.M. (1984a). Recursive parameter estimation using incomplete data. *Journal of the Royal Statistical Society B* **46**, 257–267.

Titterington, D.M. (1984b). Comments on *Application of the Conditional Population-Mixture Model to Image Segmentation. IEEE Transactions on Pattern Analysis and Machine Intelligence* **6**, 656–658.

Titterington, D.M. (1990). Some recent research in the analysis of mixture distributions. *Statistics* **21**, 619–641.

Titterington, D.M. (1996). Mixture distributions (update). In *Encyclopedia of Statistical Sciences*, Vol. 5, S. Kotz, N.L. Johnson, and D. Banks (Eds.). New York: Wiley, pp. 399–407.

Titterington, D.M., Smith, A.F.M., and Makov, U.E. (1985). *Statistical Analysis of Finite Mixture Distributions*. New York: Wiley.

Tong, H. (1990). *Non-linear Time Series.* New York: Oxford University Press.

Tsiatis, A. (1975). A non-identifiability aspect of the problem of competing risks. In *Proceedings of National Academy Sciences, USA* **72**, 20–22.

Tukey, J.W. (1960). A survey of sampling from contaminated distributions. In *Contributions to Probability and Statistics*, I. Olkin, S.G. Ghurye, W. Hoeffding, W.G. Madow, and H.B. Mann (Eds.). Stanford: Stanford University Press, pp. 448–485.

Turner, T.R., Cameron, M.A., and Thomson, P. (1998). Hidden Markov chains in generalized linear models. *Canadian Journal of Statistics* **26**, 107–125.

Uebersax, J.S. and Grove, W.M. (1993). A latent trait finite mixture model for the analysis of rating agreement. *Biometrics* **49**, 823–835.

Ueda, N. and Nakano, R. (1998). Deterministic annealing EM algorithm. *Neural Networks* **11**, 271–282.

Ueda, N., Nakano, R., Ghahramani, Z., and Hinton, G. E. (2000). SMEM algorithm for mixture models. *Neural Computation* (to appear).

Usher, J.S. (1996). Weibull component reliability-prediction in the presence of masked data. *IEEE Transactions on Reliability* **45**, 229–232.

Usher, J.S. and Guess, F.M. (1989). An iterative approach for estimating component reliability from masked system life data. *Quality & Reliability Engineering International* **5**, 257–261.

Verdinelli, I. and Wasserman, L. (1991). Bayesian analysis of outlier problems using the Gibbs sampler. *Statistics and Computing* **1**, 105–117.

Vlassis, N. and Likas, A. (1999). A kurtosis-based dynamic approach to Gaussian mixture modeling. *IEEE Transactions on Systems Man and Cybernetics Part A—Systems and Humans* **29**, 393–399.

Vlassis, N.A., Papakonstantinou, G., and Tsanakas, P. (1999). Mixture density estimation based on maximum likelihood and sequential test statistics. *Neural Processing Letters* **9**, 63–76.

Vounatsou, P., Smith, T., and Smith, A.F.M. (1998). Bayesian analysis of two-component mixture distributions applied to estimating malaria attributable fractions. *Applied Statistics* **47**, 575–587.

Wald, A. (1949). Note on the consistency of the maximum likelihood estimate. *Annals of Mathematical Statistics* **20**, 595–601.

Wallace, C.S. and Boulton, D.M. (1968). An information measure for classification. *Computer Journal* **11**, 185–194.

Wallace, C.S. and Dowe, D.L. (1994). Intrinsic classification by MML–the Snob program. In *Proceedings 7th Australian Joint Conference on Artificial Intelligence*. Singapore: World Scientific Publishing, pp. 37–44.

Wallace, C.S. and Dowe, D.L. (2000). MML clustering multi-state, Poisson, von Mises circular and Gaussian distributions. *Statistics and Computing* **10**, 78–83.

Wallace, C.S. and Freeman, P.R. (1987). Estimation and inference by compact coding. *Journal of the Royal Statistical Society B* **49**, 240–252.

Wang, P. (1994). *Mixed Regression Models for Discrete Data*. Ph.D. dissertation, University of British Columbia, Vancouver.

Wang, P. and Puterman, M.L. (1999). Markov Poisson regression models for discrete time series. Part 1: Methodology. *Journal of Applied Statistics* **26**, 855–869.

Wang, P., Puterman, M.L., Cockburn, I., and Le, N.D. (1996). Mixed Poisson regression models with covariate dependent rates. *Biometrics* **52**, 381–400.

Wang S.J., Woodward W.A., Gray H.L., Wiechecki, S., and Sain, S.R. (1997). A new test for outlier detection from a multivariate mixture distribution. *Journal of Computational and Graphical Statistics* **6**, 285-299.

Wasserman, L. (1999). Asymptotic inference for mixture models using data dependent data. *Technical Report No. 677*. Pittsburgh: Department of Statistics, Carnegie-Mallon University.

Wedderburn, R.W.M. (1974). Quasi-likelihood functions, generalized linear models, and the Gauss-Newton method. *Biometrika* **61**, 439–447.

Wedel, M. and DeSarbo, W.S. (1993). A review of recent developments in latent class regression models. In *Advanced Methods of Marketing Research*, R. Bagozzi (Ed.). Cambridge, Massachusetts: Blackwell Business, pp. 352–388.

Wedel, M. and DeSarbo, W.S. (1995). A mixture likelihood approach for generalized linear models. *Journal of Classification* **12**, 21–55.

Wedel, M., DeSarbo, W.S., Bult, J.R., and Ramaswamy, V. (1993). A latent class Poisson regression model for heterogeneous count data with an application to direct mail. *Journal of Applied Economics* **8**, 397–411.

Wedel, M. and Kamakura, W.A. (1998). *Market Segmentation: Conceptual and Methodological Foundations*. Dordrecht: Kluwer Academic Press.

Wedel M., ter Hofstede F., and Steenkamp, J.B.E.M. (1998). Mixture model analysis of complex samples. *Journal of Classification* **15**, 225–244.

Weldon, W.F.R. (1892). Certain correlated variations in *Crangon vulgaris*. *Proceedings of the Royal Society of London* **51**, 2–21.

Weldon, W.F.R. (1893). On certain correlated variations in *Carcinus moenas*. *Proceedings of the Royal Society of London* **54**, 318–329.

West, M. (1999). Hierarchical mixture models in neurological transmission analysis. *Journal of the American Statistical Association* **92**, 587–606.

West, M., Müller, P., and Escobar, M.D. (1994). Hierarchical priors and mixture models with application in regression and density estimation. In *Aspects of Uncertainty: A Tribute to D.V. Lindley*, A.F.M. Smith and P. Freeman (Eds.). New York: Wiley, pp. 363–386.

White, H. (1982). Maximum likelihood estimation of misspecified models. *Econometrica* **50**, 1–25.

Whittaker, J. (1990). *Graphical Models in Applied Multivariate Analysis*. New York: Wiley.

Wilk, M.B. and Gnanadesikan, R. (1968). Probability plotting methods for the analysis of data. *Biometrika* **55**, 1–17.

Williams, D.A. (1975). The analysis of binary responses from toxological experiments involving reproduction and teratogenicity. *Biometrics* **31**, 949–952.

Williams, D.A. (1982). Extra-binomial variation in logistic linear models. *Applied Statistics* **31**, 144–148.

Willse, A. and Boik, R.J. (1999). Identifiable finite mixtures of location models for clustering mixed-mode data. *Statistics and Computing* **9**, 111-121.

Wilson, S.R. (1982). Sound and exploratory data analysis. *Compstat 1982, Proceedings Computational Statistics*. Vienna: Physica–Verlag, pp. 447–450.

Windham, M.P. and Cutler, A. (1992). Information ratios for validating mixture analyses. *Journal of the American Statistical Association* **87**, 1188–1192.

Withers, C.S. (1996). Moment estimates for mixtures of several distributions with different means of scales. *Communications in Statistics—Theory and Methods* **25**, 1799–1824.

Wolfe, J.H. (1965). A computer program for the computation of maximum likelihood analysis of types. *Research Memo. SRM 65-12*. San Diego: U.S. Naval Personnel Research Activity.

Wolfe, J.H. (1967). NORMIX: Computational methods for estimating the parameters of multivariate normal mixtures of distributions. *Research Memo. SRM 68–2*. San Diego: U.S. Naval Personnel Research Activity.

Wolfe, J.H. (1970). Pattern clustering by multivariate mixture analysis. *Multivariate Behavioral Research* **5**, 329–350.

Wolfe, J.H. (1971). A Monte Carlo study of sampling distribution of the likelihood ratio for mixtures of multinormal distributions. *Technical Bulletin STB 72-2*. San Diego: U.S. Naval Personnel and Training Research Laboratory.

Wong, C.S. and Li, W.K. (2000). On a mixture autoregressive model. *Journal of the Royal Statistical Society B* **62**, 95–115.

Wong, M.A. (1985). A bootstrap testing procedure for investigating the number of subpopulations. *Journal of Statistical Computation and Simulation* **22**, 99–112.

Woodward, W.A., Parr, W.C., Schucany, W.R., and Lindsey, H. (1984). A comparison of minimum distance and maximum likelihood estimation of a mixture proportion. *Journal of the American Statistical Association* **79**, 590–598.

Woodward, W.A., Whitney, P., and Eslinger, P.W. (1995). Minimum Hellinger distance estimation of mixture proportions. *Journal of Statistical Planning and Inference* **48**, 303–319.

Wu, C.F.J. (1983). On the convergence properties of the EM algorithm. *Annals of Statistics* **11**, 95–103.

Wu, Y. and Xu, Y. (1999). Local sequential testing procedures in a normal mixture model. *Communications in Statistics—Theory and Methods* **28**, 1777–1792.

Yakowitz, S.J. (1969). A consistent estimator for the identification of finite mixtures. *Annals of Mathematical Statistics* **40**, 1728–1735.

Yakowitz, S.J. and Spragins, J.D. (1968). On the identifiability of finite mixtures. *Annals of Mathematical Statistics* **39**, 209–214.

Yip, P. (1988). Inference about the mean of a Poisson distribution in the presence of a nuisance parameter. *Australian Journal of Statistics* **30**, 299-306.

Yip, P. (1991). Conditional inference for a mixture model for the analysis of count data. *Communications in Statistics—Theory and Methods* **20**, 2045–2057.

Yiu, K.K., Mak, M.W., and Li, C.K. (1999). Gaussian mixture models and probabilistic decision-based neural networks for pattern classification: a comparative study. *Neural Computing & Applications* **8**, 235–245.

Yu, J.Z. and Tanner, M.A. (1999). An analytical study of several Markov chain Monte Carlo estimators of the marginal likelihood. *Journal of Computational and Graphical Statistics* **8**, 839–853.

Yuille, A.L., Stolorz, P., and Utans, J. (1994). Statistical physics, mixtures of distributions, and the EM algorithm. *Neural Computation* **6**, 334–340.

Yung, Y.-F. (1997). Finite mixtures in confirmatory factor-analysis models. *Psychometrika* **62**, 297–330.

Yusuf, S., Peto, R., Lewis, J., Collins, R., and Sleight, P. (1985). Beta blockade during and after myocardial infarction: an overview of the randomized trials. *Progress in Cardiovascular Diseases* **27**, 335–371.

Zepf, S.E., Ashman, K.M., and Geisler, D. (1995). Constraints on the formation history of the elliptic galaxy NGC-3923 from the colors of its globular-clusters. *The Astrophysical Journal* **443**, 570–577.

Zhang, J. (1992). The mean-field theory in EM procedures for Markov random fields. *IEEE Transactions on Signal Processing* **40**, 2570–2583.

Zhang, J. (1993). The mean-field theory in EM procedures for blind Markov random field image restoration. *IEEE Transactions on Image Processing* **2**, 27–40.

Zhuang, X., Huang, Y., Palaniappan, K., and Zhao, Y. (1996). Gaussian density mixture modeling, decomposition and applications. *IEEE Transactions on Image Processing* **5**, 1293–1302.

Author Index

Subject Index

WILEY SERIES IN PROBABILITY AND STATISTICS
ESTABLISHED BY WALTER A. SHEWHART AND SAMUEL S. WILKS

Editors
Noel A. C. Cressie, Nicholas I. Fisher, Iain M. Johnstone, J. B. Kadane, David W. Scott, Bernard W. Silverman, Adrian F. M. Smith, Jozef L. Teugels; Vic Barnett, Emeritus, Ralph A. Bradley, Emeritus, J. Stuart Hunter, Emeritus, David G. Kendall, Emeritus

Probability and Statistics Section

*ANDERSON · The Statistical Analysis of Time Series
ARNOLD, BALAKRISHNAN, and NAGARAJA · A First Course in Order Statistics
ARNOLD, BALAKRISHNAN, and NAGARAJA · Records
BACCELLI, COHEN, OLSDER, and QUADRAT · Synchronization and Linearity: An Algebra for Discrete Event Systems
BARNETT · Comparative Statistical Inference, *Third Edition*
BASILEVSKY · Statistical Factor Analysis and Related Methods: Theory and Applications
BERNARDO and SMITH · Bayesian Statistical Concepts and Theory
BILLINGSLEY · Convergence of Probability Measures, *Second Edition*
BOROVKOV · Asymptotic Methods in Queuing Theory
BOROVKOV · Ergodicity and Stability of Stochastic Processes
BRANDT, FRANKEN, and LISEK · Stationary Stochastic Models
CAINES · Linear Stochastic Systems
CAIROLI and DALANG · Sequential Stochastic Optimization
CONSTANTINE · Combinatorial Theory and Statistical Design
COOK · Regression Graphics
COVER and THOMAS · Elements of Information Theory
CSÖRGŐ and HORVÁTH · Weighted Approximations in Probability Statistics
CSÖRGŐ and HORVÁTH · Limit Theorems in Change Point Analysis
*DANIEL · Fitting Equations to Data: Computer Analysis of Multifactor Data, *Second Edition*
DETTE and STUDDEN · The Theory of Canonical Moments with Applications in Statistics, Probability, and Analysis
DEY and MUKERJEE · Fractional Factorial Plans
*DOOB · Stochastic Processes
DRYDEN and MARDIA · Statistical Shape Analysis
DUPUIS and ELLIS · A Weak Convergence Approach to the Theory of Large Deviations
ETHIER and KURTZ · Markov Processes: Characterization and Convergence
FELLER · An Introduction to Probability Theory and Its Applications, Volume I, *Third Edition,* Revised; Volume II, *Second Edition*
FULLER · Introduction to Statistical Time Series, *Second Edition*
FULLER · Measurement Error Models
GHOSH, MUKHOPADHYAY, and SEN · Sequential Estimation
GIFI · Nonlinear Multivariate Analysis
GUTTORP · Statistical Inference for Branching Processes
HALL · Introduction to the Theory of Coverage Processes
HAMPEL · Robust Statistics: The Approach Based on Influence Functions
HANNAN and DEISTLER · The Statistical Theory of Linear Systems
HUBER · Robust Statistics

*Now available in a lower priced paperback edition in the Wiley Classics Library.

Probability and Statistics (Continued)

HUSKOVA, BERAN, and DUPAC · Collected Works of Jaroslav Hajek—with Commentary
IMAN and CONOVER · A Modern Approach to Statistics
JUREK and MASON · Operator-Limit Distributions in Probability Theory
KASS and VOS · Geometrical Foundations of Asymptotic Inference
KAUFMAN and ROUSSEEUW · Finding Groups in Data: An Introduction to Cluster Analysis
KELLY · Probability, Statistics, and Optimization
KENDALL, BARDEN, CARNE, and LE · Shape and Shape Theory
LINDVALL · Lectures on the Coupling Method
McFADDEN · Management of Data in Clinical Trials
MANTON, WOODBURY, and TOLLEY · Statistical Applications Using Fuzzy Sets
MORGENTHALER and TUKEY · Configural Polysampling: A Route to Practical Robustness
MUIRHEAD · Aspects of Multivariate Statistical Theory
OLIVER and SMITH · Influence Diagrams, Belief Nets and Decision Analysis
*PARZEN · Modern Probability Theory and Its Applications
PEÑA, TIAO, and TSAY · A Course in Time Series Analysis
PRESS · Bayesian Statistics: Principles, Models, and Applications
PUKELSHEIM · Optimal Experimental Design
RAO · Asymptotic Theory of Statistical Inference
RAO · Linear Statistical Inference and Its Applications, *Second Edition*
RAO and SHANBHAG · Choquet-Deny Type Functional Equations with Applications to Stochastic Models
ROBERTSON, WRIGHT, and DYKSTRA · Order Restricted Statistical Inference
ROGERS and WILLIAMS · Diffusions, Markov Processes, and Martingales, Volume I: Foundations, *Second Edition;* Volume II: Îto Calculus
RUBINSTEIN and SHAPIRO · Discrete Event Systems: Sensitivity Analysis and Stochastic Optimization by the Score Function Method
RUZSA and SZEKELY · Algebraic Probability Theory
SCHEFFE · The Analysis of Variance
SEBER · Linear Regression Analysis
SEBER · Multivariate Observations
SEBER and WILD · Nonlinear Regression
SERFLING · Approximation Theorems of Mathematical Statistics
SHORACK and WELLNER · Empirical Processes with Applications to Statistics
SMALL and McLEISH · Hilbert Space Methods in Probability and Statistical Inference
STAPLETON · Linear Statistical Models
STAUDTE and SHEATHER · Robust Estimation and Testing
STOYANOV · Counterexamples in Probability
TANAKA · Time Series Analysis: Nonstationary and Noninvertible Distribution Theory
THOMPSON and SEBER · Adaptive Sampling
WELSH · Aspects of Statistical Inference
WHITTAKER · Graphical Models in Applied Multivariate Statistics
YANG · The Construction Theory of Denumerable Markov Processes

Applied Probability and Statistics Section

ABRAHAM and LEDOLTER · Statistical Methods for Forecasting
AGRESTI · Analysis of Ordinal Categorical Data

*Now available in a lower priced paperback edition in the Wiley Classics Library.

Applied Probability and Statistics (Continued)

AGRESTI · Categorical Data Analysis
ANDERSON, AUQUIER, HAUCK, OAKES, VANDAELE, and WEISBERG · Statistical Methods for Comparative Studies
ARMITAGE and DAVID (editors) · Advances in Biometry
*ARTHANARI and DODGE · Mathematical Programming in Statistics
ASMUSSEN · Applied Probability and Queues
*BAILEY · The Elements of Stochastic Processes with Applications to the Natural Sciences
BARNETT and LEWIS · Outliers in Statistical Data, *Third Edition*
BARTHOLOMEW, FORBES, and McLEAN · Statistical Techniques for Manpower Planning, *Second Edition*
BASU and RIGDON · Statistical Methods for the Reliability of Repairable Systems
BATES and WATTS · Nonlinear Regression Analysis and Its Applications
BECHHOFER, SANTNER, and GOLDSMAN · Design and Analysis of Experiments for Statistical Selection, Screening, and Multiple Comparisons
BELSLEY · Conditioning Diagnostics: Collinearity and Weak Data in Regression
BELSLEY, KUH, and WELSCH · Regression Diagnostics: Identifying Influential Data and Sources of Collinearity
BHAT · Elements of Applied Stochastic Processes, *Second Edition*
BHATTACHARYA and WAYMIRE · Stochastic Processes with Applications
BIRKES and DODGE · Alternative Methods of Regression
BLISCHKE AND MURTHY · Reliability: Modeling, Prediction, and Optimization
BLOOMFIELD · Fourier Analysis of Time Series: An Introduction, *Second Edition*
BOLLEN · Structural Equations with Latent Variables
BOULEAU · Numerical Methods for Stochastic Processes
BOX · Bayesian Inference in Statistical Analysis
BOX and DRAPER · Empirical Model-Building and Response Surfaces
*BOX and DRAPER · Evolutionary Operation: A Statistical Method for Process Improvement
BUCKLEW · Large Deviation Techniques in Decision, Simulation, and Estimation
BUNKE and BUNKE · Nonlinear Regression, Functional Relations and Robust Methods: Statistical Methods of Model Building
CHATTERJEE and HADI · Sensitivity Analysis in Linear Regression
CHERNICK · Bootstrap Methods: A Practitioner's Guide
CHILÈS and DELFINER · Geostatistics: Modeling Spatial Uncertainty
CHOW and LIU · Design and Analysis of Clinical Trials: Concepts and Methodologies
CLARKE and DISNEY · Probability and Random Processes: A First Course with Applications, *Second Edition*
*COCHRAN and COX · Experimental Designs, *Second Edition*
CONOVER · Practical Nonparametric Statistics, *Second Edition*
CORNELL · Experiments with Mixtures, Designs, Models, and the Analysis of Mixture Data, *Second Edition*
*COX · Planning of Experiments
CRESSIE · Statistics for Spatial Data, *Revised Edition*
DANIEL · Applications of Statistics to Industrial Experimentation
DANIEL · Biostatistics: A Foundation for Analysis in the Health Sciences, *Sixth Edition*
DAVID · Order Statistics, *Second Edition*
*DEGROOT, FIENBERG, and KADANE · Statistics and the Law
DODGE · Alternative Methods of Regression
DOWDY and WEARDEN · Statistics for Research, *Second Edition*
DUNN and CLARK · Applied Statistics: Analysis of Variance and Regression, *Second Edition*
*ELANDT-JOHNSON and JOHNSON · Survival Models and Data Analysis

*Now available in a lower priced paperback edition in the Wiley Classics Library.

Applied Probability and Statistics (Continued)

*FLEISS · The Design and Analysis of Clinical Experiments
FLEISS · Statistical Methods for Rates and Proportions, *Second Edition*
FLEMING and HARRINGTON · Counting Processes and Survival Analysis
GALLANT · Nonlinear Statistical Models
GLASSERMAN and YAO · Monotone Structure in Discrete-Event Systems
GNANADESIKAN · Methods for Statistical Data Analysis of Multivariate Observations, *Second Edition*
GOLDSTEIN and LEWIS · Assessment: Problems, Development, and Statistical Issues
GREENWOOD and NIKULIN · A Guide to Chi-Squared Testing
*HAHN · Statistical Models in Engineering
HAHN and MEEKER · Statistical Intervals: A Guide for Practitioners
HAND · Construction and Assessment of Classification Rules
HAND · Discrimination and Classification
HEIBERGER · Computation for the Analysis of Designed Experiments
HEDAYAT and SINHA · Design and Inference in Finite Population Sampling
HINKELMAN and KEMPTHORNE: · Design and Analysis of Experiments, Volume 1: Introduction to Experimental Design
HOAGLIN, MOSTELLER, and TUKEY · Exploratory Approach to Analysis of Variance
HOAGLIN, MOSTELLER, and TUKEY · Exploring Data Tables, Trends and Shapes
HOAGLIN, MOSTELLER, and TUKEY · Understanding Robust and Exploratory Data Analysis
HOCHBERG and TAMHANE · Multiple Comparison Procedures
HOCKING · Methods and Applications of Linear Models: Regression and the Analysis of Variables
HOGG and KLUGMAN · Loss Distributions
HOSMER and LEMESHOW · Applied Logistic Regression, *Second Edition*
HØYLAND and RAUSAND · System Reliability Theory: Models and Statistical Methods
HUBERTY · Applied Discriminant Analysis
HUNT and KENNEDY · Financial Derivatives in Theory and Practice
JACKSON · A User's Guide to Principle Components
JOHN · Statistical Methods in Engineering and Quality Assurance
JOHNSON · Multivariate Statistical Simulation
JOHNSON and KOTZ · Distributions in Statistics
JOHNSON, KOTZ, and BALAKRISHNAN · Continuous Univariate Distributions, Volume 1, *Second Edition*
JOHNSON, KOTZ, and BALAKRISHNAN · Continuous Univariate Distributions, Volume 2, *Second Edition*
JOHNSON, KOTZ, and BALAKRISHNAN · Discrete Multivariate Distributions
JOHNSON, KOTZ, and KEMP · Univariate Discrete Distributions, *Second Edition*
JUREČKOVÁ and SEN · Robust Statistical Procedures: Aymptotics and Interrelations
KADANE · Bayesian Methods and Ethics in a Clinical Trial Design
KADANE AND SCHUM · A Probabilistic Analysis of the Sacco and Vanzetti Evidence
KALBFLEISCH and PRENTICE · The Statistical Analysis of Failure Time Data
KELLY · Reversability and Stochastic Networks
KHURI, MATHEW, and SINHA · Statistical Tests for Mixed Linear Models
KLUGMAN, PANJER, and WILLMOT · Loss Models: From Data to Decisions
KLUGMAN, PANJER, and WILLMOT · Solutions Manual to Accompany Loss Models: From Data to Decisions
KOTZ, BALAKRISHNAN, and JOHNSON · Continuous Multivariate Distributions, Volume 1, *Second Edition*
KOVALENKO, KUZNETZOV, and PEGG · Mathematical Theory of Reliability of Time-Dependent Systems with Practical Applications

*Now available in a lower priced paperback edition in the Wiley Classics Library.

Applied Probability and Statistics (Continued)

LACHIN · Biostatistical Methods: The Assessment of Relative Risks
LAD · Operational Subjective Statistical Methods: A Mathematical, Philosophical, and Historical Introduction
LANGE, RYAN, BILLARD, BRILLINGER, CONQUEST, and GREENHOUSE · Case Studies in Biometry
LAWLESS · Statistical Models and Methods for Lifetime Data
LEE · Statistical Methods for Survival Data Analysis, *Second Edition*
LEPAGE and BILLARD · Exploring the Limits of Bootstrap
LINHART and ZUCCHINI · Model Selection
LITTLE and RUBIN · Statistical Analysis with Missing Data
LLOYD · The Statistical Analysis of Categorical Data
MAGNUS and NEUDECKER · Matrix Differential Calculus with Applications in Statistics and Econometrics, *Revised Edition*
MALLER and ZHOU · Survival Analysis with Long Term Survivors
MANN, SCHAFER, and SINGPURWALLA · Methods for Statistical Analysis of Reliability and Life Data
McLACHLAN · Discriminant Analysis and Statistical Pattern Recognition
McLACHLAN and KRISHNAN · The EM Algorithm and Extensions
McLACHLAN and PEEL · Finite Mixture Models
McNEIL · Epidemiological Research Methods
MEEKER and ESCOBAR · Statistical Methods for Reliability Data
*MILLER · Survival Analysis, *Second Edition*
MONTGOMERY and PECK · Introduction to Linear Regression Analysis, *Second Edition*
MYERS and MONTGOMERY · Response Surface Methodology: Process and Product in Optimization Using Designed Experiments
NELSON · Accelerated Testing, Statistical Models, Test Plans, and Data Analyses
NELSON · Applied Life Data Analysis
OCHI · Applied Probability and Stochastic Processes in Engineering and Physical Sciences
OKABE, BOOTS, and SUGIHARA · Spatial Tesselations: Concepts and Applications of Voronoi Diagrams
PANKRATZ · Forecasting with Dynamic Regression Models
PANKRATZ · Forecasting with Univariate Box-Jenkins Models: Concepts and Cases
PIANTADOSI · Clinical Trials: A Methodologic Perspective
PORT · Theoretical Probability for Applications
PUTERMAN · Markov Decision Processes: Discrete Stochastic Dynamic Programming
RACHEV · Probability Metrics and the Stability of Stochastic Models
RÉNYI · A Diary on Information Theory
RIPLEY · Spatial Statistics
RIPLEY · Stochastic Simulation
ROLSKI, SCHMIDLI, SCHMIDT, and TEUGELS · Stochastic Processes for Insurance and Finance
ROUSSEEUW and LEROY · Robust Regression and Outlier Detection
RUBIN · Multiple Imputation for Nonresponse in Surveys
RUBINSTEIN · Simulation and the Monte Carlo Method
RUBINSTEIN and MELAMED · Modern Simulation and Modeling
RYAN · Statistical Methods for Quality Improvement, *Second Edition*
SCHIMEK · Smoothing and Regression: Approaches, Computation, and Application
SCHUSS · Theory and Applications of Stochastic Differential Equations
SCOTT · Multivariate Density Estimation: Theory, Practice, and Visualization
*SEARLE · Linear Models
SEARLE · Linear Models for Unbalanced Data
SEARLE, CASELLA, and McCULLOCH · Variance Components

*Now available in a lower priced paperback edition in the Wiley Classics Library.

Applied Probability and Statistics (Continued)

SENNOTT · Stochastic Dynamic Programming and the Control of Queueing Systems
STOYAN, KENDALL, and MECKE · Stochastic Geometry and Its Applications, *Second Edition*
STOYAN and STOYAN · Fractals, Random Shapes and Point Fields: Methods of Geometrical Statistics
THOMPSON · Empirical Model Building
THOMPSON · Sampling
THOMPSON · Simulation: A Modeler's Approach
TIJMS · Stochastic Modeling and Analysis: A Computational Approach
TIJMS · Stochastic Models: An Algorithmic Approach
TITTERINGTON, SMITH, and MAKOV · Statistical Analysis of Finite Mixture Distributions
UPTON and FINGLETON · Spatial Data Analysis by Example, Volume 1: Point Pattern and Quantitative Data
UPTON and FINGLETON · Spatial Data Analysis by Example, Volume II: Categorical and Directional Data
VAN RIJCKEVORSEL and DE LEEUW · Component and Correspondence Analysis
VIDAKOVIC · Statistical Modeling by Wavelets
WEISBERG · Applied Linear Regression, *Second Edition*
WESTFALL and YOUNG · Resampling-Based Multiple Testing: Examples and Methods for *p*-Value Adjustment
WHITTLE · Systems in Stochastic Equilibrium
WOODING · Planning Pharmaceutical Clinical Trials: Basic Statistical Principles
WOOLSON · Statistical Methods for the Analysis of Biomedical Data
*ZELLNER · An Introduction to Bayesian Inference in Econometrics

Texts and References Section

AGRESTI · An Introduction to Categorical Data Analysis
ANDERSON · An Introduction to Multivariate Statistical Analysis, *Second Edition*
ANDERSON and LOYNES · The Teaching of Practical Statistics
ARMITAGE and COLTON · Encyclopedia of Biostatistics: Volumes 1 to 6 with Index
BARTOSZYNSKI and NIEWIADOMSKA-BUGAJ · Probability and Statistical Inference
BENDAT and PIERSOL · Random Data: Analysis and Measurement Procedures, *Third Edition*
BERRY, CHALONER, and GEWEKE · Bayesian Analysis in Statistics and Econometrics: Essays in Honor of Arnold Zellner
BHATTACHARYA and JOHNSON · Statistical Concepts and Methods
BILLINGSLEY · Probability and Measure, *Second Edition*
BOX · R. A. Fisher, the Life of a Scientist
BOX, HUNTER, and HUNTER · Statistics for Experimenters: An Introduction to Design, Data Analysis, and Model Building
BOX and LUCEÑO · Statistical Control by Monitoring and Feedback Adjustment
BROWN and HOLLANDER · Statistics: A Biomedical Introduction
CHATTERJEE and PRICE · Regression Analysis by Example, *Third Edition*
COOK and WEISBERG · Applied Regression Including Computing and Graphics
COOK and WEISBERG · An Introduction to Regression Graphics
COX · A Handbook of Introductory Statistical Methods
DILLON and GOLDSTEIN · Multivariate Analysis: Methods and Applications
*DODGE and ROMIG · Sampling Inspection Tables, *Second Edition*
DRAPER and SMITH · Applied Regression Analysis, *Third Edition*
DUDEWICZ and MISHRA · Modern Mathematical Statistics

*Now available in a lower priced paperback edition in the Wiley Classics Library.

Texts and References (Continued)

DUNN · Basic Statistics: A Primer for the Biomedical Sciences, *Second Edition*
EVANS, HASTINGS, and PEACOCK · Statistical Distributions, *Third Edition*
FISHER and VAN BELLE · Biostatistics: A Methodology for the Health Sciences
FREEMAN and SMITH · Aspects of Uncertainty: A Tribute to D. V. Lindley
GROSS and HARRIS · Fundamentals of Queueing Theory, *Third Edition*
HALD · A History of Probability and Statistics and their Applications Before 1750
HALD · A History of Mathematical Statistics from 1750 to 1930
HELLER · MACSYMA for Statisticians
HOEL · Introduction to Mathematical Statistics, *Fifth Edition*
HOLLANDER and WOLFE · Nonparametric Statistical Methods, *Second Edition*
HOSMER and LEMESHOW · Applied Logistic Regression, *Second Edition*
HOSMER and LEMESHOW · Applied Survival Analysis: Regression Modeling of Time to Event Data
JOHNSON and BALAKRISHNAN · Advances in the Theory and Practice of Statistics: A Volume in Honor of Samuel Kotz
JOHNSON and KOTZ (editors) · Leading Personalities in Statistical Sciences: From the Seventeenth Century to the Present
JUDGE, GRIFFITHS, HILL, LÜTKEPOHL, and LEE · The Theory and Practice of Econometrics, *Second Edition*
KHURI · Advanced Calculus with Applications in Statistics
KOTZ and JOHNSON (editors) · Encyclopedia of Statistical Sciences: Volumes 1 to 9 with Index
KOTZ and JOHNSON (editors) · Encyclopedia of Statistical Sciences: Supplement Volume
KOTZ, REED, and BANKS (editors) · Encyclopedia of Statistical Sciences: Update Volume 1
KOTZ, REED, and BANKS (editors) · Encyclopedia of Statistical Sciences: Update Volume 2
LAMPERTI · Probability: A Survey of the Mathematical Theory, *Second Edition*
LARSON · Introduction to Probability Theory and Statistical Inference, *Third Edition*
LE · Applied Categorical Data Analysis
LE · Applied Survival Analysis
MALLOWS · Design, Data, and Analysis by Some Friends of Cuthbert Daniel
MARDIA · The Art of Statistical Science: A Tribute to G. S. Watson
MASON, GUNST, and HESS · Statistical Design and Analysis of Experiments with Applications to Engineering and Science
MURRAY · X-STAT 2.0 Statistical Experimentation, Design Data Analysis, and Nonlinear Optimization
PURI, VILAPLANA, and WERTZ · New Perspectives in Theoretical and Applied Statistics
RENCHER · Linear Models in Statistics
RENCHER · Methods of Multivariate Analysis
RENCHER · Multivariate Statistical Inference with Applications
ROSS · Introduction to Probability and Statistics for Engineers and Scientists
ROHATGI · An Introduction to Probability Theory and Mathematical Statistics
ROHATGI and SALEH · An Introduction to Probability Theory and Mathematical Statistics, *Second Edition*
RYAN · Modern Regression Methods
SCHOTT · Matrix Analysis for Statistics
SEARLE · Matrix Algebra Useful for Statistics
STYAN · The Collected Papers of T. W. Anderson: 1943–1985
TIAO, BISGAARD, HILL, PEÑA, and STIGLER (editors) · Box on Quality and Discovery: with Design, Control, and Robustness

*Now available in a lower priced paperback edition in the Wiley Classics Library.

Texts and References (Continued)
TIERNEY · LISP-STAT: An Object-Oriented Environment for Statistical Computing and Dynamic Graphics
WONNACOTT and WONNACOTT · Econometrics, *Second Edition*
WU and HAMADA · Experiments: Planning, Analysis, and Parameter Design Optimization

WILEY SERIES IN PROBABILITY AND STATISTICS
ESTABLISHED BY WALTER A. SHEWHART AND SAMUEL S. WILKS

Editors
Robert M. Groves, Graham Kalton, J. N. K. Rao, Norbert Schwarz, Christopher Skinner

Survey Methodology Section

BIEMER, GROVES, LYBERG, MATHIOWETZ, and SUDMAN · Measurement Errors in Surveys
COCHRAN · Sampling Techniques, *Third Edition*
COUPER, BAKER, BETHLEHEM, CLARK, MARTIN, NICHOLLS, and O'REILLY (editors) · Computer Assisted Survey Information Collection
COX, BINDER, CHINNAPPA, CHRISTIANSON, COLLEDGE, and KOTT (editors) · Business Survey Methods
*DEMING · Sample Design in Business Research
DILLMAN · Mail and Telephone Surveys: The Total Design Method, *Second Edition*
GROVES and COUPER · Nonresponse in Household Interview Surveys
GROVES · Survey Errors and Survey Costs
GROVES, BIEMER, LYBERG, MASSEY, NICHOLLS, and WAKSBERG · Telephone Survey Methodology
*HANSEN, HURWITZ, and MADOW · Sample Survey Methods and Theory, Volume 1: Methods and Applications
*HANSEN, HURWITZ, and MADOW · Sample Survey Methods and Theory, Volume II: Theory
KISH · Statistical Design for Research
*KISH · Survey Sampling
KORN and GRAUBARD · Analysis of Health Surveys
LESSLER and KALSBEEK · Nonsampling Error in Surveys
LEVY and LEMESHOW · Sampling of Populations: Methods and Applications, *Third Edition*
LYBERG, BIEMER, COLLINS, de LEEUW, DIPPO, SCHWARZ, TREWIN (editors) · Survey Measurement and Process Quality
SIRKEN, HERRMANN, SCHECHTER, SCHWARZ, TANUR, and TOURANGEAU (editors) · Cognition and Survey Research
VALLIANT, DORFMAN, and ROYALL · A Finite Population Sampling and Inference

*Now available in a lower priced paperback edition in the Wiley Classics Library.